Human Virology

HUMAN VIROLOGY

FIFTH EDITION

John Oxford

Emeritus Professor of Virology at St Bartholomew's and The Royal London School of Medicine
and Dentistry at Queen Mary College, University of London

Paul Kellam

Professor of Virus Genomics, Imperial College London, and VP for Vaccines & Infectious Disease at Kymab Ltd.

Leslie Collier

Emeritus Professor of Virology, University of London

With case studies by Dr Berenice Langdon and Dr Juliette Oxford

OXFORD
UNIVERSITY PRESS

OXFORD
UNIVERSITY PRESS

Great Clarendon Street, Oxford, OX2 6DP,
United Kingdom

Oxford University Press is a department of the University of Oxford.
It furthers the University's objective of excellence in research, scholarship,
and education by publishing worldwide. Oxford is a registered trade mark of
Oxford University Press in the UK and in certain other countries

Second edition 2000
Third edition 2006
Fourth edition 2011

Impression: 2

Published in the United States of America by Oxford University Press
198 Madison Avenue, New York, NY 10016, United States of America

British Library Cataloguing in Publication Data
Data available

Library of Congress Control Number: 2016931927

ISBN 978–0–19–871468–2

Printed in Great Britain by
Bell & Bain Ltd., Glasgow

Preface to the fifth edition

Viruses have been described as the most abundant form of life on Earth. Students may feel challenged by the number of viruses and their diversity of form, function, and disease. In previous editions we have categorized human viruses according to the disease pattern and symptoms they cause. We have now changed to David Baltimore's classification, dependent upon the molecular properties of the virus and the methods of replication. Under Baltimore, the human viruses then fall neatly into seven groupings and we feel this will satisfy students of both medicine and science.

The publication of this latest edition follows soon after outbreaks of Ebola virus in West Africa, MERS CoV in Korea, and Zika virus in Brazil. Other viruses such as Chikungunya virus and Blue Tongue virus have now arrived in Europe, and with global warming a northern movement of mosquito vectors will expose us to further viruses.

Viral infections are not simply the domain of textbooks such as this: their diagnosis and treatment is now a routine part of clinical practice using molecular diagnostics. Treatment is also advancing with new antivirals, a resurgence of vaccine research, and the promise of therapeutic monoclonal antibodies. With this in mind, the current edition includes more case studies by two practising doctors; a General Practitioner, Berenice Langdon, and a hospital doctor, Juliette Oxford. Their first-hand reports illustrate the reality of the management of viral infections in a clinical practice and offer important learning opportunities. We envisage a new generation of scientist-doctors able to move from one discipline to the other.

We have been pleased by how well our previous editions have been received, and we thank those individuals who have provided the feedback that has informed the process of revision. We thank especially Marie Bush (hVIVO, formerly Retroscreen) for secretarial assistance in all aspects of revising this book. We think that this latest edition is a very worthy successor to those that have gone before.

Regretfully Leslie Collier, our first and most senior author, whose idea it was to write the First Edition of *Human Virology* in 1993, has died. We spent many happy hours together over tea and biscuits at his flat in Peto Place, London, thinking of viruses, disease, and the world in general. Leslie was a true academic, careful, precise, and quietly spoken, but he had had two other careers. As a medical officer he travelled through Italy with the British Army in the last part of the Second World War and worked on trachoma. Penultimately, he was a Director at the Lister Institute in the UK where vast quantities of vaccine were produced for the WHO Smallpox Global Eradication Campaign. Leslie invented a stabilizer for the live smallpox vaccine which meant it could be carried anywhere in the world regardless of the temperature.

We hope that medical and science students alike will be enthused by virology and our textbook, and will contribute in their turn, to the great endeavour of the 'Conquest of Viral Diseases'.

Contents at a glance

Contents in detail

Abbreviations

ACV	aciclovir		HBsAg	hepatitis B surface antigen
AIDS	acquired immunodeficiency syndrome		HBV	hepatitis B virus
APC	antigen-presenting cell		HCC	hepatocellular carcinoma
ARC	AIDS-related complex		HCV	hepatitis C virus
AZT	azidothymidine		HDV	deltavirus (hepatitis D virus)
bDNA	branched DNA		HEV	hepatitis E virus
BL	Burkitt's lymphoma		HFRS	haemorrhagic fever with renal syndrome
BMT	bone-marrow transplant		HGV	hepatitis G virus
CD	cluster of differentiation		HHV	human herpesvirus
CDC	Centers for Disease Control (USA)		HI	haemagglutination inhibition
cDNA	complementary DNA		HIV	human immunodeficiency virus
CMI	cell-mediated immunity		HLA	human leucocyte antigen
CMV	cytomegalovirus		HNIG	human normal immunoglobulin
CNS	central nervous system		HPS	hantavirus pulmonary syndrome
CPE	cytopathic effect		HPV	human papillomavirus
CSF	cerebrospinal fluid		HRIG	human rabies immunoglobulin
DAA	direct acting antiviral		HTLV	human T-cell leukaemia virus
D&V	diarrhoea and vomiting		ICAM	intercellular adhesion molecule
ddI	didanosine		ICTV	International Committee on Taxonomy of Viruses
DHSS	dengue haemorrhagic shock syndrome		IEM	immunoelectron microscopy
DNA	deoxyribonucleic acid		IFN	interferon
ds	double-stranded (nucleic acid)		IL	interleukin
EA	early antigen (of EBV)		IPV	inactivated poliomyelitis vaccine
EBNA	Epstein–Barr (virus) nuclear antigen		kb	kilobase
EBV	Epstein–Barr virus		kbp	kilobase pairs
ECDC	European Centre for Disease Control		LCM	lymphocytic choriomeningitis virus
ELISA	enzyme-linked immunosorbent assay		LTR	long terminal repeat
EM	electron microscope, microscopy		LYDMA	lymphocyte-detected membrane antigen (of EBV)
FITC	fluorescein isothiocyanate		ME	myalgic encephalomyelitis
GSSD	Gerstmann–Sträussler–Scheinker disease		MERS	Middle Eastern Respiratory Syndrome
HA	haemagglutinin		MHC	major histocompatibility complex
HAART	highly active antiretroviral therapy		MMR	measles–mumps–rubella (vaccine)
HAM	HTLV-I-associated myelopathy		MP	mononuclear phagocytic cell
HAV	hepatitis A virus		MRC	Medical Research Council (UK)
HBcAg	hepatitis B core antigen		mRNA	messenger RNA
HBeAg	hepatitis B e antigen		NA	neuraminidase

NANB	non-A, non-B hepatitis		SRSV	small round structured virus
NGS	next generation sequencing		ss	single-stranded (nucleic acid)
NK	natural killer (cell)		STD	sexually transmitted disease
NPC	nasopharyngeal carcinoma		SV	simian vacuolating (virus)
NS	non-structural		Taq	*Thermophilus aquaticus*-derived polymerase
NS1	non-structural protein 1		TaqMan	commercial (Roche) real time molecular diagnostics
ntr	non-translated regions		T_c	cytotoxic T cell
OPV	oral (attenuated) polio vaccine		T_{dh}	delayed hypersensitivity cell
ORF	open reading frame		T_h	T-helper cell
PCR	polymerase chain reaction		TK	thymidine kinase
PML	progressive multifocal leucoencephalopathy		TNF	tumour necrosis factor
RIA	radioimmunoassay		T_s	T-suppressor cell
RNA	ribonucleic acid		ts	temperature-sensitive (mutants)
RNase	ribonuclease		TSP	tropical spastic paraparesis
RNP	ribonucleoprotein		VCA	viral capsid antigen (of EBV)
RSV	respiratory syncytial virus		VEE	Venezuelan equine encephalitis
RT	reverse transcriptase		VZV	varicella-zoster virus
RT PCR	Reverse transcription polymerase chain reaction		WHO	World Health Organization
SARS	severe acute respiratory syndrome		YF	yellow fever
snRNPs	small nuclear ribonucleoprotein particles		ZIG	zoster immune globulin

Part 1

General principles

Virology: how it all began and where it will go next

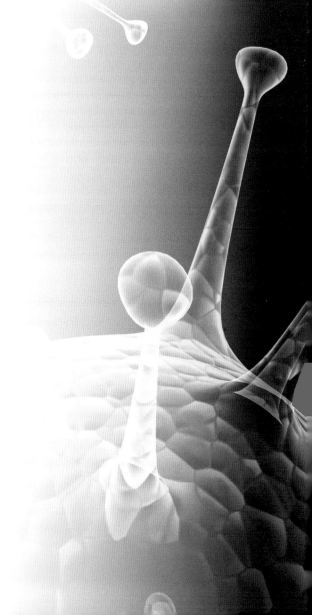

1.1 Introduction

Henry Ford (1863–1947) said that History is more or less bunk.

Although not many would agree entirely with the great industrialist's dismissal of the past, there is certainly a tendency for today's students of science and medicine to regard the history of their subjects as a waste of time. Given the sheer volume of material that has to be absorbed, digested, and regurgitated in the examinations, this feeling is understandable; but it is a pity, because history, especially that of virology, provides not only a fascinating account of technological developments in microbiology, but also the intellectual challenge of thinking about how life began. More immediate history warns us to prepare for global virus pandemics. The Spanish influenza of 1918 killed millions more young people than died in the trenches of World War I. Our research group at the Royal London Hospital, along with J. Taubenberger in the USA, has searched for the genes of this virus in lung samples of victims exhumed in England from lead lined coffins and from frozen samples from the Arctic Circle as well as formalin fixed lung specimens in museums. So sometimes historical pathology samples can give insight into current problems. Recent history shows outbreaks will continue to occur, with Severe Acute Respiratory Syndrome Coronavirus (SARS CoV), Middle East Respiratory Syndrome Coronavirus (MERS CoV), Ebola virus and Zika virus being infections of concern in the early twenty-first century.

Continued technological advances are now revealing the true extent and impact of viruses in all cellular organisms and environments. We know that viruses can infect all forms of life, that they cause huge outbreaks of disease in humans and in wild and domesticated animals and plants. In fact viruses are now recognized as 'the most abundant form of life on earth'. Bacterial viruses alone number 10^{34} and would weigh as much as one million blue whales. Given the observation that these bacterial viruses are the most numerous creatures on the planet we should also remember that 95% of bacteria with their viruses are yet to be discovered!

Viruses mostly do not leave fossils, the exception being remnants of retrovirus genes that remain in animal genomes. Our views on the origin of viruses must be based on the most slender of clues; the rest is at present speculation, based mostly on our knowledge of the behaviour of today's viruses, bacteria, and cells, but influenced sometimes by individual ideas of religion and even cosmology. For example, some believe that viruses, like all other life forms, were divinely created, while others believe that they originated in comets or elsewhere in outer space. However, most people would probably take the intuitive approach that simple organisms preceded more complicated ones, and that bacteria evolved from what were once free-living, self-replicating molecules that resemble today's viruses. Evolutionary scientists have pin-pointed RNA as being the first biomolecule to be synthesized on the planet, which places RNA viruses in early evolution. Yet we should not forget that viruses of algae, the Phycodnaviridae, may have witnessed the separation of prokaryotes and eukaryotes as long ago as 2 billion years!

Scientists can speculate endlessly about such theories and some have been stimulated to test them in the laboratory; others again are searching for clues in meteorites and in samples taken from Mars by space probes. All findings are as yet inconclusive, so we must leave this fascinating topic and take a look at the more down to earth story of how virology, rather than the viruses themselves, evolved.

1.2 How viruses were discovered

As a science, virology evolved later than bacteriology; this is not surprising, because the comparatively large size of bacteria made them visible even with the simple microscope invented by Antonie van Leeuwenhoek, a Dutch optician, who in 1673 first described their appearances. However, it was not until the nineteenth century that Louis Pasteur, Robert Koch, and others established the biological properties of bacteria and yeasts, first as the causes of fermentation and putrefaction, then as causes of disease. However, although the physical nature of viruses was not fully revealed until the invention of the electron microscope (EM), the infections they caused have been known and feared since the dawn of history. Two examples from ancient Egypt are shown in Chapters 24 (Fig. 24.5) and 7 (Fig. 7.3).

The Latin word *virus* means 'poisonous fluid', and this is just what they seemed to the first virologists. In the latter part of the nineteenth century, huge strides were being made in the study of microbes. In Pasteur's laboratory, Charles Chamberland devised a filter that would hold back even the smallest bacteria; next, Iwanowski in Russia and Beijerinck in Holland both showed that a plant infection, tobacco mosaic, could be transmitted by extracts that had been passed through a Chamberland filter, and hence could not contain bacteria. Soon afterwards, foot-and-mouth disease of cattle was also transmitted by bacteria-free filtrates, and it came to be realized that living agents, smaller than any known bacteria, but capable of multiplying, could cause a wide range of diseases in plants and animals. New viruses are still being discovered, often with new human infections originating from viruses in wild animals and birds in remote spots of the planet. MERS CoV is an example, emerging in 2013 from camels in Saudi Arabia, whilst the haemorrhagic fever Ebola re-emerged, most likely from bats, in West Africa in 2014.

1.3 How they were grown in the laboratory

Very quickly, Beijerinck realized that whatever it was that caused tobacco mosaic would grow only in living cells and could not be cultivated in the media used for bacteria. At that time, there was no way of growing cells in the laboratory, and so it was that all the early work on virus infections had to be done with intact plants or animals. Thus, Pasteur used dogs and rabbits in the development of a rabies vaccine. Now that smallpox is eradicated we can relax about this fearsome virus. Leslie Collier, our first author who died in 2011, was pre-eminent in the downfall of this virus and worked tirelessly in the Lister Institute Smallpox Laboratory. At that time the vaccine virus was still cultivated on the shaved skin of calves and sheep. He invented one of the two crucial technologies of the WHO Smallpox Eradication Campaign. His contribution was a heat stabilized vaccine whilst the other contribution was the bifurcated needle which in its Y shaped tip held by surface tension one drop of the live vaccine whilst it was scraped onto the skin. In 1938, a virus was first grown in the laboratory in suspensions of minced kidney tissue. The two scientists who developed the new cell culture later grew polio for the first time and were worthy recipients of the Nobel Prize. That very weekend in Boston they must have realized that the world could then conquer infantile paralysis. Their colleague Jonas Salk used their monkey kidney cells to grow vaccine viruses and did!

During the following decades, a great advance was made with the propagation of many viruses in developing chick embryos, including yellow fever, smallpox, and influenza but the really big breakthrough came with the discovery of antibiotics in the 1940s and 1950s. Until then, it was very difficult to keep cell and tissue cultures free from contamination with airborne bacteria and moulds, but the addition of antibiotics to the culture medium inhibited these unwanted contaminants and permitted the large-scale application of cell cultures in pharmaceutical factories to make vaccines. Nowadays cells are cultured without antibiotics in the strict air filtered factories of pharmaceutical groups.

The new millennium is witnessing a great expansion of molecular techniques for studying micro-organisms. In particular, the ability to identify, isolate, clone, and express specific nucleotide sequences has led to the identification of hitherto unknown viruses that still cannot be propagated in the laboratory by conventional methods, such as hepatitis C. This methodology is now taken to another level, for example, with the use of micro-arrays, which are in essence glass microscope slides containing hundreds of thousands of short oligonucleotides that can capture a spectrum of virus genomes, to survey samples where the identity of a virus is not known. In addition, new DNA sequencing methods, Next Generation Sequencing (NGS), and portable nanopore sequencers that allow the cheap and rapid sequencing of virus genomes are being used to identify viruses in various body fluids, infected tissues, and environmental samples, leading ultimately to more detailed understanding of the total virome. Further, large-scale sequencing (a subject of the lab of our co-author PK), together with knowledge of the date and place the sample was obtained, can be used with computational methods to inform transmission chains and epidemiological information in the new fields of phylodynamics and phylogeography. Finally, the techniques of **reverse genetics** (Chapter 3) have allowed the genetic manipulation of RNA viruses. In the laboratory polio has been completely synthesized as an infectious entity. Undoubtedly, we are on the edge of new discoveries.

1.4 Sizes and shapes

Light microscopes were well advanced by the end of the nineteenth century and the appearances of many bacteria—cocci, rods, spirals, and so forth—were already familiar. Furthermore, the vast majority of bacteria proved larger than 0.25 µm, which is about the limit of resolution of the light microscope, and so were easy to see. The reverse is true of viruses, which are smaller than this limit (Fig. 1.1). Poxviruses and the newly discovered phycodnaviruses of algae with 1200kb of DNA and whose icosahedral capsids reach 500 nm in diameter are an exception, and it is a scientific oddity that, as long ago as 1886, one of them was actually seen and measured accurately by John Buist, a Scottish microscopist. However, study of the shapes and sizes—the **morphology**—of viruses had to await the coming of the EM in 1939. Whereas the resolving power of the conventional microscope is limited by the wavelength of light, the EM is under no such constraint, functioning as it does with a beam of electrons focused by electromagnets. It is interesting that the first virus to be identified—tobacco mosaic—was also the first to be seen under the EM, appearing as regular rod-shaped particles measuring about 25 × 300 nm. [A nanometre (nm) is the unit used for expressing the sizes of viruses. It is 1/1000 of the micrometre unit (µm) used for bacteria.] At about the time that the EM started to come into its own, other methods of measuring viruses were developed. One involved the passing of suspensions through a series of filters with accurately graded pore diameters; the size of the virus was determined from the smallest pore diameter through which it could be filtered. Another was high-speed centrifugation, whereby the sizes of viruses could be estimated from their sedimentation characteristics. These were quite good methods, but were eventually

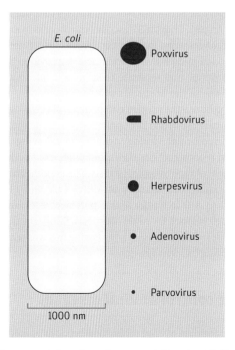

Fig. 1.1 The comparative sizes of representative viruses. The viruses are compared with a typical bacterium (*Escherichia coli*).

replaced by the less cumbersome and more precise EM technique, which had the great additional advantage of revealing the shapes of viruses; startlingly beautiful images of adenovirus emerged from RC Valentine's work in London and the virus looked like a Russian sputnik with cubic symmetry and delicate projections from each apex. It was soon found that the morphology of nearly all viruses conforms to one of two basic patterns, spherical or helical; these terms will be explained more fully in Chapter 2.

1.5 Replication

Yet again, it was the tobacco mosaic virus that yielded the first clue. In 1937, it was shown that this virus consisted not just of protein, as was first thought, but also contained ribonucleic acid (RNA). Other viruses were then found to contain deoxyribonucleic acid (DNA), rather than RNA, and in 1944, Avery and his colleagues showed that the genetic information of a bacterium, the pneumococcus, was stored in its DNA. We now know that this is true of all organisms from the bacteria upwards: viruses are unique in that their genetic material may be either DNA or RNA; in fact, the RNA viruses well outnumber those with DNA. In all viruses, the nucleic acid genome is surrounded by an outer shell of protein that functions as a protective coat during their journeys from one cell to another.

The next major event, which laid the foundation of the whole field of molecular biology, was the description by Watson and Crick and their three colleagues at King's College London in 1953 of the structure of DNA and of the way in which genetic information is encoded within it. Even this discovery did not answer the central question of how viruses replicate and thus pass their genetic information to subsequent generations. The main elements of this complicated puzzle were solved in the 1940s and 1950s, largely as a result of work on viruses that infect bacteria (bacteriophages). Chapter 3 describes the process in more detail, but in brief, it was found that, within the cell, viruses use enzymes, either encoded in their own nucleic acid or provided by the host cell, to transcribe and replicate their genetic information; they also make use of the host cell's synthetic machinery (e.g. ribosomes) to produce the protein components of the progeny virus. Some even carry RNA and DNA polymerase enzymes into the cell.

Modern molecular virology techniques allowing dissection and indeed reconstruction of viral genomes can also lead to the development of rapid diagnostic tests. The genomes of the SARS CoV virus and more recently of MERS CoV and Ebola were sequenced in a few weeks and a polymerase chain reaction (PCR) test developed for rapid diagnosis. Similarly, the genome of pandemic Influenza A H1N1/09 (swine flu) was sequenced in days following its identification.

1.6 The control of viral diseases

The understanding of viral replication brought immediate practical benefits in the form of antiviral drugs. The earliest of these antivirals, Marboran against smallpox and Amantadine against influenza, were discovered by random screening of compounds, some of which are proving remarkably successful. Aciclovir (ACV) was discovered in the 1980s and acts against certain herpesviruses, and was soon joined by a range of drugs against human immunodeficiency virus (HIV) in the 1990s. With a more complete understanding of how viruses replicate, this approach is giving way to the synthesis of drugs 'tailor-made' to attack various specific points in the replication cycle. Host cell genes and their respective proteins are now the targets of chemotherapists. Such host genes often need to be up-regulated during virus replication. So why not block these host cell enzymes and scaffolding proteins by drugs that, by definition, would not be susceptible to virus drug resistance problems? One of us (JSO) recruited volunteers to be deliberately infected with influenza in a special quarantine unit. The company (hVIVO) is now probing clinical samples such as nasal washes for cellular genes as the new target for antivirals. As many as 5000 genes aid influenza to multiply. PK, our co-author, has shown that possession of one particular human gene polymorphism in the Interferon Inducible Transmembrane 3 (IFITM3) protein can enhance the body's reaction to influenza infection and make certain groups in the world even more at risk to severe disease when infected with this global virus.

In parallel, advances in the immunology of virus infections are leading to the development of ever more safe and effective vaccines (see Chapter 30). Some viruses, such as adenovirus and pox viruses, and retrovirus, are being used as vectors to carry into the cell genes for development of virus vaccines or to correct host genetic faults. Indeed, the engineering of a virus related to rabies with the surface protein of Ebola has led to amazing progress in developing an Ebola vaccine. Sometimes, the most potent way of controlling viruses is our own immune system, and the description of an 18-year-old Frenchwoman who as a 'post treatment controller' has suppressed HIV for twelve years has reminded us of the power of our own bodies.

The Greek Goddess Hygeia was worshipped at the Parthenon in Athens 3500 years ago and her techniques of hand washing and cleanliness have been revised today in hospitals and homes and modern disinfectants formulated to kill respiratory and enteric viruses. This is now called 'implementation science'. This down to earth approach overcomes factors like human behaviour and economics which can combine to lower the impact of pure science interventions which could otherwise be effective. With HIV, more circumcision and the use of prophylactic drugs could even enhance the effectiveness of the current drug combinations.

1.7 How viruses have changed our world and will continue to do so

We have evolved for millions of years and viruses have moved with us. Some of them came out of Africa with our ancestors, most likely those causing long-term infection such as hepatitis B, hepatitis C, and herpesviruses. As nations developed and Europe dominated the world from the sixteenth century onwards, viruses travelled alongside the conquerors. Phylogeography can also be used to track the historic movement of viruses such as yellow fever and human retroviruses. For example, modern virus genetics and the construction of virus genome family trees and geographic information have enabled us to trace viruses such as HTLV-I as they spread along with the slave trade. The Spanish invaders of South America in the fifteenth century brought smallpox with them as natural infections, along with measles and influenza, all of which decimated the local Indians. Yellow Fever in South and Central America crippled economies and stopped the building of the great endeavour, the Panama Canal. Now with global warming mosquitoes can move northwards, and the African virus Chikungunya, for example, is now widespread in the Caribbean and South America and on the verge of invading North America and even the EU. Zika virus could be the same.

Gauguin called his last great painting, his scientific and artistic masterpiece 'Where have we come from, Who are we and Where are we going?' In microbial terms we have come from a world of infection but it should be clearly recognized that we have not left that world. Since Gauguin's time of the late nineteenth century only two infections, both viruses, have been truly eradicated from our world: smallpox in 1980 and rinderpest, a virus of cattle, in 2010.

Following the success of The Global Smallpox Eradication Campaign, WHO launched its 'Health for all by the year 2000' project. This was a truly global vision of the people of the world working together to eradicate disease using vaccines, hygiene, and science. But this vision has now faded a little. Minority groups are opposing vaccination and even Darwinian genetics, whilst others dislike science in general. Even the WHO vision of 'clean water for all' is being superseded by arguments within countries and between them about water ownership. One third of the population of the globe still does not have safe water. Too many countries rely today on the sewage infrastructure of 150 years ago. In some countries more citizens have mobile phones than a safe water supply and effective toilets. So the 'virology story' is not over.

Dr H. Mahler, Director General of WHO, said as long ago as in 1980, 'The present realities of the Third World are simply unacceptable. There is little joy in life nor any kind of justice for a child condemned to disease and early death because of the accident of birth in a developing country. Nor is there any rationale that can defend a system that continues to withhold the gift of health and care from 9/10ths of a nation's population. Smallpox eradication is a sign, a token of what can be achieved in breaking out of the cycle of ill health, disease, and poverty. It comes as a glimpse into the future, an intimation of a viable new order of things, in which world health, meaning health for the world, will have central significance in an upward spiral of economic and social progress.' The outbreak of Ebola in West Africa in 2014 with over 20 000 cases and 6000 deaths attests to the problems still facing us.

But as science eradicates more viruses such as polio and brings others to heel including the water- and blood-borne hepatitis viruses we all will feel proud of our efforts to fulfil his and our dreams. In turn we hope our text and the knowledge here will enthuse you and provide a start for your own contribution, either as a scientist or as a doctor, or both, in this great endeavour. And we could also remember our co-inhabitants on this planet—the animals and plants who are also very vulnerable to life threatening infection by viruses. Foot-and-mouth disease of cattle is only held under control using several million doses of vaccine per year in the world. An emerging virus of cassava, a staple crop for nearly one billion people, has been spreading at an alarming rate in East and Central Africa and will need much more virological attention. The opening of the human virus world rode upon the discovery of the first virus, that of tobacco mosaic, and perhaps now scientists of the human virology world can contribute, in return, over a century later to plant virology; perhaps to study the Mimiviruses of algae which alone exist as several million species and are key to our planet's continuing existence, with even a role in weather control.

 # Further reading

Allen, M.J. and Wilson, W.H. (2011). Viruses of Algae and Mimivirus. In Acheson, N.(ed.), *Fundamentals of Molecular Virology*, Chapter 27, pp 325–341. John Wiley & Sons, New York.

DeGoes Jacobs, C. (2015). *Jonas Salk: a life*. Oxford University Press, Oxford.

Everitt, A.R., Kellam, P. *et al.* (2012). IFITM3 restricts the morbidity and mortality associated with influenza. *Nature* **484**, 519–523.

Hayden, E.C. (2015). Teen is healthy 12 years after ending HIV drugs. *Nature* **523**, 393.

Oxford, J.S. (2012). Hygiene. In Anheier, H. K. and Juergensmeyer, M.(eds), *Encyclopaedia of Global Studies*, Volume 2, pp 838–45. Sage, Los Angeles.

Oxford, J.S. *et al.* (2002). World War I may have allowed the emergence of Spanish Influenza. *Lancet Infectious Disease* **2**, 111–14.

Patil, B.L. *et al.* (2015). Cassava brown streak disease: a threat to food security in Africa. *J Gen Virol* **96**, 956–68.

Rampling, T. *et al.* (2015). A monovalent chimpanzee adenovirus Ebola vaccine—preliminary report. *New Eng J Med DOI: 10.1056/ NEJMoa1411627*

Yarus, M. (2010). *Life from an RNA World: the Ancestor Within*. Harvard University Press, Cambridge.

Questions

1. Describe how modern virology has led to the control of some diseases and even eradication of others.

2. Write notes on:
   a. The first discovery of viruses.
   b. The widespread nature of viral diseases.
   c. The future of virology.

General properties of viruses

2

2.1 Introduction

Viruses have the following characteristics:

- They are small, retaining infectivity after passage through filters small enough to hold back bacteria (Fig. 1.1). Bacteria are measured in terms of the micrometre (μm), which is 10^{-6} of a metre. For viruses, we use the nanometre (nm) as the unit, which is a thousand times smaller (i.e. 10^{-9} of a metre). Human viruses range from about 20 to 260 nm in diameter.

- They are totally dependent on living cells, either eukaryotic or prokaryotic, for replication and existence. Some viruses do possess and indeed carry enzymes of their own, such as RNA-dependent RNA polymerase or reverse transcriptase (RT), but they cannot reproduce and amplify and translate into proteins the information in their genomes without the assistance of the cellular architecture and protein translation machinery, namely ribosomes.

- They possess only one species of nucleic acid, either DNA or RNA.

- They have a component—a receptor-binding protein—for attaching or 'docking' to cells so that they can commandeer the cells as virus production factories.

2.2 The architecture of viruses

Knowledge of the structure of viruses is, of course, important in their identification; it also helps us to deduce many potentially important properties of a particular virus. As an example, the processes of attachment and penetration of cells and, later, maturation and release differ greatly according to whether the virus possesses a lipid-containing outer envelope. Such enveloped viruses can cause fusion with the plasma membranes of host cells at some stage during their entry. The lipid bilayer protects the viral genome and acts as a permeability barrier. However viruses without this envelope tend to be more resistant to heat and are more resistant to classic hygiene practices using disinfectants. Some viruses called the 'small rounds', essentially small icosahedra, are particularly robust and a classic example is the vomiting and diarrhoea causing norovirus. There are thus important practical consequences of knowledge about virus structure.

2.2.1 Basic components of viruses

The proteins that make up the virus particle are termed structural proteins. The viral genome also codes for very important enzymes and proteins that are needed for viral replication but that do not become incorporated in the virion: these are the non-structural or NS proteins. The protein coat of a virus is termed a capsid and is itself made up of numerous capsomeres (Fig. 2.1a,b), visualized by EM as spherical virus particles, although X-ray crystallography—whereby diffraction patterns are analysed using high intensity X-rays from large synchrotrons—reveals that they are, in fact, long polypeptide

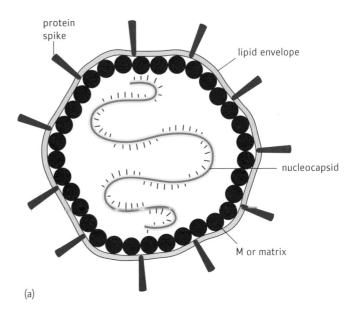

(a)

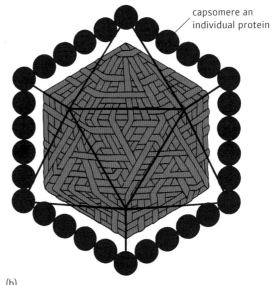

(b)

Fig. 2.1 Stylized structures of helical and icosahedral viruses. (a) A helical virus. (b) An icosahedral virus.

strands woven into complex structural patterns, much like a ball of wool. The functions of the capsid are to facilitate virus entry into the host cell and to protect the delicate viral nucleic acid outside the cell. The internal complex of protective protein and viral nucleic acid is the nucleocapsid.

As we noted above, some viruses possess an outer envelope containing lipid, derived from the plasma membrane of the host cell during their release by budding from the cell surface. In these viruses, there remains an internal capsid or core (C) made of capsomeres and containing the virus genome, but now additional virus proteins, commonly glycoproteins, are incorporated into the lipid envelope taking the form of 'spikes', protruding through a lipid bilayer. There may be a stabilizing protein membrane beneath the lipid bilayer, called the membrane or matrix (M) protein. In some enveloped viruses, such as the flaviviruses, the capsid protein has icosahedral symmetry and the external glycoprotein spike contacts the underlying icosahedron. The whole virus particle, that is the nucleocapsid with its outer envelope (if present), is called the virion.

Most of this knowledge of basic virion structure has come from electron microscopy studies. Viruses are visualized under the electron beam by negative staining using uranyl acetate. Cryo-electron microscopy, whereby the virus is frozen in liquid nitrogen combined with computer image processing, allows better preservation and visualization of the virions. A recent study which used cryo-electron microscopy observed how the core (CA) proteins of HIV fitted together as a tubular structure and even identified by meta-genesis some hydrophobic residues in the centre of the three-helix bundle. This knowledge at the level of atoms could lead to targeted chemotherapy of HIV. X-ray crystallographers have probed many viral capsomeres and we now know the precise three-dimensional configuration of the polypeptides of capsomeres of most animal viruses. More recently even complexes of viral RNA polymerases and virus replication proteins have been examined.

Perhaps the most investigated virus spike protein is the haemagglutinin (HA) of influenza and we will digress a little to use this virus as an example of modern molecular studies of a virus structural protein. The influenza HA spike protein is shaped like a 'Toblerone' chocolate bar and protrudes from the virus surface, there being about 500 spikes on each virion (Fig. 2.2). About 100 neuraminidase (NA) mushroom-like spikes also occupy space and these may cluster a little at particular surface areas. But we can conclude that there is very little space between all these spikes for access and penetration of antibodies. Biochemical studies and X-ray crystallography show each HA spike to be composed of three identical subunits with a bulb-shaped hydrophilic portion furthest from the viral membrane, while a narrower hydrophobic stalk attaches the spike to the viral lipid and protrudes through it (a transmembrane tail of 20 amino acids) to anchor the spike to the underlying membrane of matrix protein. The most exposed and distant region of the HA (called the ectodomain) contains the five antigenic sites, which often protrude from the HA, and are the very important receptor-binding site, a saucer-shaped depression near the HA tip. This ectodomain is glycosylated and hence the hydrated HAs are less prone to spurious protein interactions. Two polypeptides (HA1 and HA2), joined together by disulphide bonds, constitute the HA. These two polypeptides originate from a complete HA molecule, which during synthesis in the cell is cleaved into the two pieces by a cellular protease. An influenza virion with a cleaved HA is more infectious than one in which the HA remains as a single protein. Most viral envelope proteins are type I integral membrane proteins with the N terminus facing outwards whilst the C terminus nestles near the transmembrane tail.

It had been appreciated that the low pH environment of cytoplasmic endosomes—where the influenza virus finds itself soon after infecting an epithelial cell in the nose or pharynx—must 'do' something to the virus. In fact, it triggers a massive movement of the chain of amino acids in HA1/HA2 and the whole HA molecule becomes contorted. The central junction of HA1 and HA2 mentioned above, normally buried deep in the HA molecule near the proximal (lipid membrane) end, suddenly finds itself where, in reality, it needs to be, i.e. nearer the far or distal end, which naturally comes into contact with the lipid membrane of the endosome. The HA2 has a particular sequence of amino acids, called a fusion motif, at one end. The contortion triggered by low pH places the motif in the correct position to carry out its fusion function. Fusion of viral and endosome lipids enables the viral RNA to be released and infect the nucleus of the cell, where it will replicate.

The second spike protein of influenza, the mushroom-shaped neuraminidase (NA), has also been crystallized and studied by X-ray crystallographers. Antigenic sites were identified around the periphery of the enzyme active site on the mushroom head. This knowledge has led directly to a series of anti-influenza drugs, designed to sit precisely in the NA active site and so block enzyme action. NA is essential for release of virus from infected cells, so that these NA inhibitors are an important advance in the chemotherapy of influenza (Chapter 31). So we can see at first hand that apart from its intrinsic scientific interest, knowledge of the structure of viral capsids and capsomeres is leading to new therapies.

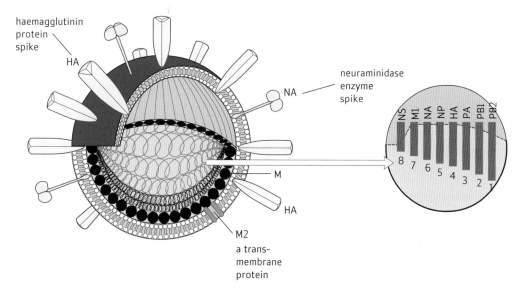

Fig. 2.2 Structural features of influenza virus.

2.2.2 Virus symmetry

The vast majority of viruses are divided into two groups (Table 2.1), according to whether their nucleocapsids have helical or icosahedral (cubic) symmetry. These terms need some explanation. A spiral staircase is a good example of helical symmetry. If one were to look directly down the centre of a spiral staircase, and if it were possible to rotate it around its central axis, the staircase would continue to have the same appearance; it would be symmetrical about the central axis. In viruses with helical symmetry, the protein molecules of the nucleocapsid are arranged like the steps of the spiral staircase and the nucleic acid fills the central core.

Table 2.1 Important structural and genome characteristics of the families of viruses of medical importance

Family name	Etymology	Representative viruses	Approximate diameter of virion (nm)	Symmetry of nucleocapsid*	Genome (kb)
DNA viruses					
Parvoviridae	'Small'	Human parvovirus B19	20	I	5 ss
Papillomaviridae		Wart viruses	50	I	8 ds circular
Polyomaviridae		BK and MC viruses		I	5.3 ds circular
Adenoviridae	Adenoid	Adenoviruses	80	I	36 ds
Herpesviridae		Herpes simplex, varicella-zoster, CMV, Epstein–Barr, Kaposi	180	I	150 ds
Poxviridae	'Pox'	Vaccinia monkey pox, molluscum, variola (smallpox)	250	C	200 ds
Hepadnaviridae	Hepatitis DNA	Hepatitis B and D	40	I	3 ds circular (partial)
RNA viruses					
Astroviridae	'Star'	Astroviruses	30–60	I	8 ss
Picornaviridae	'Small' RNA	Polioviruses, hepatitis A, common cold	25	I	8 ss
Flaviviridae	'Yellow'	Yellow fever virus, dengue, TBE, hepatitis C, West Nile	30	I	10 ss
Togaviridae		Rubella virus, Ross River, Chikungunya	80	I	12 ss
Coronaviridae	'Crown'	MERS CoV, SARS CoV	100	H	30 ss
Bunyaviridae		California encephalitis, Crimean–Congo, Hantaan, Sin Nombre	100	H	16 ss
Hepeviridae		Hepatitis E	25	I	8 ss
Orthomyxoviridae		Influenza A, B, and C	100	H	13 ss
Paramyxoviridae		Measles, mumps, Hendra, Nipah, meta pneumovirus	150	H	15 ss
Rhabdoviridae	'Bullet-shaped'	Rabies	150	H	15 ss
Arenaviridae		Lassa fever (Old World); Machupo, Pichinde (New World)	100	H	12 ss
Retroviridae	RT enzyme	HIV-1, HTLV-1	100	I	10 ss
Reoviridae		Rotaviruses	70	I	20 ds
Filoviridae	'Thread'	Marburg, Ebola	Variable	H	19 ss
Caliciviridae		Norovirus	35	I	8 ss

*H, helical; I, icosahedral; C, complex.

Cubic symmetry is more complicated, but we do not wish to delve too far into three-dimensional geometry! It is enough to say that several solids with regular sides, among them the cube and the icosahedron, share certain features of rotational symmetry, a term referring to the fact that they can be rotated about various axes and still look the same. With few exceptions, all viruses that are not helical are icosahedral, that is, they have 20 equal triangular sides. Because they belong to the same geometric group as the cube, they are often referred to as having cubic symmetry, but we shall use the term 'icosahedral symmetry'. Paradoxically these viruses do not all have the shape of an icosahedron and appear under the electron microscope as spheres. The icosahedron formation is the one that permits the greatest number of capsomeres (in fact 60) to be packed in a regular fashion to form the capsid.

Icosahedral symmetry

These viruses, such as flaviviruses, togaviruses, hepeviruses, caliciviruses, polio, hepatitis A, B, and C, and adenovirus, have a highly structured capsid with 20 triangular facets and 12 corners or apices (Fig. 2.3). The individual capsomeres may be made up of several polypeptides, as in the case of poliovirus, in which three proteins (VP1, VP2, and VP3) constitute the capsomeres. The capsomeres have a dual function, contributing both to the rigidity of the capsid and to the protection of nucleic acid from nucleases. Except for the complex poxviruses, all DNA-containing animal viruses have these icosahedron-shaped capsids, as do certain RNA viruses. The DNA-containing herpesviruses have icosahedral symmetry but, in addition, the virion is surrounded by a lipid envelope.

Helical symmetry

Examples of these viruses are the single-stranded RNA (ssRNA) viruses, such as influenza (Fig. 2.2), the parainfluenza viruses,

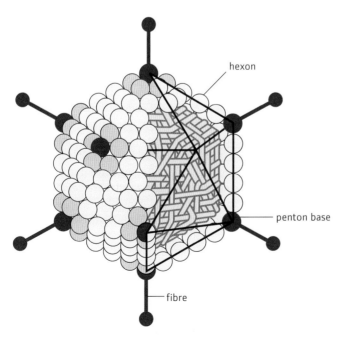

Fig. 2.3 Structural features of adenovirus.

rabies, coronaviruses, bunyaviruses, filoviruses, and arenaviruses. These viruses have the advantage of a wide variety of genome sizes: in large genome viruses the helical capsid has itself an extended length. The flexuous helical nucleocapsid is always contained inside a lipoprotein envelope, itself lined internally with a matrix protein. The lipid of the envelope is derived from the cellular membranes through which the virus matures by budding. Viral glycoprotein spikes project from the lipid bilayer envelope and often extend internally through the lipid bilayer to contact the underlying protein shell, referred to as the membrane or matrix (M) protein. This M protein may be rather rigid, as in the case of the bullet-shaped rhabdoviruses, or readily distorted, as in influenza and measles viruses. Ion channels can penetrate through the lipid and allow entry of ions into the virion interior. In the influenza virus these ions are often protons and cause a change of pH, which can trigger vital structural alterations between the four internal proteins surrounding the RNA genome and so 'activate' its infectiousness (see Section 2.2.1).

Complex symmetry

Perhaps not unexpectedly, viruses with large genomes have a correspondingly complicated architecture. Such an example is the poxvirus, which has lipids in both the envelope and the outer membranes of the virus; these viruses are neither icosahedral nor helical, and are referred to by the rather unsatisfactory designation of 'complex' viruses (Fig. 2.4). This family is literally waiting to be probed to uncover the mysteries of the internal lateral bodies and the winding of the DNA.

2.2.3 David Baltimore and virus classification dependent upon their genome

The hereditary information of the virus is encoded in the sequence of nucleotides in the RNA or DNA. This information has to be passed on to new viruses through replication of the viral nucleic acid and to direct the synthesis of viral proteins. The nucleic acid of DNA viruses does not direct protein synthesis itself. Rather RNA copies of the appropriate segments (genes) of DNA (messenger RNA, or mRNA) are used as templates to direct the synthesis of the protein. Some RNA viruses such as polio, coronaviruses, flaviviruses, and togaviruses contain a positive-strand (or positive-sense) RNA genome that acts directly as mRNA, with some of the proteins encoded by these viruses being RNA-dependent RNA polymerases which function to replicate the virus genomes and produce more virus mRNA. By contrast, negative-strand RNA viruses such as influenza, filoviruses, arenaviruses, bunyaviruses, rhabdoviruses, and paramyxoviruses possess an enzyme (RNA-dependent RNA polymerase) that copies the viral negative-sense RNA genome into a positive-stranded copy, which is then used as an mRNA to direct protein synthesis or, later, as a template from the synthesis of new genomic negative strand.

In bacteria most proteins are encoded by an uninterrupted stretch of DNA that is transcribed into an mRNA. By contrast, mammalian genes have their coding sequences (exons) interrupted

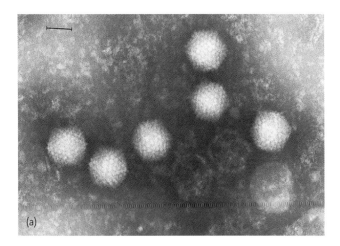

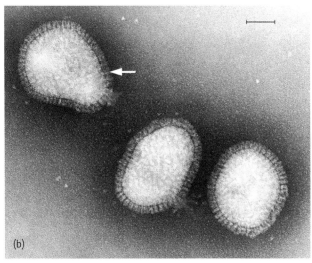

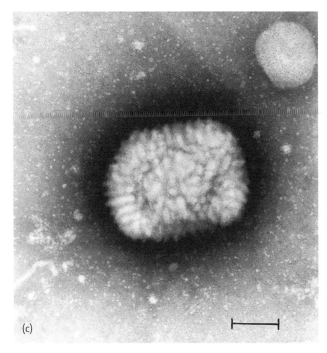

Fig. 2.4 Electron micrographs of typical icosahedral, helical, and complex viruses. (a) Adenovirus (icosahedral); (b) influenza virus (helical); and (c) poxvirus (complex). Scale bar = 100 nm.

by non-coding sequences (introns). The introns have to be removed by RNA splicing, which in essence involves a complex series of cuts followed by a 'stitching' together of the exons to form a true mRNA transcript. Splicing of precursor mRNAs can allow different proteins to be produced by the same gene and is particularly important for viruses with small genomes such as the RNA influenza or DNA parvovirus, polyomavirus, papillomaviruses, and adenovirus.

The original and rather unsatisfactory classification or grouping of viruses by their biological properties, together with the later identification of different virus genome types, led David Baltimore to propose a classification scheme based on the latter. This eponymously named 'Baltimore Classification' is now the cornerstone of molecular biology as knowledge of the virus genome type predicts properties of the viruses' replication mechanisms. All the viruses in Table 2.1 fall neatly into seven groupings. The simplest way to think about each virus genome type and Baltimore class is to determine where the virus genome inputs into the central dogma of information flow in the cell, namely DNA is transcribed into RNA which is translated into protein.

Positive-sense RNA virus genomes therefore are, in essence, messenger RNAs and so immediately amenable to direct translation in the cell. By contrast, negative-strand viruses transport into the cell an enzyme, RNA-dependent RNA polymerase, whose task is to transcribe a positive-strand RNA which can then be used as mRNA.

Single-stranded DNA (ssDNA) virus genomes, however, must be converted by virus enzymes into double-stranded DNA (dsDNA), before it is amenable for transcription. The double-stranded DNA then acts as a template for the production of more single-stranded DNA genomes.

Retroviruses and hepadnaviruses both have a reverse transcriptase (RT) in their life cycle for converting the virus RNA genome into DNA. Parvoviruses have an ssDNA genome whilst most DNA viruses have a double-stranded genome.

Conformation of genomes

Of course, viral genomes can have a number of conformations. The molecules may be double-stranded, as in higher life forms, but uniquely they may also be single-stranded; they may be linear, circular, continuous, or segmented. Quite often the viral DNA or RNA, when carefully solubilized from the virus, may be infectious, but this depends on the precise nature of the genome. Certainly, the infectivity of viral nucleic acids, either DNA or RNA, is considerably less than that of the corresponding intact virions.

We do not want students to memorize all these details of viral genes, but rather to appreciate the general principles of how viruses encode the information for manufacture of their proteins, which also enables them to commandeer a host cell and take over its machinery to manufacture virus, rather than cellular proteins.

Genome size

The sizes of human viral genomes vary greatly, but there is considerable pressure to minimize them because of the difficulty of packaging the genes into the small space within the virion.

Human cells have about 25–30 000 genes, although the diversity is increased massively by transcript splicing. In comparison, the bacterium, *Escherichia coli*, has 4000 genes. Even the largest human viruses (e.g. poxviruses) may contain only 200 or fewer genes although algal viruses can have as many as 1000 genes. The smallest human virus may have the equivalent of only three or four genes. In general, RNA viruses have smaller genomes and code for fewer proteins than DNA viruses. RNA genomes are more fragile than DNA and this feature limits their size.

It is usual to measure viral genomes in terms of the number of bases (or nucleotides) in their nucleic acid. As they are quite large, these numbers are often expressed as thousands of bases (kilobases or kb). For single-stranded genomes, the notation kb is used. For double-stranded genomes, numbers are expressed as kilobase pairs (kbp). Thus, the single-stranded measles virus genome is 16 000 bases (16 kb) in length, and the double-stranded adenovirus genome contains about 36 000 bp (36 kbp).

Virus genome and transcription maps

Figure 2.5 illustrates genome maps of two RNA viruses. Further details of particular viruses are presented in the relevant chapters. It is virtually impossible to illustrate a 'typical' viral genome or a typical transcription strategy.

Herpes and poxviruses have large DNA genomes with over 150 genes. Gene splicing is used by herpesviruses for some of their genes, but not with poxviruses, which replicate in the cytoplasm. There is precisely regulated transcription with early and

late switches, which ensure that gene products involved in DNA replication are synthesized early in the cycle and viral structural proteins much later. Both these viruses, and particularly poxviruses, have a range of unique enzymes that are carried by the viruses themselves.

The DNA hepatitis B virus has a very unusual circular genome, consisting of two strands of DNA, which is described in Chapter 28.

The smallest in terms of genome size and complexity are the circoviruses and parvoviruses, in which ssDNA replication and gene expression depend on cellular functions. Even today not much is known about the circovirus replication; however, for the parvoviruses, not unexpectedly with such a small genome, the coding regions overlap and there is splicing to produce a variety of subgenomic mRNAs.

Positive-strand RNA viruses

Poliovirus is a typical positive-strand RNA virus (Fig. 2.5a). Its RNA is polyadenylated at the 3′ end and a small virus-coded protein (VPg) is present at the 5′ end; the viral RNA is used directly as mRNA. The genome has a single open reading frame (ORF) whose primary translation product is a single polyprotein. This is cleaved to produce viral capsid proteins (VP1, VP2, VP3, and VP4), the RNA polymerases, two viral proteases, and some minor viral protein products. There are non-translated regions (ntr) at each end of the genome that have important functions. The 5′ ntr (600 nucleotides) has a significant role in the initiation of viral protein synthesis, virulence, and encapsidation, whereas the 3′ ntr is necessary for synthesis of negative-strand RNA.

By contrast, coronaviruses utilize subgenomic RNA molecules in the form of a nested set of six overlapping RNAs with common 3′ ends. Each viral mRNA is capped and polyadenylated and only the 5′ ends are translated.

Negative-strand RNA viruses

These genomes are more diverse than their positive-stranded RNA counterparts. They are not infective for cells. Furthermore, some of the viruses in this group have ambisense genomes with both negative- and positive-stranded regions of the genome encoding for proteins.

Characteristically of this group of viruses, rabies carries the RNA transcriptase in the virus itself (Fig. 2.5b). This virion-associated polymerase transcribes the single virus RNA strand by a start/stop mechanism followed by reinitiation. The five viral genes are arranged sequentially on the ssRNA genome as the nucleocapsid (N), core phosphoprotein (P), matrix (M), glycosylated membrane spike (G), and polymerase (L).

Influenza has an RNA genome composed of eight ssRNA segments, each encoding at least one protein. Using a splicing mechanism, genes 7 and 8 code for two proteins so that one protein is translated from an unspliced mRNA, whereas a smaller protein is translated from a spliced mRNA; both, however, share the same AUG initiation codon and nine subsequent amino acids.

Oddities among DNA and RNA viruses may occur, in the sense that one often sees by EM 'empty' particles that contain no

(a)

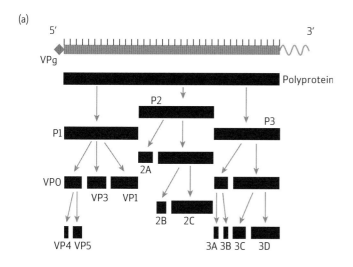

(b)

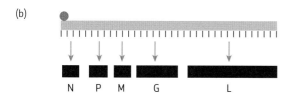

Fig. 2.5 (a) Poliovirus, a positive-strand RNA virus and encoded virus proteins. (b) Rabies, a negative-strand RNA virus and encoded virus proteins.

nucleic acid at all. These virus particles cannot, of course, replicate. Other virions may look normal by EM yet have a defective genome, lacking part of the nucleic acid needed to infect a cell; these are termed defective interfering virus particles, because although defective for their own replication, they interfere with the replication of normal viruses. This property could conceivably be harnessed to prevent virus infections.

It is useful to remember that:

- the nucleic acid of all DNA viruses except parvoviruses is double-stranded (but note that hepadnavirus DNA is partly single-stranded when not replicating);
- the nucleic acid of all RNA viruses, except reoviruses, is single-stranded.

The evolution of viruses

In the absence of fossil remains of most viruses, calculations of rates of change of viral genes must remain as estimates, although with modern use of evolutionary biology methods including molecular clocks and virus genome sequencing it is becoming increasingly possible to date important points of virus evolution, such as transmission from one species to another. An example is the estimated time of ~1900 AD when HIV emerged from a Chimpanzee to humans in Africa.

Viruses evolve rapidly because they undergo many genome duplications in a short time. Adenovirus, for example, may produce 250 000 DNA molecules in an infected cell. The host cell has evolved an editing system to test for mismatched base pairs during DNA synthesis, but RNA viruses may be unable to perform this function. With RNA viruses, the rate of divergence of RNA genomes at the nucleotide level can be as high as one nucleotide change every time a genome is copied. This is much higher than the rate for eukaryotic DNA genomes. Both viral RNA polymerases and viral RT have a very high frequency of transcription error and this, together with a lack of editing, explains why some RNA viruses, such as influenza and HIV, are so heterogeneous and exist as a quasi-species. With influenza as much as 10% of the population can be mutant. By mutation of genes coding for target proteins, these formidable RNA viral pathogens are able to evade vaccine-induced immunity, and the effects of antibody or antiviral drugs. It is a sobering thought that, because of this extraordinary genetic diversity, mutants of HIV and influenza already exist that will be resistant to antiviral drugs not as yet discovered.

RNA viruses may have originated as an RNA-only form of life whilst DNA viruses may have come from bacteria or primitive eukaryotes. Retroviruses could have evolved from mobile elements in mammalian cells.

2.3 Classification of viruses and practical applications

The precise pigeon-holing of a virus in a classification system is not only scientifically satisfying, but also of practical consequence. Two examples illustrate this assertion. For example, the AIDS virus (HIV-1) was at first thought to belong to the tumour virus group of the family *Retroviridae* (Chapter 27). When, by examination of its detailed morphology and establishment of its genome structure, HIV-1 was shown to be more related to the lentivirus group, certain features of its biology (e.g. a long latency period and absence of oncogenicity) could be filled in by reference to other viruses of the same group.

The identification of the virus causing severe acute respiratory syndrome (SARS) in South-east Asia in 2003 as a member of a previously known family of the human coronavirus (Chapter 13) brought rapid dividends, e.g. enabling diagnostic kits to be made and also giving the first intimation, compared with other members of the family, that a successful vaccine could be made.

The following are the main criteria used for the classification of viruses and these include the Baltimore categorization:

- the type of nucleic acid (DNA or RNA);
- the number of strands of nucleic acid and their physical construction (single- or double-stranded, linear, circular, circular with breaks, segmented);
- polarity of the viral genome-RNA—viruses in which the viral genome can be used directly as mRNA are by convention termed 'positive-stranded' and those for which a transcript has first to be made are termed 'negative-stranded';
- the symmetry of the nucleocapsid;
- the presence or absence of a lipid envelope.

There is a view that classification from now onwards could be based on nucleotide sequence using as a basis three automated methods of sequence comparison, namely the programme Pairwise Sequence Comparison (PASC), the more sophisticated DEmARC, or Natural Vector Representation (NVR) which records closest vector distances to other viruses. These newer methods exploit the vast number of virus sequences from Next Generation Metagenomic sequencing. On these criteria, as of 2014, all viruses can be grouped into 7 orders, 104 families, 23 subfamilies, 505 genera, and 3106 species.

Further subdivision (using the current scheme) is based on the degree of antigenic and genetic similarity. Classification lacks precision beyond this point, but antigenically identical viruses can sometimes be further categorized by differences in biological characteristics, such as virulence or cellular receptors; or in terms of molecular structure, for example, their nucleotide sequences.

Statistical comparison of the number of nucleotide sequence changes between individual members of a family enables a pictorial representative of similarity or diversity to be made, called a phylogenetic tree. The tree trunk represents the main evolutionary thrust, whereas various mutant viruses occupy branches of the tree. Viruses in the same or adjacent branches are more related to each other than those on more distant branches. An example of the usefulness of this system is shown in the case of SARS CoV virus. Phylogenetic analysis of four principal genes of the newly discovered virus showed that it was indeed new to humans and had not simply emerged from an existing grouping. Table 2.1 summarizes the genetic, biological, and structural characteristics of the families of viruses known to cause disease in humans.

2.4 The nomenclature of viruses

This topic has been in a state of flux for many years, and the names now in general use are based on characteristics that vary from family to family. Some viruses are named according to the type of disease they cause (e.g. poxviruses and herpesviruses); other familial names are based on acronyms, for example papovaviruses (*papilloma–polyoma–vacuolating* agent) and picornaviruses (*pico*, small; *rna*, ribonucleic acid); others again are based on morphological features of the virion, (e.g. coronaviruses, which have a halo or corona of spikes). Some individual viruses are named after the place where they were first isolated (e.g. Coxsackie, Marburg). More recently, in the case of hantavirus outbreaks in the USA, residents strongly objected to a newly discovered strain being named after their locality. WHO has issued guidance on the naming of new viruses and in future we may lose names such as West Nile and Ebola. Occasionally, viruses have been named after their discoverers [e.g. Epstein–Barr virus (EBV)].

The descriptions, classification, and nomenclature of viruses are specified by the International Committee on Taxonomy of Viruses (ICTV), which issues a report every few years. The complexity and expansion of the subject may be judged by the fact that these reports weigh in at nearly 3 kg. We shall not burden you with much detail, but you should know that the virus world is organized into seven orders, the *Caudovirales, Herpesvirales, Ligamenvirales, Mononegavirales, Nidovirales, Picornavirales,* and *Tymovirales* which have 3, 3, 2, 5, 4, 5, and 4 families within them, respectively. There are also 78 families currently not assigned to an order. The subsequent taxonomic levels are families, subfamilies, genera, and species, where families are printed in italics with upper case initial letters. This can produce a complex classification, but does lead to an easy to navigate structure (see: http://www.ictvonline.org/virusTaxonomy.asp). For example, measles virus is in the Order *Mononegavirales*, Family *Paramyxoviridae*, Subfamily *Paramyxovirinae*, Genus *Morbillivirus*; species Measles virus. Names used colloquially, as in 'adenoviruses cause some respiratory infections' are also printed in roman.

2.5 The range of diseases caused by viruses

We have so far dealt with viruses only as micro-organisms; we shall now start to look at them, albeit very briefly indeed in this chapter, in relation to the diseases that they cause in humans. In previous editions of *Human Virology* we adopted this more medical and biological classification but in the present book, as you will see, we have placed viruses in their virological family rather than identify them with a particular disease syndrome. We feel that adoption of the Baltimore scheme will make viruses, their biology, and disease-inducing characteristics more understandable to students.

Viruses vary greatly not only in their host range, but also in their affinity for various tissues within a given host (tissue tropism) and the mechanisms by which they cause disease (pathogenesis). Some viruses from very different families are predominantly neurotropic (Table 2.2), others replicate only in the liver (Table 2.3) or skin, and others again can infect many of the body systems. Some viruses are sexually transmitted (Table 2.4). Respiratory symptoms are an excellent

Table 2.2 Viruses causing acute infections of the central nervous system

Predominant syndrome	Viruses	Predominant neurological lesions
Meningitis	Enteroviruses, especially ECHO, Coxsackie A and B, enteroviruses 70 and 71, poliovirus	Inflammation of the meninges, with or without some degree of encephalitis
	Mumps, lymphocytic choriomeningitis, louping-ill, Epstein–Barr virus, HSV-2, VZV	
Poliomyelitis	Polioviruses; occasionally other enteroviruses	Meningitis, lysis of lower motor neurons
Meningoencephalitis	HSV-1, arboviruses	Necrosis of neurons in grey matter of brain
Encephalitis	Rabies	Varying degrees of neuronal necrosis; perivascular and focal inflammation
AIDS dementia complex (ADC)	HIV-1	Meningitis cortical atrophy, focal necrosis; vacuolation, reactive astrocytosis and microgliosis in subcortical areas; demyelinating peripheral neuropathy
Tropical spastic paraparesis	HTLV-1	Upper motor neuron lesions

Table 2.3 Viruses of different families causing hepatitis

Virus family	Main route of transmission
Picornaviridae	Enteric HAV
Hepevirus	Enteric HEV
Hepadnaviruses	Parenteral HBV
Deltavirus	Parenteral HDV
Flaviviridae	Parenteral HCV
Flavi-like viruses	Enteric GBV

example of the considerable overlap between the syndromes caused by very different viruses. Members of five families of viruses cause cough, sore throat, and runny nose (Table 2.5). As may be imagined, this can create problems in diagnosis, most of which can, however, be solved by studying the clinical picture as a whole aided by other factors such as the season (Fig. 2.6) and enlisting the aid of the virology laboratory as

Table 2.4 Families of different families causing sexually transmitted disease

Virus family	Main clinical features
(a) Localized infections	
Herpesviridae	HSV-2 more severe than HSV-1. Painful, itchy vesicular lesions on genitalia, anal, perineal areas, possibly mouth. Urethritis, proctitis, cervicitis
Papillomaviridae	Warts, anogenital mucosa, cervix, possibly larynx
Adenoviridae	Types 19 and 37 cause ulcers on external genitalia, urethritis
Poxviridae	Characteristic lesions on genitalia
(b) Generalized infections	
Picornaviridae	Hepatitis A may be transmitted by anal sex
Hepeviridae	Risk difficult to quantify, probably not high
Hepadnaviridae	Very high risk of acquisition from HBeAg-positive carriers if intercourse is unprotected
Flaviviridae	Risk difficult to quantify probably not high, recorded with Zika
Retroviridae	Male to female transmission of HIV more efficient than female to male. Spread favoured by presence of genital lesions due to, e.g. herpes, syphilis, chancroid

Table 2.5 Viruses of different families causing common respiratory infections

Virus family	Diseases
Rhinoviruses	Common cold
Picornaviridae Coronaviridae	Common cold
Herpesviridae	Pharyngitis
Myxoviridae Picornaviridae	Influenza
Adenoviridae	Pharyngitis
Myxoviridae	Bronchitis
Paramyxoviridae	Bronchitis
	Croup
	Influenza-like illness (ILI)
Myxoviridae Paramyxoviridae	Broncho pneumonia
Myxoviridae Coronaviridae Paramyxoviridae	Pneumonia

early in the illness as possible. Perhaps a little synthetic is the grouping of viruses as 'emergent'. Nevertheless this is a very important grouping and includes pandemic influenza and Ebola (Table 2.6). The factors which influence this emergence are wide and at the same time rather frightening for our future survival (Table 2.7). Emergent viruses can be from the families *Paramyxoviridae* (Hendra and Nipah, SARS and MERS), *Retroviridae* (such as HIV), *Flaviviridae* (such as Yellow Fever and West Nile and Zika virus), *Myxoviridae* (such as bird flu (H5N1)), and *Togaviridae* (such as Chikungunya).

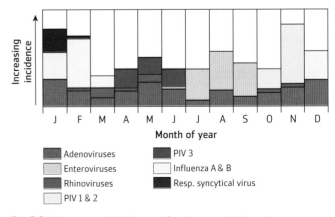

Fig. 2.6 The seasonal incidence of various respiratory viruses.

Fig. 2.7 Timeline of emerging viruses

Table 2.6 Viruses of different families which have recently emerged

Virus family	Disease in humans	Location	Reservoir
Poxviridae	Pox lesions	USA	Gambian giant rats and prairie dogs
Paramyxoviridae	Respiratory disease in Nipah	Malaysia	Bats
Coronaviridae	Respiratory disease SARS	South-east Asia	Civet cat or bat
Retroviridae	AIDS	Africa	Chimpanzee
Flavividae	Mild rash hvt teratogenic	S. America	Mosquitos
Togaviridae West Nile	Encephalitis	USA and Europe	Wild birds
Paramyxoviridae	Respiratory disease Hendra	Australia	Horses
Myxoviridae	Respiratory disease of influenza A H5N1	South-east Asia	Migrating ducks and geese, domestic ducks and chickens
Filoviridae	Ebola haemorrhagic fever	West Africa	Bats
Coronaviridae	Respiratory disease MERS	Saudi Arabia	Camels

2.6 A little 'non-human' virology

So far, in line with the intention of this book, we have considered only viruses that infect humans or other mammalian or avian species that interact with them. For the sake of completeness, however, we should mention that there are many other viruses that affect most—possibly all—of the diverse life forms on our planet. They include vertebrates, insects, plants, algae, fungi, and bacteria. Fortunately, we need concern ourselves briefly with only two of these groups.

It seems extraordinary that some of the smallest life forms on our planet should harbour even smaller living entities: but they do. As Swift (1713) wrote:

So, naturalists observe, a flea
Hath smaller fleas that on him prey;
And these have smaller fleas to bite 'em,
And so proceed *ad infinitum*.

Bacteriophages ('phage' for short), as their name suggests, are viruses that infect bacteria. They have been isolated from many species, including chlamydias, and fall into two main categories: *virulent* phages that lyse their host bacteria, and *temperate* phages that set up a non-lytic (lysogenic) cycle of replication. Their nucleic acid may be either DNA or RNA. The host range, morphology, sizes (roughly 20–2000 nm), and replication strategies of these agents vary greatly and we shall not deal with them at any length. That is not to say that they are of no importance. The following points are worth noting:

• Phages have proved invaluable as tools for studying the genetics of bacteria and for demonstrating that their nucleic acid is physically separable from their protein coats.

• Species and subspecies of various bacteria, e.g. *Staphylococcus aureus* and various salmonellas can be identified by their patterns of susceptibility to a range of phages. Rapid phage typing is particularly useful for studying outbreaks of infection with these and some other bacteria.

• Attempts have been made, notably in Eastern Europe and the former Soviet Union, to exploit the lytic properties of bacteriophages for treating enteric infections, e.g. cholera, but such measures were not very successful and have been largely superseded by the use of antibiotics. Nevertheless, in the face of widespread resistance to antibiotics, there has been a revival of interest in the possibility of using phages for therapeutic purposes.

• The presence of a phage in its lysogenic state (prophase) may give rise to phenotypic changes in the bacterial host.

• Some bacterial toxins, e.g. those of cholera and diphtheria, are encoded by prophages.

• Contamination with phages may interfere with certain industrial processes, e.g. manufacture of dairy products and antibiotics.

Table 2.7 Factors influencing the emergence of infection caused by viruses of five families

Factor	Example
Breakdown of species barriers	Probable transfer of HIV from a simian reservoir to humans. Spread of influenza viruses from animals or birds to humans, and most recently H1N1 virus from pigs to humans. Transfer of BSE to humans.
Genetic	RNA viruses, particularly HIV and influenza, have extremely high mutation rates, which favour the spread of strains resistant to immune barriers or chemotherapy.
Host factors	
Ecology and modern agriculture	New farming practices encourage human contact with rodents carrying Hantaan and other exotic RNA viruses.
Medical or surgical interventions	Immunosuppressive drugs have encouraged latent herpesviruses to emerge, such as CMV. Brain surgery can spread CJD.
Human behaviour	Changes in sexual habits have enhanced the spread of sexually transmitted diseases.
Intermediate hosts	Hantaviruses, carried harmlessly in rodents, infect humans when rodent populations expand because of increased availability of food. Recently, fruit-eating bats have been recognized as a source of novel viruses, such as Hendra and Nipah, and Ebola. Migrating birds carry potential pandemic influenza A viruses such as H5N1 (bird flu).
Population increase	Increasing population densities and urban poverty encourage the spread of water- and airborne viruses.
Global warming	Mosquitos carrying Chikungunya, formerly an African arbovirus, have moved northwards to Italy. Similarly, the veterinary virus Blue Tongue has moved out of Africa to Europe and the UK and Zika virus from Africa to S. America.
Defects in public health infrastructure	Poor control of mosquitos because of worries about the insecticide DDT or poor public health structure, or both, have allowed the re-emergence of dengue, West Nile and Zika, and other classic insect-borne viruses.

 # Reminders

- Viruses are characterized by their small size, obligate intracellular parasitism, possession of either an RNA or a DNA genome, and of a receptor-binding protein.

- Viral genomes vary in size from three to 150 genes, but most commonly have 10–15 genes. Such compression of genetic information forces viruses into strategies to extend the genetic information by splicing or utilizing different reading frames.

- The genome is protected by a coat or capsid consisting of protein subunits (capsomeres). Each capsomere is made typically of one to three polypeptides. Some viruses also have a lipid envelope. The nucleic acid is often complexed with a protein. The nucleic acid core and the capsid are together known as the nucleocapsid. The complete virus particle (nucleocapsid with envelope, if present) is termed the virion.

- The tertiary structure of many capsid proteins has been established by X-ray crystallography. The nucleocapsids of most viruses are built either like helices, with the capsids arranged like the steps in a spiral staircase around the central genome core, or like icosahedra, in which the capsids are arranged to form a solid with 20 equal triangular sides, again enclosing the genome.

- Viruses are classified into families according to the characteristics of their nucleic acid, whether DNA or RNA, the number of strands, and their polarity. Positive-stranded viruses use the viral genome as mRNA. Negative-stranded viruses must first make a transcript of the genome RNA to be used as a message and carry an RNA transcriptase in the virus particle to perform this function.

- Some families are divided into subfamilies on the basis of gene structure (DNA or RNA, single-stranded (ss) or double-stranded (ds), or circular or segmented). Further subdivisions into genera depend on antigenic and other biological properties. Phylogenetic trees can be constructed, which outline relatedness between members of a family and give indications of evolution and change.

- Viruses vary widely in the range of hosts and tissues that they can infect; members of some families cause a wide range of syndromes, whereas the illnesses due to others are much more limited in number, including STDs, neurological disease, and respiratory infection. Members of six families are so-called emergent viruses.

Further reading

Gharfour, A. (2014). Structure of the core ectodomain of the hepatitis C virus envelope glycoprotein 2. *Nature* **509**, 381.

International Committee for the Taxonomy of Viruses (ICTV) (http://ictvonline.org/virusTaxinfo.asp)

Simmonds, P. (2015). Methods for virus classification and the challenge of incorporating meta genomic sequence data. *Journal of General Virology* **96**, 1193–1206.

Zhao, G. *et al.* (2013). Mature HIV-1 capsid structure by cryo electron microscopy and all-atom molecular dynamics. *Nature* **517**, 381.

Questions

1. Discuss the classification of viruses and the Baltimore categorization of viruses.

2. Outline the range of viruses causing:
   a. Respiratory illness.
   b. STD.
   c. Neurological disease.

3 Viral replication and genetics

3.1 Introduction

In the task of reproducing themselves (replicating), viruses are at a major disadvantage compared with higher forms of life. The latter all multiply by some form of fission, so that the daughter cells start their existence with a full complement of genetic information and with the enzymes necessary to replicate it, and to catalyse the synthesis of new proteins. A virus, on the other hand is totally dependent upon the machinery of a cell and enters the cell with nothing but its own genome, a molecule of nucleic acid, which may encode for only 20 or so genes compared with 30 000 genes for a mammalian cell. Sometimes, but not always, it may not encode even a single virus enzyme to start the process of replication. This is why it must rely so heavily on the host cell for the materials it needs for reproduction involving many cellular proteins and cell pathways, and why the replication of viruses is more complicated in some respects than that of other micro-organisms. Although we now have a fairly detailed picture of the main steps, we still do not know everything about the strategies that viruses have developed over the millennia to continue their existence. In fact, detailed investigations of viral replication continue to uncover major facts about control mechanisms and genetic strategies in our own cells.

As the replication of viruses is so intimately related to cellular activities, it may be helpful to provide a brief outline of the molecular biology of the host cell, with which you are probably already familiar, and will serve as a basis of comparison between these very different life forms.

3.2 A brief résumé of the molecular biology of the mammalian cell

There are over 100 types of cell in the human body and these are mostly assembled into tissues. A typical animal cell is 20 μm in diameter. The most prominent organelle is the **nucleus**, which is enclosed by two concentric membranes that form the nuclear envelope. The nucleus contains dsDNA, which encodes the genetic specification of the cell. The nuclear pores allow certain molecules to enter and exit from the nucleus, but this procedure is strictly controlled.

The cell interior is composed of the microscopically transparent cytoplasm that is full of organelles and in which extensive traffic takes place, involving proteins and various chemical messengers, all of which may be important during viral replication. All the cell organelles are enclosed by membranes, thus enabling the cell to carry out many different processes at the same time. An important organelle is the **endoplasmic reticulum**, an irregular structure enclosed by lipid membranes. Also present is the Golgi organelle, which has the appearance of a stack of empty sacks. **Lysosomes and endosomes** are balloon-like structures in which intracellular digestion occurs. Many viruses uncoat and initiate infection in this organelle. Ribosomes are the framework upon which new proteins are made, under the direction of mRNAs.

Continual exchange of material takes place between these organelles and the outside of the cell, itself surrounded by a lipid bilayer **plasma membrane**. Underlying this is a **cytoskeleton** of filaments, such as actin, that strengthen the cell and give it a particular shape. The plasma membrane is packed with receptors for different molecules, many essential for the functioning of the cells. Viruses may use some of these receptors to enter and infect cells.

3.2.1 DNA as the carrier of genetic information

The life of the cell depends on maintaining the cell's primary function and homeostasis in a changing environment. To achieve this, the cell relies on its ability to store, retrieve, and translate genetic instructions, which are stored as **genes**. In the 1940s, DNA was identified as the carrier of genetic information. A DNA molecule consists of two strands of nucleotides held together by hydrogen bonds—a DNA double helix. The two strands of the helix each have a sequence of nucleotides that is complementary to that of the partner strand. DNA is an example of a **linear message**, encoded in the sequence of nucleotides along each strand. The DNA of a human cell is composed of 3×10^9 nucleotides and has a four-letter nucleotide 'alphabet' (A, C, T, and G). During replication this information must be copied faithfully.

3.2.2 Replication of cellular dsDNA

At replication, each DNA strand acts as a **template**, for the synthesis of a new **complementary strand**. DNA replication produces two complete double helices from the original DNA molecule. An enzyme, **DNA polymerase**, is central to this process. It catalyses addition of nucleotides to the 3′ end of a growing DNA strand by the formation of a phosphodiester bond between this end and the 5′-phosphate group of the incoming nucleotide. Virus DNA or RNAs are replicated in a similar manner—that is the virus genome acts as a template for new genomes to be produced from.

3.2.3 Transcription of DNA to form mRNAs

When a particular protein is required by the cell, the correct small portion of this immense DNA molecule is copied into RNA in a process known as **transcription**. In turn, these RNA copies (mRNAs) are used as templates to direct the synthesis of the protein. Many thousands of these transcription events occur each minute in mammalian cells. The information in the mRNA is used to make a protein, a process called **translation**. The virus can utilize this system with high efficiency. The virus can subvert the system or completely blockade translation of host-cell proteins.

3.2.4 Processing of primary RNA transcripts

The mRNA produced must be processed in the nucleus from a **primary transcript** to the final mRNA. First, the RNA is **capped** at the 5′ end by addition of a guanine nucleotide with a methyl group attached. Then a **poly(A) tail** is added at the 3′ end.

Mammalian cell genes (and many viral genes) have their coding sequences interrupted by non-coding sequences (**introns**) perhaps as long as 10 000 nucleotides. These introns are removed by **RNA splicing**. At each intron a group of small nuclear ribonucleoprotein particles (snRNPs) assembles in the nucleus, cuts out the intron, and rejoins the RNA chain. In fact, this splicing was first discovered in cells infected with viruses.

3.2.5 Translation of mRNAs into proteins

After migrating from the nucleus to the cytoplasm, the mRNAs are translated into proteins. Each group of three consecutive nucleotides in mRNA is called a **codon** and each specifies one amino acid. It follows, therefore, that an RNA sequence can be translated in three different **reading frames**. Particular codons in the mRNA signal the sites where protein synthesis starts and stops. **Initiation factors** and the mRNA interact with a small ribosomal subunit, which moves forward (5′–3′) along the mRNA, searching for the first **start codon**, namely AUG. The end of the protein-coding message is signalled by UAA, UAG, or UGA, termed **stop codons**. Viral mRNAs are translated in a similar manner but may have to compete with cellular mRNAs.

Note that genes are indicated by italics (e.g. *tax*) and their products by roman script (e.g. tax).

3.2.6 Control of gene expression

Gene expression must be controlled and a particularly important stage is at initiation of transcription. The **promoter region** of a gene attracts RNA polymerase: it has an **initiation site** and an associated 'upstream' region. Most genes also have **regulatory DNA sequences** that are required to switch genes on; they are recognized by regulatory proteins that bind to the DNA and act as activators or repressors. The process also requires the co-operation of a large set of proteins called **general transcription factors**. Some viruses code for their own transcription factors, which thereby interrupt the normal gene expression in the cells. The most recent discovery is RNA interference, the 'silencing' of gene expression by dsRNA molecules. RNA interference has evolved to silence viruses and rogue genetic elements that make dsRNA intermediates, types of RNA not usually produced by cells.

At the beginning of the twenty-first century a new mechanism for controlling gene expression was uncovered with the identification of small non-coding RNAs, now known as micro-RNAs, in animal genomes. There are hundreds of micro-RNAs in the human genome, which seem to function by inhibiting protein translation through binding to mRNAs. This mechanism for controlling protein production has not escaped the attention of viruses, with the double-stranded DNA herpesvirus in particular encoding their own, species-specific microRNAs.

3.3 Virus infection and replication in a host cell

The initial infection of a cell is a rather hit-or-miss process, depending upon chance contact, but is greatly helped if a virus enters the body at a suitable site and in large numbers. Often thousands of viruses may enter the body, and yet only two or three actually establish an infection. The remainder are destroyed by the general defences before they have a chance to infect. There follows a period of a few hours during which nothing seems to be happening. This appearance is, however, deceptive because much is going on inside the cell at the molecular level, such as **transcription** of the 'incoming' viral genes to form viral mRNAs, and their translation to produce early viral proteins, including the enzymes necessary to replicate viral DNA or viral RNA. Thus, although there are no visible signs in the cell, at the molecular level sensitive probes for the presence of viral components (i.e. those that are subtly different from the makeup of a cell) signal the subtle, but definite changes of an 'infected cell'. We now know with the help of the young sciences of genomics, proteomics, and techniques of gene arrays and transcription arrays that 5000 or more human genes may be involved and be key to influenza virus replication, for example.

There is a fundamental difference between the replication of viruses and bacteria; the latter retain their structure and infectivity throughout the growth cycle, whereas viruses lose their physical identity, and most or all of their infectivity during the initial stage of replication, which for this reason has been termed the **eclipse phase**. The next stage, the **productive phase**, is even more full of action as new virus particles are produced and released from the cell.

Hot topic Involvement of host genes in influenza infection

Many thousands of cellular proteins encoded by host cell genes are involved in virus replication, and these have been identified in cell culture experiments and in experiments involving infections in animal models. In a study carried out in London at the hVIVO (Retroscreen) quarantine unit, volunteers were infected with influenza A virus, and blood samples were taken at frequent intervals before, during, and after infection. DNA was extracted from the white blood cells and analysed by gene arrays. All the volunteers became infected but some of them produced no symptoms. The study established that transcription of over 5000 host genes was initiated but the pattern was different in the two groups.

Symptomatic volunteers involved the transcription of multiple pattern recognition receptors (PRRs), which in turn mediated responses which resulted in oxidative stress. In contrast, in the asymptomatic volunteers transcription of genes were elevated that functioned to reduce oxidative stress. Increased mRNA expression for antioxidants SOD1 and SOK1 was noted in the asymptomatic group, which may be an effective antiviral response. The asymptomatic group lacked activation of PRRs and in fact initiated a potent cell-mediated innate immune response. Also negative regulation of the inflammosome signals, especially with NCRP3 and NOD2, correlated with lack of clinical symptoms.

3.3.1 Recognition of a 'target' host cell by a virus

All viruses have on their outside a receptor-binding protein, which often has a structured pocket that reacts specifically with a corresponding receptor on a cell surface. These cellular receptors invariably have other functions and viruses simply use them for attachment. These 'virus receptors' on cells are often glycoproteins or glycolipids. Each virus family tends to have separate and unique receptors.

Once attached, which may be a more or less instantaneous process, viruses are almost impossible to dislodge. That is the 'on-rate' of virus attachment to its receptor is greater than the 'off-rate' of dissociation from the receptor. This precise 'lock-and-key' interaction explains why many viruses are restricted to a given host and, within that host, to particular cells and tissues:

For example, the AIDS virus, HIV-1, recognizes and reacts specifically with two receptors on certain T lymphocytes and other cells, and can thus attach to and infect only these cells. The primary receptor is the CD4 molecule found on immune T cells and a secondary receptor is a chemokine receptor molecule, CXCR-4, or CCR-5. (CD = 'cluster of differentiation'.)

3.3.2 Internalization of the virus

Having attached to the viral host cell, the virus must penetrate the external plasma membrane of the cell rapidly and release its genome into the cellular milieu for subsequent replication. This **internalization** is accomplished in one of three ways, but only minutes elapse before the virus finds itself inside the cell.

Fusion from without

Fusion at the cellular external plasma membrane, namely 'fusion from without', is the strategy of entry of paramyxoviruses such as measles and mumps viruses, and also HIV. Such viruses have a **'fusion protein'**, with a short stretch of catalytic hydrophobic amino acids, which mediates fusion between the lipids of the virus and the lipids of the cell membrane following receptor binding at the plasma membrane.

Receptor-mediated endocytosis (viropexis)

Viropexis is the most common cellular entry technique for viruses (Fig. 3.1). Mammalian cells have had to develop methods of attachment and entry of a range of essential molecules, such as nutrients and hormones. Viruses can exploit these existing avenues of entry. Viruses attach at special virus receptor areas on the cell membrane. The cellular protein, clathrin, which underlies the membrane, forms a so-called **coated pit** and, once the virus has attached, inversion of the cellular membrane and associated virus occurs. The virus is now in the cytoplasm, but is still bounded by the cell membrane in the vacuole known as an **endosome**, through which it has to negotiate a route to the true internal environment and often to the nucleus of the cell. The endosome is a hostile environment with many proteases and other protein-degrading components. It is likely that for viruses that enter cells via the endosome, the virus capsids and nuclear capsids have evolved to be resistant to the

inside of the endosome, protecting the virus genome until it is ready to exit the endosome and enter the cytoplasm. These endosomes offer a convenient and rapid transit system across the plasma membrane and also through the cytoplasm to the nuclear pore. Many products of cellular genes can be involved at these early stages of virus infection. Often the trigger for the virus to exit the endosome is the acidification of the inside of the endosome by pumping protons (hydrogen ions) into the endosome. This lowering of the pH causes conformational changes in the virus envelope glycoproteins and/or capsid that results in either the virus membrane fusing with the endosomal membrane and depositing the nucleocapsid in the cytoplasm, such as for influenza viruses, or the direct translocation of the virus genome from the capsid to the cytoplasm, such as for picornaviruses.

Non-clathrin-mediated endocytosis

A few viruses may enter by a third technique known as non-clathrin-mediated endocytosis or via a caveolae-assisted entry.

In all cases, quite extensive internal trafficking occurs before the virus RNA is released from the internalized virus and enters the nucleus via the nuclear pore, or alternatively starts viral gene transcription in the cellular cytoplasm. Many viruses never enter the nucleus of the cell.

3.3.3 Formation of viral mRNAs: a vital step in virus replication and a key to the classification of viruses first identified by Baltimore

The American scientist Baltimore received the Nobel Prize for his co-discovery of reverse transcriptase enzyme. But other very important discoveries and observations have come from his laboratory including a rather unique virus classification scheme. In the **Baltimore classification scheme** viruses are categorized into seven groupings by virtue of their genome type (DNA or RNA) and strategies for replicating the virus genome and for forming viral mRNA, which depend upon the sense of their genome RNA. In this context, 'sense' refers to whether the genome is homologous with the viral mRNA ('positive-sense' or 'positive-stranded') or complementary to it ('negative-sense' or 'negative-stranded').

When viruses infect cells, two important and separate events must be orchestrated, namely production of **virus structural proteins and enzymes**, and **replication of the viral genome**. Viruses have various methods of ensuring that their mRNAs are produced and then translated into viral proteins, often in preference to normal cellular mRNAs.

Before we delve into the strandedness of RNA viruses we would recall that in previous editions of *Human Virology* we veered more towards a clinical classification of viruses. Thus the diverse families causing hepatitis, respiratory disease, and diarrhoea were discussed side by side in the same chapter. Here in the new edition we have positioned viruses according to the Baltimore scheme whereby viruses are grouped according to the pathway towards mRNA and protein synthesis, which we will now consider.

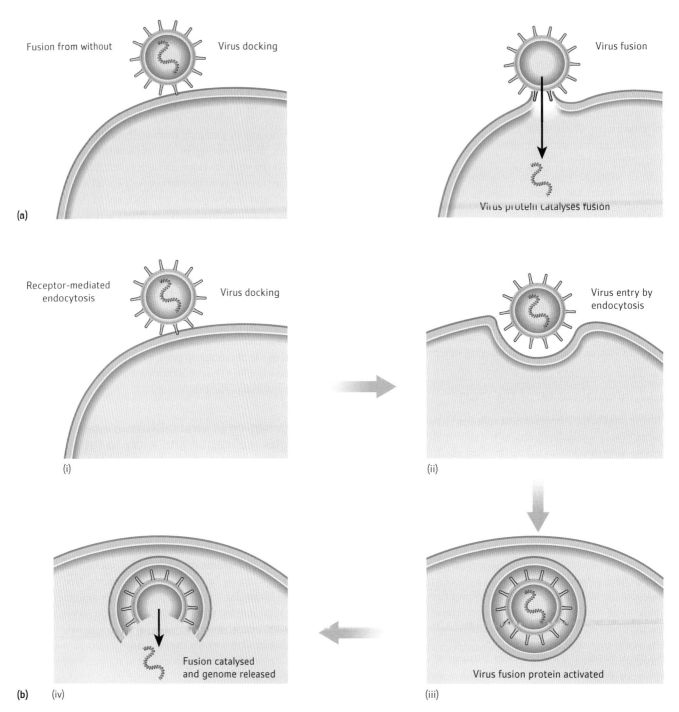

Fusion from without   Virus docking

Virus fusion

Virus protein catalyses fusion

(a)

Receptor-mediated endocytosis   Virus docking

Virus entry by endocytosis

(i)   (ii)

Fusion catalysed and genome released

Virus fusion protein activated

(b)   (iv)   (iii)

Fig. 3.1 Modes of entry of viruses into cells. (a) Fusion from without (b) Receptor-mediated endocytosis (viropexis).

Positive-stranded RNA viruses

The typical **positive-stranded parental viral RNA** (Fig. 3.2), with the addition of a poly(A) (AAA) tract at the 3′ end of the molecule and a cap at the 5′ end, is used directly as viral mRNA, from which 'early' and 'late' viral proteins are translated directly.

Polio and flaviviruses are good examples of positive-stranded RNA viruses. Another feature of these viruses is that the viral genome is itself infectious for cells, but much less so than the complete virus since naked RNA is extremely susceptible to RNAase enzymes which abound everywhere. But under

very controlled conditions polio RNA may be synthesized directly from its nucleotide constituents and directly 'forced' into cells by electroporation where the genome alone initiates virus replication.

Most positive-strand RNA viruses replicate exclusively in the cytoplasm of the cell. With the positive-stranded RNA viruses, a virus-coded RNA polymerase ('**replicase**') is translated directly from the viral genome, whereas the replicase of negative-stranded RNA viruses is carried into the cell by the virus itself. Either way, the RNA replicase synthesizes a complementary RNA

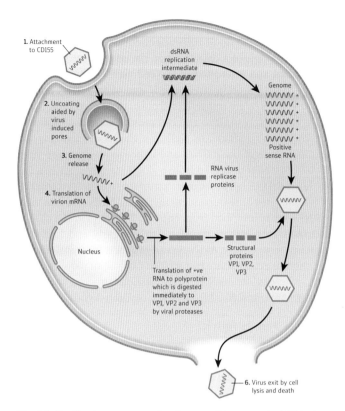

Fig. 3.2 Replication strategy of polio, a positive-stranded RNA virus.

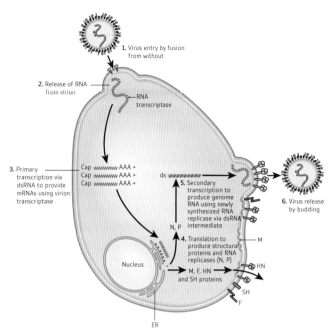

Fig. 3.3 Life cycle and replication of a negative-stranded RNA virus: paramyxovirus.

strand that serves as a template for new rounds of viral RNA synthesis. These RNA duplexes (dsRNA) are unstable and occur only as transient 'replicative intermediates'. But as we note in subsequent chapters these dsRNAs, being unique to a virus-infected cell, allow recognition of infection and through extracellular signalling gives the adjoining cells an advance warning of infection, and can trigger a huge and immediate antiviral response.

Negative-stranded RNA viruses

Some viruses, such as influenza, achieve release from the internal endosomal vacuole by internal fusion ('fusion from within') mediated by the viral HA protein. A further requirement of internal fusion with influenza is a low pH in the cytoplasmic vacuole; this triggers a movement of the three-dimensional structure of the HA protein, so allowing juxtaposition of the HA fusion sequence, normally buried deep in the HA spike protein, with both viral and cellular lipid membranes.

In the case of the **negative-stranded RNA viruses**, for example parainfluenza, influenza (Fig. 3.3), or rabies, a virus-associated RNA polymerase (transcriptase), which is carried into the cell by the virus, must first make mirror-image copies of the original negative-strand viral RNA segments. These copies, now positive-stranded, are capped at the 5′ end, polyadenylated at the 3′ end, and then function as viral mRNAs that, in turn, are translated on the ribosomes of the cell to give viral proteins.

For influenza, for example, the viral genome is in the form of eight loosely linked single-stranded RNA segments. Most transcribed mRNAs are monocistronic, i.e. they code for a single protein. However, the mRNAs of genes 7 and 8 of influenza undergo splicing, and each are now known to code for two viral proteins. The mode of transcription and replication of influenza virus is unusual as it requires co-operation with cellular RNA polymerase II ('cap snatching') and association with the cell nucleus to initiate genome replication.

RNA viruses with double-stranded RNA genomes

Similarly to the viruses with negative stranded ssRNA, dsRNA genome viruses code for an RNA-dependent RNA polymerase in the virion, which they carry into the cell. The negative strand of the double-stranded RNA is transcribed into a positive-strand RNA which then functions as mRNA.

RNA viruses with reverse transcriptase enzyme

The fourth group of RNA viruses—the **retroviruses**—have a more complex strategy of producing viral mRNAs. They have a positive-stranded RNA genome and in the case of HIV carry an RNA-dependent DNA polymerase (reverse transcriptase enzyme, **RT**) into the cell. The essentials of HIV replication and integration are illustrated in Chapter 27 (Fig. 27.6). As soon as the virus infects a cell the parental viral RNA is reverse transcribed by the virus-associated RT, which converts the incoming viral RNA genome to a DNA–RNA hybrid. The RNA strand is digested away from the hybrid and the existing DNA strand is copied to a complementary DNA to give a dsDNA molecule. With HIV this dsDNA is **integrated** into the chromosomal DNA of the host cell by a virally encoded integrase, and is now termed **proviral DNA**. Viral mRNAs are transcribed from the proviral DNA in much the same way as host-cellular mRNAs are transcribed from host-cell chromosomal DNA using host-cell RNA polymerase II. The viral messages are translated and viral proteins are synthesized.

The hepatitis B DNA genome is reverse transcribed at one point in the replication in a very different manner (see later in this section) and there is no integration of virus genome.

Single-stranded DNA viruses

All DNA viruses except poxviruses replicate their genomes in the cell nucleus, therefore poxviruses may carry or encode other appropriate enzymes in the cell. Various methods are used, depending on the configuration of the virus DNA, which may be linear and single-stranded (parvovirus), circular (papillomavirus), or linear and double-stranded (poxvirus). As with other viruses, ssDNA viruses must also produce viral mRNA transcripts soon after the infection of a cell (Fig. 3.4). A virus enzyme, **DNA-dependent RNA polymerase** generates these mRNAs. There are no cellular enzymes known which transcribe ssDNA to RNA. Therefore ssDNA virus must first synthesis the complementary DNA strand to make a dsDNA genome. Replication of ssDNA involves the formation of **a double-stranded intermediate**, which itself serves as a template for the synthesis of single-stranded progeny DNA.

Double-stranded DNA viruses

Replication of dsDNA virus genomes uses the same principles as cellular genome DNA and involves the formation of a **replication fork**. At these forks the DNA polymerase moves along the DNA, opening up the two strands of the double helix and using each strand as a template to make a new daughter strand. The replication forks move rapidly at 100 nucleotide pairs per second.

As an example of a dsDNA virus, the adenovirus genome is transcribed and replicated in the cell nucleus. Replication is mediated by a protein (P) at the 5′ end of each DNA strand. Multiple mRNAs are transcribed from both DNA strands. Splicing is extensively utilized and can provide control of different regions of the genome, as well as a means of changing the reading frame of the mRNA, and hence more than one virus protein can be transcribed from a single stretch of mRNA. For the large dsDNA herpesviruses control of virus gene transcription occurs in a time-dependent hierarchy. **Early** and **late viral mRNAs** are transcribed from either DNA strand in the case of dsDNA viruses, and are translated to give 'early' and 'late' viral proteins, respectively. As a rule early mRNAs for translation to polymerase enzymes are transcribed from input parental virus DNA, whereas late mRNAs for translation into virus structural proteins are transcribed from newly replicated viral DNA.

'Unusual' viruses with RT

This grouping includes the **hepadnaviruses** (e.g. hepatitis B) which have circular DNA genomes which are partly double stranded with a single stranded gap. These viruses encode an RT enzyme which reverse transcribes a genome-length RNA into the dsDNA for packaging into the virus particles. Upon new cell infection the gapped dsDNA genome is converted into a

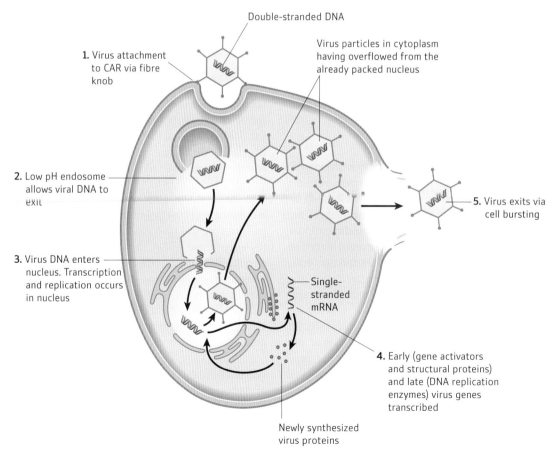

Fig. 3.4 Replication strategy of adenovirus, a double-stranded DNA virus.

covalently closed circular (ccc) DNA for transcription. The minus DNA strand is transcribed by cellular RNA polymerase II to give four mRNAs plus a 3.4kb RNA transcript, the pre-genome RNA. The viral RT reverse transcribes the pre-genome RNA into DNA and the RNA strand is degraded as the RT enzyme moves along, followed by plus strand DNA synthesis.

3.3.4 Viruses exert positive control of their replication

As viruses depend totally on the apparatus and mechanisms of the cell for replication, it is essential that the viral genome exerts control of these processes as much as possible, and so must use its genetic information to the maximum. Perhaps the most important mechanism for achieving this control is a viral code for strong positive signals to promote viral gene expression and other signals to repress expression of cellular genes. Many viruses can intervene directly to stop cellular transcription and translation. Other strategies for enhancing virus replication include **primary RNA transcript splicing**, **the use of overlapping reading frames**, or other methods of encoding multiple virus proteins in single mRNAs (Fig. 3.5).

Viral protein synthesis is completely dependent on the translation machine in the cell and so mechanisms have developed for reducing their dependence at this critical stage. Most eukaryotic mRNAs depend for initiation of translation on a 5′ terminal cap structure. Some viruses mediate initiation of translation through the internal binding of ribosomes on to mRNA at an internal ribosome entry site (IRES), a method used by hepatitis C, polio, and hepatitis A. This removes competition from host-cell caps, giving advantage for the virus. For example, two NS proteins of hepatitis C enhance IRES-directed translation.

Further control is exerted by the various properties of the viral mRNA, including its half-life and the actual flow of viral mRNAs from the nucleus to the cytoplasm, where enhancement of virus mRNA nuclear export can be achieved by virus proteins such as the protein REV of HIV. Control of viral gene replication may thus be exerted at the levels of transcription, post-transcription, or translation.

The expression of groups of virus genes is often carried out in critically timed phases. Thus **immediate early viral genes** of a virus such as herpes or adenovirus may code for regulatory proteins **and early viral genes** for genome replication proteins, whereas **late viral genes** code for virus structural proteins.

3.3.5 Synthesis of viral proteins

All viruses use the cellular ribosomes to translate viral mRNAs. The viral messages of an open reading frame are translated into the **structural proteins** that constitute the virus particle itself, or **NS proteins** (non-structural), which are enzymes or transcription factors for virus replication, but not incorporated in the virus.

The open reading frame continuity may be interrupted by the insertion or deletion of a base that causes a **frame shift** (Fig. 3.5). In other words, from the point of such a mutation the message is read as a different set of triplets and, thus, the viral

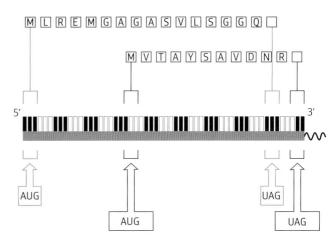

Fig. 3.5 Overlapping reading frames. The diagram shows how two or more polypeptides can be coded in a single length of nucleic acid. ∿.

protein will have a completely different amino acid sequence from that specified in the original message. This means some viruses can make very efficient use of the protein coding strand of their genomes by using the three possible open reading frames (or six in the case of ambisense genomes) to encode for different proteins, rather than just using one open reading frame as is the case for most cellular mRNAs. In this manner, the number of proteins that can be engendered from a small viral genome is increased. Several viruses use this strategy, including HIV and paramyxoviruses. Retroviruses have a special signal on the mRNA that triggers the ribosome to 'jump' and begin reading the triplet code in a different frame.

3.3.6 Post-translation modification of viral proteins

Even after synthesis of viral proteins our biochemical story is far from complete, as the viral protein must fold correctly into a precise three-dimensional structure. Important post-translation events must occur as a preparation for this folding. The initiator amino acid is removed while the polypeptide is still attached to the ribosome. Other important events may be **glycosylation** (attachment of carbohydrate), covalent attachment of lipoic acid, and addition of phosphate, sulphate, and acyl groups.

Some viruses, particularly positive-strand RNA viruses, such as polio, rhinoviruses, and flaviviruses, have a strategy whereby a very large viral polyprotein is translated initially from a single viral messenger mRNA. This **polyprotein** is then cleaved at specific sites by viral or cellular proteolytic enzymes to give a series of smaller viral proteins, some of which are incorporated into the virus while the others are non-structural proteins.

3.3.7 Virus assembly and release

The virus is now nearing the time of release and maturation. Its structural proteins have been synthesized by a host cell that appears relatively normal, or that may be irretrievably damaged. Recently the phenomenon of physical overcrowding of virus proteins at certain places of the cell plasma membrane has been noted in cells producing viruses.

Some viruses (e.g. poliovirus) assemble completely in the cytoplasm, whereas others (e.g. adenoviruses) are predominantly nuclear in location. Most enveloped viruses bud through the plasma membrane, but a few, such as rotavirus, exploit the endoplasmic reticulum membranes. Some viruses have signals on their glycoprotein spikes for specific targeting or retention.

Lytic viruses, for example polio, are released on lysis and death of the cell. Others (e.g. influenza, HIV, and measles) escape by **budding** from the cell surface.

Viral proteins are transported by the existing cell machinery and are, in the case of 'budding' viruses, inserted in the external plasma membrane of the cell; other virus structural proteins migrate to the inside of the plasma membrane. The proteins and nucleic acids self-assemble, and viral RNA or DNA is packaged as the completed virion buds by protrusion through the cellular plasma membrane. The bud is pinched off and a new virus is released.

Some viruses, such as HIV and herpesviruses, do not emerge often from the cell, but may spread to contiguous cells via connecting pores or by inducing fusion of their plasma membranes.

3.3.8 Viral post-release maturation

For some viruses, such as HIV and influenza, there is a further stage in the replication cycle, termed **post-release maturation**. Certain capsid proteins in HIV (Gag-Pol) have to be cleaved by a viral protease, which leads to changes in morphology of the interior of the new virion, quite easily detected by EM. In the case of the influenza virus, cellular proteases are needed to cleave the viral HA spike protein. Some cells do not have the correct protease and so the influenza virus does not normally initiate a successful multicycle infection in that organ: virus replication is restricted to cells of the respiratory tract, which have the correct protease. The cleavage often occurs before the HA reaches the plasma membrane and before budding, but may take place at or after budding, particularly if a protease is supplied exogenously. An example is co-infection of the lung with influenza virus and staphylococcus or streptococcus, which provide the protease and hence enhance the infectivity of the virus itself. The resulting pneumonia can be catastrophic.

3.4 Genetic variation of viruses

3.4.1 Low fidelity of reverse transcriptases and RNA replicases

Mutations involving the change by misincorporation of one nucleotide for another as well as removal or insertion of a nucleotide or a group of nucleotides (**deletion** or **insertion mutants**) are not uncommon during viral replication. Such mutations are much more frequent in RNA than in DNA viruses because of the low fidelity of transcription of RNA-dependent

RNA polymerases and RT and absence of proofreading and correction ability, compared with DNA polymerase. In fact, all RNA viruses are thought to exist as mixtures of countless genetic variants with slightly different genetic and antigenic compositions: so-called quasi-species. These mixtures of viruses exist as a dynamic equilibrium within the host, so that under a particular set of conditions one virus in the mixture is dominant, but others are still present, albeit in much lower numbers. Development of specific immunity to a dominant virus variant or use of an antiviral drug provides the selective pressure to force viral evolution allowing a minority variant to become dominant. For example a drug-resistant mutant, already present in the quasi-species and which may have arisen by chance even before the discovery of the drug itself, can emerge as the new dominant virus. This is Darwinian evolution and survival of the fittest in action! But there are some paradoxes here because in spite of the very high error rates of RNA viruses measured *in vitro* there can be great stability and conservation of sequences in nature. This does suggest strong selective constraints on sequence diversity.

3.4.2 Recombination

An important way in which viruses may vary their genomic structure is by **recombination**. This is brought about in DNA viruses by DNA strand breakage and covalent linkage of genome DNA fragments, either from within a single gene or from genes of two infecting viruses of the same family. In RNA viruses the virus polymerase can switch template strands during genome synthesis. Fortunately, such genetic interactions do not occur among unrelated viruses (such as polio and influenza), otherwise our medical and social problems would be greatly compounded. Nevertheless, this type of genetic interaction may give rise to a virus with hitherto unknown characteristics and may also give the mutant a selective advantage over its relatives. More often though, the new recombinant virus has properties incompatible with virus replication.

3.4.3 Gene reassortment

With certain RNA viruses, such as influenza and rotaviruses, in which the genome exists as separate fragments, simple exchange of genes may occur, a process known as **gene reassortment**. Such reassortant progeny viruses have characteristics that differ from those of the parental viruses. The frequency of such gene exchanges may be very high; much higher, for example, than that of true recombination. Such genetic reassortment can extend the gene pool of the virus, and allow the emergence of new and successful variants. An example is the infrequent appearance of pandemics of influenza (in 1918, 1957, 1968, and 2009), caused by reassortment of genes between human, avian, and pig influenza A viruses. A novel mutant may be created that can cross the species barrier and infect humans. The influenza A/swine virus which caused the pandemic in 2009 had genes from avian, swine, and human influenza viruses.

Reminders

- The Baltimore classification scheme designates seven viral genome coding strategies. and viruses are according assigned into families on this basis
  - dsDNA;
  - ssDNA;
  - dsRNA;
  - ss positive-sense RNA;
  - ss negative-sense RNA;
  - ss positive-sense RNA with DNA intermediate;
  - dsDNA with RNA intermediate.
- The stages of viral infection of cells are **cellular recognition** and attachment to a cell receptor, **internalization**, **genome release**, and **transcription** to form **viral mRNA, mRNA translation**, **genome replication**, **encapsidation**, and release of new virions from the cell. The complete viral life cycle characteristically takes 6–8 h and many thousands of new viruses are released from each infected cell during this initial cycle of infection.
- Primary RNA transcripts may be spliced, thus allowing several mRNAs to be coded in a single piece of viral genome. Viral messages may also be read in different **reading frames** at the translation stage, again allowing more extensive use of viral genetic information.
- **Control of viral gene expression** occurs at four levels:
  - configuration of viral DNA or RNA;
  - at transcription itself (rate of initiation, utilization of upstream transcription factors);
  - mRNA half-life, splicing of mRNA precursors, and flow of mRNA from the nucleus; and
  - at translation.
- Viruses may **bud** in many waves from infected cells (which continue to be viable) or may be released instantaneously by **cell lysis**.
- The replication of RNA viral genomes is error prone; this generates **genomic diversity**. In contrast, replication of DNA viruses is more faithful, although still more error prone that that of cellular DNA polymerases and the human genome.
- Genetic recombination and gene reassortment may both lead to genetic diversity as with the 2009 pandemic influenza A (H1N1) virus with genes from avian, swine, and human viruses.

Further reading

Cann, A.J. (2012). *Principles of Molecular Virology*, 5th Edition. Academic Press, London.

Huang, Y. *et al*. (2011). Temporal dynamics of host molecular responses differentiate symptomatic and asymptomatic influenza A infection. *PLOS Genetics* **7**.

Rima, B.K. (2015). Nucleotide sequence conservation in paramyxoviruses: the concept of codon constellation. *Journal of General Virology* **96**, 939–55.

Questions

1. Discuss the scientific basis of the Baltimore classification scheme.
2. Enumerate the various steps an RNA virus takes to infect and replicate in a cell.
3. Discuss how genetic variation occurs in viruses and give examples.

4 How viruses cause disease

4.1 Introduction

Figure 4.1 shows a child with measles, an elderly patient with severe herpes zoster ('shingles'), and another child with a malignant tumour of B-cells that can affect the jaw known as Burkitt's lymphoma (BL). Each of these people is under attack by a virus; however, the manifestations differ greatly, not only in appearance, but also in the way in which the viruses concerned caused these unpleasant effects. Measles is an acute infection of children and most patients recover quite quickly and develop

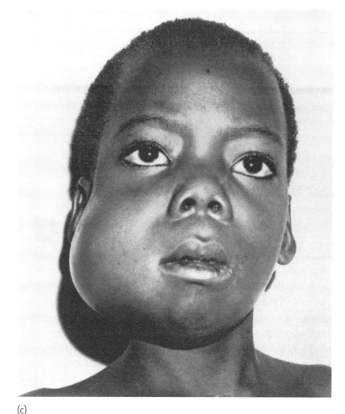

(c)

Fig. 4.1 Differing manifestations of viral infections. (a) Child with measles. (b) Elderly patient with herpes zoster. (c) Seven-year-old boy with Burkitt's lymphoma involving the right mandible induced by a herpes virus (by courtesy of Dr Joan Edwards).

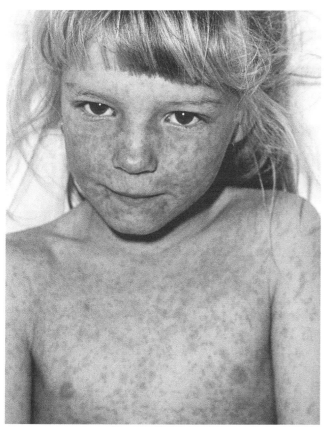

(a)

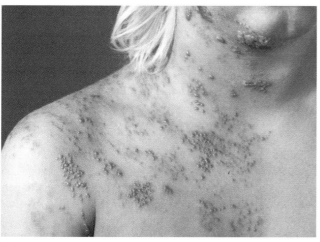

(b)

immunity against a second attack, whilst the zoster rash is caused by a reactivation of a lifelong persistent infection with a herpes virus. The lymphoma has been caused by another herpes virus, which has transformed immune B cells. This chapter is concerned with the complex interactions between viruses and hosts that result in disease, in other words, with the **pathogenesis** of viral infection. It is written primarily from the point of view of the virus; the next chapter describes the defences put up by the host.

4.2 Viral factors: pathogenicity and virulence

The terms **pathogenicity** and **virulence** are often used interchangeably, but this is not strictly correct. 'Pathogenicity' compares the severity of disease caused by **different microorganisms**: rabies virus is more pathogenic than measles. 'Virulence', on the other hand, compares the severity of the disease caused by **different strains of the same microorganism**. In practice, this may be related to the numbers of organisms needed to produce a given effect (e.g. the death of a mouse). For example, two strains of herpes simplex virus (HSV) inoculated into the skin may cause vesicular lesions, and are thus both pathogenic; but as few as 10 virions of strain A may kill the mouse, whereas 10 000 virions of strain B are needed to do so. Strain A is thus a thousand times more virulent than strain B.

What is it that makes one strain of virus more virulent than another? The development of rapid methods for determining nucleotide sequences within viral DNA or RNA is helping virologists to answer this intriguing question, which is important both from the points of view of preparing attenuated viruses for vaccines and of trying to predict whether viruses of enhanced virulence could occur either in nature or in the laboratory as the result of a small mutational change. For some RNA viruses only a few nucleotide substitutions in the viral genome are needed to control a switch from virulence to avirulence and vice versa. Poliomyelitis and influenza viruses provide two excellent examples. Only 10 **point mutations** (i.e. changes in single nucleotides) out of a total of about 7430 bases were detected in an attenuated poliovirus vaccine strain; it now appears that only three of them give rise to amino acid substitutions and, of these, only one point mutation may be responsible for attenuation of virulence. Likewise, a single change in amino acid sequence near the receptor-binding site on the HA molecule had a decisive influence on the virulence of influenza B for volunteers (Fig. 4.2). Changes at this position may also allow influenza viruses to emerge from an avian reservoir to cause a global pandemic in humans, as mutation in the receptor binding site could allow the virus to bind on to both avian and mammalian cells. Even more surprising are the details emerging

from experiments with avian influenza A virus (H5N1). As will be seen in Chapter 14, this virus is viewed as an 'emergent' virus from migrating birds and has caused deaths in poultry keepers and persons who buy from live bird markets in Asia. But the puzzling question is why it does not easily spread from human to human. Experimenters in the Netherlands and the USA have given the first explanation by the controversial experiment of deliberately mutating this influenza genome using a combination of reverse genetics and natural passage in a ferret animal model. Four mutations, two in the receptor binding site of the HA and two in the stalk of the HA, give the new mutant the ability to spread from animal to animal by the aerosol route. This experimentation has now been labelled 'dual purpose' or 'gain of function' because the results could be used to understand basic science or alternatively to create havoc by deliberately constructing viruses to facilitate human spread. Such are the dilemmas of all modern molecular virology, not just influenza research, and as a result the authorities in the USA have now severely restricted such work.

4.3 Interactions between viruses and host cells

The interactions between the virus and the cells within which it replicates are of decisive importance in determining whether infection takes place at all, the type of infection that is established, and the final outcome for the host.

4.3.1 Cellular factors

The presence of appropriate receptors on the surface of the cell determines whether virus can adsorb to it (Chapter 3). These cellular attachment factors range from lipids and proteins to carbohydrates. Examples include immunoglobulin proteins such as CD4 for HIV, SLAM (CD150 and CD46) for measles, and low density lipoprotein (LDL) for certain rhinoviruses. HIV also uses a co-receptor such as chemokine receptor CCR4 and CXCR5. Once this initial virus binding takes place and the virus gets into the cell it must replicate in order to establish infection, which may take several forms. These are described in Sections 4.4 and 4.5 below, but the general point to be made here is that none of these outcomes are possible unless the internal physical and molecular environment of the host cell is suitable for the initial replication cycle. The temperature of the cell is important. For example, respiratory viruses that replicate well at 33 °C are limited to the upper respiratory tract, where this is the prevailing temperature, whereas those that replicate well at 37 °C, but not at 33 °C predominantly infect the warmer environment of the lower respiratory tract. This simple fact of biology can have very important repercussions at the level of person-to-person spread of viruses. Thus avian influenza A (H5N1) viruses cause deep-seated infection in the lungs of those persons unfortunate enough to become infected. A simple explanation is the predilection of an avian influenza virus for the higher

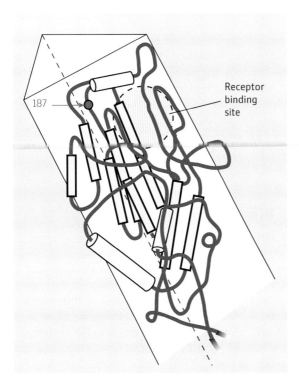

Fig. 4.2 X-ray crystallographic structure of influenza HA. The arrow indicates where substitution of a single amino acid (no. 187) causes a change in virulence in volunteers in Salisbury (J. Oxford).

temperatures of the lung (37 °C) rather than the nose and pharynx (33 °C). There is also a molecular explanation. The avian influenza HA binds to α2,3 sialic acid, which is mainly to be found in the lungs and lower respiratory tract. This implies that serious lung pathology can result from infection and that the centre of gravity of the replication factory will not facilitate easy person-to-person spread. In contrast, human influenza A viruses bind to α2,6 sialic acid receptors which predominate in the upper airways and nose, enabling easy person-to-person spread as aerosols during sneezing and coughing.

Similarly, the new Middle East Respiratory Syndrome Coronavirus (MERS) replicates deep in the lower respiratory tract rather than in the upper airways, and thus case fatality is high but there is only restricted person-to-person spread by aerosol and droplets, mostly in hospital settings.

4.3.2 Cytopathic effects

Many viruses kill the cells in which they replicate (Fig. 4.3a), sometimes with characteristic appearances or cytopathic effects (CPEs). These effects occur in the intact host, as well as in cell cultures, and are often distinctive enough to give an idea of the virus concerned, a property that is occasionally useful in the diagnostic laboratory (Chapter 29).

Cell lysis

The 'early' virus-coded proteins (Chapter 3) may **shut down synthesis of macromolecules**, particularly polypeptides, by the host cell. Later in the multiplication cycle of some viruses (e.g. adenoviruses) accumulation of large amounts of capsid protein may cause a general inhibition of both host cell and viral synthetic activities. More recently the application of techniques of gene arrays, proteomics, and genomics, to the infected cell has identified thousands of host cell genes that undergo change in the first hours of infection whereby transcription from these genes is suppressed or enhanced.

Death of the cell is followed by lysis and release of large numbers of virions. We can think of these viruses as 'bursters'. An example is poliovirus (see Chapter 7).

Many cells respond to infection by apoptosis or a programmed cell death (Fig. 4.3b), which reduces the spread of the virus. Cellular cysteine proteases such as the twelve caspases are activated either by cell surface toll like receptors or as a stress response to infection. The caspases cleave a variety of important cellular proteins and the cell self-destructs. Normally these caspases are present in the cell as inactive pro caspase. Programmed cell death is an important part of development of all multicellular organisms.

In more detail there are a plethora of ways which can be used by a virus to cause cell death. Firstly the **intrinsic pathology** leading to activation of caspases starts by activation of cellular proteins that monitor the integrity of the cell, such as p53 by virus DNA. As part of apoptosis protein oligomers form and create holes in the mitochondrial membranes, and mitochondrial proteins leak into the cytosol of the cell. These latter proteins complex to form a so-called 'apoptosome', resulting in the activation of caspases 9, 3, and 7. These latter caspases cleave cellular proteins and this leads to death of the cell.

But cell biology is never simple! Viruses have other ways of inducing cell death: for example, simply by binding to the cell surface. In this case alterations in cellular membranes can be recognized and the intracellular cytoskeleton affected leading to apoptosis signals. Alternatively, cellular signal transduction pathways can be activated by a virus binding to the cell surface via proteins such as TLR3, and intracellular signals from them

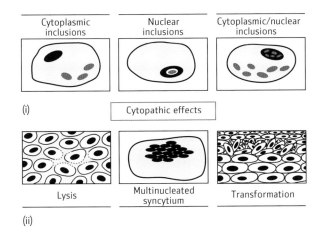

(i)

Cytopathic effects

(ii)

Fig. 4.3a Cytopathic effects of viral infection. (i) Inclusion bodies: intracytoplasmic, e.g. rabies; intranuclear, e.g. herpesviruses; or both, e.g. measles virus. (ii) The three main types of cytopathic effect: lysis, syncytium formation, and transformation.

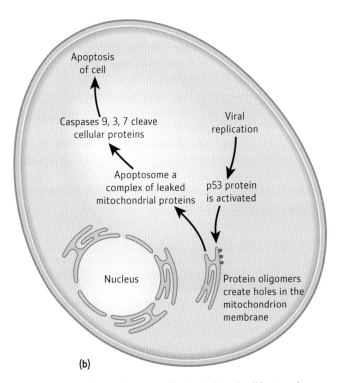

(b)

Fig. 4.3b Molecular mechanisms of virus induced cell lysis and death.

via TRIF lead to activation of caspase 8. Finally should virus RNAs be detected intracellularly then RIG-1 and MDA-5 recruit an adapter protein called IPS-1 (MAVS) and binding of the complex to mitochondrial membranes triggers caspase 3.

Some cells are connected to each other by actin-based projections called filopedia and some viruses can exploit these to move laterally from cell to cell. A number of enveloped viruses, and measles and respiratory syncytial virus are classic examples, can induce cell-to-cell fusion to form a giant cell syncytium, which can be very large indeed with multiple nuclei. The fusion is mediated by the spikes of the viruses, which can also mediate fusion of viral and cell membranes at the site of entry of the virus at the plasma membrane.

Cell fusion

We saw in Chapter 3 that virus-specified **fusion proteins** mediate the entry of certain viruses through the plasma membrane of the cell; they also cause the formation of multinucleated giant cells as we have noted above (**syncytia**, Fig. 4.3a) by **paramyxoviruses**, such as respiratory syncytial virus (RSV), parainfluenza viruses, and measles. **Herpesviruses** and some **retroviruses** also give rise to syncytia. Viruses that behave in this way tend to **pass from cell to cell**, rather than to be liberated in bursts by lysis. These are the 'creepers'. This is a potential advantage to the virus because it can evade immune responses, at least by antibody, by this intercellular spread that does not require the production and dissemination of large amounts of free virus, with all the action occurring at the junction between cell membranes of an infected and an uninfected cell.

Inclusion bodies

Inclusion bodies are eosinophilic or basophilic bodies that appear within cells as a result of infection with some—but not all—viruses and with certain bacteria (notably *Chlamydia*). Viral inclusions may be **intracytoplasmic**, **intranuclear**, or both. Some examples are given in Fig. 4.3a.

The nature of the inclusions varies with the virus concerned. They may represent aggregations of mature virions (papovaviruses, reoviruses), but are more often an assemblance standing at sites of viral synthesis or simply degenerative changes. The finding of characteristic inclusions in exfoliated cells or tissue sections stained by conventional methods was formerly much used in diagnosis, but has been superseded by the more reliable methods of molecular biology.

4.3.3 New cell-surface antigens

Another major consequence of many virus infections is the induction of **new antigens** on the cell surface. This is particularly important in the case of enveloped viruses that bud from the cell surface (e.g. herpes-, myxo-, paramyxo-, and retroviruses). As they are specified by the virus, rather than the host cell, these antigens mark the infected cells as being in a sense 'foreign', so that they are susceptible to attack by cytotoxic T cells, an important element in the immune response. Macrophages and dendritic cells, for example, can take up virus antigens and degrade them to short peptides that are then complexed with MHC-II protein and expressed on the cell surface. A roving T cell would recognize these new antigens, be stimulated, and develop into a cytotoxic T cell which can then expand in number and destroy other virus infected cells exhibiting these new peptides presented on MHC-I proteins (Chapter 5).

Some viruses induce malignant changes in cells; the important topic of the relationship between viruses and cancer will be dealt with in Chapters 21, 23, and 27.

4.4 The spread of viruses in the host

We must now move from events in individual cells to consider how viruses cause disease in the intact host.

4.4.1 Important events in pathogenesis

To cause disease, a virus has to clear a number of hurdles, that vary somewhat in type and number according both to the virus concerned and its host. The following sequence of events is typical. The virus must, sequentially:

- invade the host;
- establish a bridgehead by replicating in susceptible cells at the site of inoculation;
- overcome the local defences, for example lymphocytes, macrophages, and interferon (IFN);
- spread from the site of inoculation to other areas, often via the bloodstream;
- undergo further replication in its target area, whether this be localized (e.g. adenovirus conjunctivitis) or generalized (e.g. measles);
- exit from the host in numbers large enough to infect other susceptible hosts and thus ensure its own survival.

Some of these activities relate to specific properties of the virus itself and others to interactions between virus and host, both within individual cells and with the body as a whole, and its array of immune defences.

4.4.2 Invasion routes

Viruses gain access to the host either through the skin or mucous membranes. Examples are given in Tables 4.1 and 4.2. Please note that these lists are not exhaustive, since routes of infection will be given in more detail in the chapters dealing with individual viruses. They show:

- the diversity of routes adopted;
- the ability of some viruses to infect by more than one route.

The stratified squamous epithelium of the skin is a formidable barrier to microbes and some degree of trauma is necessary before infection can take place. Table 4.1 shows that some viruses (e.g. papilloma and some pox- and herpesviruses) cause

Table 4.1 Examples of viruses entering through the skin

Virus	Disease
Entry via abrasions	
Papillomaviruses	Warts
Poxviruses (cowpox, orf)	Vesicular or nodular lesions on milkers' fingers
Herpes simplex viruses	Herpetic lesions on face, fingers, genitalia
Ebola	Haemorrhagic fever
Entry via abrasions or inoculation with contaminated needle	
Hepadnavirus	Hepatitis B
Lentiviruses (HIV)	AIDS
Entry via insect or animal bites	
Arboviruses	Various tropical fevers
Lyssavirus	Rabies

Table 4.2 Examples of viruses entering through mucous membranes

Virus	Disease
Entry via respiratory tract	
Orthomyxoviruses	Influenza
Paramyxoviruses	Measles, mumps, parainfluenza, respiratory syncytial disease
Rhinoviruses	Common cold
Varicella-zoster	Chickenpox
Entry via gastrointestinal tract	
Poliovirus	Poliomyelitis
Other enteroviruses	Febrile illnesses affecting muscles or CNS
Rotavirus	Gastroenteritis
Entry via conjunctiva	
Enterovirus type 70	Conjunctivitis
Adenovirus type 8	Keratoconjunctivitis
Entry via genital tract	
Lentivirus (HIV)	AIDS
Hepadnavirus	Hepatitis B
Herpes simplex	Herpetic lesions of cervix, urethra
Papillomavirus	Genital warts, cancer
Hepatitis C	Liver infection

more or less localized lesions at the site of implantation, whereas others go on to produce generalized infections involving a variety of body systems. An extreme example of the latter is the filovirus Ebola (Chapter 18). This virus is pantropic, infecting every organ whilst destroying cells lining the blood vessels. Internal bleeding exhibited clinically as bloody diarrhoea and vomit causes shock and leads to death and to heavily virus contaminated body fluids. Relatives or friends of a victim contract the virus by exposure to infected tissues and body fluid especially during funeral preparations, by touching and washing the virus contaminated body. Healthcare workers, such as nurses and doctors, can have similar intimate contact and can become infected.

By contrast, many viruses entering through **mucous membranes** (Table 4.2) need little or no trauma, as they adsorb directly to the epithelial cells, in which they undergo an initial cycle of replication. The upper respiratory organs are washed by fluid produced by secretory cells of the respiratory tract. There are both upward and downward 'escalators' and as much as two litres of secreted fluid wash over the ciliated cells protecting them from infection. The fluid often has glycoprotein decoys to attach to viruses and block viral receptors. The infections caused by some of these viruses, notably those affecting primarily the conjunctiva, are localized. Most, however, are more general, and here it is important to appreciate that the body system by which a virus enters is not necessarily the one that will ultimately be mainly affected. Thus, although varicella virus enters by the respiratory tract, its main target organ is the skin; likewise enteroviruses, which, as their name suggests, do indeed invade and multiply within the enteric canal, cause disease of the CNS or muscles rather than enteritis. Also some viruses, like Ebola, can be opportunistic, invading by several routes.

In temperate climates the most common infections are those acquired by the **respiratory route.** An unstifled sneeze results in an aerosol, an ideal medium for carrying microbes into someone else's respiratory tract. The smaller the droplets, the more widely they will be disseminated and the further down the bronchial tree they will penetrate. Especially in wintertime, crowded stores, buses, and trains are ideal places in which to exchange viruses transmitted by the respiratory route and classic examples are influenza, coronaviruses, and the common cold rhinovirus. Experiments are in progress to determine the precise size of these sneeze and cough droplets. This has important practical implications because large droplets can be blocked, say in hospital, by simple paper masks whereas an aerosolized virus will penetrate through paper and therefore a denser mask is needed for protection of the nurse or doctor.

Respiratory infections are also very prevalent in the tropics; and because standards of clean drinking water are poor in many developing countries, these areas also bear the brunt of infections acquired via the **gastrointestinal tract**. This mode of infection is known as the **faecal–oral** route, which sounds revolting and is, as it means that viruses shed in faeces have got into someone else's mouth. Figure 4.4 shows a number of ways in which this can happen. Such infections are by no means confined to developing countries; they are prevalent wherever sanitary conditions are indifferent, for example, in some mental institutions.

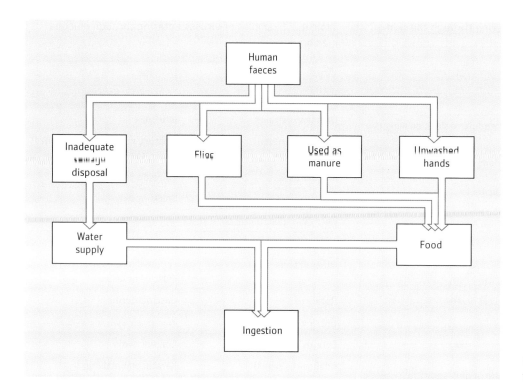

Fig. 4.4 The faecal–oral route of transmission of viruses.

Not surprisingly, viruses that infect via the alimentary tract, such as the enteroviruses, are resistant to the acid environment of the stomach, whereas those with other portals of entry (e.g. the respiratory viruses) are not.

The incidence of **sexually transmitted diseases** (STDs) has increased enormously since the end of the Second World War in 1945 and has risen still further in the first years of the new millennium. Much of the increase is due to viruses and *Chlamydia*: but the 'classical' venereal infections, gonorrhoea and syphilis, have also increased in prevalence. Although some syndromes, such as AIDS and hepatitis B, have in the past been associated with male homosexuals, in Western countries, all the STDs mentioned in Table 4.2 can also be spread by heterosexual intercourse.

The third and increasingly frequent way of acquiring infection is from **transplants**, particularly of bone marrow and kidneys. Two herpesviruses—cytomegalovirus (CMV) and Epstein–Barr virus (EBV)—are notorious in this respect. Both cause persistent but asymptomatic infections in a substantial proportion of the general population; if an organ donor is infected in this way, the recipient is liable to suffer, particularly if he or she has no pre-existing immunity, or if the immune responses have been impaired by immunosuppressive treatment. **Blood transfusions** and **blood products** such as Factor VIII are also a potential source of infection with certain viruses, such as the hepatitis viruses B and C, and possibly vCJD. However, rigorous screening, and the exclusion of high-risk donors such as drug addicts, has greatly reduced, although not eliminated, the number of infections from such sources.

Other infections resulting from **surgical treatment** have been described, but are rare. More than one person has died from rabies after receiving a corneal transplant from a donor who was incubating the disease—surely the ultimate in bad luck stories! Others again have acquired CJD, another fatal infection of the CNS, from surgical instruments contaminated with the causal agent, which is unusually resistant to sterilization.

Spread of infection from mother to foetus is a special form of transmission from one person to another and is described in Table 4.3. The pathogenesis of foetal infections is beset by uncertainties because, more often than not, there is no suitable animal model, and there is an obvious difficulty in investigating them in humans. There are two main routes by which intrauterine and intrapartum infections are transmitted. Transplacental infection results from a maternal viraemia, and may take place at any time during pregnancy, the outcome depending on the virus concerned, as occurs with human cytomegalovirus infection of the unborn foetus. There is little or no evidence that ascending infection from the cervix or vagina can penetrate the foetal membranes during pregnancy; but contact infection from these sources can certainly take place during delivery, and is facilitated by a long interval between rupture of the membranes and birth. Like bacterial and fungal infections, those caused by viruses can be classified under two main headings:

- those localized to tissues at or contiguous with the site of entry; and

- those that spread to one or more organs remote from this area.

4.4.3 Localized infections

These are infections of epithelial surfaces: the skin, the conjunctiva, and the mucous membranes of the respiratory, gastrointestinal, and genital tracts. Some examples are given in Tables 4.1 and 4.2.

Table 4.3 Modes of intrauterine and perinatal viral infection

Virus	Transplacental	During birth	Shortly after birth
Rubella	++	–	–
Cytomegalovirus	+	++	++(BM)
Herpes simplex	+	++	+
Varicella-zoster	++	+	+
Parvovirus	++	–	–
Enteroviruses	+(Late)	++	++
Human Immunodeficiency virus	+	++	+(BM)
Hepatitis B	+	++	++
Human papillomaviruses	–	++	–
Influenza A/Swine (H1N1)	+	–	+
*Zika	+	–	–

++ , Most frequent route; + less frequent route; (BM), can be transmitted via breast milk; *under current study in S. America.

Localized infections of the skin by poxviruses result in papular lesions, usually proceeding to vesicle and then pustule formation (e.g. vaccinia, orf), or in proliferative lesions of the epidermis (molluscum contagiosum and papillomaviruses). But such visual lesions on the skin may give an incomplete picture of the true pathology. Smallpox and chickenpox can infect many internal organs whilst a true pantropic virus like yellow fever, measles, or Ebola can infect virtually every organ in the body.

By contrast, infections of **mucous membranes** spread over comparatively large areas of the respiratory or gastrointestinal epithelium. The process is rapid, which means that such infections have a short incubation period, say 1–3 days. Although viral replication is restricted to these surfaces, the effects may be much more general, as anyone who has had influenza or even a bad cold knows only too well. As we will see these symptoms can reflect the huge number of chemical messengers released into circulation by cells of the innate immune system (Chapter 5).

4.4.4 Generalized infections

For various reasons, the pathogenesis of many generalized virus infections is not well understood: in some instances, there is no suitable animal model; in others, such as the viral haemorrhagic fevers (e.g. Ebola and Lassa Fever), the difficulty of working with dangerous pathogens under restricted conditions discourages extensive experimentation; and in others again, our lack of knowledge results from inability to grow the virus in the laboratory. Nevertheless, a number of viruses, notably those causing

the infectious fevers of childhood, follow a pattern of spread that was worked out by F. Fenner, in Australia, who studied a poxvirus infection of mice called ectromelia. The sequence of events, summarized in Fig. 4.5, is as follows:

(1) Virions enter through an epithelial surface, where they undergo limited replication. (2) They then migrate to the **regional lymph nodes** where some are taken up by macrophages and inactivated, but others enter the bloodstream. This is (3) the **primary viraemia**, which sometimes gives rise to

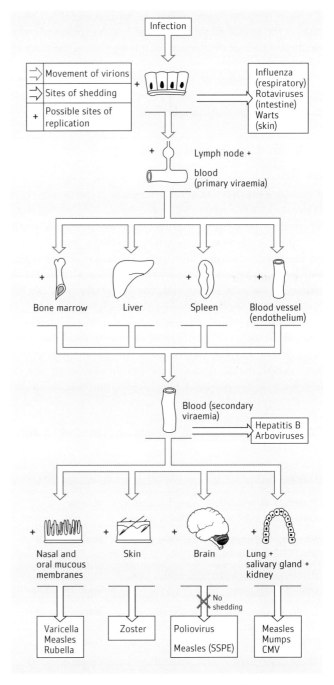

Fig. 4.5 Spread and replication of virus in a typical generalized virus infection. [Modified from White D.O., Fenner F.J. (1986). *Medical Virology*. London: Academic Press, p. 135.]

prodromal malaise and fever. (4) From the blood, the virus gains access to the large reticuloendothelial organs—**liver**, **spleen**, and **bone marrow**—in which it again multiplies. In this further cycle of amplification, a large amount of virus is produced, which again, in a manner of speaking, spills over into the bloodstream, causing (5) a **secondary viraemia**. Viruses in the circulating blood may be free, as in the case of hepatitis B, but are more often associated with lymphocytes or macrophages in which they can replicate, examples being measles, CMV, and HIV. (6) From the bloodstream, the virus finally reaches its **target organ**, the nature of which depends on the tissue tropism of the virus concerned, and mainly determines the clinical features of the illness. All this takes time, which explains why the incubation periods of infections of this type are of the order of two weeks.

In other infections, viruses reach their target organs by more direct routes, stage (3) being omitted. This may happen in, for example, certain arbovirus infections.

4.4.5 Some important target organs

The skin

A rash, or **exanthem**, features in a number of virus infections (Table 4.4). Dental students in particular should note that such an infection is sometimes accompanied by an **enanthem**, a rash affecting mucous membranes, which is of course best observed on the buccal mucosa (Table 4.5). There are several types, which are not necessary exclusive to particular viruses. Some viruses, notably the entero- and echoviruses, are notorious for causing almost any type of non-haemorrhagic rash.

Whereas the vesicular eruptions are due to replication of virus in the skin, with consequent damage to epithelial cells, other rashes may have other causes. Thus, the characteristic maculopapular rash of measles is due to the destruction of infected cells by cytotoxic T lymphocytes; purpuric rashes are

Table 4.4 Examples of virus diseases associated with a skin rash (exanthem)

Disease	Type of rash
Rubella, erythema infectiosum (parvovirus), entero- and ECHO virus infections, dengue	Macular
Measles, infectious mononucleosis, entero- and ECHO virus infections	Maculopapular
Herpes simplex, chickenpox/herpes zoster, poxvirus infections	Vesicular or vesiculopustular
Coxsackie virus infections (especially type A9 and A16), entero- and ECHO virus infections	Maculopapular
Congenital rubella	Purpuric
Viral haemorrhagic fevers	Petechial/ haemorrhagic

Table 4.5 Virus diseases associated with a rash (enanthem) on the buccal mucous membrane

Disease	Main features of the enanthem
Herpesvirus infections	
Herpes simplex (usually primary infection)	Gingivostomatitis, with vesicles and ulcers on an intensely inflamed mucosa, primarily affecting the anterior part of the buccal cavity
Chickenpox	Vesicles, rapidly ulcerating, especially on fauces and palate
Infectious mononucleosis	Petechiae occasionally seen on palate
Coxsackie A virus infections	
Herpangina	Small vesicles usually confined to posterior part of the buccal cavity (soft palate and fauces)
Hand, foot, and mouth disease	Vesicles and ulcers anywhere on buccal mucosa, but predominating in anterior part
Paramyxovirus infection	
Measles	Koplik's spots on congested mucosa near openings of parotid ducts; appear a day or two before the skin rash

Note. These and many other acute viral infections may be accompanied by pharyngitis.

often associated with a fall in blood platelets; and some haemorrhagic rashes are the result of disseminated intravascular coagulation.

The lung

Most respiratory infections, even those of the lung, result from local spread of virus, as described in Section 4.3. Sometimes, however, the lung is involved as part of a generalized infection. This is particularly so in measles, in which some degree of pneumonitis is a constant feature. In patients whose immunity is damaged, both measles and varicella viruses may cause giant cell pneumonia, characterized by the appearance of syncytia (Section 3.2); this is a dreaded complication with a high mortality.

The liver

This organ is the target for the hepatitis viruses (Chapters 7, 10, 12, and 28). It may also be damaged as part of a more general infection of body tissues by other viruses, the classical examples being the flavivirus yellow fever (YF) (Chapter 12) or the filovirus Ebola (Chapter 18).

Kidney

The kidney is rarely infected by viruses. An important exception is CMV, one of the herpes group. Characteristic inclusions are found in the proximal renal tubules, from which the virus is shed into the urine. Certain hantaviruses (Chapter 16) cause a haemorrhagic fever with nephritis.

The central nervous system

Viruses gain access in one of two ways:

- from the bloodstream, during an episode of viraemia (e.g. poliomyelitis, arbovirus encephalitis); or
- via the nerves connecting the periphery with various parts of the CNS (e.g. herpes simplex, varicella-zoster, rabies).

See also Chapter 23.

4.4.6 Immunopathological damage

We have mentioned that some skin rashes are manifestations of the immune response, rather than the result of tissue damage by virus multiplication. This is often true of pathological changes in other organs infected by viruses (Chapter 5). Particular examples are damage to hepatocytes in the liver, infected with hepatitis B virus, by cytotoxic CD8 cells of the immune system. A 'cytokine storm', a sudden outpouring of various cytokines, e.g. tumour necrosis factor (TNF), IFN-α, and interleukin (IL)-1, in response to a lung infection with influenza or SARS viruses is another important example of over-reaction of the immune system leading to immunopathological damage. In particular over-production of pro-inflammatory cytokines of the innate immune system, including interleukin-1, -6, and -12 and tumour necrosis factor alpha (TNF-α), may contribute to disease symptoms. Interleukin-1 induces fever whilst TNF-α induces fever, metabolic wasting, and lysis of infected cells.

HIV is a unique example of a virus targeting cells of the immune system itself, in this case CD4 cells, and with their destruction causing acute immunosuppression.

4.4.7 Incubation periods

Knowledge of the incubation periods of the common virus infections is important not only as an aid to the diagnosis of individual patients, but also as an essential tool in tracing the spread of outbreaks (Chapter 6). They will be given in more detail in the chapters devoted to individual viruses, but for the moment, and as an aid to memory, we shall classify them into four main groups: short, medium, long, and very long (Table 4.6).

Short means less than a week and primarily applies to viruses causing localized infections that spread rapidly on mucous surfaces. Some viruses injected directly (e.g. arboviruses transmitted by the bite of an arthropod) also, as a rule, have short incubation periods.

Medium incubation periods range from about 7 to 21 days; they are seen in generalized infections having the type of pathogenesis described in Section 5.2.

Long refers to periods measured in weeks or months (e.g. 2–6 weeks for hepatitis A and 6–20 weeks for hepatitis B). The pathogenesis of these infections has not yet been worked out, and we do not know what these viruses are doing between entering the host and producing symptoms and signs of illness.

Table 4.6 Incubation periods of representative viral diseases

Disease	Usual period	Limits
Short incubation (times in days)		
Enterovirus conjunctivitis	1–2	
Common cold	1–3	
Influenza	1–3	
Arbovirus infections	3–6	2–15
Medium incubation (times in days)		
Poliomyelitis	7–14	2–35
Measles	13–14	8–14
Rubella	14–16	14–21
Varicella	13–17	11–21
Mumps	14–18	14–21
SARS and MERS	9	8–12
Ebola	7	2–21
Long incubation (times in weeks)		
Hepatitis A	3–5	2–6
Hepatitis B	10–12	6–20
Infectious mononucleosis	4–6	2–7
Rabies	4–7	2–50
Very long incubation (times in years)		
Subacute sclerosing panencephalitis (SSPE)	6	4–8
Creutzfeldt–Jakob disease (CJD) and Kuru	20	5–30

Rabies may also have incubation periods extending for many months (Chapter 19), but in this instance we know that the time to onset depends on the rate at which the virus travels up neurons to the brain from the site of entry, and on the length of the nerves affected.

Very long incubation periods are measured in years, which is why the agents involved were originally termed 'slow' viruses. This group comprises the prions and a few 'conventional' viruses, such as polyomaviruses (Chapter 21) and measles (Chapter 17), which very occasionally cause delayed disease of the CNS. These late infections are invariably fatal. Why their incubation periods are so long remains a mystery.

The **generation time** is an important parameter, defining the time from infection of a patient to the time when that patient is infectious for others. It may vary from 3 days with influenza to 10 days with SARS CoV, MERS CoV, Ebola, and smallpox.

Together with the reproductive number (R_0), which is the number of subsequent cases developing after contact with a single infected person, the generation time can allow predictions of the size of an epidemic or pandemic. To give an example, with a new pandemic influenza virus with an R_0 of 3 and generation time of 3 days, one million persons could become infected in a 40 day period!

4.5 Patterns of disease

Details of the pathogenesis of individual virus infections will be given in more detail in the relevant chapters; this section is intended just to give you a general idea of the variations that may occur. We have classified infections into three main groups.

4.5.1 Acute non-persistent infections

Most acute virus infections resolve spontaneously; sequelae are unusual unless the CNS is invaded as, for example, in poliomyelitis and some arbovirus infections, but any very severe infection may, of course, be lethal. A comparatively rare complication of some of the infectious fevers of childhood can occur; there may be an episode of encephalitis 10–14 days after onset, by which time virus can no longer be isolated. The pathogenesis of this condition is not clear; it may be due to an autoimmune response or the consequences of rare host genetic variation.

4.5.2 Persistent infections with acute onset

In this group, chronicity of infection is often due to **latency**, resulting from **persistence in the host cell of viral DNA**; clearly, this can only happen with DNA viruses or with retroviruses that form complementary DNA (cDNA) during replication. The viral DNA may be integrated into that of the host cell or be present in **episomal** form (i.e. as a circular molecule separate from the host DNA).

The cells affected vary according to the virus concerned. Such latent infections may be established very soon after the primary infection, as with the herpesviruses, or be delayed by a matter of 2 or 3 years, as in some chronic carriers of hepatitis B. Subsequent events depend on the nature of the virus and sometimes on whether the patient's immune responses are adequate.

Note that:

- Latent infections may never cause signs of disease.
- They may reactivate on one or more occasions, causing an episode of illness.
- Infective virus may not be demonstrable while the patient is asymptomatic, but is produced in quantity during reactivations.
- Some latent infections lead to malignant disease.

With regard to the last-mentioned point, infections caused by the endogenous oncornaviruses are now of intense interest both to molecular biologists and to oncologists; they differ from the others mentioned in Table 4.7 because, being present both in somatic and germ cells, they can be transmitted vertically through many generations.

Chronic infection is the term applied to the situation in which infective virus is continually being produced, with or without integration of viral DNA into the host cell. Unlike latent infections, they may be caused both by DNA and RNA viruses and the end result for the host is often determined by an abnormal or defective response of the immune system, as may be the case in subacute sclerosing panencephalitis, a fatal, but fortunately very rare, late complication of measles; chronic active hepatitis B is another example. Other chronic, but more trivial infections are caused by the papillomaviruses responsible for warts and by the poxvirus that gives rise to molluscum contagiosum. In both these skin conditions, the viruses evade the immune response by remaining within their safe haven in the avascular epidermis. Papillomavirus genome may also integrate with host-cell DNA.

4.5.3 Insidious infections with fatal outcomes

The two types of infection in this group resemble each other only superficially. The 'slow' virus infections have already been mentioned in Section 4.4.7. The other type of infection is not known to occur in humans, but is of great interest both to virologists and to immunologists. Indeed, it was the study of lymphocytic choriomeningitis (LCM), an arenavirus infection of mice, which led the late Sir Macfarlane Burnet to the idea of **immunological tolerance** and a Nobel Prize. In brief, every member of a mouse colony in which this virus is present becomes infected at birth and continues so for the rest of its life. In such mice, the

Table 4.7 Latent virus infections

Virus	Site of latency
Herpesviruses	
Herpes simplex types 1 and 2, varicella-zoster	Neurons in dorsal root ganglia
Epstein–Barr virus	B lymphocytes
Cytomegalovirus	Lymphocytes; macrophages
Human herpesvirus type 6	Probably lymphocytes
Hepadnavirus	
Hepatitis B	Hepatocytes
Polyomavirus	Renal medulla, brain, lymphoid cells
Papillomavirus	Epithelium
Retroviruses	
Endogenous oncornaviruses	Somatic and germ cells
Lentiviruses	
Human immunodeficiency viruses	T lymphocytes, macrophages, brain cells

immune system fails to recognize the virus-infected cells as 'foreign'. Some antibody is produced, but forms virus–antibody complexes; these are not dealt with in the normal way and a proportion of infected mice eventually die because of their deposition in the kidney.

4.5.4 Symptomless viral infections

With many, if not most, infections, there can be equivalent numbers of infectees who show no symptoms whatsoever. This ratio can be as low as 1:1 with influenza and as high as 200:1 with poliovirus. The scientific basis of this diversity has yet to be explored. In some viruses there may be genetic resistance but a more likely explanation is pre infection with a related or the same virus and hence immune memory to the pathogen. Regardless of the mechanism an important remaining question is whether virus can spread from asymptomatic ('silent') carriers and cause disease in others. We would note here that typical clinical symptoms of viral infections from influenza to smallpox to Ebola fever—nausea, muscle pain, and lethargy—result from virus induced release of pro-inflammatory cytokines such as

tumour necrosis factor alpha (TNF-α) and interleukins 6 and 12. These symptoms occur regularly and transiently in humans without the development of a full infection: could these represent subclinical infections that for some reason are inhibited before causing more obvious pathology?

4.6 Shedding of virus from the host

Microbes only live to fight another day by getting out of one host and into another; the means of escape is therefore very important. Viruses may be shed from the **primary site of multiplication** or, in the case of generalized infections, from the target organ. Viruses can escape readily in one way or another from all the main body systems, with the exception of the CNS, in which they are effectively bottled up. It is important to remember that many viruses may be shed from clinically normal people: they include herpes simplex (saliva), CMV (urine, breast milk), and the viruses that infect the gut (faeces).

We shall see in Chapter 6 what happens when viruses escape into the community.

 ## Reminders

Viruses gain access to the host through **skin** and mucous membranes, and via the respiratory, gastrointestinal, and genital tracts.

- Their virulence or lack of it may be determined by very limited number of mutations in the genome.

- Viral infection of cells may result in **lysis**, **fusion**, and **syncytium formation**, the appearance of **inclusion bodies**, or, in some instances, **transformation** to cells with characteristics of malignancy. Apoptosis and self-destruction occurs as well.

- Infections caused by viruses are **localized** at or near the site of entry or **generalized** to involve one or more target organs. In generalized infections a common pattern of spread during the incubation period is **site of entry, local lymph nodes, primary viraemia, liver, spleen, bone marrow, secondary viraemia, and target organ**.

- In humans, viral infections follow various basic patterns: **acute non-persistent, acute followed by persistent latent infection**, and **chronic with continued shedding of virus**.

- Very rarely, polyomavirus, measles virus, and prions cause long-lasting, fatal disease of the CNS.

- During the course of infection, viruses are **shed** from the primary site of multiplication, or—with the exception of the CNS—from target organs, and are then free to infect other susceptible hosts.

- The R_0 number is the average number of secondary cases generated by one primary case in a susceptible community. Along with the **generation time** (days between infection and ability to infect others) it is useful in predicting the course of an epidemic.

- Most viruses cause silent infections without signs in as many patients who show clinical signs and in some, like polio, the ratio can be as high as 200:1.

 ## >> Further reading

Dahiya, S., Nonnemacher, M.R., and Wigdahl, B. (2012). Deployment of the human immunodeficiency virus type 1 protein arsenal: combating the host to enhance viral transcription and providing targets for therapeutic development. *Journal of General Virology* **93**, 1151–72.

Gatherer, D. (2014). The 2014 Ebola virus disease outbreak in West Africa. *Journal of General Virology* **95**, 1619–24.

Killingley, B., Enstone, J., Booy, R., Hayward, A., Oxford, J., Ferguson, N., and Nguyen Van-Tam, J. (2011). Potential role of human challenge studies for investigation of influenza transmission. *Lancet Infect Dis* **11**, 879–886.

Lawlor, H.A., Schickli, J.H., and Tang, R.S. (2013). A single amino acid in the F_2 subunit of respiratory syncytial virus fusion protein alters growth and fusogenicity. *Journal of General Virology* **94**, 2627–35.

 # Questions

1. Describe entry mechanisms of viruses both at the whole human body level and into cells themselves.

2. Write short notes on:

   a. Acute infections.

   b. Latent infections.

   c. R_0.

   d. Immunopathology.

3. Give three examples of the spread and replication of a virus in the body. Link these events to clinical signs.

Resistance of the human body to virus infections

5

5.1 Introduction

The scientific investigation of immunology started about a century ago with Metchnikoff's studies of phagocytosis of foreign particles, including micro-organisms. This early interest in what we now refer to as **cell-mediated immunity (CMI)** was soon overtaken by research on **antibody-mediated (humoral) immunity**, which was for long regarded as the primary defence against microbial disease. We now know that, although the presence of antibody is important in preventing virus infections, cellular immunity plays a major role in the immune response once infection has been established. We also know how interconnected and interdependent these two systems are and the central role of the 'hormones' of the immune system: cytokines and chemokines and interferons.

The relationship between viruses and the immune system is much more intimate than it is for most bacteria: viruses often modify the cells within which they replicate, thereby rendering the cells essentially 'foreign' and susceptible to attack by sensitized lymphocytes; furthermore, some viruses can multiply within the very lymphocytes and mononuclear phagocytes that are important components of the immunological defences. A good example is HIV, which replicates in and destroys the CD4+ 'helper' lymphocytes that are so important to the integrity of the immune system and indeed act as the conductors of the immune orchestra (Chapter 27).

So far, the emphasis of these opening chapters has been almost entirely on the ingenious means adopted by viruses to invade and damage their hosts. On its part, however, the animal world, including our own species, has developed efficient and highly complex mechanisms for combating infection by these intracellular parasites, which can be considered under the headings of **innate** ('general' or 'non-specific') immunity and **adaptive** (or 'specific') immunity. The latter term refers to the array of immune responses directed against particular microbes, but it is not always possible to make a hard and fast distinction between this type of resistance and the more general defences, which will be dealt with first.

By 'innate' we mean those defence mechanisms with which we are born and which form the first line of protection against microbial invasion. They can be regarded as 'built-in' defences, and fall into two categories: those that protect the individual (Sections 5.2.1–2.5) and others, genetically diverse, that determine the resistance or susceptibility of populations (Sections 5.2.6 and 5.2.7). Innate resistance mechanisms differ from adaptive mechanisms in having no immunological 'memory' to a given pathogen (Section 5.4.2).

5.2 General non-immune factors in resistance

5.2.1 Mechanical and chemical barriers

The importance of the skin, otherwise known as epithelial surfaces, as a barrier to infection was mentioned in Chapter 4 (Section 4.4.2). But these huge surfaces of the body are exposed to infection, including the nose, throat, lungs, gut, genitals, and eyes. The retrograde movement of mucus by epithelial cilia in some tissues such as the respiratory tract acts to prevent infection and damage to these cells by influenza and paramyxoviruses which may open the way to secondary bacterial infection. The gastrointestinal tract is, to some extent, protected against ingested viruses by the low pH in the stomach, although viruses that regularly infect by this route are resistant to acidity; this applies to most enteroviruses.

The secretions from mucous membranes, e.g. eyes, mouth, and respiratory, genital, and gastrointestinal tracts, offer a means of transporting various elements of the immune systems (cytokines, antibodies, lysozyme, etc.) to where they are needed. These 'wash out' defences are sometimes underrated. But we excrete over two litres of fluid a day in the respiratory tract and therefore pure 'washing' can be very important. Most of this fluid is secreted from globular cells lining the respiratory tree and the cilia escalator carries the fluid upwards to the throat where it is swallowed. Also, many viruses can enter via abrasions or small cuts in the skin including poxviruses, herpes, and the more recently highlighted Ebola. But the skin still represents a very important barrier.

5.2.2 Fever

A high temperature is naturally regarded by patients as an unpleasant effect of a virus infection, but a body temperature much above 37 °C is inimical to the replication of a number of viruses and is thus an important defence mechanism. The febrile response seems to be triggered by soluble factors, notably IL-1, TNF, and IFNs, all of which are produced by macrophages (see Section 5.3.2).

5.2.3 Age

This factor is an example of the way in which the general overlaps with the particular, as age-related resistance is, in part at least, mediated by immune responses. An infant is sent into the world with a useful leaving present from its mother in the form of a package of immunoglobulin type G (IgG) antibodies directed against infections from which she has suffered. IgG antibodies to these predominantly viral infections, supplemented by immunoglobulin type A (IgA) antibodies in colostrum and breast milk, helps to tide the baby over the first six months or so of life, after which its susceptibility to viral infections increases. The protection conferred by maternal milk is a good reason for breastfeeding, especially in developing countries where the energetic sale of manufactured substitutes is not in the best interest of babies born into a particularly hostile environment. Another great advantage of breast milk is that it is bacteriologically sterile. But 'the life of microbes' should never be underestimated. Several viruses, HIV particularly, can use milk as a vehicle to transmit infection to the new baby.

The most obvious effect of ageing on the immune system is atrophy of the thymus gland, which starts in adolescence and continues for the rest of life. As might be expected, this process is accompanied by failures of function in both B- and T-cell immunity, and in disturbances of cytokine responses. Classical effects of a waning immunity are the increased risk in the over 60-year-olds of mortality and serious illness from viruses such as influenza.

Virus infections in childhood are not usually serious, but become increasingly so with the advance of age; for example, poliomyelitis usually causes mild or even subclinical infection in children, but adults are often hit harder and the incidence of paralytic disease is higher. Some virus infections in elderly people can be very severe, herpes zoster (shingles) being a case in point.

5.2.4 Nutritional status

Poor nutrition may exacerbate the severity of some virus infections, an often-quoted example being measles in African children, which has a much higher mortality rate than in developed countries. However, assessment of the importance of malnutrition is complicated by other factors such as intercurrent infections, particularly malaria, which is in itself immunosuppressive. In developed countries more attention is paid to diet and, for

example, to attempts to control the microflora of the gut using natural products such as yoghurt.

5.2.5 Hormones

It is well known that treatment with steroids exacerbates the severity of herpes simplex and varicella-zoster infections, but their precise role in natural resistance or susceptibility is unknown. The severity of hepatitis E (Chapter 10) may be exacerbated by pregnancy, presumably because of hormonal influences, but again, the mechanism is not yet understood.

5.2.6 Human genetic factors

In experimental animals there is clear evidence that genetic factors influence resistance, or conversely, susceptibility to virus infections. Thus, some highly inbred lines of mice are killed by very small inocula of HSV, whereas others withstand enormous doses with no sign of illness. In this case, resistance is dominant, and is mediated by only four genes: with other viruses, susceptibility may be the dominant genetic factor. The genes involved are sometimes, but not always, part of the major histocompatibility complex (MHC; Section 5.5.4).

Case study A single human gene can increase susceptibility to severe influenza

The influenza pandemic of 2009, with wide outbreaks in all countries of the world, provided researchers with a unique chance to search for correlates between human genes and outcome of infection. Early laboratory work had shown that the replication of many viruses that enter the cell by the acidic endosome route was abrogated by the interferon-inducible transmembrane 3 protein (IFITM3). A gene knockout mouse model was used to test the hypothesis directly showing that IFITM3 was crucial for defence of the animal against influenza, in that mice which had the gene removed suffered much more severe influenza than control 'intact' mice; in fact all mice without IFITM3 succumbed to influenza infection. It is much more difficult to obtain human data but a group of researchers from the Sanger Institute near Cambridge including Paul Kellam, our co-author, and his team achieved this by analysing IFITM3 alleles of hospitalized patients and found that a minor CC genotype of IFITM3 enhanced influenza virus disease severity. This allele is rare in Caucasians but

much more common in Han Chinese. The CC genotype was later found in 69% of Chinese patients with severe pandemic influenza. The CC genotype gave a six-fold greater risk for severe infection than the CT and TT genotypes. These figures translate into a population-attributable ten-fold extra risk for severe influenza infection in certain Chinese populations compared to Northern Europeans.

Technically IFITM3 protein acts as a virus restriction factor for influenza, West Nile, and Dengue virus amongst others and mediates cellular resistance to the viruses in the acidic endosome. The expectation is genotyping of Han Chinese and other Asian patients early in infection with influenza, and possibly Dengue, could predict those who might progress to severe disease. This test can be performed simply by Single Nucleotide Polymorphism (SNP) typing at very low cost, enabling the targeted use of antiviral therapies into a subgroup of patients, or the stratification for vaccination before an influenza season.

5.2.7 Species resistance

The host range of many viruses is restricted, probably because the cells of resistant species do not possess appropriate receptors. The best understood example is poliovirus, the receptors for which are present only in humans and other primates. Modern molecular biology allows transfer of human genes to mice,

for example, and when the human cell receptor for poliovirus is transferred the mouse becomes susceptible to poliovirus infection when the virus is injected into the mouse. Others, notably the human immunodeficiency viruses and some hepatitis agents, are equally selective; by contrast, others, such as rabies, are capable of infecting most or all warm-blooded animals. Influenza virus is another classic example of a species restriction and this is very fortunate because it means in practical terms that the movement of avian influenza A viruses from the bird reservoir to humans is rare because humans have $\alpha2,6$ sialic acid receptors in the upper airways and avian influenza viruses prefer $\alpha2,3$ sialic acid receptors normally present in birds (Chapter 14).

5.3 Early innate immunity

The ways in which the above responses are mediated are, for the most part, extremely important but still poorly understood. We are on firmer ground when we examine what weapons are brought into play by local non-specific immune defences in the earliest phases of infection. These include **soluble factors**, such as the interferons (IFNs) (Table 5.1), cytokines (Table 5.2), chemokines, complement, and C-reactive protein, phagocytic cells, and particularly **natural killer (NK) cells**, all of which are important components of the **innate immune system**.

5.3.1 Natural killer cells

About 10% of circulating lymphocytes are larger than the average lymphocyte, have electrondense granules in their cytoplasm, and are known as 'natural killer' cells or 'large granular lymphocytes'. NK cells circulate and in fact are primed to leave the blood at any time and enter infected tissue. They mature in

Table 5.1 Some properties of IFNs

IFN	Derived from	Principal functions	Type	Receptor
IFN-α	Most nucleated cells, especially fibroblasts. 12 species, produced by infected leucocytes and macrophages	Antiviral enhances MHC class 1 expression	I	IFN α/β R
IFN-β	One species, produced by fibroblasts and epithelial cells	Antiviral	I	IFN-α/β R
IFN-γ	T_h1 cells, NK cells macrophages	Immune stimulation Enhances MHC class 2 expression	II	IFN-γR
IFN-λ	Three species produced by fibroblasts, macrophages, and leucocytes	Antiviral	III	IL-28 receptor and IL-10 receptor

Table 5.2a Innate and adaptive immunity

Innate	Adaptive
Rapid response	Slower response
Little or no memory	Highly specific immunological memory

Table 5.2b Representative cytokines

Cytokine	Examples of activity
IL-1α and β	Inflammatory; pyrexia
IL-2	T-cell activation
IL-4	B- and T-cell activation
IL-5	B- and T-cell activation; eosinophil differentiation
IL-6	A pro-inflammatory cytokine and an anti-inflammatory myokine (a cytokine released by muscle cells)
IL-8	A chemokine: attracts polymorphonuclear leucocytes
IL-9	Mast cell growth factor
IL-12	Induces T_h1 cells
IL-13	Suppresses T_h1 cells
IFN-α	Antiviral
IFN-β	Antiviral
IFN-γ	Antiviral; activates macrophages; inhibits interleukin 6 and T_h2 cells
IFN-ω	Under investigation
IFN-τ	Under investigation

the bone marrow and are short-lived cells with a half-life of only 1 week. They have no immunological specificity but can kill cells that they recognize as 'foreign'. NK cells are thought to recognize infected cells as foreign or non-self, through the lack or low-level expression of MHC class I on the cell's surface. MHC class I usually presents virus peptides to killer T cells and viruses therefore often down-regulate MHC class I on the cell surface but in doing so render themselves visible to killing by NK cells. NK cells can also recognize and kill cells coated with virus-specific IgG, a process known as antibody-dependent cell cytotoxicity (ADCC). Their activity is greatly enhanced by various cytokines such as IL-12, IL-15, IL-18, IL-2, and CCL5, and NK cells are recognized as a very important defence mechanism to limit virus disease severity.

NK cells kill by excreting cytotoxic proteins such as perforin, which can make holes in the cell membrane, aiding penetration of the cell by granzyme B, a molecule which activates apoptosis and programmed cell death.

5.3.2 Macrophages

These cells are phagocytic cells, highly efficient at removing dead or dying cells and removing cell debris. The name derives from the Greek for big eaters, from makros, meaning 'large' and phage, meaning 'eat'. There are two main groups of macrophages, called M1 and M2. M1, so-called 'killer', macrophages are activated by LPS and IFN-γ, and secrete high levels of IL-12 and low levels of IL-10. In contrast the M2, so-called 'repair', macrophages function in constructive processes like wound healing and tissue repair. M2 is the phenotype of resident tissue macrophages, and these sentinel cells reside in places of pathogen entry to the body such as the lungs, intestine, and skin. Macrophages also function as antigen-presenting cells, however unlike dendritic cells they do not travel and must themselves be activated and then restimulate T cells. Macrophages themselves have a lifespan of months. However, once activated they produce cytokines such as TNF-α and can kill virus infected cells.

5.3.3 Cytokines

Interactions between the various cells of the innate and adaptive immune system are mediated by a substantial variety of small (about 20 kDa) glycoproteins referred to collectively as cytokines. The cytokines act as chemical messengers and are rapidly produced upon virus infection. They are produced in virus-infected cells and can, after secretion, act at local and distant sites as both autocrine (affecting the cell that produced them) and paracrine (affecting other cells). These cytokines bind to surface receptors and can activate cellular genes via signal transaction pathways. Cytokines are a focus of huge research effort at the present time. They contain four subgroups of molecules, of which the most important as far as viral infections are concerned are the interferons (IFNs) (Section 5.3.4). Another category of cytokines are the chemokines such as IL-8, RANTES, and IP-10 which recruit macrophages, T cells, B cells, or eosinophils to sites of ongoing infection and inflammation. A third group of important cytokines such as IL-1 and IL-6, IL-12 and TNF-α are pro-inflammatory cytokines inducing fever whilst at the same time stimulating B cells or T-helper type 2 (T$_h$2) cells. IL-1 stimulates T cells and induces fever, IL-6 stimulates B and T cells, IL-12 stimulates NK cells, and TNF-α itself activates neutrophils and triggers acute fever.

In contrast the fourth group of cytokines includes anti-inflammatory cytokines such as IL-4, IL-10, and transforming growth factor beta (TGF-β) which inhibit macrophages and T cells. These cytokines push the immune system to its normal resting state where it can be prepared again to counter further infections.

The number of known ILs runs well into double figures (Table 5.2b), and they play an important part in regulating both T-cell and B-cell activities. For example, the binding of antigen to an immature CD4 cell is followed by differentiation of the latter into one or other of two main subsets of helper cell, T helper 1 (T$_h$1) or T helper 2 (T$_h$2). These subsets are distinguished by the patterns of ILs that they secrete. Thus, T$_h$1 cells secrete IL-2 and IFN-γ, with stimulation of T cells leading to CMI, whereas IL-4, -5, -6, and -10 are secreted by T$_h$2 cells (see also Section 5.5) leading predominantly to B-cell and antibody mediated immunity.

5.3.4 Interferons

The discovery by Isaacs and Lindenmann in 1957 that virus-infected cells produce a soluble factor that protects other cells from virus infection seemed to herald a new era, as the substance was effective against a wide range of viruses and apparently non-toxic. At a time when rapid strides were being made in the chemotherapy of bacterial infections, IFN seemed like a virus 'penicillin' and an answer to virologists' prayers. These early hopes were soon to turn to disappointment, but now, with better scientific understanding and with molecular cloning technology to produce large quantities, IFNs have made something of a comeback.

The interferons are proteins with molecular weights of about 20 kDa, produced by many cell types including leucocytes or fibroblasts in response not only to viral infection, but also to stimulation by natural or synthetic dsRNA, and some bacteria (e.g. *Chlamydia*). IFN types I (α and β), II (gamma), and III (lambda) differ in the way they are produced, in amino acid sequence, and in mode of action (Table 5.1). These IFNs have different receptors.

These interferon molecules are not virus-specific, so that IFN induced by one virus is effective against others; on the other hand, they are animal **species-specific**, so that IFN produced by, say, a guinea-pig, is ineffective on mouse or human cells. IFN-gamma differs in a major way from the alpha and beta IFNs as it is produced by a subset of T memory cells in response to stimulation by an antigen previously encountered.

IFNs do not kill viruses, nor do they act like antibodies. The mechanisms of IFN induction are complex and indirect (Fig. 5.1). The three mechanisms of virus blocking are summarized in Fig. 5.2. Most DNA and RNA viruses induce IFNs but dsRNA viruses such as reoviruses are the best inducers. Also, double-stranded RNA produced during the replication of RNA viruses is a potent signal for IFN stimulation.

The genes required for transcription events leading to the production of IFN are, not unexpectedly, heavily suppressed in cells not infected with viruses. Firstly the pattern recognition receptors (PRRs) of the cell, or detector proteins, predominately TLR2, 3, 4, 7, 8, and 9 (Table 5.3 and Fig. 5.1), and also RIG-1 and MDA-5, are activated by binding to viral dsRNA and viral proteins. They also can be activated internally in the virus-infected cell. Internally these PRRs initiate binding to the so called 'adapter proteins' (Table 5.4) such as MyD88 and TRIF and IPS-1. There now happens a very important step in this complex scenario: the recruitment of cellular protein kinases by the adapter proteins. Important examples of the protein kinases are the IKK enzyme complex and the TBK-1 kinase. IKK phosphorylates a protein that normally binds to and deactivates NF KB. NF KB is released and moves to the nucleus to act as part of the interferon induction signal transduction cascade.

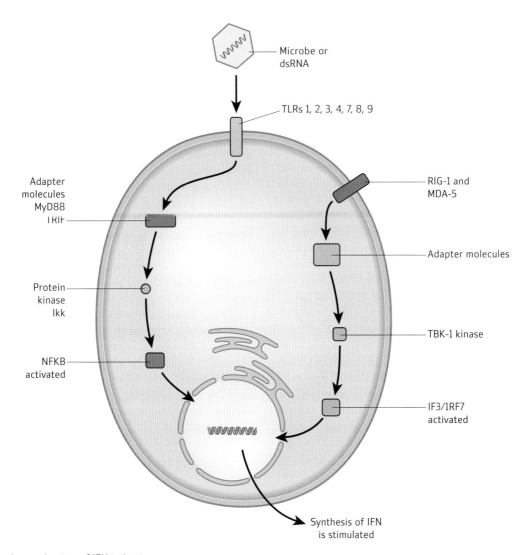

Fig. 5.1 Molecular mechanism of IFN induction.

In a parallel manner the kinase TBK-1 phosphorylates IRF-3 and IRF-7 (Interferon Regulatory Factors) which are also part of the interferon induction cascade and which also move to the nucleus. Excitingly for molecular biologists, but sometimes dismaying for students, we note that there are other pathways which lead to the activation of kinases. The transcription of IFN genes and hence ultimately to interferon itself is stimulated by these signal transduction cascades.

These IFNs are transported out from the cell and induce an **antiviral state** in neighbouring cells. Interaction with the type I IFN, for example, and receptor IFNBRI induces numerous genes in the cells that are directly or indirectly antiviral. Hundreds of human genes are up-regulated by IFN, and these are known collectively as Interferon Stimulated Genes (ISGs), but only a handful have been fully characterized.

Many investigators have to date focused on three systems induced in the interferon-treated cell. Firstly, the formation of the Mx proteins, secondly an important **protein kinase** PKR (double-stranded RNA-dependent protein kinase), and thirdly **2–5A oligo synthetase**. The Mx proteins block virus RNA polymerase, particularly influenza, and also block transport of influenza virus RNP complexes across from cytoplasm into the nucleus. The activated protein kinase PKR is capable of phosphorylating, and thus inactivating, a factor that initiates synthesis of proteins. The activated 2–5A oligo synthetase, in turn, produces 2′-5′ oligoadenylates, which activate a latent ribonuclease leading to the degradation of cellular and viral RNA. The end result is blockade of virus replication. Elucidating the role of different ISGs and how they inhibit different properties of virus infection is becoming a rich source of understanding the cell biology of virus infection and identifying new targets for broad spectrum antiviral drug therapies.

As well as blocking viral replication, IFNs have profound additional effects on cells, some of which also help indirectly to control infection. One of the most important is **enhancement of the display on cell surfaces of histocompatibility antigens**, which are essential to antigen-driven activation of T cells.

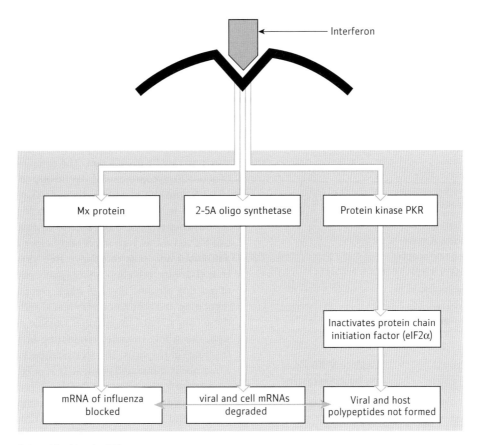

Fig. 5.2 Mechanisms of virus blocking by IFN.

Table 5.3 Complex and growing nomenclature of cellular dsRNA and DNA recognition proteins (PRPs)

Abbreviation	Name	Function
TLRs 1–10	Toll-like receptors	Bind to viral dsRNA or ssDNA and some proteins and recruit 'adapter proteins'
RIG-1	Retinoic acid inducible gene (1) (a RNA helicase)	Acts similarly to TLRs
MDA-5	Melanoma differentiation associated gene 5	As above
DA1	DNA-dependent activator of interferon regulatory factors	Binds to dsDNA (not usually in cytoplasm in uninfected cells) and recruits adapter proteins
AIM-2	Absent in melanoma 2	As above
NLRP3 (a group of 20 proteins some of which also recognize dsRNA)	Nucleotide binding domain leucine rich repeat containing family with pyrin domain containing 3 protein	As above

Others involve modulation of both B- and T-cell activities, including enhancement of the cytotoxicity of the NK and cyto-toxic T cells (CTLs). IFN-α helps regulate antibody production and has a role in B-cell activation by switching production from IgM to IgG2a producing plasma cells.

In addition to their antiviral effects, certain IFNs can inhibit cell division and this property, in conjunction with the immuno-modulating activities just described, has been used to some effect as an adjunct to immuno- or chemotherapy of cancer.

IFN molecules activate macrophages and this is a major way these crucial cells are stimulated. The macrophages are induced to form nitrogen oxide because of IFN-generated nitrogen oxide synthetase. Finally and perhaps most importantly, IFN enhances the synthesis of interleukin 12 in antigen-presenting cells and can simultaneously block interleukin 4 which would otherwise induce expansion of T_h2 cells. Thus the cytotoxic T_h1 system is favoured, which is active in killing virus-infected cells.

How important are interferons in preventing or treating virus infections in patients?

Although the effects of IFN are clear in experiments using cell cultures, it is quite difficult to disentangle them from all the other components of the immune response patients or in intact animals. IFNs are generated very quickly—within hours of the start of infection—and help to hold the fort in the interval before clonal proliferation of T and B cells and antibody production get

Table 5.4 The complex and growing nomenclature of cellular adapter proteins and the protein kinases and transcription factors which are recruited

Abbreviation	Name	Function
Adapter protein		Adapter molecules which then recruit protein kinases
MyD 88	Myeloid differentiation primary response gene 88	
TRIF	Toll interleukin-1 receptor domain containing adapter inducing IFN-β	
IPS-1	Interferon-β promoter stimulator	
Protein kinases		
IKK	IKK kinase	Leads to NFKB signal transduction cascade
TBK-1	Tank binding kinase	Phosphorylates IRF-3 and IRF-7
IFN Transcription (regulatory) factors		
IRF-3	Interferon regulatory factor 3	After phosphorylation these molecules move to the nucleus and induce expression of interferon
IRF-7	Interferon regulatory factor 7	

under way. This view is supported by the finding that treatment of animals with anti-IFN serum exacerbates virus infections; furthermore, some people who have a natural defect in IFN production are abnormally susceptible to upper respiratory tract infections.

With hindsight, the early attempts to use IFN for treatment were doomed to failure, because the amounts available were minuscule by today's standards. In fact, the first demonstration of the activity of IFN in volunteers at the MRC Common Cold Unit in Salisbury, UK used almost the entire world supply of the molecule of that time! Now that genes coding for IFN can be cloned in yeast or other cells, very large quantities can be produced and have been used with success in treating chronic hepatitis C; and hepatitis B, often in combination with directly acting antiviral (DAA) drugs. This will be dealt with in Chapter 31, but at this point it is interesting to note that part of the therapeutic effect against the virus may be due to IFN-mediated enhancement of MHC display on infected liver cells, resulting in their more efficient destruction by cytotoxic T cells. IFNs are also used for treating various malignancies, including Kaposi's sarcoma.

5.3.5 Complement

The complement system (originally discovered as a 'complement' to microbial effects of antibodies) is an important component of the innate immune response to infection and is present in all vertebrates. In fact, from an evolutionary viewpoint, sea urchins several hundred million years ago used this system.

In humans genesis of the complement system starts in the first trimester of pregnancy and persons with a defect in the system succumb rapidly to infection. The complement system consists of about 20 glycoproteins of MW 80 000–200 000, of which nine (C1–C9) are the most investigated. They are serum proteins produced in the liver and can reach high concentrations in serum. They recognize and bind to viruses. In this manner they can be viewed as very important helper molecules which aid antibodies and cells to remove free virus and virus-infected cells. The nine proteins, appropriately stimulated, are capable of reacting sequentially. The whole complement system is self-regulating and is initiated very quickly after infection. Some of these components are enzymes capable of catalysing the next step, the whole cascade terminating in lysis or opsonization (coating) of the virus. This 'classical' pathway to destruction of a microbe is stimulated when an IgM or IgG antibody binds to a viral antigen. The 'alternative pathway' is triggered by the spontaneous hydrolysis of complement component C3, which is cleaved into two molecules, C3a and C3b, which initiates the next step in the cascade.

However, the complement system is activated, the end result is cleavage of C3 into its two fragments C3a and C3b. To give an idea of the speed of the cascade, if C3b does not find a target, an amino group or a hydroxyl group on a microbe, in a few microseconds, it is neutralized. C3b can trigger a second cascade (C5–C9), which forms a complex able to attack virus membranes or membranes of virus-infected cells, the so-called Membrane Attack Complex (MAC). Cleavage of C3 and C5 can also lead to molecules that release histamine. This results in increased blood flow and attraction of leucocytes to a site of infection.

A third activation pathway in the complement system is lectin activation via mannose binding lectin (MBL). MBL can bind to HIV and influenza, for example, but not to uninfected cells, so there is some discrimination. MBL, which is plentiful in blood, binds to mannose on a pathogen and the complex acts as a complement convertase enzyme to process complement C3 into C3b. The C3b binds to the pathogen triggering the rest of the complement cascade.

5.4 The adaptive immune system

The two fundamentals of the specific adaptive immune responses are as follows:

- The ability to distinguish between 'self' molecules that belong to the body itself and are fortunately ignored by the immune system; and 'non-self' molecules (e.g. those of microbes), against which the system is capable of reacting.

- The ability to induce memory, whereby an enhanced immune response is evoked to an antigen previously encountered.

Specific immunity is also referred to as adaptive or acquired immunity, as opposed to the non-specific innate mechanisms with which we are born and which we have just described above. From the viewpoint of patient therapy acquired immunity may be 'passive', or preformed when large amounts of antibody is injected for prophylaxis of, say, rabies or Ebola, or 'active', as in the full antibody response to an infection or vaccine.

Let us now take a look at the main components of the specific immune response. It will be brief because the adaptive immune system is complicated; nevertheless, a knowledge of the basic features is essential. It should be remembered that many of its activities apply also to microbes other than viruses.

5.4.1 Antigens

Antigens are molecules, or groups of molecules, capable of eliciting an immune response. Within limits, the larger the molecule the stronger the immune response to it. Although antigens are usually large molecules, such as proteins, only small sequences of, say, 6–12 amino acids actually induce the formation of antibodies which react with them, or are presented to T cells to allow specific infected cell killing. These sequences are known as **antigenic determinants**, or **epitopes**.

5.4.2 Immunological memory

This is mediated by lymphocytes, each of which is capable of responding to a single antigen specific for that particular lymphocyte cell. It is astonishing that during the course of evolution we have developed a range of lymphocytes that can react to virtually any one of the myriads of molecules capable of evoking an immune response. There are two main classes of lymphocytes, B and T cells, that mature in the bone marrow and thymus, respectively. Both types react with antigens by means of cell-surface receptors. Although responsible for antibody- and cell-mediated immunity (CMI), respectively, the two main classes of lymphocyte should not be thought of as occupying separate compartments of the immune response. On the contrary, they co-operate closely with each other in the same milieu and with dendritic cells and macrophages that process antigens and 'present' them to the lymphocytes. The immune system can be viewed as a large, finely tuned, and carefully conducted orchestra.

5.4.3 Activation of naïve B cells and antibody-mediated immunity

The interaction of B lymphocytes with antigen is summarized in Fig. 5.3.

B lymphocytes originate from stem cells in the **bone marrow**, enter the circulation, and mature mainly in the spleen and lymphoid tissue. Their surface receptors are **antibody molecules** that react on encountering their specific antigen. Activation

requires four signals. On interaction with the specific antigen the antibody receptors cluster and induce intracellular signalling (Stage 1). This is the first signal. The foreign antigenic material is also processed in a way not yet fully understood, and presented on the B cell surface in the form of many short peptides that are recognized and interacted with by helper T cells which provide signal 2 (Stage 2, Fig. 5.3). Cytokines produced by antigen-presenting cells and helper T cells stimulate B cells, providing signal 3. Pathogen-associated molecular patterns and/or danger signals stimulate pattern recognition receptors (PRRs) on B cells giving signal 4. These signals stimulate the B cell to proliferate, thus forming a clone of plasma cells with the same antigenic specificity (Stage 3). Sometimes B cells can have a T-cell-independent activation.

Many of the cells in the B cell clone differentiate into **effector cells**, in this instance large and very productive **plasma cells** that secrete high yields of antibodies capable of combining with that antigen and that antigen only. The immunologist Macfarlane Burnet received the Nobel Prize for his work on this clonal expansion. Some of the sensitized B lymphocytes persist for long periods as **memory cells** that respond rapidly to further encounters with the antigen by clonal proliferation and production of more of the appropriate antibody. These memory cells can last a lifetime. The amount of antibody generated is controlled by T cells.

The five classes of immunoglobulin—IgA, IgM, IgG, IgD, and IgE—are each produced by particular clones of plasma cells; only the first three seem to be important in virus infections. Each consists of one or more units made up of four polypeptide chains, two 'heavy' and two 'light', which together form a Y-shaped molecule (Fig. 5.4). The tips of the two prongs of the 'Y' are the **variable regions**. Within the population of lymphocytes, the extremely wide variation of amino-acid sequences in these regions ensures that there will be a few cells capable of reacting with any conceivable antigen encountered for the first time. The binding site has space for an epitope of up to 13 amino acids. The variable regions thus confer upon each immunoglobulin molecule its individual **antigenic specificity**. The constant region of the heavy chains at the base of the Y is called the Fc, and determines the immunoglobulin isotype and binds to complement and to Fc receptors on macrophages. There is an immense library of 'variable' region exons, which can be further mutated and also randomly selected by recombination during lymphocyte development. As we noted above, clonal expansion happens once a B cell has interacted with a viral peptide.

IgA antibodies

IgA antibodies are produced by lymphoid tissue underlying the mucous membranes at whose surfaces it acts; they are found in secretions of the oropharynx, gastrointestinal, and respiratory tracts, and are thus important in defending against viruses that enter by these routes (Chapter 4). The IgA secreted at mucous surfaces consists of two immunoglobulin units (a dimer) attached to a 'secretory piece' that aids its passage through cells and onto the mucosal surface. IgA is also produced during lactation, particularly in the colostrum, and it is the specific

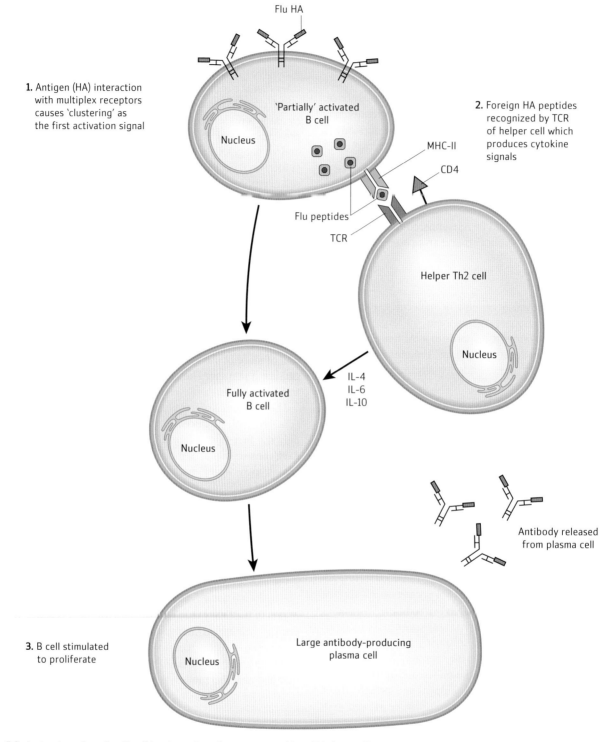

1. Antigen (HA) interaction with multiplex receptors causes 'clustering' as the first activation signal

Flu HA

'Partially' activated B cell

Nucleus

2. Foreign HA peptides recognized by TCR of helper cell which produces cytokine signals

MHC-II

CD4

Flu peptides

TCR

Helper Th2 cell

Nucleus

Fully activated B cell

Nucleus

IL-4
IL-6
IL-10

Antibody released from plasma cell

3. B cell stimulated to proliferate

Nucleus

Large antibody-producing plasma cell

Fig. 5.3 Activation of a naïve B cell by clustering of receptors and by a T helper cell.

antibodies provided in this form by the mother that help to protect against infections in early infancy (Section 5.2.3).

IgM antibodies

IgM antibodies are the first to be produced in systemic infections and are particularly avid (tightly binding) in combining with antigen and complement. IgM is a large molecule and cannot cross the placenta; hence, specific IgM antibody in a foetus or neonate indicates intrauterine infection. Production of IgM antibody is a fairly short-term process, lasting for a few weeks, and the molecules have a short life span of 3–4 days. **The finding of a specific IgM is thus evidence of a recent or current**

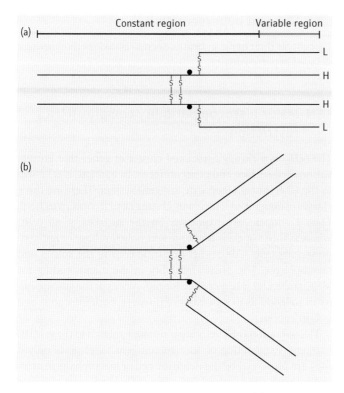

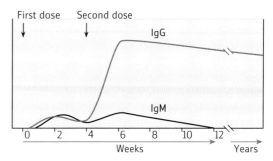

Fig. 5.5 Primary and secondary antibody responses to spaced doses of an inactivated vaccine.

Fig. 5.4 Structure of an immunoglobulin molecule. (a) Basic structure: two identical 'heavy' polypeptide chains (H) are linked to two identical 'light' chains (L) by disulphide bonds. (b) The molecule opens at the hinge region. This is the position in which it combines with an antigen.

infection and is used for diagnosis in particular infections such as rubella or Zika during pregnancy.

Eventually, a negative feedback mechanism operates whereby the manufacture of IgM is switched off and replaced by production of specific IgG antibody, which is the most abundant immunoglobin.

IgG antibodies

By contrast with IgM, IgG antibodies continue to be produced for very long periods—often during the entire lifespan—and thus afford **long-term protection** by binding viruses and neutralizing their proteins or enabling phagocytosis of free virus particles. As their presence is evidence of past infection, such antibodies are useful in epidemiological surveys (Chapter 6). IgG antibodies cross the placenta and, as mentioned above, provide protection during the first few months of life, thus supplementing the action of IgA antibodies.

The time course of antibody production can be observed most clearly during immunization with a non-replicating antigen such as an inactivated vaccine (Fig. 5.5). The **primary response** to the first injection results from the initial clonal expansion of B lymphocytes on first encounter with the antigen: it is comparatively slow and of low magnitude. By contrast, the **secondary responses** to subsequent injections given some weeks later reflect the presence of pre-existing memory cells that are now available to undergo a greatly amplified and rapid clonal

expansion; the resultant antibody response takes place very quickly (i.e. within a day or so), and much greater antibody quantities are produced. This clear-cut picture is blurred when a replicating antigen such as live poliovirus vaccine is given, or during the course of a naturally acquired virus infection: in this instance the primary stimulus continues for some time and other immune mechanisms relating to recovery are brought into play. The response picture also becomes complicated following vaccination or infection with viruses that undergo antigenic variation, such as the antigenic drift and change that occurs with influenza virus. In this case, older-aged people may have some immune memory of previous infections but the antibody still cannot give full protection against new virus variants because of antigenic drift.

IgE antibodies

These mediate allergies by attaching to mast cells, which are 'activated' and release histamine causing allergic symptoms.

5.4.4 Mode of action of antibodies

There are several possible ways in which a specific immunoglobulin can act against a virus, the exact mechanism depending on the virus concerned:

- It can neutralize by agglutinating the virions and thus stop them attaching to susceptible cells or by blocking the receptor binding site, namely the virus docking protein. Some antibodies can even block the functioning of an internal viral protein when, for example, it is expressed on the cell surface like the influenza M2 protein. Antibodies may even enter a cell attached to a virus and together trigger the antiviral state.

- Antibody may act as an opsonin, combining with virus particles and increasing the ability of macrophages to phagocytose and destroy them. Macrophages can bind to specific antibodies by the Fc receptor and are therefore able to destroy infected cells expressing on their surfaces viral antigens recognized by such antibodies.

- Antibody plus complement can combine with viral antigen expressed on the surface of an infected cell, and together lyse the cells. This effect is known as antibody-dependent-cellular cytotoxicity (ADCC).

- Antibody can interfere with the uncoating mechanism after the virus has entered the cell.

Memory B cells can show startling longevity—12 years for mumps, rubella, and Rift Valley fever, and up to 70 years for yellow fever, vaccinia, and measles. We shall refer to the use of antibody therapy elsewhere (Chapter 30), but with the availability of the new molecular cloning methods we are noting the use of yeast cells, mammalian cells, and tobacco plant cells as substrates to produce human monoclonal antibodies for therapy of some virus infections such as Ebola, rabies, and hepatitis B.

5.5 Activation of T cells and cell-mediated immunity

Like B cells, **T lymphocytes** originate from stem cells in the bone marrow, but then migrate to the thymus where they mature and acquire their specific antigen receptors. These cells are responsible for CMI; there are various subsets, some that operate by secreting ILs (of which more than 20 are involved in the immune response), some helping the production of antibodies, and others with cell killing responses. Like B cells, T lymphocytes undergo clonal expansion when appropriately stimulated, with the production of both effector and memory cells.

5.5.1 Helper (T_h) and regulatory (T_{reg}, T_s) T cells

These cells either promote or inhibit the activities of other T cells and the production of antibody by the B cells; they are known, respectively, as **T-helper** (T_h) and **T-suppressor or regulatory T cells** (T_s or T_{reg}) cells, and express the CD4 cell surface marker. T-helper cells can develop into T helper 1 (T_h1) T cells, which promote adaptive immune responses against intracellular pathogens, or T helper 2 (T_h2) T cells, which promote adaptive immune response to extracellular pathogens, whereas T_{reg}s maintain tolerance to self antigens and control autoimmune reactions.

5.5.2 Cytotoxic T cells (CTLs)

These cytotoxic T cells (CTLs) express CD8 on their cell surface; they are particularly important in virus infections, as they recognize virus-specified antigens presented on the surface of infected cells by the MHC class I molecules, which they attack and lyse. This system is carefully controlled so that autoreactive cytotoxic T cells do not arise and kill uninfected cells. Activation and action of CTLs involves three stages for safety (Fig. 5.6): a CTL with receptors recognizing a foreign virus peptide, an activated dendritic cell (APC) presenting the virus peptides, and an activated T_h1 helper cell. In fact T-helper cells can be redundant at the beginning of a virus attack and become more essential during the control of an infection. Cytotoxic T cells kill using perforin and granzyme, the first protein making holes in the cell membrane and allowing subsequent entry of granzyme B. A CTL cell is abundant in these two enzymes and so can kill a number of cells infected with virus before the CTL dies by apoptosis (Fig. 5.7).

5.5.3 Delayed hypersensitivity

T_{dh} cells, as their name indicates, secrete cytokines that mediate delayed hypersensitivity.

5.5.4 Major histocompatibility complex (MHC) restriction

The MHC is a closely associated cluster of genes that encode various cell-surface and other antigens. In humans, it is known as the HLA (human leucocyte antigen) system. There are three classes of antigen, of which only classes I and II concern us here. Class I antigens are expressed on the surfaces of all cells, whereas those of class II are present only on those of antigen-presenting cells (APCs), i.e. dendritic cells in the spleen and macrophages. The great variability in the alleles coding for MHC molecules is responsible for differences in the susceptibility of individuals to a number of diseases, both infectious and non-infectious.

The operations of T cells are intimately related to these MHC antigens. Recognition of 'foreign' antigens takes place only if they are presented by the dendritic cell or macrophage in association with the correct MHC class II molecules, or all other cells if presented by MHC class I molecules: thus CTLs and T_{reg} cells act only when antigen is presented with MHC class I molecules identical with their own, whereas the activities of T-helper lymphocytes are restricted by the need for homologous MHC class II molecules on the APC cell surface complexed with antigenic peptide.

Initially the T-cell receptor (TCR) of a T cell recognizes a viral peptide presented by an MHC molecule and transmits a signal to activate the T cell. This signalling is expedited by a co-stimulatory molecule called CD3. There is also lateral movement of TCRs to cell surface clusters to form an immunological synapse. At this point the activation signal of CD3 is stronger. As you would expect there are important safety mechanisms to prevent uncontrolled T cell activation, such as the elimination in the thymus of T cells that recognize self peptides derived from normal cellular proteins.

Activated T-helper cells also produce T_h1 cytokines such as TNF, IFN-γ, and IL-2 and help to activate the CTL cells. The IFN-γ also keeps macrophages in action and stimulates B cells to produce IgG3 antibodies. Finally IL-2 can restimulate NK cells to re-engage in killing virus-infected cells. There is considerable benefit in involving stimulation of both the innate and adaptive immune systems to engage a virus infection. Figure 5.6 illustrates an important interaction between a T cell and an APC.

A cytotoxic T cell (CTL) is stimulated by reaction with viral antigen on the surface of the APC presented by MHC class II (Stage 1, Fig. 5.6) to undergo clonal proliferation. Later, in association with MHC class I molecules, CTL cells attach to viral peptides on the surface of infected ('target') cells and lyse them (Fig. 5.7). These activities are assisted by T-helper cells acting through IL-2.

Another subset of T cells that mediates delayed hypersensitivity (T_{dh}) reacts with viral antigen on the surface of the APC in

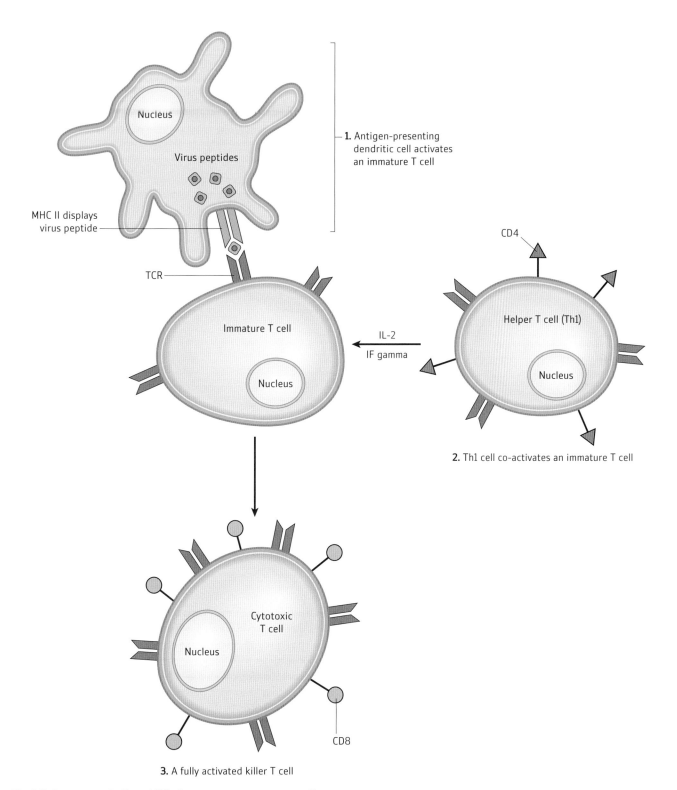

Fig. 5.6 Activation of a T_h and CTLs by an antigen-presenting cell.

association either with class I or class II MHC molecules; these, too, are assisted by T-helper cells. The stimulated T_{dh} cells secrete chemotactic cytokines that attract more T cells and some macrophages to help at the site of infection.

As in the B-cell response, T_{reg} cells stand by to moderate over-energetic behaviour by the T cells. The central role of

T lymphocytes in immunity to virus infections has been likened to that of a conductor orchestrating the entire immune response.

Apart from mopping up antigen–antibody complexes, polymorphonuclear leucocytes are relatively unimportant in virus infections. This can be inferred readily from tissue sections or

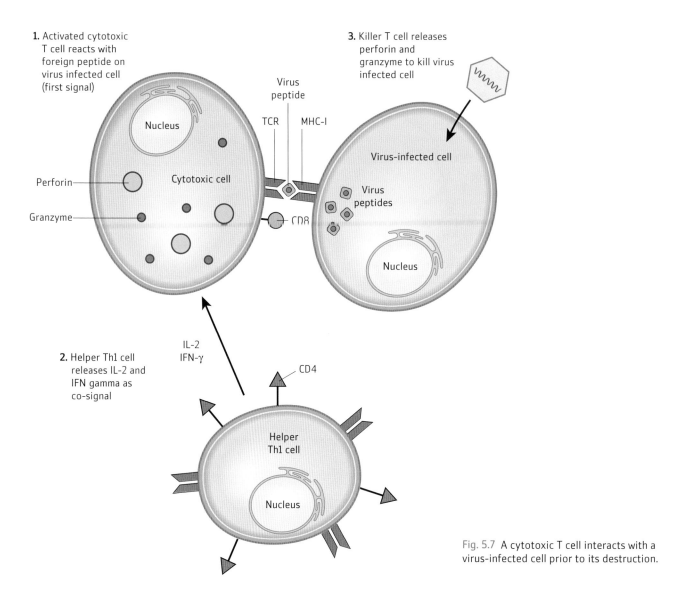

1. Activated cytotoxic T cell reacts with foreign peptide on virus infected cell (first signal)

Virus peptide

TCR   MHC-I

Nucleus

Virus-infected cell

Perforin

Cytotoxic cell

Virus peptides

Granzyme

CD8

Nucleus

3. Killer T cell releases perforin and granzyme to kill virus infected cell

IL-2
IFN-γ

2. Helper Th1 cell releases IL-2 and IFN gamma as co-signal

CD4

Helper Th1 cell

Nucleus

Fig. 5.7 A cytotoxic T cell interacts with a virus-infected cell prior to its destruction.

exudates in which, by contrast with bacterial infections, the overwhelming majority of cells are lymphocytes. An example is shown in Fig. 22.7 (a).

5.6 Harmful immune responses

5.6.1 Enhanced T-cell responses

The destructive potential of the immune system is considerable and can sometimes be dangerous to the host. This was well illustrated during the outbreaks of hepatitis B that used to occur in renal dialysis units before the hazards posed by this infection were properly understood and guarded against by vaccination of staff and virological screening of blood products. The immunity of the patients was impaired by their illness and by treatment with immunosuppressive drugs; therefore, in those patients who contracted hepatitis B a chronic virus carrier state developed (Chapter 28) rather than symptomatic acute infection. By contrast, members of staff, healthy but similarly exposed to large amounts of virus, mounted vigorous cytotoxic T-cell responses, resulting in autoimmune

destruction of their own virus-infected hepatocytes and acute illness that was, on average, much more severe than that suffered by their patients. A classic example of the danger of enhanced immune response along with antigenic cross-reactions of virus proteins is seen with the tropical mosquito-borne disease dengue. The virus exists as four serotypes. Infection with one type can result in enhanced infection by the other types leading to illness called dengue haemorrhagic shock syndrome (DHSS) in some people, which is clinically much worse than the first infection.

It is now recognized that in many serious respiratory infections such as SARS CoV, MERS CoV, and pandemic influenza, the symptoms of acute infection may be the result of a cytokine hyper-reaction or over-reaction involving an outpouring of inflammatory cytokines as a result of virus infection.

5.6.2 Depressed T-cell responses

Immune responses to virus infections are complicated by the ability of some viruses to infect and damage the very cells that mediate these reactions. The best example is that of AIDS

(Chapter 27), in which the HIV infects and destroys the T-helper subset (CD4 T cells), which, as we have seen, is crucial in the activation of B cells and of other T lymphocytes. The resultant havoc in the immune system opens the way to so-called opportunistic infections by other viruses and bacteria and sometimes to tumour formation, resulting in serious illness and, without treatment with antivirals, death.

Measles is also a good example of the importance of an intact B- and T-cell response. In children with a defect in CMI they are ill but there is no rash, which, as explained in Chapter 4, is the result of cytotoxic T cells destroying virus-infected cells in the skin. The presence of a rash can thus be a good sign that the patient is recovering; its absence can be sinister, because a possible outcome is uncontrolled viral replication, resulting in giant-cell pneumonia, and death. Measles is still a major cause of childhood mortality in Africa and developing countries where vaccination coverage is low.

Another example is infection with Epstein–Barr Virus (EBV), a herpesvirus that causes infectious mononucleosis (Chapter 23) and then becomes permanently latent in B cells. Because of efficient immune surveillance by T cells, the latent infection is usually without effect on the health of the host; but an upset of this balance by immune suppression (e.g. in renal transplant patients) can result in transformation of the infected B cells by the virus, with uncontrolled proliferation and death from lymphoma.

5.6.3 Viral infection of mononuclear phagocytes

Many viruses can replicate within these cells, which, as mentioned in Section 5.3.2, play a major role in regulating immunity. Because of the diversity of viruses involved, of the types and functions of mononuclear phagocytes (MPs), and of the cells with which they interact, the influence of viral infection of MPs on the immune response is extraordinarily complex. From the above description of the immune system you will appreciate that interference with MPs by viral infection can, among other activities, affect phagocytosis, antigen presentation, interactions with T cells, chemotaxis, and antibody-dependent cytotoxicity.

The age of the MP may also be important: newborn mice are highly susceptible to herpes simplex because the virus multiplies readily in their macrophages; however, a pronounced increase in resistance to disease develops about 3 weeks after birth and is related to a decreased ability of the virus to replicate in these cells.

Genetic resistance to certain viruses may be partly mediated by MPs. In addition, viruses such as influenza and CMV infections may depress the function of lung macrophages and increase susceptibility to bacterial infection. Finally, arginase secreted by macrophages may interfere with the replication of HSV, which has a requirement for arginine; this, however, is an unusual type of effect.

The precise role of these and other factors in the immune response cannot as yet be quantified, but they do provide some pointers to the part played in pathogenesis by viral infection of these important cells.

5.6.4 Immune complex disease

In certain chronic infections of animals, immune complexes between virus and antibody lodge in small blood vessels, particularly in the kidneys, causing glomerulonephritis. This sequence of events does not seem to be an important feature of infections of humans; nevertheless, immune complexes may be the trigger for the disseminated intravascular coagulation seen in some of the viral haemorrhagic fevers (Chapter 18). They may also be implicated in dengue haemorrhagic shock syndrome (DHSS), which occurs in people, mainly children, with antibody to dengue virus that subsequently become infected with another serotype (Chapter 12).

5.7 Virus countermeasures to inhibit innate immunity and acquired immune responses

Viruses have evolved elegant mechanisms to evade and modulate many aspects of the host immune response. Indeed, the attention that different viruses pay to interfering with certain immune functions highlights their importance in controlling virus infection. Depending on their life cycle different viruses have different immunological constraints that must be overcome for a successful infection. At the simplest level the error prone nature of virus RNA polymerases and reverse transcriptases allow the accumulation of genetic change in the virus genome. Some of these changes will be deleterious and lost through purifying selection, Darwinian laws of natural selection, and survival of the fittest. Others will be neutral, providing neither an advantage nor disadvantage to the virus. Importantly, some mutations will be advantageous, often allowing the virus to escape antibodies through mutation of antibody binding sites and T-cell epitopes (antigens) in virus proteins.

More specific methods of immune evasion are encoded by proteins of different virus genomes. Often the more a virus is able to cause a persistent infection or replicate over a prolonged period of time then the more immune modulatory functions are found encoded by the virus. This requires the virus to be able to pack more genes into its genome that are not purely involved in replicating the genome and making a new virus particle. Therefore, extensive immune modulation is often the preserve of large DNA viruses such as the herpesviruses and poxviruses.

There are two classes of virus immune modulatory genes, those that have sequence similarity to immune function genes of the host genome and those that do not have sequence similarity. Using these genes, viruses are able to inhibit components of the complement system and to produce inhibitors of cytokines, such as soluble receptors for tumour necrosis factor (TNF) and interferons (IFNs) that soak up the TNF and IFN produced in the local tissue environment of the virus infection. Conversely, various viruses produce their own chemokines and cytokines, such as viral IL-6 and viral IL-8 to modify the immune response. Finally, viruses also inhibit intracellular antiviral pathways with virus proteins thereby inhibiting intracellular IFN signalling pathways, MHC antigen presentation, and apoptosis. Rabies P protein, rotavirus (NSP1), and HSV ICP0 inhibit IRF-3, whilst reovirus capsid protein, vaccinia SK-1 protein, HIV TAT protein, and hepatitis C NS5A bind to dsRNA in the infected cell and so block activation of PKR. Why, when, and what immune modifying proteins are deployed by a virus is often still not well understood especially in the context of their effect on the virus infection. Some will be required to establish an initial infection, some to sustain replication, and others to allow onward transmission in the face of an active immune response. Ultimately, one of three outcomes occurs: containment and clearance of the virus, death of the host, or a mutual stand-off where both human and virus co-exist.

5.8 The role of the immune arms in resistance to infection versus recovery from infection

At the start of this chapter it was emphasized that the mechanisms whereby we resist infection differ from those involved in recovery.

By and large, **resistance to invasion by viruses** is mediated by **antibodies**. These may be:

- acquired passively by infants from the mother given passive immunization (e.g. immunoglobulin to protect against rabies (Chapter 19) or hepatitis A (Chapter 7) or Ebola (Chapter 18)); or

- acquired actively as the result of an earlier encounter with the same or mutated virus such as influenza. In this case, the comparatively small amount of existing antibody (IgG or IgA) is supplemented rapidly by clonal proliferation of antibody-secreting B lymphocytes.

Recovery from infection, once acquired, is a more complicated affair and the responses brought into play depend to some extent on the virus concerned. In Chapter 4 cytopathic viruses were classified as 'bursters', lysing cells with the liberation of large numbers of virions into the circulation, or as 'creepers', moving directly from cell to cell. During the viraemia resulting from 'burster' infections, such as poliomyelitis or Yellow Fever (YF), the virions are susceptible to destruction by antibody, notably the IgM induced early in a primary infection. Antibody thus plays a larger part in recovery from 'burster' than from 'creeper' infections, the latter being caused by enveloped viruses, such as herpes and influenza. 'Creepers' have a weak spot because they may induce specific viral antigens on the surfaces of infected cells, thus evoking powerful cell-mediated responses.

Finally, experimental evidence for the relative roles of antibody and CMI in resistance and recovery is powerfully supported by observations of infections in patients with deficiencies in these responses.

 Reminders

- The mechanisms of **innate immunity** come into play very quickly in response to microbial infections, and act as a stop gap until the adaptive mechanisms come into play. IFN, cytokines, NK cells, and other molecules such as complement system form part of innate immunity. It is a powerful system and now the subject of great interest to researchers.

- A highly diverse system of molecules known as **cytokines** form a network of intercellular messengers that help to regulate the response to infection and act as a bridge between the innate and acquired immune systems.

- Toll-like receptors and other pattern recognition receptors (PRRs) which bind to pathogen associated molecular patterns (PAMPs) and activate adaptor molecules, kinases, and finally transcription factors (IRF and NF KB) are key components of the innate immune system and activate IFN genes. IFNs act via many cellular mechanisms including Mx

proteins, protein kinase (PKR), and 2-5A oligo synthetase which block virus replication.

- **Adaptive immunity** is characterized by (1) its ability to distinguish between 'self' and 'non-self' molecules, and (2) memory, whereby an enhanced response is evoked to an antigen previously encountered; memory can last for years.

- The two main components of adaptive immunity to virus infections are (1) **antibodies** produced by B lymphocytes in cooperation with APCs and T lymphocytes, and (2) **CMI** conferred by a variety of cells, mostly T lymphocytes, including particularly helper, suppressor, and cytotoxic cells.

- The operations of T cells are intimately related to the **MHC** antigens; thus, CTLs and T_{reg} cells act only when antigen is presented with class I molecules identical with their own, whereas the activities of T-helper lymphocytes are restricted by the need for homologous class II molecules on the presenting cell surface.

• As a generalization, **protection** against virus infections is largely mediated by antibody, whereas in recovery CMI is relatively more important, especially in the case of the enveloped 'creeper' viruses.

• Immune responses can sometimes be **harmful**; for example, several manifestations of virus infections are caused by the cytotoxic effects of T cells or by immune complexes of antigen with antibody.

 Further reading

Acheson, N.S. (2011). *Fundamentals of Molecular Virology*, 2nd Edition, John Wiley & Sons Inc, New York.

Everitt, A., Kellam, P. *et al.* (2012). IFITM3 restricts the morbidity and mortality associated with influenza. *Nature* **484**, 1–6.

Sompayrac, L.M. (2012). *How the Immune System Works*, 4th Edition. Wiley-Blackwell, Hoboken, NJ.

 Questions

1. Discuss the way IFNs are transcribed in cells.

2. What are:
    a. Chemokines?
    b. Cytokines?
    c. NK cells?

3. Describe the molecular basis of cell-mediated immunity.

6 Viruses and the community: the science and practice of epidemiology

6.1 Introduction

In Chapters 2 and 3, we considered viruses at the molecular level, after which we went on to discuss them in relation to individual cells, to organs of the body, and thence to their effects on individuals. It is now time to stand back and look at a much broader canvas—the way in which viruses affect whole communities. This is the province of **epidemiology**, a word of Greek derivation meaning 'upon (i.e. affecting) the people'; the corresponding term for disease in animal communities is **epizootology**.

The story of the way in which microbes and other parasites affect communities is one of a constantly shifting balance of power between parasite and host, both of which, during their evolution, evolve quite elaborate means of attack and defence, and hence of survival. The results of such battles, which determine the effects of microbes on individuals and hence on the community, are the outcomes of a highly complex interplay of factors. These were neatly summarized by the American epidemiologist John Paul under the headings of *seed*, *soil*, and *climate*, the first referring to the parasite, the second to the host in which it grows, and the third not just to the weather, but to all the other environmental factors involved. Underlying everything is Darwin's theory of natural selection and survival of the fittest. Viruses are the most abundant life form on Earth, and they are subject to evolutionary pressures and can mutate to fit perfectly and to become dominant. Two billion people in the world are infected with hepatitis B and each harbours a population of 1 billion viruses. During an influenza pandemic involving 1 billion people, each will produce many millions of virus particles. Therefore, we are talking huge numbers upon which evolutionary pressures can be brought to bear, allowing viruses that are superoptimized to their immediate environment to emerge.

6.2 Definitions

6.2.1 Prevalence

The number of cases of a given disease—clinical or subclinical—recorded **at a particular time** and expressed as a proportion of the population under study.

A useful ratio is that of case/infection which expresses the proportion of infections resulting in overt disease. It is commonly expressed as the number of cases per 100 infections. Similarly case fatality is expressed as a case/fatality ratio and/ or percentage. Thus polio has a case fatality ratio that ranges from 2–5% among children under 5 years of age to 10–30% among adults and a case/infection ratio of <1. Measles has a case fatality of 0.1% and a case/infection ratio of 95.

6.2.2 Incidence

The number of cases recorded **during a particular period** (e.g. 1 year). This measurement is often given in terms of **an attack rate** (i.e. the number of cases per thousand or per hundred thousand of the general population (or of a subgroup within it) during the period in question).

6.2.3 Endemic

This term refers to a disease that is constantly present at a significant level within a community. Herpes simplex is an example. Endemicity may be high or low. The corresponding term for infections of animals is **enzootic**.

6.2.4 Epidemic

An epidemic is an unusual increase in the number of cases within a community. In this context, 'unusual' is defined arbitrarily: 100 cases of measles in London might not be regarded as an epidemic, but 100 cases of poliomyelitis certainly would be. The corresponding term for infections of animals is **epizootic**. Localized epidemics are usually referred to as **outbreaks**. Members of The Royal College of General Practitioners in so-called spotter practices around the UK, and comparable groupings in Europe, release figures each week of the incidence (number of cases per 100 000 in the population) of respiratory infections such as influenza, and also gastroenteritis. An influenza epidemic is in progress when the incidence reaches 150 clinical cases per 100 000 persons in the community per week.

6.2.5 Pandemic

This is an epidemic involving several continents at the same time. An excellent example is influenza A virus, where global pandemics occurred in 1918, 1957, 1968, and 2009. Pandemic implies speed. HIV could be referred to as pandemic but although 30 million persons have been infected around the world this has taken 30 years and not 18 months as with influenza.

The corresponding term for infections of animals is **panzootic**.

6.2.6 Incubation period, reproductive number (R_0) and generation time

A traditional and still extremely useful piece of information is the **Incubation Period** of a viral infection: the interval between acquisition of the infection to the onset of illness. Another useful measure for mathematical analysis of epidemics is the **Reproductive Number** (R_0) of a virus, defined as the number of persons actually infected from one case of the disease. R_0 varies from a high of 13 for measles to a low of around two for SARS and smallpox, and 1.5 for Ebola. However, a corollary piece of data is the **Generation Time** of a virus—the time between infection of a person and their ability to transmit virus to others. This can be short, viz. 2 days with influenza, or as long as 8–9 days with SARS and smallpox and 12 days with Ebola. A combination of high R_0 and short generation time can spell trouble in the community, especially with a newly emerging virus, such as influenza, whereas long generation time alongside low R_0 gives time for the possible use of quarantine to stop an outbreak, as used in the past in the smallpox era and, more recently, with SARS CoV, MERS CoV, and Ebola. When the figure for $R_0 < 1$ then virus transmission cannot be sustained.

6.3 What use is epidemiology and surveillance?

Medical case story   Cough

It is often surprising how the most intelligent and able of people will present to a doctor with the most trivial of illnesses. However, we all have to learn about managing our own health sooner or later, and for some people this doesn't begin until their twenties.

Your next patient, Mr L, knocks and enters. He is a 26-year-old man, extremely well dressed, with dark cropped hair and a bright-eyed look. You know from the notes that he works as a solicitor. He is alert and lively in his movements, and yet responsive and deferential to your greeting and friendly gesture toward a chair. As he sits down he meets your eye comfortably.

Mr L, 'I've come in because I've had a cough for two weeks,' he says.

Your attentive expression becomes a little less lively, but you attempt to nod in a friendly way. He is the fourth young man with a cough you have seen today.

You have been following the Royal College's weekly returns service and know that the rate of respiratory illness is up at 238 per 100 000 in your area—unusual for this time of year (usually 98 per 100 000).[1] You know it's not just you: all the doctors in your part of the country are having to deal with an unusually high number of coughs.

Mr L continues, 'Everyone at work told me they'd had enough of my coughing. They said I might need antibiotics'.

This is a very typical story. Your patient has not come in now because he is especially ill. He has come in now because his symptoms are having an impact on his life.

He continues, 'Last night I couldn't sleep I coughed so much, and this morning I vomited'.

'After coughing,' you clarify. He nods, 'I was worried because the cough has been going on so long'.

You are now on very familiar ground. You don't need to work things out from first principles here. It seems obvious that he has a self-limiting viral cough.

With luck you'll complete this consultation nicely within your allotted 10 min. This may sound a little uncaring, but you have a duty to care efficiently for all the patients you see this morning. The small amount of time saved now will allow you to catch up a little from your previous consultation, and prevent you running so late for your last four patients, all young children.

You also hope to change his future consulting behaviour. After today's experience he will know to self-manage his next mild viral infection and not seek antibiotics inappropriately.

Mentally you now chalk him up on your 'cough challenge' board. You and a fellow doctor have an informal challenge to give as few antibiotics out as possible for patients with coughs. This is in line with the government directive trying to cut antibiotic use, the overuse of which has already generated worrying bacterial resistance.[2][3] Also, you do not like to give unnecessary treatment for anything, no matter how cheap or free from side effects it is.

First you need to confirm your initial diagnosis. What is the differential diagnosis of cough?

You bring to mind the main infectious causes of coughs:

Mild coughs and upper respiratory tract infections

- Adenoviruses
- Parainfluenza
- Rhinoviruses
- Respiratory syncytial virus (RSV)
- Corona viruses
- Influenza

Pneumonia

- Streptococcal pneumonia
- Mycoplasma pneumonia
- *Haemophilus influenzae*
- *Moraxella catarrhalis*
- Legionella
- *Staphylococcus aureus*
- Influenza

Other differentials of cough, seen more with a chronic history, are asthma, chronic obstructive pulmonary disease (COPD), post-nasal drip, and gastric reflux.

For this young patient with a short history of infectious cough, the main concern is to exclude pneumonia. What questions would you ask to exclude this serious illness?

First you ask if he has green sputum and particularly increased sputum: he does. He has not noticed feeling hot and cold (which would indicate a raised temperature), and has no breathlessness. In fact, he is reasonably well, he is still at work, and has a good appetite. He is a non-smoker and has no history of asthma.

The history is entirely reassuring. If this was a telephone consultation you would now have enough information to advise this patient safely without asking him to come to the surgery for an examination. However, as this patient is in front of you, you take his temperature and reach for your stethoscope: besides, patients expect their chests to be listened to, and will feel that the doctor has been thorough if they have been examined carefully.

As you put your stethoscope in your ears, you signal to him that he is unlikely to be given antibiotics by saying; 'We usually try to save antibiotics for people with pneumonia, but let's have a listen first'.

His aural temperature is 36.7 °C, his respiratory rate is 16 breaths/min, and his chest has good air entry with no wheezes or crackles. Whilst you are trying to examine his chest he mentions he has had a touch of diarrhoea and a slight sore throat as well. These multiple mild symptoms again conform to a typical picture of a viral illness.

After he has dressed, you both sit down again and you explain your findings. You conclude by saying, 'I think you have a viral infection. Antibiotics don't work on viruses, so I am not going to give you antibiotics because they wouldn't work for you'.

He is keen to have advice on managing his symptoms and wants to know when he will be better.

You explain that on *average* a cough will last about 3 weeks, some people will get better sooner than this and for some it will last longer. However, if his goes on for 6 weeks in total, he should come back and you will do a chest X-ray. You agree that coughs at night are very disturbing and suggest a night cough mixture. These have mild sedatives in them to help with sleep.

You explain that coughing can trigger the gag reflex and this can cause vomiting. People often worry about this symptom, but it is not dangerous, just annoying and very uncomfortable. A long bout of coughing can sometimes be helped by a drink of water, or by holding the breath (which is one physiological effect of drinking water, in fact). The main ingredient of cough mixtures is syrup. In fact, 'simple linctus' is *just* syrup—so honey is just as good.

If the cough doesn't go, or if it gets worse, or if he has any pains in his chest (pleuritis), is breathless or unable to take fluids, he should come back. However, you don't think any of that will happen. You anticipate that his cough will simply improve slowly over the next 2–3 weeks.

'Your own immune system will fight it off, naturally.'

Notes

[1] Royal College of General Practice Research and Surveillance Centre Weekly Returns Service.

[2] European Antibiotics Awareness Day. Department of Health, Public Health. Available at: www.gov.uk/government/collections/european-antibiotic-awareness-day-resources

[3] Antimicrobial Resistance and Prescribing in England, Wales and Northern Ireland. Available at: www.hpa.org.uk

---

★ Learning Points

- **Virological:** Common causes of mild upper and lower respiratory illness include: adenovirus, parainfluenza virus, rhinovirus, and coronavirus. There is only symptomatic treatment for these viruses. Antibiotics do not help.

- **Clinical:** It can be important to ask 'Why has the patient come in *now*?' This patient has come in because his symptoms are impacting on his life, not because he is feeling at his worst. Your advice will help address his symptoms at work.

- Personal: Your assessment of a patient will vary according to all sorts of things, *including* whether you like your patient or feel interested in their illness. Your goal (and it's not easy) is every single day to give every single patient an equally thorough assessment no matter whether you like them, feel interested, feel tired, or even upset from a previous patient.

You will have appreciated from the clinical description above that there is much more to this science of epidemiology than dry tables of statistics; it is the key to five major clinical and scientific activities, many of which are aided by the application of mathematical models:

- Predictions of trends in diseases; knowledge of the behaviour of an infection in the past helps to predict the course of an epidemic. For example, determining if cases of Ebola are going to double in the next few weeks. Of course, too much attention to the past can lead to wrong conclusions. Ebola caused thousands of deaths in West Africa in 2014, but the same virus caused only a few hundred deaths in yearly outbreaks from 1976 to 1996. But Africa itself has changed with more movement of people to urban areas.

- A guide to the introduction, improvement, or modification of control measures; for example, immunization programmes, control of insect vectors, or improvements in hygiene.

- An evaluation of the success of control measures, locally, nationally, or worldwide.

- An aid to diagnosis—a knowledge of what infectious diseases are currently prevalent is very useful to the physician as a diagnostic pointer. Such knowledge is particularly useful to doctors in the respiratory virus season (see Medical case story: Cough).

The fifth area is new and developing and involves virus genetics. We are now entering an age where the relatively inexpensive techniques of large-scale sequencing of virus genomes, when linked to two pieces of information about the sample, namely the date the sample was taken and the place the sample was taken (geographically), allows investigators to trace movement of viruses locally and around the globe. This has been undertaken for HIV-1 and 2, influenza viruses, and West Nile disease virus, to name but a few. To make full use of the virus genome data, techniques that utilize the date/time information, called phylodynamics, and place information, called phylogeography, are used. Phylodynamics is a term that describes the interplay between evolution and epidemiology when they occur on the same timescale, as they do in many virus infections, whereas phylogeography focuses on geographical movements of evolutionarily and therefore transmission chain related viruses. Clearly these two methods can be combined. The core method in phylodynamics and phylogeography is the phylogenetic analysis of viral genome sequences sampled over time, which reveal events and processes, recorded in the genomes as mutations by the inevitable errors that accumulate as the virus replicates. As these errors are transmitted from host to host they give us a footprint for reconstructing the epidemiological connections between their hosts. When this information is linked to dates and places phylodynamics and phylogeography reveal the details of virus spread over time, a valuable adjunct to traditional epidemiology. For HIV-1, the reconstructed genetic history of the HIV-1 group M pandemic showed how HIV spread across the African continent and around the world from Kinshasa in the 1920s. This suggested that at this time and place a 'perfect storm' of factors existed, namely rapid urban growth, strong transport links through railways that developed during Belgian colonial rule, and changes to the sex trade. These combined to see HIV emerge and spread across the globe over the next 60 years until enough HIV disease was present in Africa and in risk groups in the rest of the world for AIDS and HIV to be recognized. Similarly, HIV-2 has zoonotic origins in sooty mangabey monkeys in West Africa and sequence analysis of the viral pol and env genes provided data for phylogeographic analysis. Guinea-Bissau, Côte d'Ivoire, and Senegal acted as the main sources at the early stage of the epidemic around 1950 followed by four routes of dispersal along colonial ties namely Guinea-Bissau and Cape Verde to Portugal, Côte d'Ivoire and Senegal to France. Within Europe the virus moved from Portugal to the UK and Luxembourg where there was a large Portuguese population. Phylodynamics and phylogeography can be used in real time, as in West Africa, where these methods are aiding transmission chain analysis of Ebola to help bring the number of Ebola cases to zero. In the future, these methods together with routine virus genome sequences from clinical samples will become important tools in the control of virus infections.

6.4 Epidemiological and surveillance methods

It is not possible to discuss epidemiological techniques in detail here. For our purposes, it is enough to know that for infectious diseases the two principal methods of surveillance are clinical and microbiological, both of which must be tied in with an adequate system of data collection and processing. Increasingly, mathematical modelling is used to mine the data set and to offer advice on the course of epidemics in real time.

6.4.1 Clinical observations

Well before viruses were discovered, clinical observations alone, carefully recorded, made major contributions to epidemiology. Careful clinical observation is still most important and has led to the discovery of emerging acute respiratory viruses such as MERS CoV and SARS CoV and the retrovirus HIV. In the final years of the smallpox eradication campaign whole communities in India were asked to look out for persons with pox lesions and printed pamphlets were issued with photographs of smallpox lesions versus chickenpox lesions. Small rewards were offered for accurate diagnosis. Following a smallpox case a ring fence vaccination strategy was employed (Chapter 24). As more of the world has mobile phones and data sharing through social network platforms increases, it is likely the same principle can be applied, but with electronic information and data capture, rather than paper pamphlets.

In many countries approaching eradication of polio, children who were partially crippled and using crutches were counted, and these observations gave the first indication of an ongoing problem. One of the most striking clinical observations of a disease was P. L. Panum's study of measles in the Faroe Islands in 1846. During that year there was a major epidemic in this isolated community, which, not having been exposed to the infection since 1781, consisted almost entirely of non-immune people. Within the next 6 months, more than 6000 people in a total population of 7782 contracted measles, an extraordinary attack rate of over 770 per 1000. Panum noted that none of the elderly people who had had measles in the previous epidemic acquired it on this occasion. By meticulously recording the dates of contacts and of the onset of disease, this young Danish doctor established that measles is infective for others only at about the time of appearance of the rash. He found the incubation period to be 13–14 days and confirmed by personal investigation that cases with significantly longer or shorter incubation periods either were suffering from something other than measles, or had had a contact outside this period.

Nearer our own time, N. M. Gregg noticed that during 1941 there was a high incidence of cataract in newborn babies in Sydney, and that this abnormality was often associated with deformities of the heart. He then made the acute observation that the mothers of most of these infants had had rubella (German measles) while pregnant during an epidemic in the previous year. We should remember that this idea was ridiculed at the time. The making of this association, purely by clinical methods, was fundamental to the recognition that several virus infections acquired during pregnancy may damage the foetus. Many countries today use trained doctors in the community to look out for respiratory infection and report their findings to the Central Collaborating Service. Incidence is then reported as the number of respiratory virus cases (influenza-like illness or ILI) per 100 000 persons per week. The importance of this basic clinical public health work cannot be overemphasized. Should polio return, for example, the first cases would be picked up by acute clinical observation.

6.4.2 Laboratory studies of antibodies and viral genomes

Good clinical observations are useful, provided that an adequate **case definition** is established at the outset; this means laying down the criteria by which a case is accepted or rejected as suffering from a particular disease. Purely clinical studies do, however, have the following disadvantages:

- Despite the use of case definitions, some syndromes, such as respiratory infections or diarrhoea, are often difficult to diagnose accurately on clinical grounds alone.
- They cannot reveal the extent of very mild or subclinical infections.
- They cannot usually provide accurate information about the prevalence of a particular viral infection in the past (which is often a good guide to the state of immunity to that virus possessed by the community as a whole).

For these purposes, laboratory investigations are brought into play, and are especially useful in viral infections, most of which induce a firm and long-lasting antibody response. As in general diagnostic work, the tests fall into two main categories:

- detection of virus or quantification of virus RNA and DNA in acutely ill patients;
- detection of specific antibody in infected and/or convalescent people.

The molecular techniques of reverse transcription (RT) and polymerase chain reaction (PCR) have been commercialized, along with kits to detect most viruses in our community. As we note in Section 6.3, virus genome sequencing, which is now becoming cheaper and more widely available, can also be used to probe clinical samples for entirely new viruses. These molecular methods are extremely rapid and can give preliminary results within a working day. Following clinical observations of persons with acute respiratory illness in South-east Asia (Severe Acute Respiratory Syndrome or SARS), the virus was isolated, identified as a member of the family *Coronaviridae*, and the genome sequenced in a few days. Later a Middle East variant of the virus (MERS) was identified in the same manner. From these genome sequences rapid RT PCR-based tests were developed for the use as front line diagnostics.

6.5 Serological epidemiology and surveillance

This expression was also coined by John Paul, who wrote that serum antibodies detected by classic virus neutralization tests in the laboratory 'represent footprints, either faint or distinct, of

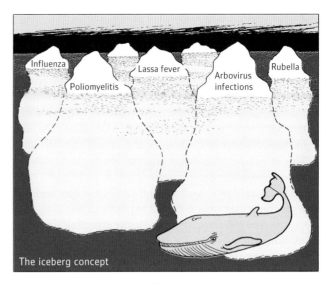

Fig. 6.1 The iceberg concept of infections in communities.

an infection experienced in the remote or recent past'. During the 1930s it was found, even with the crude methods then available, that antibodies to certain infections were present in much higher proportions of the study populations than was warranted by the amount of clinically apparent disease. This situation can be likened to that of an iceberg (Fig. 6.1), in which the part above water represents clinical illness, and that below, the prevalence of antibody in the population. The simultaneous discovery that the prevalence of antibody varies with age provided a powerful tool for studying the history of viral infections in communities.

Now look carefully at Fig. 6.2, which contains much more information than at first appears. These are the results of early studies on antibody to a poliovirus, carried out during 1949–51 in three locations differing greatly in socio-economic condi-

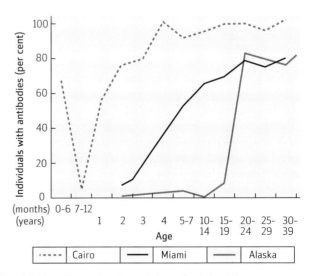

Fig. 6.2 Prevalence of poliomyelitis antibody in different populations. Percentages of individuals with antibodies to poliovirus type 2, surveyed during 1949–51. (Reproduced with permission from Paul, J.R. and White, C. (eds) (1973). *Serological Epidemiology*. Academic Press, New York, p. 10.)

tions. The first, in Cairo, included young infants and clearly shows that those aged less than 6 months still had maternal antibody, which was lost after this period; thereafter, however, it rapidly reappeared, reflecting an active immune response to early infection with the virus. This finding was surprising, because epidemics of *clinical* poliomyelitis were unknown—or at least unreported—in Egypt; it represented hitherto unsuspected infection—the part of the iceberg under water. The early acquisition of antibody is characteristic of societies with poor hygiene in which transmission of infections by the faecal–oral route occurs readily.

In Miami, a world apart both geographically and economically, hygienic conditions were such that exposure to the virus came later in life, and poliomyelitis tended to occur in clinically apparent outbreaks at intervals of 7–10 years. (We explain in Chapter 7 how at certain times and places poliomyelitis is more severe in older than in younger people.)

In the much more isolated Alaskan population, there had been an epidemic of poliomyelitis 20 years previously, but none since. Note that the percentage of people with antibody is insignificant in those aged less than 20 years, but rises steeply to nearly 80% in those aged 20 years or more. If Dr Panum had had the benefit of serological methods, he would have found a very similar pattern 20 years after the measles epidemic in the Faroes in 1846.

Sero-epidemiology therefore produces a wealth of data and is still being undertaken by government and WHO laboratories worldwide to monitor potential gaps in immunization levels and susceptibility of communities to new emerging viruses.

6.5.1 Monitoring an immunization programme

In addition to studying patterns of disease, serological tests are invaluable for monitoring the success of mass immunization programmes. Here, it is important to know what proportion of people in the target population is immune to the disease in question, whether as a result of the immunization itself or of natural infection. The true extent of the influenza pandemic of 2009 was not known until sero-surveys were carried out around the world when it was seen how many silent infections had occurred. In the UK such a survey showed that most of the population had been infected with the virus over two years. However, when sero-surveys were combined with other methods such as RT PCR for infection diagnosis, it was found that only about 23% of infections produced symptoms, so 77% of people were infected but without overt symptoms of disease. The iceberg analogy therefore also applies to influenza.

In many countries, sera from women attending antenatal clinics are routinely screened for rubella antibody; if negative, the patient is advised that she should be immunized soon after delivery, to avoid the risk of rubella in subsequent pregnancies, with consequent damage to the foetus (see Chapter 11). Following the unfounded allegation that MMR vaccine is linked to autism, childhood vaccination rates plunged and have only just recovered. Laboratory serological monitoring is a key element of the campaign and can identify, for example, ethnic and other minority groups where vaccination rates may be extremely low.

6.5.2 Processing data

The collection of epidemiological information both from studies of serology and virus genes is pointless unless there is an efficient system for collating it and applying the results for the improvement of public health. Such systems exist at the local, national, and international levels. It is not possible here to explain in detail the machinery for processing epidemiological information, but an important principle is that much of the traffic is two-way; data collected by clinicians and in laboratories are processed centrally, and the results are ultimately returned to these 'front-line' workers for action at the local level.

On the global level, much information is collected by the World Health Organization, an outstanding example being the monitoring, through 140 national centres around the world, of the incidence of influenza and identification of the prevalent strains of virus, an activity of prime importance to vaccine manufacturers. There are other dedicated WHO laboratories to search for polio virus, and insect-borne viruses. These laboratories also act as centres of scientific excellence for virological training.

In the USA, the surveillance and investigation of infective diseases is managed centrally by the Centres for Disease Control (CDC) in Atlanta, Georgia. In the UK, these functions are the responsibility of Public Health England (PHE) and equivalents in Scotland and Wales, and throughout Europe by the ECDC in Stockholm. Certain infections must by law be reported to the appropriate authorities.

These sources of intelligence cater primarily for healthcare professionals. It should also be the job of governments to provide up-to-date, accurate information and guidance about current epidemiological problems in their respective countries (e.g. outbreaks of food poisoning, influenza immunization, AIDS, vCJD, and so forth). Such information is—or should be—disseminated by notices to doctors in general practice, and to the press.

6.5.3 Can social media help in disease surveillance?

Twitter has over 500 million accounts and over half the active users post or view messages each day. Facebook has nearly 1 billion active users and WordPress holds over 15 million blogs. Social media is being viewed as a 'new frontier in disease surveillance'. Already software programs of digital surveillance platforms like Health Map and Biocaster trawl and scrape the internet searching for advance warning clues of outbreaks. As an example there is an indication that the 2009 influenza pandemic could have been identified a week ahead of the official reports. The ubiquitous mobile phone, numbering over 3 billion subscriptions in the developing world alone, can be used to track population movements and to give early warnings of disease outbreaks.

6.5.4 Mathematical modelling

Knowledge of the age structure in a community, population diversity, and movement by road, rail, and air, together with details of the virus such as the reproduction number R_0, sero-susceptibility, and generation time allows mathematical modellers to produce predictions of speed of spread of infection. Even more important, preventative measures can be placed into the model to give information on timing, and the number of people requiring vaccination to blunt the spread of disease. An analysis of emergence of bird flu (H5N1) in Thailand in 2005 predicted that a global pandemic could be obviated, or at least slowed, by the prophylactic use of antivirals in the local region. These models have also been used to predict the impact of MERS CoV, SARS CoV, influenza, and Ebola. A word of warning is that the modellers often publish data with large uncertainties, particularly when looking further forward in time and moreover with 'worst case' scenarios. This can lead to a 'cry wolf' attitude from the public who remember dire warnings from the influenza pandemic of 2009 which were, in reality not fulfilled. But conversely, mathematical estimates of the doubling time of cases in the 2014 Ebola outbreak alerted the world to a potential health disaster in West Africa.

6.6 Factors in the spread of viral infections

At this point, we must take stock of the 'seed, soil, and climate' factors mentioned at the opening of this chapter. First, the seed.

6.6.1 Characteristics of the virus

Table 6.1 lists the main features of viruses that determine how they are transmitted and, hence, their potential for spread within communities.

First, how well does the virus survive in the environment on its way from one host to another? Enteroviruses such as hepatitis A and polioviruses can remain viable for weeks in water or sewage, an obvious advantage for waterborne agents. They are also unusually resistant to acid pH, which helps them to survive transit through the stomach on their way to their site of replication in the small intestine. Such survival of polio in sewage can be used to monitor spread of the infection in a community, especially now that the disease is so rare. In Israel in 2014, for example, there were no clinical cases of polio but virulent polio virus was detected in sewage. This warned practitioners to enhance the vaccination campaign whilst increasing clinical surveillance for clinical cases. It should be recognized that very extensive laboratory work is needed to analyse thousands of sewage samples month by month.

The existence of a **reservoir of infection** in a mammal or a bird clearly has many implications for the mode of spread. Such infections are called zoonoses, among which are many infections by toga and flaviviruses (Chapters 11 and 12) and by those causing certain haemorrhagic fevers (Chapters 15 and 18). If—as is often the case—a vector such as a mosquito or tick is also involved, many more factors complicate the picture; these include its feeding and breeding patterns, range of mobility, length of time for which the virus persists

Table 6.1 Properties of viruses that determine transmissibility

Property	Particular features	Examples
Survivability outside the host	Resistance to ambient temperatures, drying, ultraviolet light (in sunlight), pH	Enteroviruses, e.g. polio and Loxsackie viruses
Existence of an alternative host	If so, direct transmission? Via arthropod vector?	Rabies virus, arboviruses, e.g. yellow fever, Zika
Portal of entry	(see Tables 4.1 and 4.2 in Chapter 4)	
Evasiveness	Rapid multiplication on a mucous membrane before immune response can be mounted	Viruses infecting the respiratory tract, e.g. rhinoviruses, and conjunctiva, e.g. some adenoviruses
	Variability in antigenic structure, thus evading immune response to a previous infection	Influenza A and B viruses, human immunodeficiency virus (HIV)
Pathogenesis	Incubation period: short, medium, or long	See Chapter 4
Route by which virus is shed	Respiratory secretions	Viruses causing childhood fevers, e.g. measles, mumps, rubella; those causing respiratory infections such as influenza
	Conjunctival secretions	Conjunctivitis viruses, e.g. some adenoviruses, enterovirus 70
	Skin, epithelial mucosae	Warts, herpes simplex and zoster
	Faeces and diarrhoea	Entero- and rotaviruses; Ebola
	Blood transfusion, contaminated needles or instruments	Cytomegalovirus, hepatitis B and C, HIV

within it, and whether the virus is transmitted to its offspring. The transmission of such infections to humans clearly depends on the degree of exposure of the latter to the vector, which in turn may be conditioned by the place of work or recreation. Current examples of animal reservoirs are the identification of fruit bats as a source of Ebola and camels as a source of MERS CoV. There are continuing cases of 'bird' influenza A (H5N1) in Egypt and South-east Asia. These influenza A viruses (H7N9, H5N1) reside in migrating swans and geese and are transferred via bird droppings to domesticated chickens and ducks from where they can enter the human population.

The routes by which viruses enter and are shed from the body are obviously very important in transmission and what happens to them within the host may be equally so. Much depends on how well the virus is able to evade the host's defences, in terms of either its site of replication or its ability to undergo mutations that give rise to new, antibody-resistant, strains. The ways in which viruses spread within their hosts were discussed in Chapter 4; they determine the incubation period of a given infection, and thus the rapidity with which it can be transmitted from person to person. To take two extremes, conjunctivitis caused by enterovirus type 70 has an incubation period of about 2 days, and causes explosive epidemics that sweep through whole communities like wildfire, whilst the AIDS viruses, with incubation periods measured in months or even years, spread correspondingly more slowly and insidiously.

Virus transmission by aerosol or droplet is perhaps the most dangerous method of spread, employed by MERS CoV, SARS CoV, influenza, common cold, and paramyxoviruses. In contrast, water can be treated and filtered, food heated, sex avoided, but breathing is continuous. This aerosol transmission can be so important that groups of scientists have recently mutated bird influenza A (H5N1) to quantify how many virus mutations are required to allow the virus to spread by aerosol from its natural host and then to transmit by the same route between mammals. The answer was startling—only four mutations at critical points in two viral proteins gave the virus this new property.

6.6.2 Characteristics of the host and the environment

We have combined the 'soil' and 'climate' factors in Table 6.2 because one cannot consider host species apart from their environments. We have not given examples of how the various factors operate because there are so many that isolated instances would give a misleading impression. However, there is one very important point. The environment does, to a great extent, determine what sort of virus infections are most prevalent in given geographical areas; this means that there is a considerable difference between the patterns of infection in developed and developing countries (Table 6.3). A more recent theory is that there may be a low proportion of 'super spreaders', This super spread may arise by physical activity of the patient such as sociability or by excretion of higher levels of virus.

6.7 Herd immunity

This expression signifies the proportion of the population that is immune to a given infection, whether as a result of natural infection or artificial immunization. Most acute viral infections

Table 6.2 Characteristics of the host and environment that influence the pattern of viral infections

The host
• Age
• Sex
• Ethnic group and genetic factors
• Occupation and economic status
• Nutrition
• State of immunity
• The environment
• Geographical location
• Urban or rural setting
• Existence of zoonotic infections/vectors
• Socio-economic status/state of hygiene/overcrowding
• Seasonality—some infections peak in the winter or 'raining' season

induce firm and long-lasting immunity. Should the immune status of a community reach over 90% then otherwise susceptible individuals in the group are usually protected by a sea of immune fellow citizens, preventing the virus from finding seronegative people to form an active transmission chain.

There are five main methods by which viral infections are propagated in communities:

1. The respiratory route, e.g. influenza, colds, measles, rubella, mumps.

2. Exceptionally, as with Ebola, intimate contact with bodily fluids such as vomit, blood, semen, or diarrhoea can be a source of spread from person to person, the virus entering via skin abrasions or via the eye.

3. Sexual transmission is common and examples include herpes simplex virus type 2, HIV, hepatitis B, hepatitis C, and rarely some haemorrhagic filovirus infections such as Marburg and Ebola.

4. Many viral infections are spread by the faecal–oral route (e.g. poliomyelitis, hepatitis A, gastroenteritis viruses).

5. The bites of infected arthropods spread viruses (e.g. tick-borne encephalitis, yellow fever, dengue fever and Zika) in most countries of the world.

In all of these modes of spread, the degree of herd immunity is of paramount importance in determining patterns of endemicity and epidemicity. From what has been said, it will readily be appreciated that such patterns are governed by extremely complex factors, so much so that they are often difficult to predict even with the most elaborate mathematical models. Nevertheless, two simplified examples will give a good idea of the general principles.

Table 6.3 High-prevalence viral infections: comparison between developed and developing countries

Developed countries

- Upper respiratory tract infections
- Paramyxoviruses
- Influenza
- Herpesviruses
- Papillomaviruses
- Human immunodeficiency virus (in certain population groups)

Developing countries

All the above, plus

- Poliomyelitis
- Gastroenteritis viruses
- Hepatitis A
- Hepatitis B
- Yellow fever, haemorrhagic fevers and encephalitis due to arbo-, filo-, and arenaviruses
- SARS
- MERS
- Rabies

6.7.1 An epidemic in an isolated community

First, let us go back to the 1846 measles epidemic in the Faroe Islands. The situation faced by Dr Panum is shown schematically in Fig. 6.3. The box represents the isolated community, all of whom, except for a small minority of elderly people, had never been in contact with this infection and were thus fully susceptible. The arrow at the left represents the cabinet-maker who imported the infection on returning from a trip to the mainland. The incubation period was 2 weeks, and on the assumption that those who became infected were immune after another 2 weeks, the increase in herd immunity can be plotted, and is shown as the shaded area. (Needless to say, epidemics do not behave as regu-

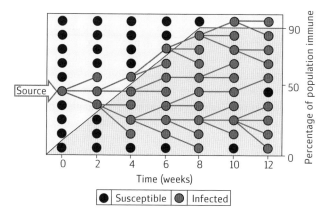

Fig. 6.3 Spread of measles in a highly susceptible population in the Faroe Islands, 1846

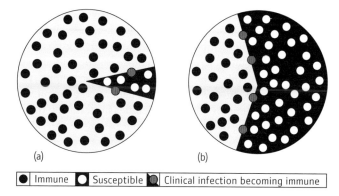

Immune | Susceptible | Clinical infection becoming immune

Fig. 6.4 Herd immunity in poliomyelitis. (a) 'Cairo' model. Herd immunity is high and clinically apparent infection is rare: no danger of epidemic. (b) 'Miami' model. Herd immunity is only moderate and there are sporadic cases of clinically apparent infection: an epidemic is probable

larly as this in real life. Successive waves of infection soon get out of synchrony.) Note that some infections are 'dead-end' (i.e. are not transmitted to anyone else). The epidemic petered out when there were too few susceptible people left to keep the chain reaction going. In practice, not all the population must be immune for this to occur; it can happen when the proportion reaches 70–80%. Other historical examples include measles outbreaks in Africa transported during colonization and smallpox in the Americas.

6.7.2 Endemic infections

Look again at Fig. 6.2, in which the age-specific distributions of antibodies to poliovirus type 2 in different geographical areas during 1949–51 are compared. In Cairo, the majority of the population had been infected during infancy, an age when most polio infections are asymptomatic. As immunity following exposure to this virus is virtually lifelong, it follows that the proportion of susceptibles in the total population was very small (Fig. 6.4a). There may have been the odd sporadic case of clinically apparent disease, but the pie chart shows that there are certainly not enough susceptible people to sustain an epidemic. By contrast, the proportion of immune people in the various age groups in Miami did not much exceed 50% until early adolescence, leaving a substantial proportion of children unprotected during a period of life when poliomyelitis is liable to cause severe paralytic illness. The situation in this age group is shown in Fig. 6.4b, from which it can be seen that there has been a build-up of susceptibles to a point at which an epidemic could be initiated. Before the introduction of polio vaccine, such epidemics did, as mentioned in Section 6.5, occur at 7–10-year intervals in this community.

6.8 Hospital-acquired or nosocomial infections (HAIs)

Infections are classed as hospital-acquired if they become manifest more than 48 h after admission or less than 48 h after discharge. They are also known as **nosocomial infections**. HAIs present us with the paradox that some people who enter hospital to have their illness cured acquire another, possibly worse infection from other patients, their environment, or even medical or surgical equipment used for their treatment. Various factors contribute to this state of affairs, including, for example, the increasingly rapid turnover of staff and patients, the complexity of modern instruments and the difficulties in sterilizing them, and the emergence of antibiotic-resistant pathogens. All this is compounded by the fact that the immunity of some hospital patients is impaired by their illness.

6.8.1 Sources of infection

Viral infections account for a substantial proportion of illnesses contracted in hospitals. The chances of acquiring one are the result of several factors including:

- good staff discipline, especially in relation to hand washing with hot water and soap and disinfectant;
- cleanliness of the wards;
- length of stay;
- whether surgery is involved;
- prevalence of the virus;
- transmissibility of the virus;
- state of immunity of the patient;
- age of the patient (risk greater for very young or very old people); and
- care and cleanliness of apparatus such as air-conditioning systems and water.

Table 6.4 lists the most important viruses causing HAI. Note, however, that the same virus may have several routes of entry. A special situation arises when a patient has a reactivation of herpes zoster, particularly if immunocompromised, in which case he or she may shed virus, and be a danger to other patients, and a potential hazard to the medical and nursing staff. To avoid this problem, the immune status of the attendants should be ascertained before, rather than after, they take up their duties and any without antibody should be offered varicella-zoster virus (VZV) vaccine. The patient should be transferred to an isolation facility forthwith.

There have been several instances of patients acquiring rabies from transplant donors (Chapter 19).

There is more attention paid now to the vaccine immune status of healthcare workers (HCW) who must have received the hepatitis B vaccine and be shown to have protective levels of antibody. More pressure is now being placed on HCW to be regularly immunized with influenza vaccine to reduce their chance of introducing new viruses to the ward or hospital viruses to their own family.

A relatively new phenomenon has been the spread of certain viruses with high fatality such as MERS CoV, SARS CoV, and Ebola from patients in the hospital environment to doctors and nurses treating them. Unfortunately these three viruses may

Table 6.4 Examples of hospital-acquired infections (HAIs) caused by viruses

Body system	Virus
Respiratory tract	Influenza
	Parainfluenza
	Respiratory syncytial virus
	SARS
	MERS
Gastrointestinal tract	Noroviruses (Norwalk)
	Rotaviruses
	Ebola in West Africa
Skin	Varicella-zoster
Contamination with blood	Hepatitis viruses
	HIV
	Filoviruses such as Ebola and Lassa fever
Generalized infections	Measles
	Rubella

be first identified in the community by the death of a healthcare worker. Hospitals are a major source of spread of the haemorrhagic fever Ebola in West Africa, and at the beginning of the 2014 outbreak as many as one third of deaths were recorded in HCW. In the polio epidemics of the 1950s in the USA and Europe a constant hazard for medical and nursing staff was contracting the virus from the predominately young patients.

6.9 The periodicity of epidemics

In the absence of interference with the natural order (e.g. by introducing vaccines, improvements in hygiene, or campaigns against arthropod vectors), the pattern of acute infections may alternate between endemic and epidemic. Epidemics create highly immune populations, which, however, steadily become diluted with susceptibles as new babies are born into the community. When the proportion of non-immune people reaches a 'critical mass', contact with an infectious person may start a 'chain reaction' epidemic; water- or vector-borne diseases may also occur in epidemic form as a result of waning immunity in the general population. In areas unaffected by immunization, epidemics of acute infections, such as measles or rubella, may recur at regular intervals. Figure 6.5 shows such a pattern, and also illustrates the effect of introducing immunization. There is another type of periodicity, superficially similar but basically different. Here, regular annual epidemics result from infections transmitted from an external source—in this case food—and not from person to person. The peaks are due not so much to a build-up of susceptibles, but to the increased opportunities for bacteria to multiply and be transmitted during the warmer months.

6.10 Control measures

So far, we have described the factors involved in the spread of infective agents in the community and the methods for collecting epidemiological data. Clearly, such information is of little value unless it can be used both to combat outbreaks as they occur and to provide guidelines for preventing them in the future. Such control measures fall into two main groups: those directed respectively at individuals and at the community (Table 6.5). Almost by definition, most of them cannot be implemented adequately in developing countries, although within recent years there has been a welcome increase in the number of mass immunization programmes in the less affluent areas supported by international charities such as GAVI and by the WHO childhood immunization campaigns.

Epidemiologists attempt to quantify the effect of vaccines and hand washing and social distancing and epidemics by designing **prospective** and **retrospective** studies. In a prospective study, two population groups are studied and one for, example, has been vaccinated. Incidence rates in both groups are recorded. Incidence rates in both groups can be compared and a relative risk (RR) calculated as a ratio of the two rates. Large numbers of subjects are required and followed for at least one year. In a retrospective study smaller numbers in the two groups are analysed for infection rate and relative risk.

6.10.1 Individual measures

It goes without saying that personal hygiene is important for everyone, but particularly for those whose occupations carry the risk of spreading infections to others. They include food handlers, healthcare workers, and laboratory staff. **The single most important measure is probably hand washing**, especially after defaecation and, for clinical personnel, between contacts with patients with infective illnesses. The CDC have issued very important guidelines on the technique of hand cleansing which takes at least 20–30 seconds. We should remember the pioneer Dr Himmelweis who insisted that his gynaecology students in Vienna used hand hygiene and so reduced puerperal fever in his wards. At first he was ostracized by his medical colleagues and made a figure of fun.

Infections transmitted by blood and body fluids, such as hepatitis and HIV and, of course, bacterial STD, can be prevented by avoiding both unprotected penetrative intercourse and the sharing of syringes and needles used to inject drugs.

The control of dangerous infections such as the haemorrhagic fevers Ebola (Chapter 18) and Lassa fever (Chapter 15) in the community at large is possible only when cases are few in number (e.g. importations from an infected to a non-infected country) and where there are adequate containment facilities for nursing such patients. Many communities and indeed small hospitals in West Africa where Ebola and Lassa fever spread did not have access to safe water for the purposes of hand hygiene.

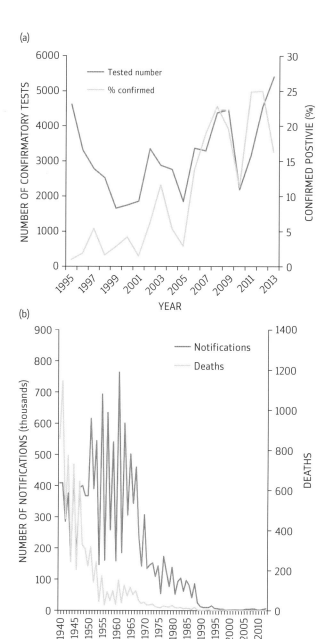

Fig. 6.5 The periodicity of measles epidemics. Typically, this pattern occurs when the infection is transmitted from person to person and confers long-lasting immunity. The effect of introducing mass immunization in 1970 is apparent. (Data from The Office for National Statistics, see http://webarchive.nationalarchives.gov.uk/20140505192926/http://www.hpa.org.uk/web/HPAweb&HPAwebStandard/HPAweb_C/1195733811358 (accessed 19 Feb. 2016) and https://www.gov.uk/government/publications/measles-deaths-by-age-group-from-1980-to-2013-ons-data/measles-notifications-and-deaths-in-england-and-wales-1940-to-2013 (accessed 19 Feb. 2016).

Finally, travellers can protect themselves against infections not prevalent in their own countries by immunization against, for example, yellow fever, hepatitis A, and some encephalitis viruses. Given the number of citizens of all ages now travelling around the world this branch of 'travel medicine' is flourishing.

Table 6.5 Control of infection in individuals and communities

Individual measures	Community measures
Personal hygiene, especially hand washing	Surveillance of food/milk/water supplies and blood products
Personal behaviour, e.g. 'safe sex', avoidance of injected drug abuse	Maintenance of hygiene and avoidance of overcrowding in residential institutions
Isolation of patients with dangerous infections ('quarantine')	Where appropriate, control of insect vectors and immunization of animals
Appropriate individual immunizations for travellers	Routine immunization programmes
'Social distancing' with avoidance of crowded theatres and also avoiding infected persons in the family	Dissemination of information to professional colleagues to the public

6.10.2 Community measures

Constant **surveillance of the production and distribution of food and water**, together with mass immunization, are among the most important elements in protecting the health of the community at large. The rapid transport of foodstuffs within and across national boundaries is a potent means of spreading microbial infection. Objections are often raised to the intrusion of government into our daily lives, but there is no doubt that appropriate legislation is essential for ensuring that the production and distribution of food, milk, and water meet satisfactory standards all the way from production to the point of sale.

Although contamination of blood and blood products such as plasma, immunoglobulins, and clotting factors is comparatively rare and affects much smaller numbers of people, the consequences for individuals may be very serious. Virtually all such episodes involve viruses, rather than bacteria, and a high degree of technical expertise may be needed to identify them. It is far better that blood products originate from healthy, unpaid

donors, rather from those giving blood for money, who are more prone to infections such as HIV and viral hepatitis, and may be drug addicts. Furthermore, where paid donors are used, screening facilities may well be inadequate.

Hygienic measures in institutions such as those housing the mentally subnormal may be difficult or impossible to maintain, with resultant outbreaks of gastrointestinal and respiratory infections.

In some areas, often tropical or subtropical, **certain infections are transmitted to humans from animals**, with or without the participation of an insect vector. Control measures often involve counter-attacks on the animals or insects involved. Some of these infections are caused by arboviruses (Chapters 11, 12, and 16), and a number fall into the 'exotic and dangerous' category (Chapters 15 and 18). Table 6.6 gives some examples. New influenza pandemics arise from the transfer of virus from asymptomatic migrating geese and ducks to domesticated chickens and pigs and onwards to humans. But it has to be admitted that influenza surveillance at this human/bird interface is very weak. The world's poultry population exceeds 21 billion and yet only 1000 influenza nucleotide sequences are commonly analysed. Weak surveillance and epidemiology can give a sense of false security. In recent years we have appreciated the central role of bats in the emergence of new viruses. These are very prevalent creatures and are a major group of mammals. In many countries they are an important source of animal protein for human consumption and at the same time carry diseases such as Ebola, Hendra, and Nipah viruses. The meat is safe after cooking but infection occurs during hunting or gathering or in market places.

A most important weapon in the control of infective illnesses is the **mass immunization** of susceptible people, the great majority of whom are young children (Chapter 30). By this means alone, smallpox was eradicated from the planet by 1977, and we are in sight of doing the same for polio and measles, which have already been eliminated from many countries. But surprising to many is the post-polio eradication scenario: vaccination will continue for at least five years using killed polio vaccine until we are absolutely certain that true eradication has happened.

It should be clear that none of the control measures described in this section could be properly implemented without a constant and efficient exchange of information. This important topic was dealt with in Section 6.5.3.

Case study Surveillance, case fatality ratio (CFR), and generation times of pandemic influenza A (H1N1)

A new variant of influenza A (H1N1) virus was identified in Orange County, California around 23 April 2009. Mexico, meanwhile, was experiencing an unusual outbreak of community-acquired pneumonia in young persons and declared a local emergency on 25 April. A virus cause was suspected in Mexico because such bacterial pneumonia is normally detected in older persons. Throat swabs were analysed, typed, and an influenza A virus

found which was identified as A/California/7/2009. Initial data from Mexico suggested a high case fatality of around 4%, not so distant from the related influenza A (H1N1) virus of 1918, the so-called Spanish influenza which caused 80 million deaths that year. This was a major factor in the WHO decision to declare an influenza pandemic. Also by June, although estimates of case fatality had reduced, nevertheless the new virus had been found to cause unusual and severe pathology in animal models and was spreading quickly and widely in both hemispheres. This is a classic example of global spread by air transport, not by catching virus on a plane, which is rather infrequent with most aircraft equipped with HEPA filters for cabin air, but by travellers and holiday makers carrying the virus homewards as infectees often with subclinical or very mild symptoms. A pandemic situation was declared by WHO on 6 June.

Early in the outbreak it was quite apparent that the infection was moving amongst children and young adults and that anyone older than 60 years was immune because a very similar virus had circulated in the early 1950s. The virus had a R_0 of 2.5 and a generation time of two days. In the more recent Ebola outbreak in West Africa, 14 000 cases were detected over a year from December 2013, whilst, by contrast, from 2009–2010 in the USA over 40 million cases of influenza were detected. In the UK it soon became apparent that 75% of hospitalized patients had no prior underlying medical problem, whilst non-white and pregnant patients were over-represented in the severely ill. In common with the yearly influenza epidemics, vulnerable 'at risk' people were those with diabetes, respiratory problems, asthma, and heart and kidney disease. Rather new was the number of obese persons who became seriously ill. The rates of severe or fatal illness in patients with severe obesity (body mass index >35) were increased 5–15 times. A combination of compromised respiratory function alongside immunological changes could contribute to this unfortunate outcome. A higher case fatality was soon noted in aborigines in Australia and First Nation individuals in Canada and the USA. In the UK a higher mortality was noted in children of groups who had immigrated from the Indian subcontinent, and was attributed to proven socio-economic factors rather than genetics per se. However, recent studies have identified a variant of the gene IFITM3 that correlated with extra severity of the disease.

Overall the case fatality rate with the virus approximated to 0.02% but this still equated to 15 or more million years of life lost. In comparison, in the 1918, 1957, and 1968 pandemic comparable figures were 6.3, 2.6, and 1.6 million years of life lost in the world. In a striking re-evaluation of the original WHO figures of 18 500 deaths worldwide, more recent calculations show as many as 575 500 deaths worldwide with the corresponding increase in years of life lost in the 2009 pandemic.

Table 6.6 Examples of control measures for infections transmitted from insects and animals

Disease	Transmitted by	Control measures
Yellow fever and Zika (Ch. 12)	Mosquitoes	Anti-mosquito measures: insecticides; spraying or elimination of breeding grounds; use of mosquito nets and insect repellents. Immunization (for yellow fever)
Dengue (Ch. 12)	Mosquitoes	Anti-mosquito measures
Viral encephalitides (Chs 11, 12, 16, 20)	Mosquitoes, ticks	Anti-mosquito and tick measures
Crimean and Omsk haemorrhagic fevers (Ch. 16)	Ticks	Anti-tick measures
Lassa and other arenavirus infections (Ch. 15)	Contact with infected rodent urine or faeces	Rodent-proofing of living quarters
Hantavirus haemorrhagic fevers, pulmonary, and renal infections (Ch. 16)	Contact with rodent urine	Rodent-proofing of living quarters
Filoviruses: Marburg and Ebola (Ch. 18)	Contact with infected monkeys or bats. Thereafter the viruses are transmitted by contact with human blood and body fluids.	Staff training in isolation, and barrier nursing using PPE. Change funeral practices in Africa to avoid relatives touching a fever victim. Quarantine cases
Rabies (Ch. 19)	All warm-blooded animals, especially dogs, foxes, bats, raccoons	Immunization (humans, dogs, foxes). Quarantine where appropriate
Poxviruses, e.g. cowpox, orf, monkeypox (Ch. 24)	Cows, monkeys	Avoid contact with skin lesions
Pandemic influenza	Contact with chickens or pigs thereafter person-to-person spread.	Control live bird markets in Asia
SARS	Initially contact with bats and civet cats	Control live animal and bird markets in Asia
MERS	Initially contact with bats and camels	Control access to camels in the Middle East

Planning for the next influenza pandemic and other disease outbreaks

WHO has issued an updated document 'Pandemic Influenza Risk Management' and a particular change has been identification of a continuum of four phases, namely interpandemic, alert, pandemic, and transition. The current potentially pandemic influenza A (H5N1) and (H7N9) viruses are both at the 'alert phase'. Importantly, the WHO recognized that a framework to manage a range of disasters is already in existence in many countries (Emergency Risk Management for Health, ERMH) and that the new pandemic influenza guidance should fit more easily into these existing plans. A practical consequence is an uncoupling of global phases from risk management decisions and actions at the country level. As part of the new plan WHO has devised a PIP (Pandemic Influenza Preparedness) framework, to encourage sharing of viruses and a more equitable access to vaccines and antivirals. The PIP was activated in 2011 at the Sixty Fourth World Health Assembly and is incorporated in the new guidance. Such preplanning for influenza and SARS aided the world community to quickly react to MERS in 2012 and to a lesser extent to Ebola in 2014.

The new WHO plan has several layers of disease severity to incorporate 'mild' and 'moderate' clinical presentations and to make the plan 'living', flexible, and accessible.

Case study Application of contact tracing, quarantine, and hygiene to stop the spread of Ebola in West Africa

Ebola (Chapter 18) is a haemorrhagic fever spreading infrequently from two species of fruit-eating bats in Central and West Africa. The virus has a low R_0 of around 1.5 and spreads directly from person to person after intimate contact with blood, vomit, or diarrhoea. The virus has a lipid envelope that is easily destroyed by detergents, disinfectants, and hot water and soap. However, these are all in short supply or non-existent in slum townships and villages in Liberia, Sierra Leone, and Guinea. Additionally there is lack of knowledge about hygiene and a belief in a sizeable minority of the population in traditional healing and exorcism. The local custom of burying family members nearby and preparing the body for burial by washing at home add to the contagion problems. There is an experimental vaccine for contacts of a clinical case but no proven antiviral therapy.

In the past and in the 2014 outbreaks the classic public health and epidemiological methods of contact tracing, case management, surveillance, and quarantine have been applied successfully. In addition public awareness and community engagement and national legislation and regulation have been brought into play.

Alongside the low R_0 of the virus, the incubation period averaging 12 days is long, as is the generation time (12 days). There is no person-to-person spread until symptoms of temperature rise and vomiting begin to appear. As an example, an individual who was in contact with Ebola patients in Liberia travelled by plane to Texas during the incubation period of the disease and became ill after several days in the USA. He was turned away from the A&E at the local hospital but returned there two days later with clinical signs of Ebola and was admitted to a security ward. Meanwhile his home was visited by the police, who were not dressed in personal protective equipment (PPE). Essentially this is a Tyvac set of overalls with a separate hood, visor, and incorporated filter, two pairs of rubber gloves, and rubber boots. Training on taking off the PPE is essential to avoid self-contamination of any specks of blood and vomit on the suit. The subsequent contact tracing in Texas quickly drew six close contacts into quarantine at home and this had to be legally enforced. A further ring of 40 or so persons were contacted as well as over 600 travellers on the planes from Liberia to Europe and then onwards to the USA. All close contacts were quarantined for 21 days, the longest incubation time measured for the infection to date. The only transfer of infection was to two nurses at the hospital at the first visit.

This case illustrates perfectly the great efforts needed by public health specialists to abort community outbreaks, but also shows how effective these measures can be. Obviously blood was taken from the immediate contacts for rapid molecular testing by RT PCR, whilst the others were phoned regularly and monitored their own temperatures and symptoms at home.

Identification of the source of the 2014 Ebola outbreak

This is an excellent example of a public health response in Guinea to contain the Ebola outbreak. The objective was to obtain early samples of virus and trace the genesis of the outbreak. Samples and clinical data were obtained from the first twenty patients. The team generated 48 complete genome sequences, searched for sequence evolution, and composed phylogenetic trees. Data was gathered through interviews with patients, affected family attendees at funerals, and also hospital staff.

The suspected first case was a child who died in Meliandou on 6 December 2013. Her sister, mother, and grandmother had died by 1 January 2014. A nurse transferred the infection to Guectaedou hospital on 5 February and the virus spread out via funerals and other healthcare workers in the next six weeks. A high degree of virus genome nucleotide sequence similarity of the viruses plus the epidemiological evidence of links indicated that there was a single introduction of virus into the population. An early piece of important clinical data was that haemorrhage was not marked, but that fever and vomiting were very common. Case fatality was above 70%.

Public health measures can be undertaken on a more international scale to involve a restriction of flights to affected countries together with exit or entry screening of travellers. However, in the case of large outbreaks, where international help is required, as in West Africa in 2014, flight bans impede the response and therefore are not favoured by WHO. Exit and entry screening is essentially a form-filling exercise but can be useful. Diseases like Ebola, SARS, and MERS have a long incubation time and so temperature monitoring at exit or entry to an airport may not be successful in detecting cases. It is worth noting that airport entry screening in South-east Asia at the time of the SARS outbreak in 2003, and involving two million total body temperature scans, did not detect a single case of SARS. Likewise in Canada an analysis of 750 000 detailed entry questionnaires failed to detect a single case of SARS and nor did temperature screening at the airport. Key elements of the public health response are to identify cases early by RT PCR and to segregate them from the community. Meanwhile contact tracing is carried out and suspects closely monitored clinically, and also by RT PCR.

Reminders

- Epidemiological techniques are essential for predicting health trends in the community, and as a guide to the implementation and evaluation of control measures.

- Mathematical and computer predictions are more used now to predict the speed of spread of infection in the community. Key pieces of information are the reproductive number (R_0) of a virus, namely how many persons are infected from a single case of the disease, and the generation time, being the time between infection and infectiousness for others.

- The use of virus genome sequence information and the methods of phylodynamics and phylogeography have become widespread in infection control and epidemiology.

- Careful clinical observations, making use of adequate case definitions, are important in establishing patterns of endemic and epidemic disease, but usually require supplementation by laboratory tests.

- Serological surveys of communities must be well designed in terms of their objectives and collection of adequate information about the study population. They show that for many viral infections there are high ratios of antibody positivity—the 'footprints' of past infections—to clinically apparent disease.

- Social media and Twitter are new sources of information and may give an early warning of epidemics.

- The control of infection in communities depends on the implementation of control measures both by individuals and the community. They include maintenance of personal hygiene, immunization, surveillance of food and water supplies, control of insect and animal vectors, and dissemination of epidemiological information at local, national, and international levels. Many factors affect the transmission of viral infections, among which are survivability outside the host, the existence of reservoirs of infection in alternate host species, the need for an arthropod vector, the routes by which the virus enters and is shed from the host, and incubation period.

- Hospital-acquired infections pose their own epidemiological problems. It is most important that hospitals have written control of infection policies in place, and that the medical and nursing staff implement them meticulously. Staff should be trained in barrier nursing and the use of PPE for haemorrhagic fever.

- The prevalence of the various viral infections differs with socio-economic status, and between developing and developed countries.

- The degree of herd immunity is important in determining the balance between endemicity and epidemic disease. In turn, shifts in this balance may result in recurrences of epidemics of a particular disease at regular intervals.

Further reading

Everitt, A.R.*et al.* (2012). IFITM3 restricts the morbidity and mortality associated with influenza. *Nature* **484**, 519–25.

Faria, N.R.*et al.* (2012). Phylogeographical footprint of colonial history in the global dispersion of HIV type2A. *Journal of General Virology* **93**, 889–99.

Févre, E.M., Bronsvoort, B.M., Hamilton, K.A., Cleaveland, S. (2006). Animal movements and the spread of infectious diseases. *Trends Microbiol* **14**, 125–31.

St Louis, C. and Zorlu, G. (2012). Can Twitter predict disease outbreaks? *British Medical Journal* **344**, e 2353.

Van Tam, J. and Sellwood, C. (2003). *Pandemic Influenza*, 2nd Edition. CABI, Boston, USA.

WHO Ebola Response Team. (2014). Ebola Virus Disease in West Africa—the first 9 months of the epidemic and forward projections. *New England Journal of Medicine* **371**, 1481–95.

? Questions

1. Write notes on:

   a. Characteristics of host and environment which influence infection.

   b. R_0, Generation Time and Incubation Period.

   c. What is the 'Iceberg concept'. Give some examples.

2. Write short notes on epidemic and pandemic viruses versus those which are endemic. Illustrate with specific examples.

3. Can Ebola be controlled without a vaccine?

Part 2

Specific viruses

Group 1 Positive-sense single-stranded RNA viruses

7

Picornaviruses: polio, hepatitis A, enterovirus, and common cold

7.1 Introduction

The family *Picornaviridae* (Fig. 7.1) is one of the largest in numerical terms, whilst from a purely molecular viewpoint the picornavirus was the first animal virus to be cloned and sequenced, and then to be reconstructed from chemical synthesis of the nucleotide sequence (Fig. 7.2). A picornavirus was also the first human virus to have its three-dimensional structure analysed using X-ray crystallography. In clinical terms poliomyelitis was one of the first virus diseases to be recorded; an Egyptian tomb carving of the Nineteenth Dynasty shows the dead man to have had a foot-drop deformity typical of paralytic poliomyelitis (Fig. 7.3). A world polio eradication campaign, spearheaded by WHO, is the largest public health project the world has ever known and should result in this virus becoming eradicated.

Another globally important virus in the family is hepatitis A, formerly enterovirus type 72, and this can also be controlled by vaccinations. Of course rhinovirus infections are known to us all, as it causes the common cold.

7.2 Properties of the viruses

7.2.1 Classification

The family *Picornaviridae* (Fig. 7.1) contains six genera of which five infect humans (Table 7.1). Its name derives from the small size of these viruses (*pico* = small) and their RNA genome. Picornaviruses are found in several mammalian species and in birds, but we shall describe only those that infect humans.

Poliovirus was the first human virus of the whole family to be isolated in 1908. Much later, Enders, Weller, and Robins earned a Nobel Prize in 1948 for describing the growth of this neurotropic virus in non-neural tissue, in this instance derived from monkey kidney. Soon after the isolation of the polioviruses, similar viruses were discovered that paralysed infant mice. On

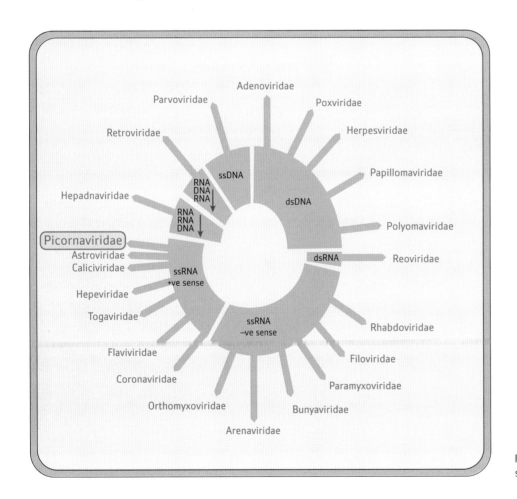

Fig. 7.1 Baltimore classification scheme and picornaviruses.

Egyptian tomb image	Clinical description	Virus transmission established	Virus cultivated in monkey kidney cells	Killed polio vaccine licensed (Salk)	Live vaccine licensed (Sabin)	Molecular cloning of polio cDNA	X-ray crystallography of virus	WHO sets goal of global eradication	Chemical synthesis of polio RNA	Last case of polio in India	Virus on the edge of eradication
1300 BC	1840	1909	1948	1955	1961	1981	1985	1988	2010	2014	2015

Fig. 7.2 Timeline for polio.

Fig. 7.3 Egyptian tomb carving of the Nineteenth Dynasty. The 'foot-drop' deformity is characteristic of residual paralysis due to poliomyelitis, a disease hopefully soon to be confined to history.

the basis of the pathological changes in mice, two groups were defined, termed Coxsackieviruses A and B after the town of Coxsackie, New York, where they were first isolated.

Although, historically, the enteroviruses were classified into serotypes by neutralization tests, numerous difficulties and disputes arose, and now molecular sequencing is the gold standard for identification and classification (Table 7.2). The molecular scheme is based on shared or distinct nucleotide sequences of

Table 7.1 The *Picornaviridae*

Genus	Main syndromes
Enterovirus	Infections of the central nervous system, heart, skeletal muscles, skin, and mucous membranes. Polio, Coxsackie, and ECHO viruses.
Hepatovirus	Hepatitis A
Rhinovirus	Common colds
Aphthovirus	Foot-and-mouth disease of cattle (rarely in humans)
Cardiovirus	Encephalitis and myocarditis
Parechovirus	Diarrhoea

Table 7.2 Members of the genus *Enterovirus* that infect humans

Group	No. of serotypes
Poliovirus	3
Coxsackie A	23
Coxsackie B	6
Echovirus	32
Enterovirus	5*

*Numbered 68–72

the VPI capsid protein. Hepatitis A virus was first cultivated in monkey kidney cells in 1979. Volunteer experiments confirmed the nature and pathology of hepatitis in 1983.

7.2.2 Morphology

Picornaviruses are only 18–30 nm in diameter and icosahedral, with a regular protein capsid composed of 60 protomers each containing a single copy of four structural proteins called VP1, VP2, VP3, and VP4. Picornaviruses are non-enveloped (Fig. 7.4). VP4 lies on the inside of the virus particle and the virus's external facing 'shell' is made up of VP1-3. The entire

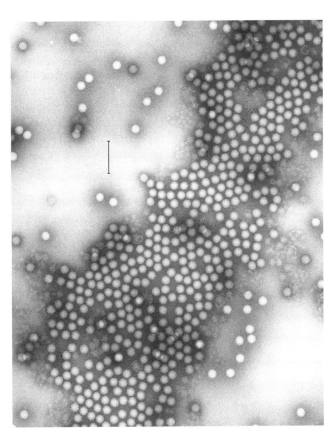

Fig. 7.4 Electron micrograph of a cluster of picornaviruses (poliovirus). (Courtesy of Dr David Hockley.) Scale bar = 100 nm.

protomer structure is wedge shaped allowing the strong icosa-hedral shape of the virus particle to be formed.

X-ray crystallography has helped to pinpoint functional areas in the virion, the most interesting being the 'canyon' or cleft where the cell receptor binding site is located. There is great interest in this saucer-shaped depression as a target for antiviral drugs, particularly against rhinoviruses. Immunogenic sites are present on the exposed external parts of the capsid. These antigenic areas are binding sites for neutralizing antibody and are a key feature in the immunogenicity of poliomyelitis vaccines.

7.2.3 Genome and polypeptides

Picornaviruses are positive-strand RNA viruses with a genome of 7.5 kb. The RNA genome is essentially an mRNA molecule and is sufficient if introduced into cells to establish an infection and to produce new virus particles. The genome RNA has a covalently attached protein termed VPg at the 5′ end; it is polyadenylated at the 3′ end (Fig. 7.5). The genome encodes for a single large open reading frame (ORF) for translation into one very large polyprotein. The protein coding region is flanked by a 750-nucleotide-long 5′ untranslated region and a 3′ untranslated region ending in a poly(A) tail (Fig. 7.5). The 5′ non-coding region contains sequences that control genome replication and virulence.

7.2.4 Replication

Much of the following detail comes from experimental work with polio, by far the most well studied member of the family. Poliovirus enters target cells via interaction between the canyon in the poliovirus capsid and the host cell adhesion protein CD155 (Fig. 7.6). CD155 binding also induces virus RNA uncoating. There are considerable changes in the virion at this moment with the loss of the structural protein VP4 for example. It is thought that these conformational changes also include exposing the lipophilic terminus of VP1 which is now positioned to insert itself into the cell membrane, possibly forming a pore through the cell membrane. The virion RNA now has the opportunity to move into the cytoplasm of the cell.

Other members of the family also enter cells by attaching to the cellular adhesion protein CD155. Coxsackie viruses bind to CD55 and human rhinoviruses to ICAM-1 and LDLR, whilst echoviruses attach to the integrin VLA 2.

The initial event of poliovirus replication is the association of the virus genome with cellular ribosomes. The 5′ VPg protein is removed from the virus genome by cellular enzymes as the virus enters the cell. This leaves the 5′ untranslated region (UTR) of the virus genome free to associate with a ribosome. The 5′ UTR contains an RNA secondary structure called the internal ribosome entry site (IRES), which recruits the cellular ribosomes and positions them adjacent to the AUG start codon for the translation of the virus polyprotein. The polyprotein is cleaved by proteinases which are themselves part of the virus polyprotein into virus proteins P1, P2, and P3. All picornaviruses have 2A, 3C, and L proteinases but cleavage pathways are somewhat different within the family. The polypeptide P1 of polio undergoes further cleavage to give the four viral capsid proteins (VP1, VP2, VP3, and VP4). P2 codes for three (non-structural) NS proteins including a protease, whereas P3 codes for four proteins including the RNA polymerase. The production of the virus RNA polymerase allows the initiation of genome replication, which usually starts within 1 h of infection of a cell. RNA replication is primed by VPg. The parental positive-sense RNA strand is transcribed into a negative-sense strand, which serves repeatedly as a template for transcription into progeny positive-sense RNA strands (Fig. 7.6). The dsRNA, termed a replicative intermediate, has only a fleeting existence. Assembly of viral RNA nucleic acid and structural proteins takes place in the cytoplasm. Most viral proteins from the P2 and P3 regions are involved in picornaviruses RNA replication on cellular membranes and vesicles which serve as a scaffold. During poliovirus replication the translation of cellular mRNAs is effectively shut down by the viral 2A protease cleaving and inactivating the cellular cap binding complex termed eIF-4G which is needed for binding of cellular mRNAs to ribosomes. This complex is not required for polio mRNAs as they contain the 5′ UTR IRES. The cell normally dies after a viral replication period of 4 h and large numbers of new virus particles are released by cell death and lysis.

7.3 Clinical and pathological aspects

7.3.1 Poliomyelitis

This term is derived from the Greek *polios* (grey) and *muelos* (marrow), and refers to the propensity of the virus to attack the grey matter of the spinal cord. The shortened form 'polio' is often used to indicate the paralytic form of the disease. Hopefully the world should see the very last cases of clinical polio in the next few years.

Clinical features

The incubation period is usually 7–14 days, with extremes of 2–35 days (Fig. 7.7). The course is variable; there may be

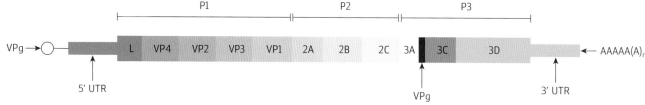

Fig. 7.5 Genome structure of polio.

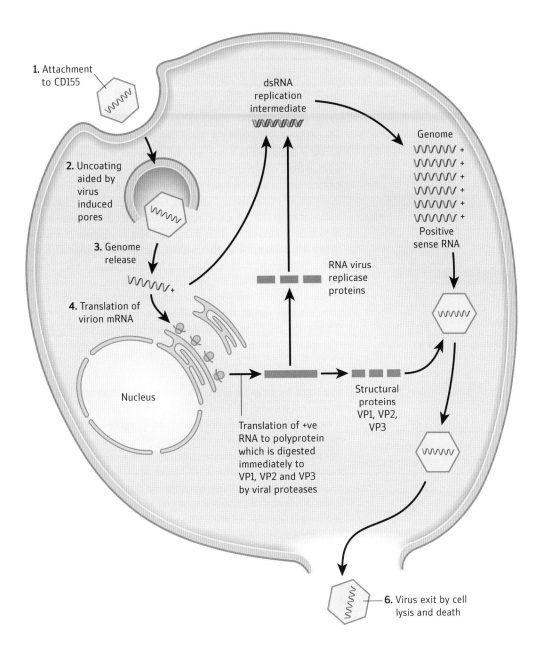

1. Attachment to CD155

dsRNA replication intermediate

Genome
+
+
+
+
+
+
Positive sense RNA

2. Uncoating aided by virus induced pores

3. Genome release

4. Translation of virion mRNA

RNA virus replicase proteins

Nucleus

Translation of +ve RNA to polyprotein which is digested immediately to VP1, VP2 and VP3 by viral proteases

Structural proteins VP1, VP2, VP3

6. Virus exit by cell lysis and death

Fig. 7.6 Replication cycle of polio (see text). VPg and polyA tail omitted for clarity.

asymptomatic infection with as many as 200 asymptomatics compared to one symptomatic person. There may be a minor illness with malaise, fever, and sore throat, or a major illness heralded by a meningitic phase. Each of these, except the major illness, may resolve without sequelae. In those progressing to the major illness, there may be a few days of apparent well-being before the onset of the meningitic phase, giving a biphasic presentation.

The major illness onset is abrupt, with headache, fever, vomiting, and neck stiffness. The meningitic phase often concludes in a week. However, in a minority of persons, about 1%, paralysis sets in. Paralysis is caused by virus entering the central nervous system and replicating in the motor neurones of the spiral cord. Its extent ranges from part of a single muscle to virtually every skeletal muscle, in which case severe impairment of respiration

may demand the use of artificial ventilation. Paralysis is of the lower motor neuron type with flaccidity of affected muscles giving the clinical signs of paralysis. In bulbar poliomyelitis involvement of cranial nerves results in paralysis of the pharynx, again bringing difficulty with respiration.

When there is paralysis, its full extent is apparent within 72 h of the first signs, although a firm prognosis cannot be made for about a month, by which time most reversible neuronal damage will have disappeared and the residual permanent damage can be assessed (Fig. 7.8). Fortunately, there is often great improvement during this anxious waiting period. In bulbar polio the outlook is good if the patient is still alive by the tenth day or so, as the pharyngeal muscles then begin to show signs of recovery. Case fatality in paralytic cases is 5–10% in children and 15–30% in adults.

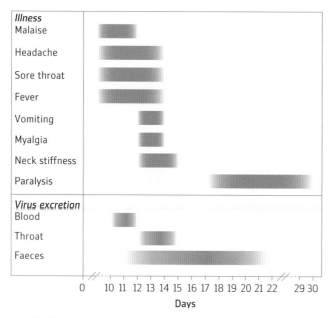

Fig. 7.7 Clinical and virological events during infection with polio.

Post-polio syndrome

Decades after paralytic polio, survivors of the acute paralytic attack may show muscle weakness, fatigue, and pain. There is current controversy as to whether the slowly progressive atrophy of muscles is caused by persistent infection with polio virus or whether it is more physiological and caused by age-induced decline of neurons in the absence of compensatory neurons which were destroyed much earlier during the acute infection.

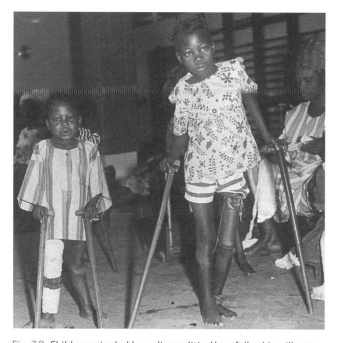

Fig. 7.8 Children crippled by poliomyelitis. Hopefully, this will soon become a disease of the past.

Pathogenesis

Poliovirus may be acquired by the respiratory route, but the faecal–oral mode of transmission is much the more important. Virus from faeces is stable in lake or river water for days at 30 °C but is susceptible to UV light. Virus is also resistant to mild detergents and disinfectants. The virus replicates in the lymphoid tissue of the pharynx and gut, including Peyer's patches. A viraemic phase is followed in a minority of patients by extension to the CNS, within which the virus spreads along axons. Lytic infection of neurons, with secondary degeneration of axons, is the primary cause of paralysis. Typically, the anterior horn cells of the spinal cord are the worst affected, which accounts for the lower motor neuron type of limb paralysis. The virus does not replicate within the muscles themselves, rather widespread atrophy starts because muscle is no longer innervated. In severe cases, various centres within the bulb and brain may be attacked, with respiratory paralysis sometimes leading to death. More transient damage results from inflammatory infiltration in the CNS. Pregnancy increases the incidence of paralysis as does intramuscular injection, e.g. of other vaccines or antibiotics.

Pathology

The most remarkable feature is the high selectivity of the virus for the nervous tissue. In addition to the lytic destruction of neurons there is an inflammatory reaction, an initial polymorphonuclear response giving way to lymphocytes. There is perivascular infiltration with lymphocytes ('cuffing') characteristic of inflammatory reaction in the CNS, microglial proliferation, and oedema. Inflammatory foci in lymphoid tissue are also present.

Experimental observations suggest that the virus is often disseminated throughout the CNS to a much greater degree than is suggested by the clinical signs.

Polio vaccine

The discovery that the three polio types could be grown on monkey kidney cells, coinciding with large outbreaks of disease in American children in the 1940s and the affliction of the US President Franklin Roosevelt, encouraged scientists to develop vaccines and citizens to donate to the 'March of Dimes' to finance the research. Very large sums were raised to fund the vaccine research of Jonas Salk and also Albert Sabin. Hopefully, the world is suffering its last cases of polio but even after total eradication the childhood vaccination campaign will continue using the inactivated polio vaccine for some time to come.

Inactivated vaccine (IPV)

By the mid-1950s, Jonas Salk's group in the USA had demonstrated that a poliovirus suspension from the supernatant fluid of infected monkey kidney cell cultures, treated with formalin to destroy its infectivity, induced immunity in susceptible persons. This IPV contains the three serotypes and is given as three injections at monthly intervals, often combined with diphtheria, tetanus, pertussis, and *H. influenzae* type b as a combined vaccine for children. It is also advised for travellers, particularly to Africa, Afghanistan, Pakistan, and the Middle East, and for laboratory personnel.

The modern vaccine production process uses Vero cells, a continuous line of monkey kidney cells, as a cell substrate on which to grow virus. New vaccines are being developed where the 'starter' virus for production of IAV will be an attenuated poliovirus. This will reduce the chance in the post-polio era of accidents in vaccine factories leading to escape of virulent polio.

Live attenuated vaccine (OPV)

While Salk's researches were still in progress, Albert Sabin and co-workers were using a different approach; they were trying to attenuate highly virulent polioviruses by repeated subcultures (passages) of virus at low temperature in monkey kidney cells.

These 'attenuated' strains were used to prepare a live vaccine, which was given by mouth (oral polio vaccine, or OPV). This, too, was given in three doses at monthly intervals, but the reason is quite different from that on which the similar schedule for IPV is based (Table 7.3). The first dose of the IPV induces the classical primary immune response to a non-replicating antigen, followed by enhanced (secondary) responses to the second and third doses (Chapter 5). By contrast, the virus in the attenuated vaccine replicates in the lymphoid tissue of the gut; when the first dose is given, only one of the three serotypes (say type 2)

secures a foothold to the exclusion of the others, and induces immunity to type 2 strains only. At the next dose, either type 1 or type 3 wins; similarly, the third dose provides immunity against whichever of the three serotypes is left. Molecular analysis of wild type and attenuated polio strains identified the stem loop V region of IRES as a determinant of virulence. Not surprisingly, therefore, the Sabin attenuated viruses have point mutations in this region.

Sabin live attenuated vaccine can 'back' mutate or 'revert' in an immunized child to viruses of increased neurovirulence, and can cause vaccine-associated polio in 1 case per 2.5 million vaccines. Reversion happens in the gut and the virus is shed in the faeces such that these viruses (mainly polio type 3) can then circulate leading to polio outbreaks in the unvaccinated. Therefore, only inactivated poliomyelitis vaccine (IPV) will be used after polio has been eradicated and is being recommended for the final push of eradication.

Where is the world today as regards polio vaccine status?

There remain niches of the polio virus in Nigeria, where the vaccine has not been used for religious reasons, in war-conflicted Afghanistan and Pakistan, and, more recently, in war-torn Syria.

Virus monitoring will continue worldwide to detect infantile paralysis using visual surveillance for acute flaccid paralysis, which at a basic level means searching for children on crutches. Although cases of Guillain–Barré and traumatic neuropathy will also be included by error, the direct observation test is still very useful. Virology monitoring of polio virus in sewage will also continue worldwide. The importance of the global network of 145 laboratories testing sewage plants and outlets for live polio cannot be overemphasized. A laboratory may be presented with 50 000 samples per year and this screening can be very intense.

Here is a word of caution about the post-polio era: vaccination will have to be continued for billions of children for some years until the world can be declared 'polio free'. For example, in the 'post-polio' era some immunosuppressed persons in the community will continue to excrete polio virus and may become life-long excretors of virulent polio, a threat not to themselves, but to others.

7.3.2 Clinical picture of enterovirus disease

Table 7.4 lists the diseases caused by these agents; there is much overlap between the syndromes caused by the various groups and it may help you to remember them better if we classify them in terms of the body systems affected rather than of the virus groups involved.

Table 7.3 Comparison of inactivated and attenuated (oral) poliomyelitis vaccines (IPV and OPV)

	IPV	OPV
Stimulates IgG antibody in blood	Yes	Yes
Stimulates IgA antibody, hence local immunity, in gut	No	Yes
Duration of immunity	Medium	Long
Cost	High	Low
Route of administration	Injection	Oral
Skilled staff needed	Yes	No
Inadequate response in some tropical countries	No	Yes
Involuntary transmission to susceptibles within the community	No	Yes
Possible mutation to neurovirulence	No	Yes, but very rare
Contraindicated in immunodeficiency states or pregnancy	No	Yes

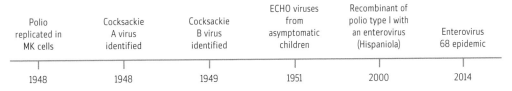

Polio replicated in MK cells	Cocksackie A virus identified	Cocksackie B virus identified	ECHO viruses from asymptomatic children	Recombinant of polio type I with an enterovirus (Hispaniola)	Enterovirus 68 epidemic
1948	1948	1949	1951	2000	2014

Fig. 7.9 Time line for enterovirus.

Table 7.4 Syndromes caused by Coxsackieviruses, echoviruses, and enteroviruses which are surging at present, namely 70, 71, and 68

System affected/syndrome	Coxsackieviruses A†	Coxsackieviruses B	Echoviruses	Type 70	Type 71	Type 68
			Enteroviruses†			
Central nervous system	+	+	+			
Meningitis	+	+	+	+	+	
Paralytic disease	+	+	+	+		
Encephalitis	+	++	++	++	+	
Myalgic encephalomyelitis		+				
Heart/skeletal muscle	+	++	+			
Gastrointestinal tract			++			
Respiratory tract	++	++	+			+
Skin, mucous membranes	+	+	+			
Rashes	+		+			
Herpangina	++					
Hand, foot, and mouth disease	++	+		++	+	
Conjunctiva	+*		++	+		
Pancreas		+				
	+	++	+			

*A variant of Coxsackievirus A24.
†Note that the + scores compare the frequencies of occurrence only within the same row and not overall.

Most of the enteroviruses, both disease causing and 'orphan', are recovered from stools from children. Primary infection occurs in childhood. In general there is positive relationship between symptomatology and age. Rashes caused by enteroviruses are common in childhood. Similarly, young children are featured in outbreaks of encephalitis and other CNS complications as with enterovirus 71. Poorer economic status and poor sanitation correlate with outbreaks. Adults are over represented amongst cases of severe disease including carditis, meningitis, encephalitis, and paralysis.

Most student textbooks list the serotypes predominantly involved in each syndrome, but there is no need to memorize them in such detail; we shall merely indicate the broad relationships. Table 7.4, therefore, gives only a rough idea of the quantitative involvement of the viruses in the various syndromes. The 'plus' scores compare the frequencies of occurrence only within the same row, and not overall.

Pathogenesis of enterovirus infections

Enteroviruses are mainly transmitted either by the respiratory or the faecal–oral route. Those causing conjunctivitis are spread by eye secretions, probably via contaminated clothing, handkerchiefs, etc., or contaminated hands. Infections of the mucosae of the upper respiratory tract, conjunctiva, and gut result from direct viral invasion, whereas the pathogenesis of generalized

infections, e.g. those involving the CNS, heart, or muscles, has a viraemic phase, and is similar to that of poliomyelitis (Section 7.3.1); it follows the pattern of acute infection described in Chapter 4, Section 4.4.4. Destruction of tissue is in the main due to direct lysis of cells, but some damage may be caused by immunopathological mechanisms, especially in the case of Coxsackievirus infections.

Paralysis caused by novel recombinant between enterovirus and poliovirus in Hispaniola

Outbreaks of paralytic poliomyelitis in the island of Hispaniola (Dominican Republic and Haiti) in 2000 were shown to be caused by a novel virus with genes of polio type 1 and another enterovirus. Most of the patients were children who had not been vaccinated with polio or in whom full vaccination had not been completed. Furthermore, there was a low vaccination rate in the local community. The non-poliovirus was found to have contributed to the non-structural (NS) proteins in the new virus.

7.3.3 The enteric hepatitis A virus (HAV): classification, morphology, and replication

The virions have cubic symmetry and are 27 nm in diameter; uniquely the virus exists in an enveloped form. However high resolution X-ray analysis of the virus shows unique features and places hepatitis A as a 'link' virus between modern picornaviruses. HAV differs from other picornaviruses, however, in that the nucleotide and amino sizes of several HAV proteins are different, the virus replicates slowly and without cytopathic effects in cell culture, and that despite being an RNA virus only one dominant serotype exists with one antibody binding neutralization site being immunodominant. There are several genotypes and nucleic acid sequencing can be used to investigate special outbreaks.

The nucleic acid of HAV is single-stranded RNA with positive polarity approximately 7.5 kb in length. It codes for four polypeptides: VP1, VP2, VP3, and VP 4.The genome can be divided into three sections:

- a 5' non-coding region, which is capped by a viral protein Vpg;
- an ORF coding all the viral proteins;
- a short 3' non-coding region.

Little is known about the mechanism of entry into cells and whether a specific receptor is involved. It is assumed that as it lacks a virion pocket structure that uncoating is unique. It is presumed that, in common with other picornaviruses, genome replication occurs entirely in the cytoplasm where the genome RNA can act as mRNA directly and as a template for negative-strand genome, which subsequently leads to more positive-sense genome production. The incoming viral RNA strand directs the synthesis of a large viral polyprotein, which is then cleaved into segments. Translation is a crucial step because synthesis of new picornaviral RNA cannot begin until the virus has translated an RNA-dependent RNA polymerase. The virus genome encodes four structural proteins, VP1–4, and seven non-structural proteins, 2A, 2B, 2C, and 3A, B, C, and D. 3D encodes for the virus RNA-dependent RNA polymerase.

HAV assembly is complex, with formation of non-infectious 'provirions', which require a 'maturation' cleavage of one of the structural polypeptides. The new virions are usually released by an infection-mediated disintegration of the host cells. Unlike other members of the family it is unable to shut down host protein synthesis. Hepatitis A, exceptionally in the family, can establish a persistent infection with much reduced cell destruction.

7.3.4 Clinical and pathological aspects of hepatitis A virus infections

Clinical features

The incubation period is 2–6 weeks. Many infections are silent, particularly in young children. In adults (Fig. 7.10) clinical illness usually starts with a few days of malaise, loss of appetite, vague abdominal discomfort, and fever. The urine then becomes dark and the faeces pale; soon afterwards jaundice becomes apparent, first in the sclera and then in the skin; if severe, it may be accompanied by itching. The patient starts to feel better within the next week or so and the **jaundice** disappears within a month. Hepatitis A is nearly always self-limiting, but relapses have been reported. The severity of illness is less in

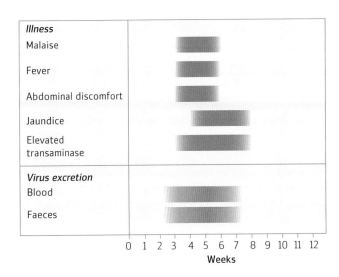

Fig. 7.10 Clinical and virological events during infection with hepatitis A virus

children than in adults. Complications, such as fulminant hepatitis, fortunately rare, are seen mainly in older people. Mortality from HAV is about 1 in 1000 people infected.

Epidemiology

The main mode of spread is that of other enteroviruses, i.e. transmission is by the **faecal–oral route** (Chapter 6). Like them, HAV survives for long periods in water and wet environments. Large quantities of virions are excreted in the **faeces** for several days before and after the onset of jaundice, but after a week the patient's stools may be regarded as non-infectious. Other routes of transmission include transfusion of blood or blood products such as factor VIII and factor IX inadvertently collected during the viraemic phase; sharing of needles by drug abusers; and **sexual contact**. Homeless people living in shelters and hostels can also be a source of infection because of poor hygiene.

The geographical distribution of hepatitis A follows the patterns described for other enteroviruses, notably poliomyelitis. Thus, it is widespread in countries where sewage treatment and hygiene generally are inadequate, such as India and Eastern Europe, most persons acquiring it as a subclinical infection in early childhood. In more developed areas, outbreaks are liable to occur in institutions where personal hygiene is poor. It occurs both in endemic form and as epidemics, some of which have been traced to infected shellfish. More recently outbreaks have occurred in countries like Norway where sero-prevalence is very low (under 10%) and have been attributed to faeces-contaminated frozen berries.

Pathogenesis and pathology

After ingestion the virus replicates in cells of the epithelial crypts and reaches the liver via the blood. The virus replicates primarily in the **hepatocytes**, from which it passes through the bile duct to the intestine, and is shed in large quantities in the faeces. There is necrosis of hepatocytes, particularly in the periportal areas, accompanied by proliferation of Kupffer and other endothelial cells. Damage is likely to be immune modulated.

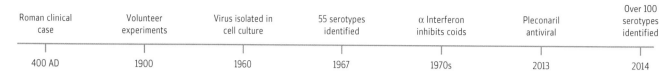

Roman clinical case	Volunteer experiments	Virus isolated in cell culture	55 serotypes identified	α Interferon inhibits coids	Pleconaril antiviral	Over 100 serotypes identified
400 AD	1900	1960	1967	1970s	2013	2014

Fig. 7.11 Timeline for common cold virus.

Temporary damage to liver function is indicated by **elevated liver enzymes** in the blood. However, unlike infections with HBV and HCV, there is no tendency to chronicity, cirrhosis, or malignant change.

Immune response

Specific IgM appears during the prodromal phase, is present at high titre by the time jaundice is apparent, and persists for several months. **IgG neutralizing antibody** is detectable for many years after infection and protects against further attacks. Cytotoxic T cells lyse virus-infected hepatocytes, and these, together with IFN and other cytokines, are important in the **cell-mediated immune response**, which in turn seems to be a major factor in pathogenesis.

Laboratory diagnosis

During the acute phase, liver dysfunction is indicated by **raised serum bilirubin and transaminases and a depressed prothrombin level**. Genome detection by RT PCR is a rapid and sensitive method.

Hepatitis A vaccine

For practical purposes there is only one virus serotype.

There are four monovalent vaccines licensed for persons over 12 months of age where virus is grown in human diploid cells inactivated with formalin and adjuvanted with aluminum hydroxide. These viruses are highly protective, resulting in up to 97% decline in infection in communities. The vaccines can be used even after an outbreak has started. Combined hepatitis A and B vaccines are also available. For the former vaccines, two doses are sufficient, whereas for the combined vaccine a three-dose schedule must be completed. A live attenuated vaccine has been licensed in China. The virus is cultivated on human diploid cells and has been attenuated by classic passage. WHO recommend incorporation of vaccine into the childhood schedule for higher endemic regions of the world.

General measures

Control of infection in the community depends on maintenance of hygiene, a counsel of perfection that is often very difficult to achieve.

A food handler with hepatitis A must be kept away from work for 2 weeks after the onset of jaundice. Fellow workers should not be given immunoglobulin as prophylactic as this may merely mask an attack without completely preventing it, an obviously dangerous situation; they should be kept under surveillance and asked to report any illness during the next 12 weeks.

In **hospitals**, patients should be nursed with appropriate precautions against the spread of an enteric infection, with particular attention to **safe disposal of faeces** during the infective period.

7.3.5 Rhinoviruses

A landmark in the study of the virology of the common cold came in the 1960s (Fig. 7.11) when, after years of patient investigation, David Tyrrell and his colleagues at the MRC Common Cold Research Unit in Salisbury isolated viruses in a simple cell-culture system. A chance observation that an alkaline pH in the culture medium favoured the growth of the viruses, together with using roller cultures at 33 °C, were the key factors in their success.

By conventional neutralization tests rhinoviruses fall into more than 150 serotypes. However, the more recently applied molecular techniques define different patterns of genetic relationships between the viruses. Genomic analysis has uncovered three groups of viruses, namely HRVA, B, and C, which have, respectively, 74, 25, and (at least) 50 serotypes. HRV-C viruses are difficult to grow in cell culture.

These genetic differences may have practical implications since some antiviral agents inhibit only common cold viruses in particular groups. Also it is now obvious that there is a low rate of recombination in the RNA genome alongside mutational changes particularly in VP1.

In many properties, rhinoviruses resemble other members of the *Picornaviridae*. They differ, however, in their inability to withstand acid conditions, and for some members in their low **optimum temperature for growth (33 °C)**. The latter characteristic has been thought to be an evolutionary adaption to the comparatively cool environment of the nasal mucosa. But the application of RT PCR for diagnosis has detected many more deep-seated infections than previously found.

Clinical and epidemiological aspects

Replication is most often restricted to cells of the upper respiratory tract. The infection is spread by aerosol, by hand contact, and from the surface of cups and plates. The incubation period is 2–3 days (Fig. 7.12). Inflammation and copious exudation from the upper respiratory tract lasts for a few days. The symptoms are nasal congestion, sneezing, sore throat, and often headache and cough. There is rarely a fever.

There can be serious sequelae in chronic bronchitics and asthmatics, in whom attacks may be precipitated by infection with a common cold virus. These complications cost $60 billion in the USA alone each year in lost work days and medical care.

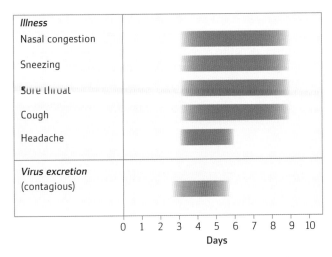

Fig. 7.12 Clinical and virological events during infection with common cold.

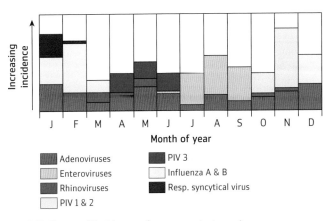

Fig. 7.13 Seasonal incidence of some respiratory viruses.

Rhinovirus colds occur throughout the year in all countries of the world (Fig. 7.13). It is thought that several virus serotypes co-circulate for a year or so, to be displaced by a new group; one individual may experience two or three infections per annum. Children are more susceptible than adults and one of the hazards of living in a large family is the risk of contracting colds from the younger members. The frequent occurrence of more than one cold per season in the same person is due to the large number of serotypes, which confer little or no cross-protection.

Although corona- and rhinovirus infections are often trivial in themselves, the economic losses due to absences from work are enormous, a fact well recognized by the general public. Most medical students and doctors will have been asked the question 'Have you found a cure for the common cold yet?', which brings us to the following question.

Antiviral chemotherapy or vaccines to prevent the common cold?

The potential antiviral effects of IFN were first established in volunteers infected with common cold viruses at the Salisbury experimental unit. This single experiment in 1973 used up almost all the world's supply of the new drug. Nowadays, with the arrival of recombinant DNA technology, a range of IFN molecules is available for clinical use. However, even cloned and highly purified IFN is not without side-effects, a particularly unfortunate one being the production of a stuffy nose after intranasal use. Therefore, the early promise of IFN as an effective antiviral compound for the common cold has not yet been realized. A number of synthetic antiviral molecules, such as enviroxime and dichloroflavan (see Chapter 31), have good antiviral activity against certain rhinovirus serotypes in the laboratory, but only marginal effects have been noted in volunteer experiments. Capsid binding molecules such as pleconaril, vapendavir, and pirodavir have reached clinical testing and almost licensure. These drugs bind to the hydrophobic pocket of capsid proteins and tend, paradoxically, to stabilize the virion and interfere with attachment to cellular receptors and early uncoating.

Little progress has been made with vaccines because of the antigenic diversity of the rhinoviruses. Data from X-ray crystallography and whole genome nucleotide sequencing may now help to delineate common amino-acid sequences among the capsid proteins of rhinoviruses, which could be used as peptide vaccines. However, the effective control of the common cold lies very much in the future.

 Reminders

- The family *Picornaviridae* comprises six genera of small RNA viruses, of which three, *Enterovirus*, *Rhinovirus*, and *Hepatovirus*, frequently infect humans.

- The picornaviruses are icosahedral, about 25 nm in diameter, non-enveloped, and have a positive-sense RNA genome 7.5 kb in size with a covalently attached protein (VPg) at the 5' end. The genome extends its information base by coding for a long polyprotein, which is subsequently cleaved into 20 viral proteins.

- Polioviruses can specifically shut down cellular mRNA translation by a viral protease that inactivates the cellular cap binding complex eIF–4G. The virus is therefore cytocidal for cells, including neuronal cells.

- The viruses of the *Enterovirus* genus cause a wide range of illnesses, including infections of the CNS, the heart, skeletal muscles, liver, skin, and mucous membranes, including the conjunctiva. Some serotypes such as enterovirus 71 have become epidemic causing aseptic meningitis and flaccid paralysis.

- Enteroviruses are transmitted predominantly by the faecal–oral route, but sometimes via the respiratory tract; conjunctival infections are spread by contact with infective secretions.

- Rhinoviruses and common cold viruses exist as over 100 serotypes and cause upper respiratory tract infection all the year around in every country of the world. Prevention of poliomyelitis is accomplished by immunization with **IPV** or, in the recent past, live attenuated OPV. Both are highly effective. Hepatitis A virus causes jaundice but there is a carrier state and no malignancy unlike hepatitis B and C.

- There is an effective hepatitis A vaccine for travellers to endemic areas of the world.

Further reading

DeCroes Jacobs, C. (2015). *Jonas Salk: a life*. Oxford University Press, Oxford.

Falah, N., Montserret, R., Lelogeais V., Schuffenecker, I., Lina, B., Cortoy, J-C., and Violot, S. (2012). Blocking human enterovirus 71 replication by targeting viral 2A protease. *J Antimicrob Chemother* **67**, 2865–9.

Guzman-Herrador, B.Jensvoll, L., Einöder-Moreno, M., Lange, H., Myking, S., Nygård, K., *et al.* (2014). Ongoing hepatitis A outbreaks in Europe 2013-2014: imported berry mix cake suspected to be the source in Norway. *EuroSurveillance* **19** (15), 20775. DOI: http://dx.doi.org/10.2807/1560-7917.ES2014.19.15.20775

Hovi, T., Shulman, L.M., van der Avoort, H., Deshpande, J., Roivainen M., and De Gourville E.M. (2012). Role of environmental poliovirus surveillance in global polio eradication and beyond. *Epidemiol Infect* **140** (1), 1–13.

Kew, O. *et al.* (2002). Outbreaks of poliomyelitis in Hispaniola associated with circulating type 1 vaccine derived polio virus. *Science* **296**, 356–8.

Meng, F.Y., Li, J.X., Chu, K., Zhang, Y.T., Ji, H., Li, L., Liang, Z.L., and Zhu, F.C. (2012). Tolerability and immunogenicity of an inactivated enterovirus 71 vaccine in Chinese healthy adults and children: an open label, phase 1 clinical trial. *Hum Vaccin Immunother* **8**, 668–74.

Minor, P.D. (2012). The polio-eradication programme and issues of the end game. *Journal of General Virology* **93**, 457–74.

Wang, X.*et al.* (2015). Hepatitis A virus and the origins of picornaviruses. *Nature* **517**, 85–8.

Wimmer, E. and Paul, A.V. (2011). Synthetic poliovirus and other designer viruses: what have we learned from them? *Annu Rev Microbiol* **65**, 583–609.

Questions

1. Describe the detailed morphology of a picornavirus like polio, the genome, and the replication strategy. Why is it called a positive-strand virus?

2. What are the relative merits, as well as the problems, of the two polio vaccines?

3. Write short notes on Coxsackie, ECHO, and enteroviruses. Describe the range of diseases caused by these viruses.

4. How does hepatitis A cause disease and how can the infection be prevented?

Astroviruses: gastroenteritis agents

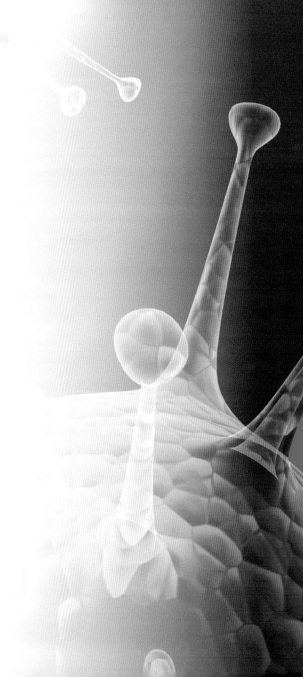

8

8.1 Introduction

These viruses identified by electron microscopists 40 years ago (Fig. 8.1) are the second most common cause of gastroenteritis in young children after rotavirus (Chapter 20). Indeed rotavirus may often be a co-pathogen. Multiple serotypes circulate simultaneously, especially in developing countries. It is also likely that many persons have a subclinical infection. These viruses (Fig. 8.2), spread by the faecal–oral route, also cause gastroenteritis in animals such as calves, piglets, dogs, cats, and mink (genus *Mamastrovirus*).

8.2 Properties of the viruses

8.2.1 Morphology

The family name (Fig. 8.2) is coined from the characteristic negative contrast EM of astroviruses where the exterior surface of this small rounded 30-nm icosahedral non-enveloped virion has a curious five- or six-pointed star appearance (Greek *astron* meaning a star). More recently three-dimensional views of the virion have been constructed by a combination of cryo-electron microscopy and electron density maps. Thirty dimeric spikes emerge from the capsid extending 5nm or so (Fig. 8.3). But a very wide variety of these so called 'small round viruses' are seen, often dependent upon the condition of the clinical sample, and indeed many virions do not display the 'star' like morphology at all.

8.2.2 Genome

As with caliciviruses the genome is a linear, positive-sense, single-stranded RNA molecule of 6.4–7.3 kb (Fig. 8.4). The RNA is polyadenylated at the 3′ terminus and a virus encoded VpG protein caps the 5′ end of the genomic RNA. The genome encodes for three ORFs; ORF1a and ORF1b are located towards the 5′ end of the genome. ORF1a encodes a non-structural protein, while ORF1b encodes the virus RNA-dependent RNA polymerase. The 3′-located ORF2 encodes the major capsid structural protein. There is also a subgenomic RNA which is produced in large quantities.

Electron micrograph of astrovirus	Cell culture cultivation	5 Serotypes	EIA test	Astovirus genome sequencd	15 Serotypes
1975	1981	1984	1988	1993	2014

Fig. 8.1 Timeline for astroviruses.

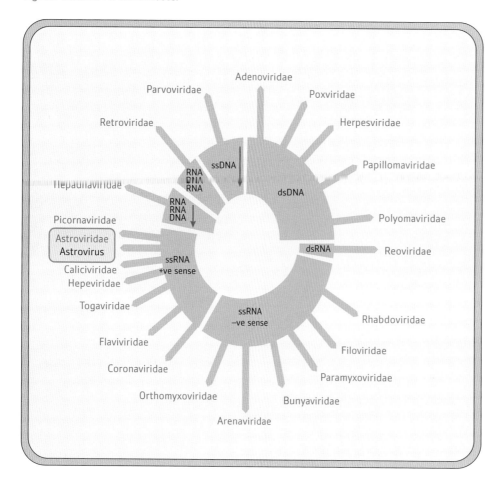

Fig. 8.2 Baltimore classification scheme and astroviruses.

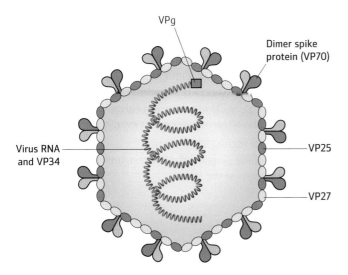

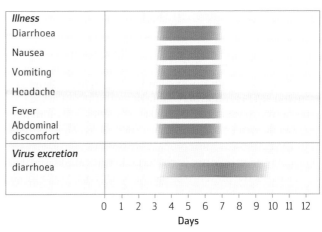

Fig. 8.5 Clinical and virological events during infection with astrovirus.

Fig. 8.3 Structural features of astroviruses. Note the 'star' appearance is often not seen by electron microscopy. Furthermore, ambiguity is present as regards assignation of VPs to structure with the exception of the 30 dimers protruding from the surface (VP70).

8.2.3 Replication

Astroviruses enter cells by adsorption to an as yet unidentified receptor and the virus is internalized by endocytosis. The virion RNA is infectious and so translation of the incoming genome starts the replication cycle. The virus replicates in the cytoplasm where, following internalization, the NS proteins from ORF1a and 1b are translated, and the new virus polymerase directs the synthesis of negative-sense RNA genome that serves as the template for more positive-sense genome production. In addition, a subgenomic mRNA is synthesized that encodes for a viral polyprotein, which is cleaved into capsid proteins. New virions self-assemble in the cytoplasm and may be seen as crystalline arrays of multiple virions, whence they are released by cell lysis. These viruses can be cultivated in the laboratory in trypsin-treated human colon carcinoma or monkey kidney cells.

8.3 Clinical and epidemiological aspects

After an incubation of 3–4 days the clinical attack involving watery diarrhoea, nausea, and abdominal discomfort lasts 3–4 days (Fig. 8.5). Virus is excreted in huge numbers at around

10^{10} virions per gram of diarrhoea. Astroviruses cause outbreaks of watery diarrhoea in all countries of the world, predominantly in children where they are second in clinical importance after rotaviruses (Chapter 20). The modes of transmission by the faecal–oral route resemble those of caliciviruses (Chapter 9). Food-borne transmission is less important since most infections are in children who do not prepare food. The exception is sewage-contaminated seafood such as oysters. Mussels, on the other hand, are more often cooked and the virus is destroyed.

The viruses are somewhat resistant to heat although 70% methanol reduces infectivity. The virion is resistant to chlorine and stable in water in streams and rivers. In temperate climates, there is a well-marked peak of incidence in the winter. The virus is detected, for example, in drinking water, rivers, and effluents from water treatment plants.

Most heroic adult volunteers deliberately infected with virus have a subclinical infection. There are 15 serotypes of the virus and there is little evidence of cross-protection immunity.

Data on pathology has been gleaned from studies of human volunteers given astroviruses and also from limited human field samples. Overall the virus is considered to infect columnar epithelial cells in the lower part of the gastrointestinal tract.

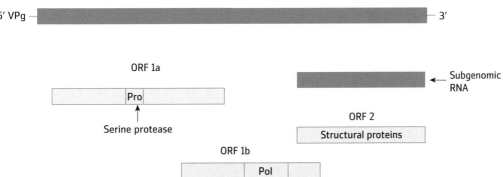

Fig. 8.4 Genome structure of astroviruses.

8.4 Laboratory diagnosis

We have already pointed out (Chapter 6) that detection of a micro-organism in a clinical specimen does not necessarily mean that it is causally related to the patient's signs and symptoms. This is nowhere more obvious than in acute gastroenteritis. Under the EM, viruses, sometimes of more than one family, can be found in faecal specimens. Their association with disease is most probable in the case of rotaviruses, but as far as astrovirus is concerned, epidemiological factors, such as the presence of the same virus in a high proportion of people involved in the same outbreak, must also be considered.

None of this makes the task of the virology laboratory any easier. There are two approaches to diagnosis:

- EM of faecal extract. This is sometimes called the catch-all method, because it is potentially capable of demonstrating all the viruses present in the specimen; astroviruses were first identified by EM but these days electron microscopes are rare and electron microscopists even more so.

- Specific tests such as RT PCR using primer pairs targeting the 3′ end of the genome, the RNA polymerase gene, and the capsid gene at the 5′ end.

 # Reminders

- **Astroviruses** are 30 nm spherical virions with a five- or six-pointed star morphology.
- The genome is positive-sense ssRNA, 6.4–7.3 kb in size.
- These viruses cause outbreaks of diarrhoea, mainly in young children, with peak prevalence in the winter.

- The virions are heat resistant but can be destroyed by 70% methanol.
- Hand and surface hygiene are important techniques for prevention and also controlling outbreaks.
- There are many subclinical cases.

 ## Further reading

Kapoor, A. *et al.* (2015). Multiple novel astrovirus species in human stools. *Journal of General Virology* **90**, 2965–72.

Wang, Y., Li, Y., Yu, J., *et al.* (2013). Recently identified novel human astroviruses in children with diarrhoea, China. *Emerging Inf. Diseases* **19**, 1333–5.

 # Questions

1. Write about the natural history of astroviruses. Why are infections mainly detected in children?

Calciviruses: norovirus causing vomiting and diarrhoea

9

9.1 Introduction

The caliciviruses are global in prevalence and depend on faecal–oral spread coupled with poor hygiene. They are notorious in causing serious gastroenteritis outbreaks on cruise ships, and in restaurants, hospitals, schools, and military camps. The clinical picture of projectile vomiting and severe diarrhoea does nothing to reduce anxiety in the community. Heroic volunteering (Fig. 9.1) enabled the virology and precise clinical details to be carefully correlated. Other members of the family infect cattle, pigs, mice, and cats. Only recently has the clinical impact been recorded in immunosuppressed patients. In developing countries 200 000 deaths per year in the under-fives are caused by noroviruses and nearly 1000 in the USA. They remain, however, a rather under-investigated family because of the lack of cell-culture systems.

9.2 Properties of the viruses

9.2.1 Classification

Noroviruses are members of the family *Caliciviridae* (Fig. 9.2) which infect a very wide range of hosts, including mammals, birds, reptiles, and even fish. There are currently seven species in this family, divided among 5 genera, namely the lagoviruses, neboviruses, noroviruses, spoviruses and vesiviruses (Fig. 9.2). Of these, two, *Norovirus* (formerly 'Norwalk-like viruses') and *Sapovirus*, affect humans, causing outbreaks of gastroenteritis. Undoubtedly noroviruses head the list of gastroenteritis pathogens important to humans. It is quite possible that the family is zoonotic with a potential to move across the species barrier.

Filterable virus used in volunteer studies	Volunteer transmission	Norwalk (Ohio) outbreak	'Small round viruses' seen in stools by EM	Virus genome cloned	Sapporo virus outbreak in Manchester	Experimental norovirus vaccines
1950	1971	1972	1976	1990	1993	2011

Fig. 9.1 Timeline for norovirus.

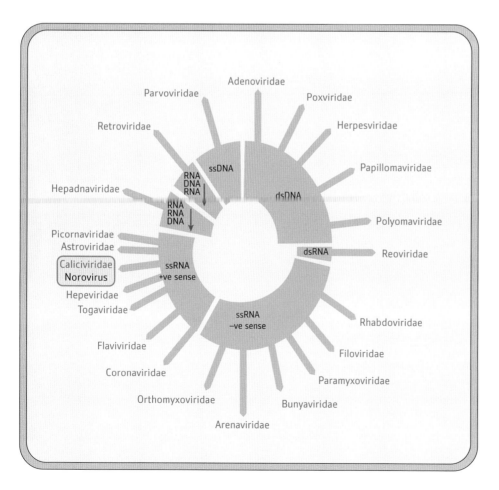

Fig. 9.2 Baltimore classification scheme and norovirus.

9.2.2 Morphology

The family *Caliciviridae* comprises non-enveloped, icosahedral viruses 27–40 nm in diameter with 32 cup-like indentations (Latin *calix* = cup), on the surface of the virions (Fig. 9.3). A less classical appearance is sometimes seen under the EM, but the family are known collectively as 'small round structured viruses' (SRSV). To date (Fig. 9.1), human caliciviruses have not been cultivated in the laboratory.

9.2.3 Genome

The virus genome is a polyadenylated positive-sense single-stranded RNA (ssRNA) 7.3–7.5 kb in size (Fig. 9.4). The 5′ end of the virus genome has the virus protein VPg attached and the 3′ end is polyadenylated. There are short UTRs at either of the genome and these have important functions in viral replication, translation, and pathogenesis. The genome encodes three open reading frames (ORFs) (Fig. 9.4).The virus non-structural proteins are encoded towards the 5′ end of the genome whereas the structural proteins are encoded towards the 3′ end. In the case of norovirus the first ORF is translated as a large polyprotein that is subsequently cleaved into seven non-structural proteins

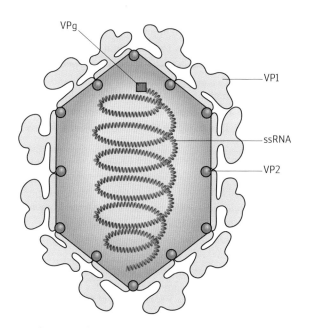

Fig. 9.3 Structural features of norovirus.

(NS1–7). Of these NS5 is the virus genome cap protein VpG, NS7 is the virus RNA polymerase (Pol), and NS3 is a nucleoside triphosphatase (NTPase) and RNA helicase. The two remaining norovirus ORFs encode for the virus capsid protein VP1 and a structural protein of unknown function, VP2, respectively.

9.2.4 Replication

The virus attaches to cells in the upper intestinal tract using carbohydrate receptors and enters via endocytosis. Thereafter, *Calicivirus* (norovirus) replication is similar to other positive-sense RNA viruses. VPg acts as a cap substitute and recruits cellular factors which normally initiate translation of cellular mRNAs including e1F4F, e1F4E. The large polyprotein encoded by ORF1 is cleaved after translation by the virus-encoded protease NS6. Norovirus RNA replication starts closely aligned to complexes of host membranes in the cell cytoplasm. The viral RNA polymerase is NS7 encoded by ORF1. dsRNA, the replication intermediate, associates with defined regions of the perinucleus. These molecules are joined by viral NS proteins and form a replication complex. VP1 and cellular tubulin are also thought to associate. Therefore this so-called replication platform is composed of host membranes of the secretory pathway, the endosomes, the endoplasmic reticulum, and the trans Golgi network. The mRNA becomes a template for the double-stranded replicative form and the subsequent synthesis of positive-sense genomic and subgenomic RNA. A norovirus protein, namely p22, inhibits excretion of cellular proteins and can stop normal trafficking to the Golgi.

Necessarily, positive-strand RNA viruses have a number of problems and features in common, including unwanted stimulation of the innate immune system and interactions with cell proteins and pathways. The replication of viruses is now considered to involve many cellular genes and their products. Norovirus, like others, can to its advantage disrupt host-cell protein excretion and reduce the expression of HLA antigens on the cell surface. In this way the host immune response is diminished.

Virus assembly is considered to involve VP1 with its self-assembling properties whilst VP2 has a highly basic nature allowing it to interact with virus RNA. Finally, exit from the cell may involve norovirus-initiated apoptosis, down-regulation of survivin, and up-regulation of caspase. The error-prone virus polymerase leads to virus sequence diversity and quasi-species within infected individuals.

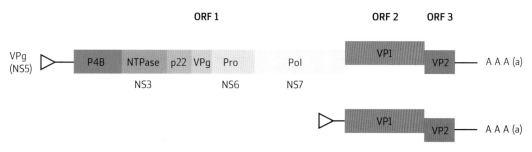

Fig. 9.4 Genome structure, encoded proteins, and open reading frames of calicivirus. ∧/∨/∧, poly(A) tail.

9.3 Clinical and epidemiological aspects

Medical case story Gastroenteritis

A gastroenteritis consultation is all about giving the patient the information they need in order for them to manage for themselves.

You call your next patient in, but four people enter instead of one: a mum, a dad, and two 15-month-old boys.

'This is Joe,' says the mum, 'and Jem has the next appointment.' Joe and Jem are twins. The mum and dad sit down with a boy on each lap. One is crying and one is lethargic.

'We have an older boy too,' the mum begins, 'and he had diarrhoea and vomiting 10 days ago, but he was better within a few days. And then these two caught it and have been ill ever since. They started vomiting first and had temperatures. Then they started having diarrhoea. They've eaten nothing for days.'

You sit for a moment stunned into silence by the thought of having to sift through a diarrhoea story not once, but twice.

The mother has given a clear description of infectious gastroenteritis, which you also know is circulating at high levels in the community. However, the concern here is not the diagnosis, which is obvious, but assessing how well each child is coping. You focus on the mum and the child in *her* lap.

'Which one is this?' you ask. 'Joe,' says mum.

'Let's focus on Joe first,' you say and then plough in. You want exact days of illness, exact numbers of vomits and stools, and exact millilitres drunk. You note it down as you go so as not to get the two children muddled up.

It soon becomes apparent that neither parent is quite sure which child has had or done what. The dad is helpful at trying to remember.

In each child you check skin turgor by pinching a skin fold on the tummy, capillary refill by pressing their fingertips, and peripheral warmth by feeling their fingers and toes. There are no signs of clinical dehydration in either, but these are late and non-specific signs.

You are concerned about both boys. You are aware that Jem is the sicker of the two. However, neither child has been replacing lost fluids. You wash your hands, which gives you a private moment to think. You envisage referring *two* boys at once, with profuse diarrhoea, to the on-call paediatric registrar and the thought makes you quail (especially as you happen to know that your local hospital has a huge notice outside saying in large black capital letters; 'NOROVIRUS OUTBREAK. If you have diarrhoea and vomiting do not enter the hospital,' and then a number to call for advice).

However, the children have not had a proper trial of oral fluids at home yet. You decide that the parents need to push fluids at home first. You can then review them later that day by telephone.

First, you explain to the parents that they should be very careful with hand hygiene and food production. Viral gastroenteritis is spread faecal–orally. Vomit is also very infectious: when it splashes it aerosols nicely, and floats around the room.

As soon as they get home both children should be given a dose of paracetamol to bring their temperatures down and make them less likely to vomit. (Temperatures in themselves can cause vomiting.) This should take about half an hour. Then they will be ready to accept fluids. You explain that any fluids will do: water, juice, squash: in babies less than 6 months old milk is fine, too. However, electrolyte replacement fluid (dioralyte) is the best, if they will take it. (It tastes rather salty.) It replaces the salts lost in vomit and diarrhoea, and is absorbed almost immediately in the upper part of the bowel. Whichever fluid the children take, the parents need to keep an exact record.

The formula for calculating the amount of fluid needed over 24 h is based on the patient's weight and is as follows:

- 100 ml/kg for the first 10 kg of weight.
- 50 ml/kg for the next 10 kg of weight.
- 25 ml/kg for the next 10 kg of weight.
- 1 ml/kg for any further kg of weight.

For example, a 60-kg adult needs about 1800 ml in 24 h. A 15-month-old child weighing approximately 7 kg will need about 700 ml in 24 h. Typically, a toddler might have 250 ml of milk morning and evening, and a 300-ml beaker of juice during the day.

An older child no longer has milk feeds. Typically a 4–10-year-old might have a small cup of 120–150 ml each time they have a meal or a snack. This would be about 5–6 times in a day, so 700–900 ml.

So, in fact, in all children you are recommending about 70 ml (2 fluid oz) an hour. (This assumes they are awake and drinking for 10 h in a day.) If a child is really in need, and not taking fluids when asked, you can ask the parent to give 5 ml orally every 5 min, with a syringe. This is very intensive work for the parent, but a hospital admission can sometimes be avoided by this method.

You explain these details very carefully to the parents and offer 5-ml syringes, and suggest sucky bottles to encourage the boys to drink spontaneously. By the time you telephone in four hours' time, both children should have had at least 300 ml. You also tell them that Jem is more ill and in particular need of their care. It may not be possible to achieve the intake they need, in which case one or both boys may need a hospital admission, with a view to nasogastric fluids or a drip.

'Do they need any medicine?' says the mum. You explain that there is no drug treatment for viral gastroenteritis. The children's own immune system will have to fight it off.

'They can have diarrhoea for many days, as long as they replace their lost fluids,' you say. Treatments like kaolin with morphine, or immodium do not get them better more quickly, they simply block up the bowel. You do not recommend them, particularly in children.

The parents are effusively thankful: they didn't know about the fluids and they want to avoid hospital if at all possible. You check they are both off work as this is a job for two. 'Of course, of course,' says the dad.

Later that day you telephone and confirm they have taken the 300 ml. The dad says Jem is less lethargic. Both are watching telly holding their beakers.

Over the course of the week you field further telephone calls from the mum as the diarrhoea drags on. You explain viral gastroenteritis can last anytime from 12 h to over 10 days. Sometimes, you say, the child seems better with a normal appetite, temperature and energy, but the diarrhoea continues as if the bowel has to have a bit more time to heal up.

Mum introduces the bland foods you suggested: crackers, rice cakes, dried toast. She does not take them back to their nursery until the diarrhoea has completely settled for 48 h.

Eventually you don't hear from them again and assume all is well.

Joe	
Vomiting	Started 7 days ago, three times, none since.
Diarrhoea	Started 7 days ago at eight times a day. Yesterday three times, still watery.
Urine	Hard to tell.
Fluids	They share a beaker. Maybe one-third of a beaker this morning.
Energy	Crying and wriggling around. Alert, miserable.

Jem	
Vomiting	Started 6 days ago and also 4 days ago. None since.
Diarrhoea	Started 6 days ago, 8–10 times a day. Yesterday six times, very watery.
Urine	Hard to tell.
Fluids	Refusing fluids today. Yesterday 'not much'.
Energy	Lying in dad's lap, passive, lethargic.

★ Learning Points

- **Virological:** Infectious viral gastroenteritis can be caused by noroviruses (also known as Norwalk-like viruses or small-round-structured-viruses (SRSV)), rotaviruses, enteric adenoviruses, and astroviruses. Stool can be sent for electron microscopy for viral identification, but is only done if the diarrhoea is unusually prolonged (over 10 days), or severe, mainly to exclude bacterial causes. Treatment is symptomatic and supportive with fluid management, careful monitoring of hydration levels and gradual reintroduction of bland carbohydrates.

- **Clinical:** Parents quite often give a joint history for siblings, but it is important to assess one child at a time.

- **Personal:** A lot of the work of a doctor is teaching and education, patient by patient. And each time you see another case, you have another chance to improve your teaching technique.

So called serotypes and genogroups infect humans, some named after the places where they were first identified in the 1970s; the list includes the Norwalk, Hawaii, Taunton, and Snow Mountain viruses. The aetiological role of some of these viruses was proved by remarkably heroic volunteers, who developed gastroenteritis after drinking filtrates of faeces from patients. The two norovirus genotypes infecting humans are themselves subdivided into 15 genetic clusters. A sample from Sapporo in Japan showed a virus which is morphologically similar but antigenically distinct. Details of pathology related to gene structure are awaited.

9.3.1 Pathology

Electron microscope studies show a shortening of microvilli in the jejunum, cytoplasmic vacuolization and atrophy of villi. There is increased mononuclear infiltration in the lamina propria. Abnormal gastric motor function may result in the nausea which is not uncommon in patients and experimentally infected volunteers.

There are several modes of transmission including the **faecal–oral route**, inhalation of **aerosols** from vomit and point source outbreaks from **contaminated food or water**. Shellfish are a frequent source of infection.

Caliciviruses such as norovirus cause outbreaks of vomiting, sometimes projectile, and diarrhoea. **The incubation period is 12–48 h** (see case study), and the onset is often sudden (Fig. 9.5). There may be abdominal pain, nausea, and low-grade fever, but the stools do not contain blood or mucus. The clinical problems subside by the third day.

Noroviruses are prevalent worldwide and are the major cause of outbreaks of diarrhoea and vomiting (D&V), many of which occur in closed institutions such as cruise ships, army camps, and hospitals, and even restaurants. Norovirus infections tend to occur mainly in the winter, whereas other calicivirus epidemics are seen all the year round. In the UK, for example, laboratory-confirmed norovirus has increased from 400 or so cases in 1986 to 8000 in 2010, but this data would be the small tip of a very large community iceberg. In developing countries, noroviruses cause

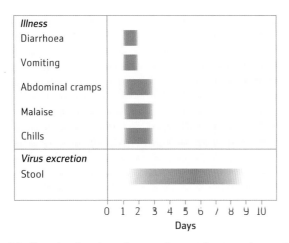

Illness	
Diarrhoea	
Vomiting	
Abdominal cramps	
Malaise	
Chills	
Virus excretion	
Stool	

0 1 2 3 4 5 6 7 8 9 10
Days

Fig. 9.5 Clinical and virological events during infection with norovirus.

an estimated 200 000 childhood deaths a year, whilst even in a country like the USA with relatively high levels of hand hygiene and a regulated and chlorinated water supply this virus family is a leading cause of gastroenteritis deaths. There is some evidence that infection is associated with neutralizing enterocolitis, seizures in infants, and disseminated intravascular coagulation.

9.3.2 Diagnosis

RT PCR kits are now widely available, and are a sensitive and rapid approach to detecting virus in stools; virions can be present in numbers such as 10^{10} per gram of diarrhoea. EM using specific antisera (immunoelectron microscopy, IEM) remains a key secondary diagnostic method.

9.3.3 Management of outbreaks

Control is difficult because asymptomatic infection can occur, and indeed can be quite common, and excretion of virus may be prolonged in some persons for 1–2 weeks after symptoms have disappeared. Moreover, there are very large quantities of virus in diarrhoea and vomit. However, virus-excreting food handlers can now be identified quickly and removed from contact with food. The viruses are hardy and resistant to many disinfectants. 70% alcohol can be used on a small scale. High levels of chlorine are required, twofold in excess of the levels in chlorinated water (10 mg/l) of chlorine. It is more resistant to chlorine than polio and rotavirus but is destroyed at pasteurization temperatures of 70°C. Vaccines are under development and further adults have volunteered for this wonderful endeavour.

Reminders

- Norovirus, a member of the family of caliciviruses, is a small, non-enveloped positive-sense RNA virus, with 7.5kb coding potential.

- Virus replicates in the cytoplasm in complexes of cell membranes and virus is released by cell lysis.

- After a short incubation of 24 hours the important clinical symptoms are vomiting and diarrhoea with stomach cramps.

- There is a high secondary attack rate (50%) and the virus notoriously causes outbreaks on cruise ships, and in hospitals and army camps.

- Members of the family are global pathogens affecting both developed and developing countries.

- There is no vaccine at present although volunteer experiments are underway.

- Surface hygiene, hand washing, and social distancing are important control elements, as well as self-imposed quarantine.

Further reading

Ayukekbong, J.A., Mesumbe, H.N., Oyero, O.G., Lindh, M., and Bergström T. (2015). Role of noroviruses as aetiological agents of diarrhoea in developing countries. *Journal of General Virology* **96**, 1983–99.

Repp, K.K. and Keene, W.E. (2012). A point source norovirus outbreak caused by exposure to fomites. *Journal of Infectious Diseases* **205**, 1639–41.

Thorne, L.G. and Goodfellow, I.G. (2014). Norovirus gene expression and replication. *Journal of General Virology* **95**, 278–91.

Questions

1. Describe the replication as well as genome structure of the calicivirus norovirus.

2. Summarize the clinical features of norovirus infection relating virus replication to symptoms where possible.

Hepatitis E

10

10.1 Introduction

Hepatitis E is the sole member of the genus *Hepevirus*. Higher hygiene levels in developed countries probably encouraged the relative demise of the virus which is now more common in less developed countries alongside Hepatitis A. Nevertheless the virus causes 20 million infections per year with 3 million acute illnesses. A particular secondary source of infection can be from meat products, especially derived from pigs. Recently the virus has been detected in pooled plasma products and the infection in the EU is probably wider than previously considered. Around one quarter of the UK population have antibody to the virus.

10.2 Properties of the virus

10.2.1 Classification

HEV was originally classified in the *Caliciviridae* family (Fig. 10.1). However, the HEV gene order is not identical to caliciviruses and the HEV genes that can be compared are more closely related to rubella virus. For these reasons HEV was reclassified as the only member of the genus *Hepevirus* in the new family *Hepeviridae*.

HEV strains can be divided into four major genetic groups; genotype 1 in Asia and North Africa, genotype 2 in Mexico and South America, genotype 3 in North and South America, Europe, and Asia, and finally genotype 4 in Asia.

10.2.2 Morphology

The virion is icosahedral, 27–34 mm in diameter.

10.2.3 Genome

HEV is a non-enveloped virus with a single-stranded positive-sense RNA virus, 6.3–7.2 kb in length, containing a short 5′ UTR, three open reading frames (ORF1, 2, and 3), and a short 3′ UTR, terminated by a poly(A) tract. ORF1 encodes non-structural proteins such as RNA helicase and RNA-dependent RNA polymerase (Fig. 10.2). The ORF 2 and 3 proteins are translated from a single subgenomic RNA. The ORF2 protein is the viral capsid protein and the ORF3 protein is a small phosphoprotein essential for viral infectivity. Because HEV does not grow in cell culture, and because it is only distantly related to other viruses, little is known about its replication.

10.3 Clinical and pathological aspects

10.3.1 Clinical features

The course of HEV infection is generally similar to that of HAV with jaundice, enlarged and tender liver, fever, and nausea and vomiting (Fig. 10.3). The main differences are:

• The incubation period is rather longer (about 6 weeks).

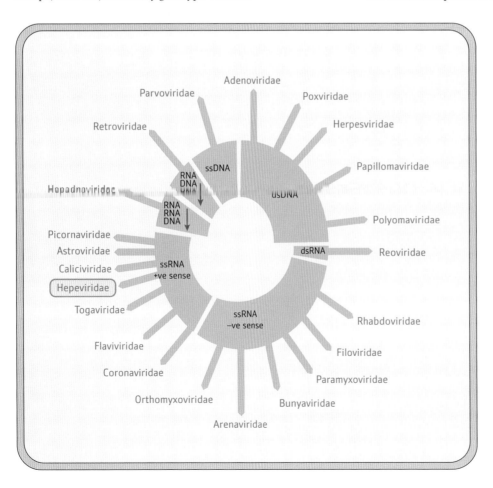

Fig. 10.1 Baltimore classification scheme and hepatitis E.

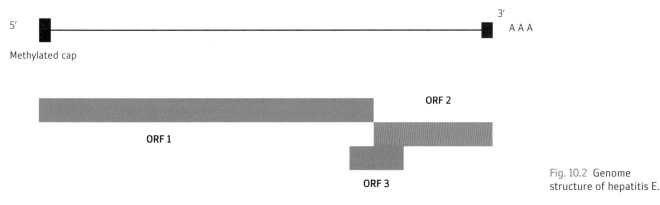

Fig. 10.2 Genome structure of hepatitis E.

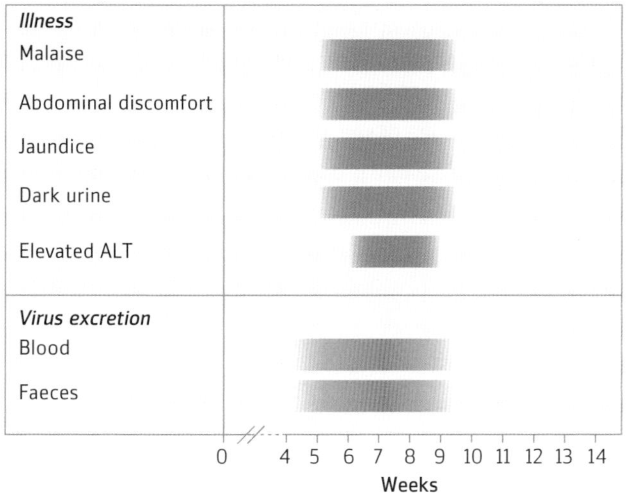

Fig. 10.3 Clinical and virological events during infection with HEV.

- Infection is generally acquired in adolescence or adulthood, rather than in infancy. Person to person transmission rates are low.
- In women becoming infected during the later stages of pregnancy the mortality rate is approximately 20%.
- Fulminant disease is more common than in HAV infections.

10.3.2 Epidemiology

In 1955, following a breakdown in the water supply and sewage systems of Delhi caused by floods, there was a huge epidemic of water-borne hepatitis affecting more than 30 000 people, which affected pregnant women particularly severely (Fig. 10.4). HAV was blamed at first, but years later, retrospective serological tests excluded this possibility (and showed the value of keeping specimens for long periods). In 1989 the cause was identified as

a new agent, HEV. Since then outbreaks have occurred in most developing countries. The mode of spread is not so well understood as that of HAV, but certainly involves the **faecal–oral route**. Raw or uncooked shellfish can be a source of virus, as well as undercooked pork. Subclinical infections are common and may be as high as seven times the clinical incidence. Based on sequence analysis HEV isolates have been classified into four genotypes. Types 1 and 2 have caused water-borne outbreaks, whilst types 3 and 4 are more sporadic and most likely zoonotic from pigs, deer, wild boar, and moose. In the UK for example, the number of cases has increased by 39% between 2011 and 2012 and a particular source has been sausage meat. Screening of 225 000 blood donations in the UK by RT PCR detected 79 viraemic donors and this blood had been used to prepare blood components: 43 recipients of this contaminated material showed evidence of infection.

10.3.3 Pathogenesis and pathology

The pathology of HEV infections broadly resembles that of HAV disease. However in some immunosuppressed patients the infection can become resistant.

10.3.4 Immune response

Apart from the fact that antibodies to HEV can be detected in the serum, little is yet known about the humoral and cell-mediated responses to HEV. New serological tests now make epidemiology studies more accurate.

10.3.5 Laboratory diagnosis

The diagnosis depends on finding serum IgM antibody to the virus. An efficient cell culture system has not been developed, although there are DNA clones of HEV which replicate in primates and pigs. RT PCR molecular diagnosis is available in more specialized laboratories.

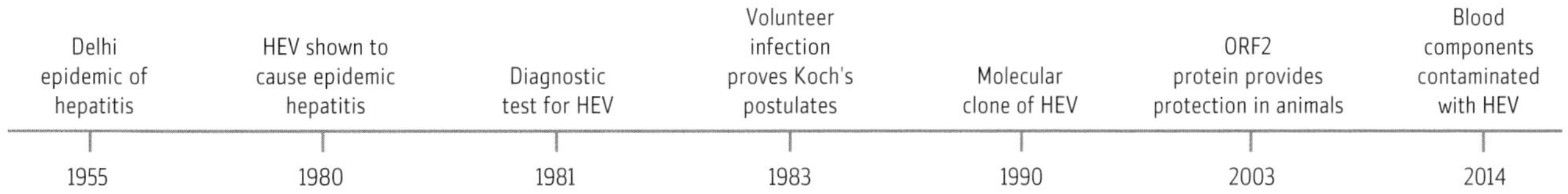

Delhi epidemic of hepatitis	HEV shown to cause epidemic hepatitis	Diagnostic test for HEV	Volunteer infection proves Koch's postulates	Molecular clone of HEV	ORF2 protein provides protection in animals	Blood components contaminated with HEV
1955	1980	1981	1983	1990	2003	2014

Fig. 10.4 Timeline for HEV.

10.3.6 Control

Immunization

Recombinant vaccines have shown efficacy in experimental clinical trials whilst a Chinese-produced vaccine has been licensed there.

General measures

These depend on the maintenance of a clean water supply, and generally resemble those used to control HAV, including quality hand washing and the avoidance of water and/or ice of unknown purity. Regulatory requirements for routine testing of blood products are being drawn up.

 Reminders

- HEV is a member of the new family *Hepeviridae*.
- The virion is a non-enveloped icosahedron.
- The genome is a positive sense ssRNA.
- Clinically and epidemiologically, HEV infections resemble those due to HAV, but are severe in pregnant women and are more likely to cause fulminant disease.

- Apart from a serum IgM antibody test which is available, the most satisfactory diagnostic method is RT PCR used to screen blood in transfusion clinics.
- There is no vaccine or specific treatment.
- Transmission via blood products is not insignificant.

 Further reading

Hewitt, P.E. *et al.* (2014). Hepatitis E virus in blood components: a prevalence and transmission study in SE England. *Lancet* **384**, 1766–73.

Lin, J. (2014). Novel hepatitis E like virus found in Swedish moose. *Journal of General Virology* **95**, 557–70.

Nagashima, S. *et al.* (2014). HEV utilizes an exosomal pathway for virion release. *Journal of General Virology* **95**, 2166–74.

Smith, D.B. (2014). Consensus proposals for classification of the family Hepeviridae. *Journal of General Virology* **95**, 2023–33.

Widdowson, M.A., Jaspers, W.J., van der Poel, W.H., Verschoor, F., de Roda Husman, A.M., Winter, H.L., *et al.* (2003). Cluster of cases of acute hepatitis associated with hepatitis E virus infection acquired in the Netherlands. *Clin Infect Dis* **36**, 29–33.

 Questions

1. Outline the genome structure of hepatitis E virus.

2. Why is this virus spreading?

Togaviruses: mosquito-borne Chikungunya and teratogenic rubella

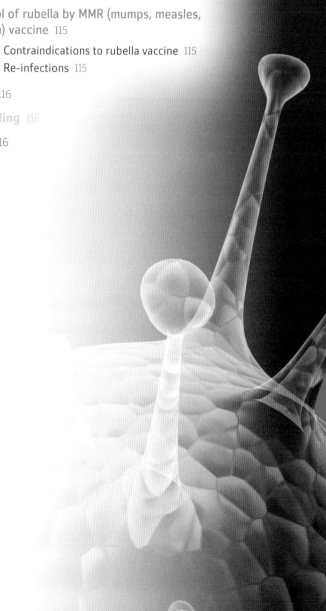

11

11.1 Introduction

The family *Togaviridae* derives its name from the closely fitting envelope (Latin *toga* meaning gown and cloak) surrounding the virions. The *Togaviridae* includes two genera: *Alphavirus* and *Rubivirus* (Fig. 11.1). The *Rubivirus* genus has a single virus and it is not arthropod-borne, but causes rubella and foetal infections. The *Alphavirus* genus has 28 viruses divided into 7 complexes, most of which are spread by mosquitoes and have a history dating back 200 years. In general, the Old World alphaviruses, such as Ross River, Sindbis, Semliki Forest, and Chikungunya, cause febrile illness with rash and arthralgia, whilst New World viruses, such as Western equine encephalitis (WEE), Eastern equine encephalitis (EEE), and Venezuelan equine encephalitis (VEE), are neurovirulent, causing headaches and encephalitis.

11.2 Properties of the togaviruses

11.2.1 Morphology

The viruses are 70mm in diameter and are spherical particles (Fig 11.2) with an icosahedral nucleocapsid (C) internal structure.

The outer virion surface is made up of 80 flower-like structures, each of three subunits of E1 and E2 glycoproteins, and

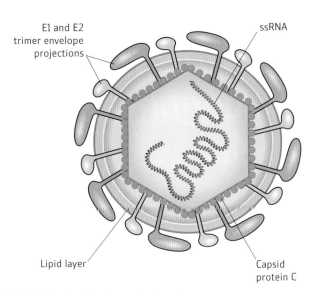

Fig. 11.2 Structural features of a togavirus.

these cover the surface, much like flaviviruses, and contact the underlying nucleocapsid.

11.2.2 Genome

Alphaviruses have single-stranded positive-sense RNA genomes 11.7 kb in size (Fig. 11.3). The genome is 5′ capped and has a 3′ poly(A) tail. The RNA codes for a single ORF flanked by 5′ and 3′

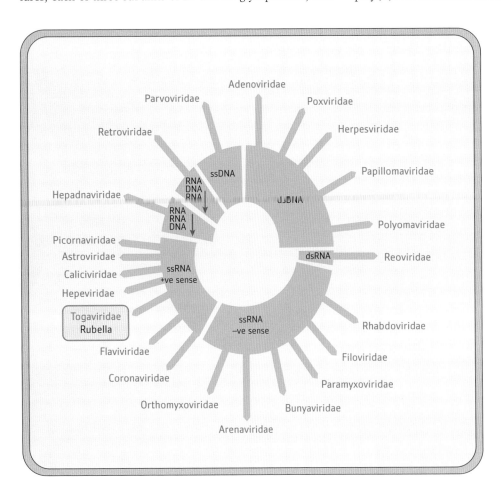

Fig. 11.1 Baltimore classification scheme and togaviruses.

The 5′ end of the genome encodes non-structural proteins (nsP1-P4) whilst structural proteins C, E2, 6K and E1 are encoded at the 3′ end. The protein 6K is only present in small numbers in the virion as a signal sequence for E1. The three black regions are non-coding, and involve replication promoters, and the internal promoter for synthesis of subgenomic RNA.

Fig. 11.3 Genome structure and encoded proteins of alphavirus.

non-translated regions (NTRs). Following infection, two mRNAs are produced that encode either the structural or non-structural proteins.

11.2.3 Replication

The alphaviruses have an extremely broad host range for infection. They can infect a variety of different cell types and different species from fish to mammals and insects. It is not clear how the viruses achieve this broad host range, but a combination of multiple receptor binding sites in the envelope E2 protein and ubiquitously expressed receptors are both likely to contribute. The viruses attach to cellular receptors such as a laminin-like receptor or heparin sulphate and enter by receptor-mediated endocytosis. The genome is released into the cytoplasm of the cell by fusion mediated by a low pH in the endosomes and phagolysosomes. Much like influenza, the spike glycoprotein complex E1-E2 exposes a hydrophobic fusion sequence at low pH. Virus replication takes place exclusively in the cytoplasm. The virus genome serves as the mRNA, but as two separate modules. Non-structural proteins (nsP1–4) are encoded at the 5′ end of the virus genome. These are produced as an abundant polyprotein P123 that is cleaved into nspP1–3 and a less abundant P1234 polyprotein encoding nsP1–4 that is produced by ribosomal read-through at low efficacy of the stop codon at the end of nsP3. The virus protease nsP2 cleaves the non-structural proteins into their functional units. The non-structural protein nsP4 encodes the virus RNA-dependent RNA polymerase. The nsP1–4 replication complex directs the production of virus negative-sense anti-genome that serves as the template for the production of new virus genomic RNA. This replication occurs in cytopathic vacuoles on the surface of endosomes and lysosomes. The negative strand also serves as the template for the production of a sub-genomic length mRNA that encodes for the structural polyproteins C, E3, E2, 6K, and E1, which are again cleaved into constituent proteins. The subgenomic mRNA is about three times more abundant than full-length genome RNA and, therefore, leads to the production of abundant structural proteins. The virus assembles and buds through specific sites in polarized cells.

11.3 Clinical and pathological aspects of alphaviruses

Anyone doing medical, nursing, and volunteer work in Africa, South America, or South-east Asia will be struck by the many villagers who complain of fever and aches and pains in the joints. The names of these viruses sometimes graphically describe the clinical symptoms: African O'nyong-nyong for example means 'weakening of the joints' and Chikungunya 'that which bends up'. Epidemics of these viruses can be massive, involving millions, but mortality is low. Alphaviruses can also cause severe arthritis, myositis, and itchy maculopapular rashes (Table 11.1 and Fig. 11.4).

Table 11.1 Some important mosquito-borne togaviruses

Geographical distribution	Viruses	Vertebrate host	Symptoms in humans
Africa, Asia	Semliki Forest		Rash, arthralgia
S America	Venezuelan equine encephalitis (VEE)	Rodents, horses, birds	Encephalitis
India, SE Asia, West Indies	Chikungunya	Primates	Haemorrhagic fever, severe joint pain
SE Asia, Australia, Oceania	Ross River	Marsupials, rodents	Arthritis, rash
USA	Western and Eastern equine encephalitis (WEE and EEE)	Birds, horses	Encephalitis

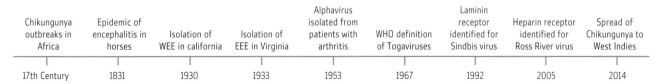

Chikungunya outbreaks in Africa	Epidemic of encephalitis in horses	Isolation of WEE in california	Isolation of EEE in Virginia	Alphavirus isolated from patients with arthritis	WHO definition of Togaviruses	Laminin receptor identified for Sindbis virus	Heparin receptor identified for Ross River virus	Spread of Chikungunya to West Indies
17th Century	1831	1930	1933	1953	1967	1992	2005	2014

Fig. 11.4 Timeline for the family *Togaviridae*.

Medical case story Chikungunya

Ross is a fit and adventurous 28-year-old man. Studying for a master's in zoology, he has just returned from a 6-month trip in Uganda, where he spent most of his time living in the jungle and studying monkeys. He has been referred to the infectious diseases clinic by his GP with symptoms of joint pain. Given his recent exotic excursion you can understand why the GP immediately thought his symptoms could be related to some wild and wonderful disease that no one can pronounce. But you remind yourself that common things are common, and that the travel history may be a coincidence or a red herring.

You listen with awe as he describes his life there; sleeping on a raised platform in a makeshift shelter with an improvised mosquito net, drinking from a nearby stream, eating a variety of bush meats and rice for dinner. Whilst out there Ross had suffered from the odd infected mosquito bite, and had been bed-bound once or twice by diarrhoea and vomiting. He had lost weight, he had lost count of the number of scratches from the jungle and insect bites he had sustained, and he had even been bitten by a monkey.

But one illness episode stood out. Ross had been there 4 months when he began to feel unwell with a fever one evening whilst sitting around the fire. He decided to go to bed, and crawled under his mosquito net, ensuring that all the edges were tucked in, and fell asleep only to be woken in the night in a drenching night sweat. The next day he didn't really feel like eating, and his friend brought him stream water which he had diligently filtered for Ross before popping in a couple of chlorination tablets. Ross had a headache all day, and the next day developed a fine rash which disappeared if he pressed on it. Then he began to ache all over, especially the small joints in his hands.

Ross tried to think through all the possible diagnoses; could it be malaria, or dengue, or something more dramatic like meningitis? He had tried his best to mitigate the risks by getting every vaccination he could before arriving in Uganda, and fastidiously took his anti-malarials. But when you are out in very remote areas you have to think carefully about the triggers that would make you trek for 6 hours to the nearest road and hitch a lift 5 hours into the nearest big town to seek medical attention. You have to be unwell enough to seek help but not too unwell that you wouldn't make the journey. Ross reflected that his conundrum is faced by millions of people around the world every day who live in remote areas with few resources. No wonder your first thought is to seek some kind of local remedy or advice. The various local people in the camp all had their own ideas and suggestions, and one even presented him with a homemade remedy which Ross thanked him for politely but did not take.

Over the next two days the fever continued and at times Ross felt he could see things crawling up the walls, and noises sounded very amplified and scary. His friend kept on bringing him water but other than that Ross couldn't remember much. On the fifth day, he felt better and over the next week continued to gain strength. The only symptom that remained was the joint pain which mainly affected his hands and toes, but also occasionally his hips and back. He said there was no visible swelling or redness although the joints felt swollen and tight inside. For a fit and active man, this constant pain affected him badly and when he eventually returned home he made an appointment with the GP straight away.

You examine his various joints. He seems in pain when he moves them and his range of movement is affected, but there is not much to see. His blood pressure and urine dip are fine, and there are no positive clinical findings including lymphadenopathy, conjunctivitis, or rashes. He is sexually active, but denies any pain or abnormalities such as penile discharge.

Your list of differential diagnoses range from malaria, or dengue, to first presentation of inflammatory joint disease, or a possible reactive arthritis in the context of a sexually transmitted disease, although admittedly the timings aren't correct. Ross agrees to a barrage of tests including a routine full blood count, renal and liver function tests, HIV testing, autoimmune antibodies, and a spare blood tube of viral serology which gets stored in the laboratory and can be used at a later date for any outstanding ideas. You also arrange for Ross to have X-rays of his various joints and make a referral for him to be seen by one of your Rheumatology colleagues in order to get a second opinion. For the time being Ross will have to wait for results and continue taking simple pain relief.

Later in the day you have a discussion with your consultant to run through the patients you have seen in clinic and to double check you haven't missed anything.

'Why don't you do an arbovirus screen? It sounds like he had ample opportunity to get bitten by a mosquito or a tick. There's plenty of dengue, Chikungunya, and Rift Valley fever around that area of the world. The results would be rather academic because it sounds like he's over the worst of it and there's no treatment anyway, but one is always curious.'

In fact the consultant was correct. A month later the results came back as IgG positive of Chikungunya which means that Ross has been infected in the past and has developed an immunological response, although there is no timescale of infection and no guarantee he won't get it again. You receive a letter from the Rheumatologist who has seen Ross in clinic quite recently and comments that his joint pains are improving, there is no evidence of rheumatoid arthritis, and that the joint pains are likely to have been reactive. You write to Ross informing him of the Chikungunya results and enclose some information on the virus because you know he will immediately turn to the internet for answers. You say you are pleased to hear things are improving but reinforce the fact that should he return to similar parts of the world he would need to be strict with insect bite prevention.

★ **Learning Points**

- Chikungunya is an arbovirus that is transmitted by the *Aedes* mosquito in tropical parts of the world. It is recognized as an emerging virus, infecting more and more people in countries not typically associated with the virus, possibly due to overcrowding and climate change.

- Symptoms start 3–7 days after a mosquito bite and typically consist of temperatures, joint pain, rash, and fatigue. The joint pains can continue for months or even years.

- Chikungunya is often underdiagnosed or misdiagnosed as malaria or dengue fever.

- Diagnosis is usually based on clinical presentation, as laboratory results may take several weeks to come back, by which point the illness has passed. The virus rarely causes severe disease except in those with multiple comorbidities.

- There is no treatment apart from symptomatic relief, but infection can be reduced by using mosquito repellent, nets, and by draining mosquito breeding sites.

11.3.1 Pathology

The natural hosts are wild mammals and birds while humans and horses are dead-end hosts being unable to transmit to each other. A typical vector, for example, *Aedes albopictus*, is widespread in the world allowing rapid infection of humans. The objective of modern molecular virology is to allocate symptomatology to different virus genes. In experimental models of this family of viruses single nucleotide changes in the 5′ untranslated regions of the genome correlate with varying survival times in laboratory animal experiments. In a similar manner, single base changes in the E2 glycoprotein affects neurovirulence in laboratory experiments. Many human cells are susceptible, for example, to Chikungunya virus including myeloid cells as well as fibroblasts and epithelial cells and this would contribute to the broadness of clinical symptoms and pathology. For many members of this family, signalling through type I IFN receptor plays a role in both controlling infection and causing virus-induced arthritogenic disease. Most of these viruses alter the levels of inflammatory cytokines and chemokines which contribute to the clinical symptoms.

11.3.2 The recent spread of Chikungunya virus

The virus was first detected in East Africa in the 1950s. A large epidemic began in 2004 with around 6.5 million cases, primarily in Africa and Asia, but is now involving 40 countries. A significant outbreak has been recorded around the Caribbean since December 2013 with over 300 000 cases in the Dominican Republic, 60 000 in Haiti, 70 000 in Guadeloupe, and 50 000 in Martinique. The virus could now spread into N. America quite easily, and several hundred returnees to the USA have already

been infected. An explanation is that a specific mutation in the virus has allowed it to infect the Asian tiger mosquito (*Aedes albopictus*) which itself has been expanding in the USA and worldwide. Previously the virus had commonly infected *A. aegypti* but this mosquito is killed in winter weather.

11.3.3 Alphaviruses as vectors for vaccines or even as insecticides

For experimental purposes the alphavirus RNA is firstly converted in the laboratory to cDNA, which can then be cloned into a bacterial plasmid, modified at will by molecular technology, and then transcribed back into infectious RNA. A range of foreign genes can be introduced into the alphavirus genome, including influenza HA and, more recently, Ebola surface glycoprotein. Such an alphavirus can also be modified to produce a protein toxic to insects, which could kill the mosquito vector or even produce an interfering RNA to other human pathogens which are mosquito-spread, such as dengue!

11.4 Teratogenic rubella virus

Rubella (Fig. 11.6) is derived from the Latin *rubellus*, meaning 'reddish', and refers to the pink rash that is seen in most patients. Its popular name, German measles, probably derives from the fact that it was first described in Germany during the eighteenth century. Rubella is predominantly an infection of children, in whom it causes a mild febrile illness. Were this all, we should spend but little time on it; however, as a potent cause of foetal abnormality, rubella has a major claim on our attention.

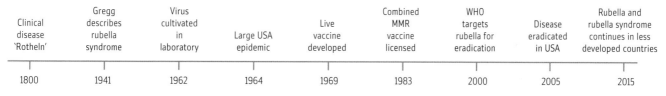

Clinical disease 'Rotheln'	Gregg describes rubella syndrome	Virus cultivated in laboratory	Large USA epidemic	Live vaccine developed	Combined MMR vaccine licensed	WHO targets rubella for eradication	Disease eradicated in USA	Rubella and rubella syndrome continues in less developed countries
1800	1941	1962	1964	1969	1983	2000	2005	2015

Fig. 11.5 Timeline for rubella.

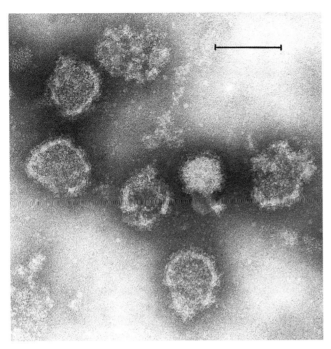

Fig. 11.6 Electron micrograph of rubella virus (courtesy of Dr David Hockley). Scale bar = 100 nm.

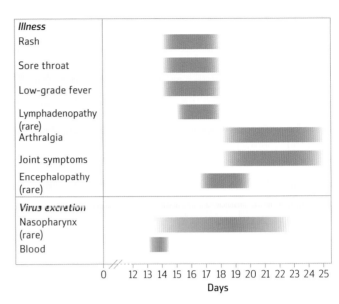

Illness															
Rash															
Sore throat															
Low-grade fever															
Lymphadenopathy (rare)															
Arthralgia															
Joint symptoms															
Encephalopathy (rare)															
Virus excretion															
Nasopharynx (rare)															
Blood															

0   12 13 14 15 16 17 18 19 20 21 22 23 24 25

Days

Fig. 11.7 Clinical and virological events during infection with rubella.

11.5 Clinical and pathological aspects of rubella

11.5.1 Clinical features

Rubella in children

The virus is spread from respiratory secretions and later produces a viraemia, as well as being present in stool samples (Fig. 11.7). The onset, after an incubation period of 8–12 days, is marked by slight malaise, a small rise in temperature—usually to less than 38 °C—and sometimes suffusion of the conjunctivae. Ohotty enlargement of the **lymph nodes** in the suboccipital region, behind the ears, and in the neck and axillae is characteristic. The **rash** usually appears a day or two after the onset of symptoms, but is by no means a constant finding; its appearance is not diagnostic, as it may resemble those seen in other viral infections. It consists of small pinkish macules, rarely exceeding 3 mm in diameter, and usually more discrete and regular in appearance than the eruption of measles. The face and neck are first affected, followed by the trunk. Circumoral pallor is sometimes present, but is not diagnostic. Unlike the pronounced enanthem seen in measles, there may be some very small erythematous spots on the soft palate, but often the buccal mucosa appears normal. A purpuric rash has been described, but is rare.

The vast majority of patients get better within a few days. Post-infection **encephalitis** is rare, affecting one in several thousand patients, more often in adults, and the prognosis is generally good.

Rubella in adults

The main difference between rubella in children and in adults is that in the latter, **polyarthritis** is a not infrequent complication. It most often affects the hands and wrists, but may also involve the larger joints of the limbs. There may also be some myalgia. Rubella arthropathy predominantly affects post-pubertal women. It usually clears up fairly quickly, but may persist for months or even years.

11.5.2 Epidemiology

An important point to appreciate is that half the persons infected with rubella show no symptoms and yet can be a source of infection to others. In the USA the largest outbreak was recorded in 1962–65, with 12 million cases, but after the licensing and use of the vaccine in 1969 incidence there dropped rapidly and by 2004 the virus was no longer endemic in the USA.

Because rubella is of significance mainly in relation to foetal infection, an important epidemiological aspect is its prevalence in women of childbearing age. Prior to mass vaccination rubella was present in all countries, periods of endemicity alternating with epidemics at irregular intervals. Outbreaks were seasonal, often in the spring months in the Northern hemisphere. Its distribution now varies from country to country. Serological surveys show that, in most populations, over 70% of women have been infected, but in others, notably rural areas in the Indian subcontinent, the figure is lower.

Effective immunization programmes radically alter the prevalence, and where they are in force, rubella is now a rare infection. However foci of unvaccinated citizens such as in Hispanics and in the Amish in the USA allow the virus to survive even in this highly vaccinated population.

11.5.3 Pathogenesis and pathology

The portal of entry is the respiratory tract; the pathology of foetal infections may involve slowed growth rates of foetal cells infected by the virus and apoptosis induced by viral protein. Infection of cells in the laboratory shows depolymerization of actin filaments and disruption of cytoskeleton structures.

Immune response

Specific IgM antibody appears within a few days of the rash, and is followed soon after by IgG. The titre of IgM increases rapidly, reaching a peak about 10 days after onset and thereafter declining to undetectable amounts over several weeks or months. The rapid appearance of specific IgM antibody is invaluable for diagnostic purposes. IgG antibody peaks at about the same time as IgM, and persists for many years, as does IgA antibody, which appears in the serum and nasopharyngeal secretions.

The cell-mediated response precedes the appearance of antibody by a few days, reaches a peak at about the same time, and is also detectable for many years.

The E1 glycoprotein spike of the virus is a major target for both antibody and T cells.

Laboratory diagnosis

Because acute rubella is often difficult to diagnose on purely clinical grounds, laboratory diagnosis assumes considerable importance. RT PCR is now used widely with clinical material from a throat swab.

Tests for rubella infection, past or present, can be considered under three headings:

- screening tests for rubella antibody, to ascertain the immune status of women of childbearing age;
- tests for acute infection in pregnancy;
- tests on infants for congenital infection.

Screening tests on women

In the UK and elsewhere, women attending antenatal clinics are routinely screened for rubella antibody. If the result is negative, immunization soon after the birth is advised in order to protect subsequent pregnancies. Screening is also advised for women of childbearing age at particular risk of infection, e.g. teachers and clinic and hospital staff in contact with children. Such tests should be done irrespective of a history of past infection or immunization, which may be unreliable. As many sera must be processed, the requirement is for a reliable test that can be done on a mass scale and that is not too labour-intensive. Commercially available ELISA tests for IgG antibody are used because they can be automated.

11.5.4 Intrauterine infection and the congenital rubella syndrome

Of all the distressing situations encountered in medical practice, the loss of a wanted pregnancy or the birth of a malformed infant are among the most poignant; the latter may indeed carry the

prospect of many years of misery and hardship for both child and parents. Some of these tragedies are caused by infection with viruses, bacteria, and even protozoa; it must, however, be stressed that microbial infection is responsible for only a small proportion of the overall total of miscarriages and damaged infants; in the UK, the incidence of birth defects is about 20 per 1000 live births.

The diagnosis of congenital infections has in the last few years been greatly facilitated by ultrasonic examination, the ability to take samples of blood, amniotic fluid, and chorionic villi from the foetus, and the application of the newer laboratory tests, such as PCR.

The discovery by Gregg, in Australia 75 years ago, of the congenital rubella syndrome (CRS) was the first recognition that foetuses can be damaged by viruses. Congenital rubella is the result of a primary infection of the mother during the first 16 weeks of pregnancy. In the absence of maternal antibody, the virus readily crosses the placenta and the time at which this happens is an important factor in determining the outcome. By and large, the earlier in gestation infection takes place, the worse the result; during the first month, the teratogenic rate approaches 100% (Fig. 11.8). Full-blown congenital rubella, called the 'expanded rubella syndrome', has features in common with those of congenital infections by HSVs, CMV, and toxoplasmosis, a parasitic disease. Many systems are affected and the prognosis is very poor. Figure 11.8 gives an idea of the relative frequency of some of the major lesions; the chart is compiled from figures obtained during a major

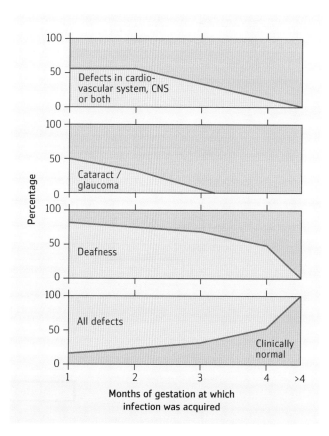

Fig. 11.8 Congenital defects in rubella-infected neonates.

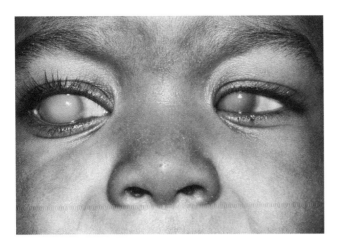

Fig. 11.9 Congenital rubella syndrome: bilateral cataracts in a 5-month-old baby.

epidemic of rubella in New York during 1964, well before the introduction of rubella vaccine. The virus also causes microcephaly.

It shows, for example, that of 106 infants born after their mothers contracted rubella during the second month of pregnancy, only five (4.7%) were clinically normal; 58% had defects of the cardiovascular system or CNS; 29% had eye defects (Fig. 11.9); and 72% were later shown to have sensorineural deafness. Clearly, the inner ear is susceptible to damage over a longer period than the cardiovascular system or CNS. It is interesting that the curve for clinically normal infants is a mirror image of that for deafness, perhaps implying that aural damage is the lesion best correlated with infection *in utero*.

Some infants who appear normal at birth are later discovered to have hearing defects, with or without some degree of mental retardation.

Rubella diagnosis in the pregnant mother

Because the clinical diagnosis of rubella is not always easy, laboratory tests are essential as a guide to the important decisions that must be made following a history of contact with rubella, or of a rubella-like illness, within the first trimester of pregnancy. Tests for specific anti-rubella IgM and IgG antibodies are made and action is taken according to the scheme shown in Fig. 11.10. These tests must be done irrespective of a history of past immunization, which is not always reliable.

The plan looks a little complicated, but is, in fact, quite logical; it is based on:

- the time course of the appearance of IgM and IgG antibodies; and
- the possible results of tests, arranged according to whether they are first done during, or outside the extreme ranges of the incubation period of rubella (10–21 days). There is one trap to beware. In addition to a known contact, say 14 days previously, there may have been another unsuspected contact some days later; if this seems a possibility, weekly tests for IgM antibody should be continued until the end of the longest possible incubation period has passed.

Diagnosis of the neonate

Congenital rubella is normally diagnosed by demonstrating specific IgM antibody in the cord or peripheral blood. Virus isolation or quantification using RT PCR is used only for confirming a diagnosis of prenatal rubella by testing the tissues of an aborted foetus and monitoring the shedding of virus or virus RNA from a congenitally infected baby.

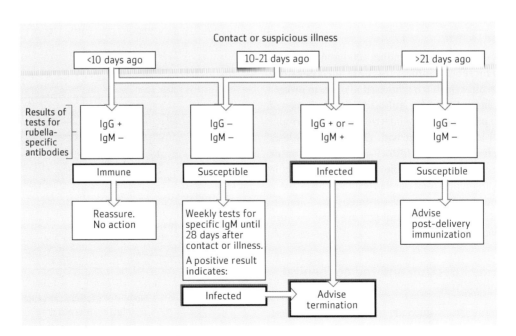

Fig. 11.10 Action following exposure to rubella early in pregnancy.

11.6 Control of rubella by MMR (mumps, measles, rubella) vaccine

Rubella vaccine is the only one not given primarily for the benefit of the recipient, but to protect another, in this case, a foetus. The live vaccine is given by subcutaneous injection.

Rubella immunization was started in the USA during 1969, and in the UK a year later, following a rubella epidemic in 1962–65 when over 12 million cases occurred in the US alone, with 20 000 children with congenital defects.

Early immunization policies reduced the incidence of congenital rubella, but less so in the UK than in the USA. Figure 11.11(a) shows the antibody profiles by age group of people in a UK population during the period 1969–85, rubella immunization of teenage girls but not boys having been introduced in 1980. The antibody profiles are similar in pre- and post-vaccination years. Figure 11.11(b) shows a similar analysis of a population in Finland which, in 1982, adopted a policy more like that of the USA, vaccine being given to both boys and girls aged 14 months and 6 years. Note the sharp increase in seropositivity in the younger age groups from 1984 onwards. A typical immunization schedule is:

- Boys and girls aged 12–15 months: first dose of MMR vaccine.

- Boys and girls aged 3–5 years: if no documentary evidence of previous vaccination, second dose of MMR before entering primary school (i.e. at same time as diphtheria–tetanus–polio booster).

- School leavers are offered MMR vaccine if they have not previously been immunized against rubella or measles.

- Non-immune women, before pregnancy or after delivery: single rubella vaccine.

It is now accepted that the immunity resulting from vaccination, although not as solid as that following natural infection, is sufficiently durable to allow a programme of this type. Its aim is the eventual elimination of rubella, measles, and mumps. Now that the MMR vaccine acceptance rate has reached 90% again, this goal is well within sight. In the UK in 2015 there were only a handful of confirmed cases of rubella on the records.

The disease is still common in Africa, India, and some parts of South-east Asia, causing over 100 000 babies to be affected yearly. These countries act as reservoirs of virus and, moreover, a threat to travellers who have not been immunized.

11.6.1 Contraindications to rubella vaccine

The main contraindications are those applying to other live vaccines, i.e. a febrile illness, allergy to one of the constituents (rare), and defective immunity. The important additional contraindication is **early pregnancy**; women are advised not to become pregnant a month before to a month after receiving rubella vaccine. That said, no instance of foetal damage has been recorded following inadvertent vaccination shortly before pregnancy or during the first trimester, although serological evidence of infection *in utero* has been obtained in a small proportion of the babies going to term. There is very little indication for terminating a pregnancy if rubella vaccine is given inadvertently.

11.6.2 Re-infections

Re-infections, both after second attacks of rubella and immunization, have been recorded and confirmed serologically. As far as is known, such episodes, even if in early pregnancy, pose little or no risk to the foetus, probably because they are not accompanied by the viraemia characteristic of primary infections.

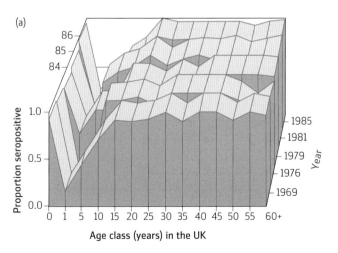

(a)

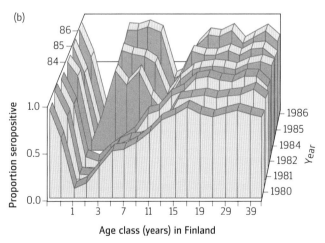

(b)

Fig. 11.11 The effects of different immunization policies on herd immunity to rubella. For explanation, see text. (Adapted with permission from Anderson, R.M. and May, R.M. (1990). Immunisation and herd immunity. *Lancet* **335**, 641–5.)

Reminders

- Togaviruses are **enveloped positive-sense ssRNA viruses** of 10–11 kb. The RNA codes for a single ORF flanked by two non-translated regions.

- The virus replication complex is in the cytoplasm of the cell and budding occurs into the Golgi.

- The virion is spherical with a fringe of flower-like structures, each with three subunits. There is an icosahedral internal structure.

- Many togaviruses are spread by mosquitoes and cause a range of diseases including arthritis, rash, joint pains, and encephalitis.

- Rubella (German measles) is a worldwide infection of childhood. It is spread by the **respiratory route** and causes a

mild febrile illness with **lymphadenopathy** and **rash**. In older people, especially women, it is liable to cause **arthropathy**.

- **Infection during early pregnancy with rubella damages the foetus and causes a triad of blindness, deafness, and mental retardation called the 'rubella syndrome', and also microcephaly.**

- For this reason, efforts are made worldwide to prevent the infection of women of childbearing age by:
  - **mass immunization** of the child population with MMR vaccine;
  - **serological screening** of certain categories of adult women and immunization of those without rubella antibody.

Further reading

Babigumira, J.B. *et al.* (2013). Health economics of rubella: a systematic review to assess the value of rubella vaccination. *BMC Public Health* **13**, 406.

Bajak, A. (2014). US assesses virus of the Caribbean. *Nature* **512**, 124–5.

Manore, C.A. *et al.* (2014). Changing epidemiology of a mutated chikungunya virus. *Journal of Theoretical Biology* **356**, 174–91.

Taylor, A. (2015). Mouse models of alphavirus induced inflammatory disease. *Journal of General Virology* **96**, 221–38.

World Health Organization (2011). Rubella vaccines. WHO position paper. *Vaccine* **29**, 8767–8.

Questions

1. Outline the morphology, genome and replication strategy of an alphavirus.

2. Summarize the range of symptoms caused by alphavirus and their geographical distribution. Could these viruses extend more widely into the EU?

3. Describe the clinical nature and pathology of the 'congenital rubella syndrome'.

4. Write notes on rubella vaccine and effectiveness.

5. Describe the clinical nature and pathology of the 'congenital rubella syndrome'.

Flaviviruses: yellow fever, dengue fever, and hepatitis C

12.1 Introduction

There are at least 70 serotypes of flavivirus. The prototype virus of the family causes yellow fever (Latin: *flavus* = yellow). Molecular clock analysis dates the sojourn of these viruses from 10–100 000 years. Yellow fever (YF) particularly is a classic disease of antiquity and applying the term 'white man's grave' to West Africa resulted from its impact on colonizers, whilst in classic literature yellow fever has a walk-on part in Coleridge's 'Ryme of the Ancient Mariner' and Wagner's 'The flying Dutchman'. Although Carlos Finlay, as early as 1881, viewed the mosquito as the source and scourge of YF, it took heroic volunteer experiments coordinated by Walter Reed in 1900 to prove this.

Other medically important flaviviruses are tick-borne encephalitis virus (TBE), dengue virus and West Nile virus, Japanese encephalitis, Murray Valley encephalitis virus and Zika virus (Fig. 12.1). There are a number of flaviviruses of rodents and bats with no known vector to transmit to humans. Hepatitis C virus (HCV) is a well-researched member of the family. Of course HCV does not have the typical mosquito transmission of the other flaviviruses, but is blood-borne. Zika virus is known to be teratogenic.

12.2 Properties of flaviviruses

12.2.1 Genome

Viruses of this family have a single-stranded positive-sense linear RNA genome around 10-12kb in size, capped at the 5' end but without a poly A tail at the 3' end (Fig. 12.2). For hepatitis C, the genome has a complicated secondary structure and functions as an internal ribosome entry site (IRES), which mediates viral protein translation in a cap-independent manner, in an identical way to picornaviruses.

12.2.2 Morphology

The virion has the appearance of a golf ball with an icosahedral nucleocapsid structure within an outer lipid envelope (Fig. 12.3). There are no projections of the glycoproteins and rather they appear to lay flat on the surface of the virion.

The nucleocapsid is made up of the C or capsid protein whilst 180 copies of M and E glycoproteins constitute the outer layer embedded through the host-derived membrane (Fig. 12.3).

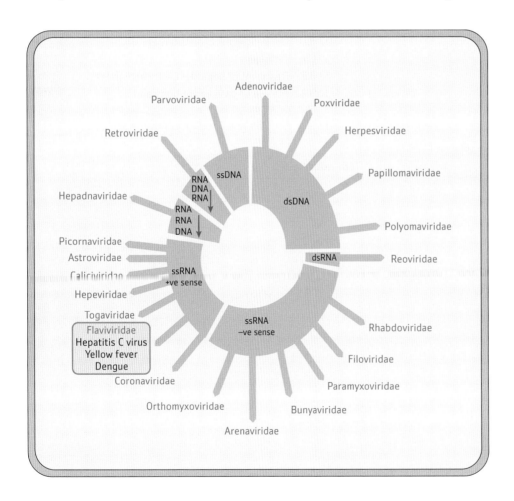

Fig. 12.1 Baltimore classification scheme and flaviviruses.

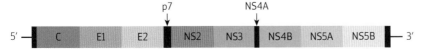

Fig. 12.2 Genome structure and some encoded polyproteins of flaviviruses.

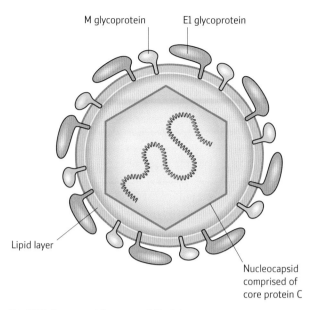

M glycoprotein   E1 glycoprotein

Lipid layer

Nucleocapsid comprised of core protein C

Fig. 12.3 Structural features of flaviviruses.

The E protein is a type I transmembrane protein which binds to cellular receptors and later mediates fusion of host cell and virus membranes. The capsid protein C encapsulates the RNA. The function of M, a small envelope glycoprotein, is not well understood. There are 7 non-structural (NS) proteins. NS1, NS2A, NS4A, and NS5 are RNA replicase components, NS2B and NS3 are virus proteases, whilst finally NS4B has an unknown role.

12.2.3 Replication

The viral E protein binds to cell receptors, possibly DC-SIGN, laminin binding protein, or α-β 3 integrin. Hepatitis C can utilize SR-B1 (scavenger receptor class B type one), CD 816 (tetraspaninprotein), and low density lipoprotein receptor (LDLR). Entry is achieved by endocytosis via clathrin-coated vesicles which fuse with endosomes (Fig. 12.4). Acidification of endosomes induces structural changes of the E dimer, which in turn causes fusion of virus and endosome membrane.

After fusion the RNA genome is released into the cytoplasm and the RNA is translated into a single polyprotein. Cleavage of the polyprotein results in the three structural proteins (E, prM, and C) and the seven non-structural proteins (P7, NS2, NS3,

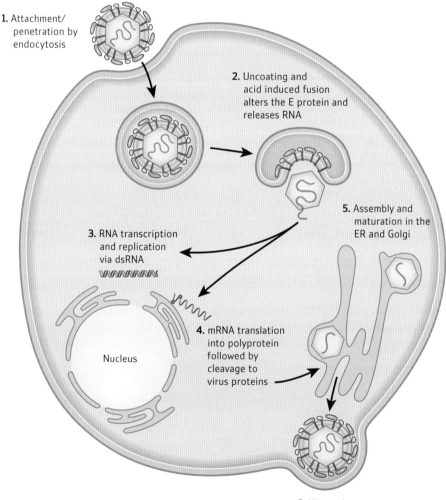

1. Attachment/ penetration by endocytosis

2. Uncoating and acid induced fusion alters the E protein and releases RNA

3. RNA transcription and replication via dsRNA

4. mRNA translation into polyprotein followed by cleavage to virus proteins

5. Assembly and maturation in the ER and Golgi

Nucleus

6. Virus release by exocytosis

Fig. 12.4 Replication cycle of flavivirus.

NS4A, NS4B, NS5A, NS5B). NS4B and NS5A are part of an RNA-dependent RNA polymerase which transcribes the input virus RNA genome into an antisense negative-strand RNA, which in turn acts as a template for the infectious positive-strand RNA genome. NS3 is a bifunctional protease/helicase.

The virions are assembled inside the cell and the viral membrane is formed from intracellular membranes. Since all assembly phases are taking place on the same internal membrane the new RNA genome can interact with C protein as the start of virus assembly per se. The viral RNA is encapsidated and virions bud into endoplasmic reticulum and then move through the Golgi along the exocytosis pathway as immature virions. The virions are released by budding. In the case of HCV as many as 10^{12} viral copies/ml of virus load are released as a viraemia.

12.3 Clinical and pathological aspects of the flaviviruses

12.3.1 Yellow fever

As its name implies, this infection is characterized by **jaundice**, the result **of mid-zone necrosis of the liver**; the flavivirus responsible also damages the kidney and heart, and bleeding from the gastrointestinal mucosa may cause '**black vomit**' and melaena (Fig. 12.5). This is a very serious infection with mortality around 20%.

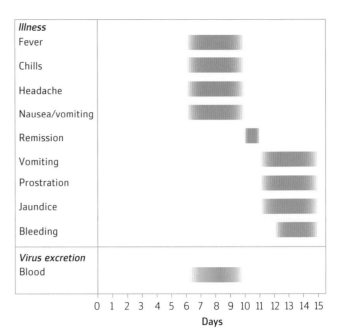

Fig. 12.5 Clinical and virological events during infection with yellow fever.

The virus originated in Africa, from whence the mosquito vector, *Aedes aegypti*, was spread to the New World during the seventeenth century in sailing ships (Fig. 12.6); the unsuspected

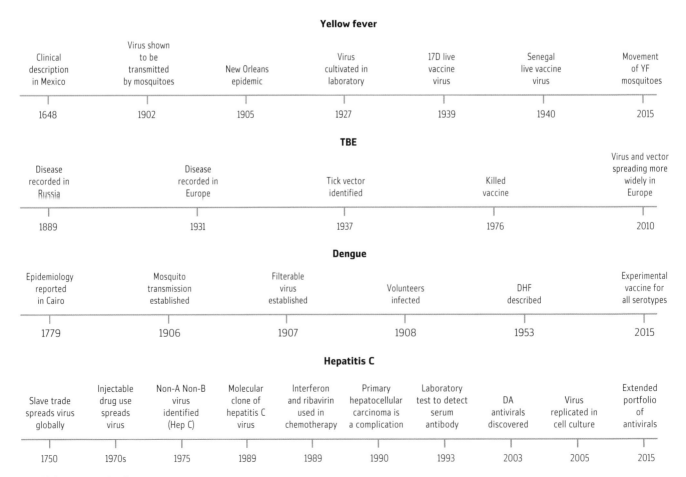

Fig. 12.6 Timeline for flavivirus.

importation of a deadly disease along with the slaves that many of these boats carried might be regarded as poetic justice. The virus is endemic in Central Africa and South America where it exists mainly as an infection in monkeys and only rarely transmits to humans (Fig. 12.7a).

There are two forms of YF, which differ only in their epidemiology. The so-called **urban** variety is unusual in that the only hosts are humans (Fig. 12.7b).

Urban YF has caused huge epidemics when introduced into populations with no immunity. During the nineteenth century its ravages in newly arrived foreign workers completely halted construction of the Panama canal for some time; there were major outbreaks in Ethiopia and West Africa during the 1960s, with thousands of deaths.

By contrast with urban YF, which as its name implies mainly affects inhabited places, the **sylvan** (or sylvatic) form occurs in

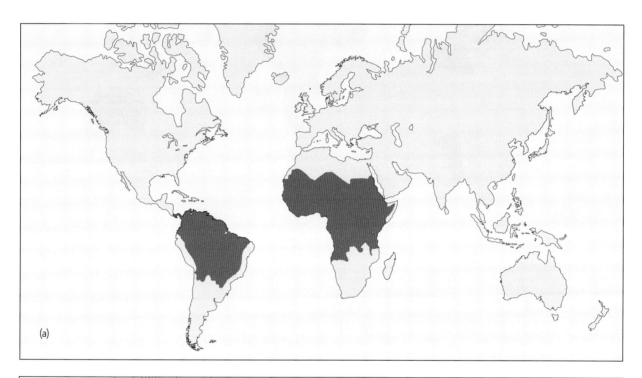

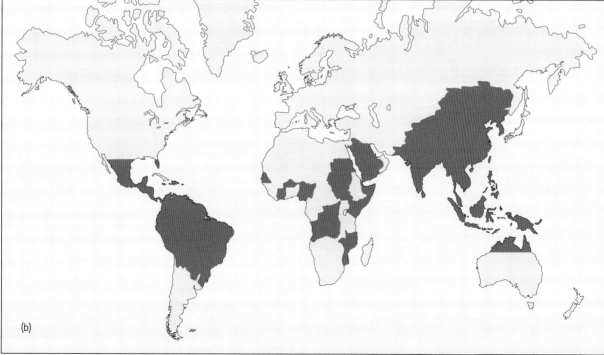

Fig. 12.7 Geographical distribution of (a) yellow fever (b) dengue and (c) West Nile.

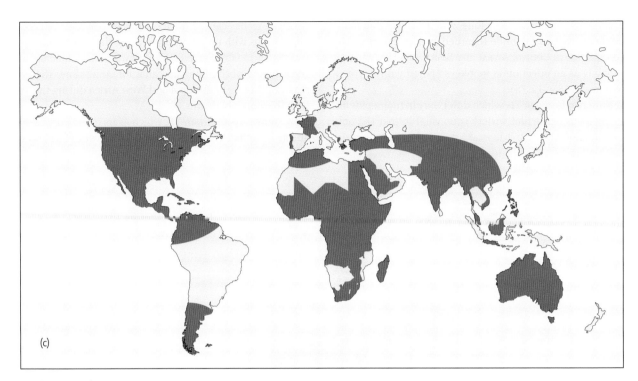

Fig. 12.7 (*Continued*)

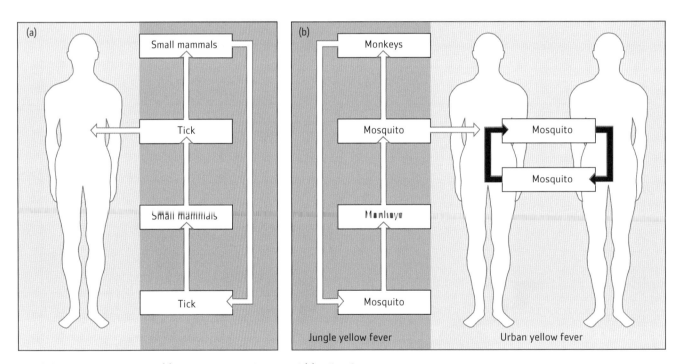

Fig. 12.8 Transmission cycle of (a) tick-borne encephalitis and (b) yellow fever.

forested areas; here, the epidemiology more closely follows the classical arbovirus pattern as there is an animal reservoir, namely monkeys, among which the virus is spread by a variety of mosquitoes; men—and here the term is used to imply gender,

rather than species—are infected when they venture into forests, usually to hunt; women are much less often the victims.

YF is still prevalent in Africa and Latin America (Fig. 12.7c); why it does not occur in the Indian subcontinent, where there is no shortage of susceptible mosquitoes, is something of a mystery.

The live attenuated yellow fever vaccine 17D

As long ago as the 1930s Theiler and Smith in the USA passaged YF virus in fertilized hens eggs 17 times and tested the resultant live but hopefully attenuated virus for residual virulence in monkeys and themselves! Fortunately what is known as classic 'Pasteurian' passage selected out attenuated virus mutants. There are 68 nucleotide changes and 32 amino acid substitutions in the attenuated virus (of a total of some 10 862 nucleotides) compared to the virulent wild-type virus. The E protein has 12 mutations and appears to be the dominant virulence gene.

The vaccine is given to travellers visiting endemic YF regions and is highly protective—immunity lasts for more than 10 years. Current production of vaccines is 15 million doses per year and in endemic regions the vaccine can be used to ring fence an outbreak or, better, as a prophylactic vaccine given with measles vaccine to children at 9 months of age.

The application of reverse genetics to the RNA genome of YF has allowed researchers to place mutations in the virus genome and these studies could expose further mutations to make the live vaccine even safer.

Medical case story Yellow fever

The sorts of incidents that cause you to learn your next virological or medical fact can be surprising. Often it might be a newspaper report of a foreign epidemic or a local hospital outbreak, for example. Sometimes a friend mentions in passing an interesting diagnosis from their childhood, and sometimes it is a story, maybe a recent Booker prize winner, or perhaps even a film.

One afternoon, after a week of night shifts, you are idly flicking television channels on mute. You come across a black and white film, and there, on the screen, is a beautiful woman in a strikingly styled crinoline dress. It is sunny outside and spring-like, but for you it is too late: you turn up the sound and are sucked into the story.

You don't know it yet, but this is a classic film from 1938. It is only when the credits go up at the end you find out that it is called *Jezebel*. The story is taking place in Southern America, in a period known as 'the antebellum'—that is, 'before the war'.

You may be thinking you are simply watching a romantic drama, but you would be wrong. Your medical education, in relation to one of a group of lethal viruses causing haemorrhagic fever, is about to be advanced.

The story unfolds; Jezebel is waiting for her fiancé to return. Eventually he does, but not to see her. He has come back for the sole purpose of warning the city authorities of an outbreak of Yellow Jack.

Yellow Jack? What's that? You are completely distracted from the romance. Just then the camera pans in close for a shot of our hero. You hear the whining noise of an insect and you see him slap his hand. Is it a mosquito-borne illness?

From the panic that is depicted next as people escape the city, you also gather it is a serious illness. The sequence is then intercut with a scene of the hero sweating and then collapsing in the middle of a public space, and no one helping him: he has a fever. Finally, you are told that all the victims are to be quarantined on an island.

So what was the real name of this illness? Where is the story set? And is it true?

You find out that the story is based on an epidemic that really happened and the illness is now known as yellow fever.

Sailors would be infected with yellow fever in West Africa and then arrive back with a cargo of slaves to New Orleans, where they would be bitten by local mosquitoes. The mosquitoes then carried the infection to the local population, and river traffic carried it right up into the Mississippi valley. Memphis had yellow fever epidemics in 1828, 1855, 1867, and 1873. However, the worst epidemic of all was in 1878 when a mild winter had allowed the mosquitoes to multiply. The first death was reported at the beginning of August and over the next 2 weeks 25 000 people fled the city, just as you have just seen being depicted in the film.

The deaths continued until October when the mosquitoes died off in the first frosts. Of the white population who remained in Memphis, 90% caught yellow fever, and, of these, 70% died. Only 7% of the black population who caught it died. There *were* quarantines imposed, as featured in the film, but they had little or no effect.[1]

You turn to your tatty old medical dictionary given to you on your first day at medical school. It doesn't have 'yellow jack', but it does have:

'... yellow fever. The virus causes degeneration of the tissues of the liver and kidneys. Symptoms include chills, headache, aches and pains, fever, vomiting, a reduced flow of urine (which contains high levels of albumin), and jaundice. Yellow fever often proves fatal. The disease can be prevented by vaccination'.[2]

The history page you have found is more graphic and remarks 'the trademark of the disease was the victims' black vomit composed of blood and stomach acids'.

Your microbiology book says the virus is a flavivirus. 'If suspected in a patient who has travelled to Africa or South America, malaria should be excluded first'. This, it explains, is because malaria has a similar non-specific presentation of fever, headache, and aches and pains, and of course is far more common.

If malaria is excluded and haemorrhagic fever is suspected—mainly from a travel history—then the patient should be isolated and referred to an infectious diseases specialist.[3]

A differential diagnosis might be:

Disease	Transmission
Malaria	Mosquito
Typhoid	Faecal–oral
Typhus	Louse
Influenza	Droplet spread
Haemorrhagic fevers	
Lassa fever	Person to person
Yellow fever	Mosquito
Dengue haemorrhagic fever	Mosquito
Ebola	Person to person
Marburg disease	Person to person

You find yellow fever is not a nineteenth-century phenomenon. Every year 2 000 000 cases of yellow fever are still recorded and of these 30 000 people die (15%).[4]

Yellow fever vaccination protects patients from 10 days after the jab and lasts for 10 years. Strangely, yellow fever is not in Asia, although the right mosquitoes live there. This explains why yellow fever vaccination is sometimes required before crossing the border in some countries, and why yellow fever jabs can only be given at WHO recognized vaccination centres. These centres are given a stamp to document the jab on an international yellow fever vaccination card.

You feel obscurely comforted that yellow fever is mosquito-borne. You would not be at risk of contracting it from a patient as there are no mosquitoes (allowed) where you work. As you mull over the differential diagnoses, you realize that would not be the case for, say, Ebola (see the film *Outbreak!*).

Notes

[1] http://www.tenesseeenclyclopedia.net/entry.php?rec=1545 (accessed 19 Feb. 2016).
[2] *Chambers Medical Dictionary* 1993.
[3] Humphreys, H., and Irving, W. (2004) *Problem-orientated clinical microbiology and infection*, 2nd Edition. Churchill Livingstone, London.
[4] www.netdoctor.co.uk

★ Learning Points

- **Virological:** Yellow fever is a flavivirus transmitted by mosquito bite. Patients present with temperature, headache, aches and pains, vomiting, and jaundice. If suspected, exclude malaria and contact the infectious diseases team at your hospital. Treatment is supportive. Yellow fever is prevented by vaccination.

- **Clinical:** In a non-specific presentation of temperature, headache, and aches and pains, consider asking about immunizations, recent travel, and, if relevant, malaria prophylaxis.

- **Personal:** When you come across a patient with an illness that is new to you it can trigger 'just in time' learning. This is a good way of remembering information. New facts are linked to the patient in your mind and often put into practice straight away. This is as opposed to 'just in case' learning—for example, lectures, courses, and textbooks, which are, of course, just as essential.

Viral lifecycle: yellow fever

Following a mosquito bite the onset of symptoms is rapid. Viraemia is coincident with the acute non-specific symptoms. In this biphasic disease a remission may precede a more ominous clinical picture, including haemorrhages, or remission may lead to complete recovery. In others a rise in serum transaminases is paralleled by prolonged clotting, delirium, stupor, and coma. Deaths are not uncommon at days 12–15 of illness.

12.3.2 Dengue

It is undoubtedly the most important mosquito-borne viral disease in the world today in terms of mortality and morbidity and is widespread geographically in South-east Asia, Indonesia, sub-Saharan Africa, and Central and South America (Fig. 12.7b). The common mild form, dengue fever, is an old disease from over 200 years ago. The disease is maintained in urban areas by cycling between humans and female *Aedes aegypti* mosquitoes, but in rural areas monkeys may form a reservoir of potential importance. The severe form, dengue haemorrhagic fever (DHF), occurred after the Second World War causing epidemics involving millions of people.

The clinical features are usually similar to those of the fever/rash syndrome with severe joint and muscle pains, lymphadenopathy, and altered perception of taste.

Medical case story Dengue

Barney had experienced the time of his life in Darwin, Australia where he had spent 8 weeks on a medical elective. The shadowing placement had been intense and, although he had come half way around the world, his supervisor was keen that he spent a lot of time on the wards and less time on the beach. He was here to learn about tropical diseases after all, and where better than a hotspot for melioidosis! Darwin also boasted a very small number of cases of Ross River virus, Murray River encephalitis, and dengue fever but usually only during the rainy season.

On the way back home, Barney had booked to stop over in Bali for a week to have a holiday with his friend Andrew, who had also been on a medical elective placement. In Bali they did all the usual tourist things like visiting temples, snorkelling, tours to more remote villages and rice paddy fields. At night they stayed in various standards of accommodation from an air conditioned hotel with an infinity pool to a poorly resourced hostel with no mosquito nets. Every evening they ate out, often enjoying street foods that smelled so enticing and tasted so good. Whilst they were there they drank bottled water, although they washed their hands and brushed their teeth in water from a tap. Andrew was ill for one day with diarrhoea and vomiting but Barney felt fine most of the time. Neither of them had thought about travel vaccinations or malaria prophylaxis for Bali because they persuaded themselves that they were only just passing through.

Andrew and Barney were on the same flight home. They had been out partying on their last night in Bali and were feeling terrible when they made it to the airport. Barney was feeling especially unwell and had a really sore headache. The flight home was long and arduous with an 8 hour stopover in Dubai, so they tried to sleep most of the way. On arrival Andrew had recovered from his hangover but Barney was feeling worse; he was starting to get hot and feverish with occasional cold chills, and the aching had spread to his whole body. Andrew accompanied him home because he was worried about his friend, and as they entered the hallway and took the bags off their backs, Andrew noticed that Barney had a rash on his neck. It was quite faint and red, and on closer inspection was more widespread than previously thought. Reassuringly, it blanched with pressure.

Andrew decided to play doctor; after all he had nearly graduated and felt he had learned a thing or two on elective.

'Do you have any neck stiffness Barney?'

'No, I don't have meningitis if that's what you're thinking?'

'I'm just worried about you. I think you should get an appointment with the GP.'

'I'm fine, probably just jetlag. I want to go to bed now Andrew. Thanks for helping me back home.'

Barney rolled into his cold bed, unwashed, unfed, his baggage emptied on the floor. Later on he was woken by a text message from Andrew: 'You alright mate? Just thinking, could it be malaria or something? We got bitten loads out there. Give us a shout if you need anything.'

He barely registered the message, he wasn't in the mood for thinking. Just sleeping and piling up the duvet because he was so cold.

The next morning when he woke up he felt worse. He had soaked his sheets in the night with sweat and he was feeling really light headed, sick and sore all over. He phoned Andrew who decided to take him to the nearest hospital where luckily he knew there were infectious disease specialists.

The Infectious Disease Registrar who reviewed him in the Emergency department knew all the right questions to ask, not only where Barney had been in Australia and Bali but also about his stopover in Dubai. She also asked about which vaccinations and malaria prophylaxis he had been given and looked askance when told he hadn't bothered. She even asked about any sexual encounters he had made, or tattoos he had had done.

The doctor made a quick note of his observations; temperature of 39.5 °C, heart rate of 105, blood pressure normal at 122/60, and respiratory rate up a bit at 20 with normal saturations on air. On examination Barney looked unwell and dehydrated, his rash was similar to previous but more widespread over his trunk, his chest was clear, and his abdomen was soft with no palpable liver or spleen.

She took some blood tests including a blood cultures, a malaria screen which would need to be repeated over the next two days, and a couple of bottles for viral serology—one to test for arboviruses like dengue, and the other to store for any bright ideas that might arise over the next few days.

Barney stayed in hospital for a week on a drip and regular paracetamol. The doctors in the unit quickly excluded malaria on repeat blood films and the presumption was that he had dengue fever based on where he had been travelling, the symptoms he had, and the incubation period. They did regular blood tests to check his full blood count, coagulation screen, liver, and kidneys in case numbers started to change. This could indicate dengue shock syndrome with severe bleeding, organ failure, and plasma leakage. Everyone knew this would be extremely rare especially given this was the first time that Barney had caught dengue, but they were a little excited to be dealing with a real tropical infectious disease instead of the usual diarrhoea and cellulitis.

After Barney was discharged he was followed up in the clinic. The results were back and he did indeed have dengue fever. He was cautioned about visiting countries where dengue is endemic because if he were to catch dengue again he was more likely to develop the haemorrhagic shock form. He was also given advice about protecting himself against mosquito bites, the vector of dengue but also other infectious diseases common amongst travellers.

★ **Learning Points**

- Dengue is an RNA virus from the arbovirus family, and is commonly spread by the female *Aedes* mosquito, but can also be transmitted via blood transfusion.

- It is endemic in tropical countries and is often underreported as it can be mistaken for malaria. Numbers are growing due to rapid urbanization and overcrowding and it is thought to infect over 390 million people per year.

- With global warming, the mosquito and hence the virus it carries is widening her territory and cases have been found in Europe, China, and the USA.

- The virus has an incubation period of 4–10 days and the illness usually lasts for 2–7 days. Typical symptoms include a high fever, aches and pains, a rash, diarrhoea, and vomiting.

- Treatment consists of symptomatic relief in the form of intravenous fluids and paracetamol. NSAIDs and ibuprofen should be avoided.

- More severe complications of dengue fever usually only affect those who have had dengue in the past. This includes plasma leakage with pleural effusions, ascites, and shock, organ dysfunction, and coagulopathy.

- There is no treatment for dengue fever although progress is being made to develop a vaccine which is in the early stage of clinical development. It is very important to prevent mosquito bites using nets and insect repellent, but also to control mosquito populations by locating and draining breeding grounds.

Children may suffer from **dengue haemorrhagic shock syndrome** (**DHSS**). This is a dangerous complication with a mortality of 4–12%, and it is worrying that it appears to be on the increase, notably in South-east Asia. It affects 50 million people annually. In Malaysia, for example, over 6000 cases per annum are not unusual in a busy year. A brief febrile illness is followed by collapse with shock, low blood pressure, and haemorrhagic signs.

There are four serotypes of dengue virus (numbered 1–4), and it is thought that DHSS may be the result of **immune enhancement** due to a second infection with a heterologous serotype: the virus forms complexes with pre-existing non-neutralizing antibody to the first virus and then, via Fc receptors, gains access to monocytes, in which it multiplies very easily and becomes widely disseminated. The symptoms arise from hypocalcaemia shock associated with haemoconcentration. Thus a cytokine storm response can be associated with DHF and virus-infected macrophages are thought to contribute to this response via activation of caspase-1, an important innate immunity gene, which in turn activates pro-inflammatory cytokines.

Treatment is quite effective; it involves replacement of fluid loss, correction of electrolyte balance, and transfusion of whole blood if haemorrhage is severe. There are experimental live vaccines but to date they have not succeeded in giving protection against all four serotypes.

12.3.3 West Nile

This flavivirus, like others, has a wide geographical distribution, causing outbreaks in Africa, Asia, and in Europe (Greece, Italy, France, and Spain). There are five virus lineages which are geographically distinct. Birds and culicine mosquitoes are the main alternate hosts. The virus is endemic in Africa and the Middle East and probably was imported into the USA via a sick bird. It is transmitted from birds to humans by *Culex pipiens*, which is a hybrid mosquito feeding off birds and mammals. From a small beginning in two human cases in New York in 1999, the virus spread rapidly carried by birds and by 2002 there were over 4000 cases with 284 deaths in 44 states in the USA, and in 2014 over 17 000 cases of severe neurological disease (Fig. 12.7c).

The disease is characterized by fever after an incubation period of 1–6 days. Often the patient has a fever, headache, backache, and generalized myalgia for a week. Nearly half of the patients have a rosealar rash involving the chest, back, and arms, while many people also suffer with pharyngitis, nausea, and diarrhoea. There are many subclinical infections, perhaps as many as 250 per clinical case. The case fatality rate is about 1–3% and the elderly are particularly at risk.

12.3.4 Louping ill

This virus is of some interest because it is the only arbovirus infection seen in the UK. It is a tick-borne flavivirus causing serious infection of the CNS of sheep and cattle in upland areas of northern England, Devon, Wales, Ireland, and Scotland, and has been known for over 100 years. Distribution is closely associated with the presence of the vector, *Ixodes ricinus*, the hard tick. There are four genetic lineages. The occasional infection of a human, usually a farmer, characteristically presents as a biphasic febrile illness with meningitis; complete recovery is the rule. There have been 44 cases published in the UK since the first patient was recorded in 1934.

12.3.5 Tick-borne encephalitis (TBE)

There are more than 13 000 clinical cases of TBE each year in the EU. The transmission cycle involves small mammals as a reservoir and persons are infected via ticks, particularly in the summer periods (Fig. 12.8a). The infection produces a variety of symptoms from common febrile illness to severe meningio-encephalitis. The pathology in the brain is characterized by large inflammatory events with cytokines, IFN-γ, TNF-χ, IL-6, and chemokines, and infection of astrocytes. There is a well-established vaccine.

12.4 Epidemiology of flaviviruses

This section should really be entitled 'epizootology' because these infections, as you will have read in the previous sections, occur primarily in animals and are, thus, examples of vector-mediated zoonoses. Replication of the viruses in these hosts causes **viraemia**, thus permitting their spread to other hosts by the bites of blood-sucking arthropods, which are referred to as **vectors** of infection.

The large variety of vectors, viruses, and mammalian, avian, and reptilian hosts makes for complex and variable epidemiological situations. Here we can give only a generalized picture of the factors that affect transmission.

The ability of the vector to spread infection is determined by its feeding preferences, range of mobility, and whether the virus concerned can be transmitted to the next generation, thus enabling it to 'overwinter' between one breeding season and the next. Unfortunately, arboviruses do not kill their insect vectors, thus ensuring their own survival.

The **reservoir** is usually a wild bird or small mammal, which acts as an **amplifier of infection**. Sometimes, for example in equine encephalomyelitis, mosquitoes transmit infection from wild to domestic animals, in this instance horses, causing a local epizootic that presages an epidemic in a nearby human population. Another example is seen in Japan, where epidemics of Japanese B encephalitis have been predicted by testing locally kept pigs for recent seroconversion.

Humans may also acquire arboviruses by entering an area harbouring infected arthropods, either for work or recreation. Their susceptibility will depend on their state of immunity: by contrast with new arrivals, locals who have had subclinical infections in the past are protected.

Given this mode of transmission, it is apparent that direct person-to-person infection cannot take place unless there is actual transfer of blood, e.g. by a 'needlestick' injury to a medical attendant during the viraemic stage of an illness. In Crimean-Congo haemorrhagic fever there have been several deaths among theatre staff that mistakenly operated on such patients.

Many arboviruses do not cause significant illness in humans. Some affect both humans and animals, others animals only, and others again are known only because they have been isolated from arthropods.

12.5 Pathogenesis of flaviviruses

After the virus is introduced into a subcutaneous capillary from the saliva of an infected arthropod there is an incubation period of a few days during which it replicates in the lymphatic system and endothelium; the first signs of illness are usually malaise and fever caused by the subsequent viraemia. Characteristically, signs of infection of the target organs follow 4–10 days later, resulting in a **biphasic illness**.

Immunity to re-infection is mediated by the antibody response, which may also confer some protection against related viruses, but its role in recovery is not clear.

12.6 Introduction to hepatitis C

This clinically important virus is the leading cause of liver disease globally and is estimated to infect more than 180 million people. As many as three million people are infected each year. Yearly deaths approximate to 500 000. Chronic HCV leads to chronic liver disease with liver fibrosis and cirrhosis, and can lead to hepatocarcinoma (HCC).

The virus emerged from Africa around 180 years ago and initially spread via the slave trade (Fig. 12.6). The natural reservoir may be rodents or bats. A triumph of molecular virology in 1989 allowed the identification of the virus by direct cDNA cloning of the viral genome, but a paucity of cell-culture systems still impedes research.

12.7 Clinical and pathological aspects of hepatitis C virus infections

12.7.1 Variants of hepatitis C virus

There are eleven major genotypes or **clades** (designated 1–11) with many subtypes (designated a, b, c, etc. within a genotype), resulting in over 100 strains of HCV. In terms of nucleotide identity, the genotypes may differ by 30% whilst 15% nucleotide variation distinguishes the subtypes. The major genotypes are related to the various risk groups, response to antiviral therapy, and geographical areas of prevalence. For example, the response to treatment is worse in the widely distributed genotype 1b infection than in other genotypes.

Unfortunately, there appears to be no antigenic cross-protection between the various genotypes, which mitigates against the development of a simple monovalent vaccine.

12.7.2 Clinical features of hepatitis C

Mode of onset

The **incubation period** is about **8 weeks** (Fig. 12.9). Only 10–20% or so of those infected have symptoms, e.g. anorexia and nausea, which resemble those caused by other hepatitis viruses; frank jaundice is uncommon. Therefore, the term acute hepatitis C is misleading. When jaundice does occur symptoms

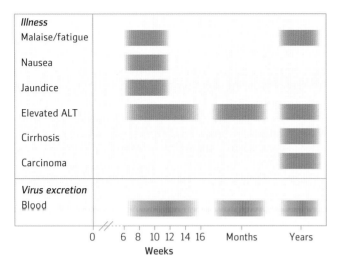

Fig. 12.9 Clinical and virological events during infection with hepatitis C.

and biochemical changes are identical to other forms of hepatitis. Alanine aminotransferase levels begin to increase shortly before symptoms and a 10-fold elevation can be detected.

Persistent infection with hepatitis C

Hepatitis C becomes chronic in about 80% of those infected. After many years, sometimes as long as four decades, **cirrhosis of the liver** may supervene in 10–20% of patients, but death from this cause alone is rare. Even so, the development of cirrhosis is sinister because it is often a precursor of HCC, which develops in 1–5% of those with chronic infection and, of course, has a very poor prognosis.

In a minority of HCV infections, liver disease is accompanied by glomerulonephritis and various forms of vasculitis, of which some at least are caused by deposition of immune complexes. HCV is also involved in the pathogenesis of type 2 cryoglobulinaemia, which often gives rise to a purpuric rash.

Medical case story Hepatitis C

The government has sent every doctor in the country a leaflet on hepatitis C. You gather that the main drive of the leaflet is to increase hepatitis C testing. This is because hepatitis C can now be treated with some degree of success and that the classical treatments can be superseded with a new collection of directly acting antivirals (DAA).

But who would you test? Apparently, not just your known risk groups of intravenous drug users and prostitutes. You need to consider screening other people, who may have had just a single tattoo or a one-off piercing, perhaps a very long time ago. You particularly need to screen anyone who has ever injected drugs, even once. You are looking to identify that unlucky group of people who have never cleared their infection, and who therefore still have the hepatitis C virus circulating in their blood stream. This chronic infection means they stay persistently infectious to others. It also means they are at risk of chronic hepatitis, cirrhosis, and even hepatocellular carcinoma.

The classic treatment still used by many countries is pegylated interferon and ribavirin, and more recently a number of DAA drugs.

That afternoon, 38-year-old Laura Carny comes in. She has hepatitis C. She is 5 months into a course of pegylated interferon and ribavirin, and she has come to ask for your advice about her enlarged groin lymph nodes. Are the lymph nodes due to her treatment or to something else, she asks?

Now you wish you had read a bit further in the leaflet sent to every doctor in the UK, because you don't know one thing about pegylated interferon and ribavirin and the new anti-hepatitis C drugs.

However, Laura is, in fact, an expert on these treatments. She was taught about them in detail by the hepatitis nurse at the tertiary liver centre at the time her treatment was initiated. She kindly talks you through the facts.

She says ribavirin is the antiviral drug appropriate for her genotype 1a hepatitis C. She takes it twice a day, and will do so for 12 months. The side effects are depression and anaemia, and it can also have teratogenic effects on unborn babies. Contraception is of paramount importance with this drug.

The pegylated alpha interferon 2b is administered as one injection a week. Laura has been trained to self-administer this and knows all about storage and disposing of sharps. This is particularly important for someone with a known blood-borne virus.

The side effects of interferon are primarily flu-like symptoms, although these have improved now that the interferon has been attached to PolyEthylene Glycol (that is, has been 'PEGylated'), which increases its half-life (the amount of time the molecule stays active in the body). Other side effects include nausea, indigestion, hair loss, headaches, insomnia, and weight loss, to name but a few.

You find the letters which the nurse has sent you in the notes, confirming these facts. Each letter ritualistically starts, 'Upon review in clinic today Laura reported experiencing the following symptoms: muscular aches and pains and constipation', or 'tiredness and abdominal pain and difficulty sleeping,' and so on.

You check the British National Formulary (BNF), but there is no mention of enlarged lymph nodes as a side effect.

On examination, she does have enlarged groin nodes and you take a fuller history, and bring her back for swabs and a smear. These turn out to be normal and her symptoms settle.

The next time she sees you 3 months have passed. She describes uncontrollable anger and depression, and wanting to scream at her 12-year-old son who keeps running off. You didn't know she had a son. You restart her sertraline antidepressant and chase up the family counselling. She then drops from view. You forget about her.

Case notes

Eight years ago

November	Alcohol problem since age 15. Has 8-year-old son. Heroin until aged 21 then gave up. Treatment; refer cognitive therapy.
December	Abstaining from alcohol well. Liver function tests done.
Late December	Depression. Attending Alcoholics Anonymous.

Seven years ago

January	Virology sent as liver function tests abnormal. Right upper quadrant pain.
Late January	Hepatitis C positive. Worried that may have had this before gave birth to son. Very worried. Liver function tests now normal.
February	Son 9 years old has hepatitis C. Refer tertiary liver centre.
March	Housing letter.
April	Appointment liver centre. Letter from liver centre, liver biopsy; mild disease. Monitoring. No treatment advised.
Late April	Housing letter.
May	Depressed.
August	Vaccinated hepatitis A and B.
October	Letter from liver centre, still monitoring.

Six years ago

January	Discussed contraception. Advised coil.

Three years later you have a day of study leave and you are doing a presentation on hepatitis C. You remember Laura Carny, and you bring up her notes and review her history.

You now review her collection of hospital outpatient letters (see Case Notes table). The liver unit monitored her 6-monthly, from that April 7 years ago. This continued for 2 years. Five years ago you find a letter where she appears to push for treatment insisting that she can't manage as a single parent with her fatigue and right upper quadrant pain. The letter advises; 'Although the response rate to combination therapy is around 40–50% for those with more advanced disease, current evidence indicates a lower response rate in people with mild disease on biopsy ... we will check the viral response at 3 months and only continue therapy for the full 12 months if she is successfully controlling the virus at this interim analysis point.'

Three months into treatment you read; 'Viral load check today. The minimum drop required is 99% or a 2 log drop'. The next letter reveals that Laura has achieved this viral response. You are then interested to find a letter that mentions your own consultation of 3 years before '... has experienced mood swings. Laura's GP has restarted her sertraline'.

Laura completes her course 7 months after your original consultation (3 days after Christmas, in fact). You read 'bloods reveal virus detectable. Patient very disappointed.' A letter 18 months after this describes Laura 'working at the hepatitis C trust. Declined further treatment (protease inhibitor; Telapravir).'

You then look up her son's letters to see if he has started treatment yet and find the letter, to the right of this paragraph, dated 6 months ago, transferring his care to the adult liver team.

As you sift through the story, writing brief notes for your presentation, you remember the leaflet with the pierced nose woman on the front, and contrast that bohemian image with the havoc wreaked by hepatitis C on Laura Carny's life.

Master Robert Carny, 16 years old.

Final diagnosis: chronic hepatitis C.

Robert was started on treatment 2 years ago. He was a non-responder and the treatment was stopped after 6 months.

He did have quite a lot of side-effects related to his treatment. He had lots of behavioural and psychological difficulties for which he has been seen by child psychiatry.

I have enclosed a copy of Robert's last clinic letter, but if you need further information, please do not hesitate to contact me.

Yours sincerely ...

★ Learning Points

- **Virological:** Hepatitis C virus is spread mainly by blood and blood products and less efficiently by sexual contact. Patients can present with pyrexia, malaise, nausea, and jaundice, but in fact 80% of seroconversions are subclinical. Chronic carriage occurs in 80% of those infected and can lead to chronic hepatitis, cirrhosis, and hepatocellular carcinoma. The classic treatment is with pegylated interferon and ribavirin but now we have a number of newly discovered directly acting antivirals (DAA) called telaprevir, boceprevir, asunaprevir, daclatasvir, rimeprevir, and sofosbuvir. There is no vaccine at present.

- **Clinical:** Have a low threshold for hepatitis C testing on any patient with a history of tattoos, piercing, and IV drug use, because treatment will clear the infection in 40–50% of carriers.

- **Personal:** It is natural for patients to become experts on their own illness and you will find that this will alter the usual dynamics of a consultation. Some patients also volunteer to help others with the same infection, as shown here and by the many patient self-help groups (www.hepctrust.org.uk and www.britishlivertrust.org.uk).

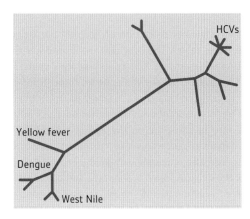

Fig. 12.10 Phylogenetic tree of hepatitis C within the flavivirus family.

12.8 Epidemiology of hepatitis C

The groups at risk are broadly similar to those listed for hepatitis B, but their relative proportions are different. Most cases of hepatitis C **are intravenous drug abusers** but many may fall into a now middle-aged grouping who only briefly experimented with drugs as long as 40 years ago. Transmission via inadequately sterilized needles and body-piercing, tattooing, and circumcision have also been implicated.

Sexual transmission and congenital infections are less important for transmission; and now that blood can be screened for HCV, infections from this source and from blood products are much less frequent than they used to be. Organ transplants have also transmitted HCV infections.

12.8.1 The geographical distribution

The geographical distribution, as measured by serological surveys, is lowest in northern and western Europe (<1%), the USA, and Australia, and highest in Japan and the Middle East (4%), reaching an extraordinary 20% in Egypt. The high incidence in Egypt is attributed to public health programmes using inadequately sterilized syringes in the 1960s. In contrast the high incidence in certain immigrant groups, like Bangladeshis in the UK, is not yet understood. There is a pronounced variation in the prevalence of the seven various genotypes and 70 subtypes in different regions. Genotypes 1–3 are most widely distributed in the world. Genotype I is mainly detected in the USA and Europe, types II and III in the USA. Type IV is more common in North and Central Africa.

12.8.2 Pathogenesis

In most instances there is a slowly progressive asymptomatic hepatitis, with persistent viraemia. Those with chronic active disease are liable to proceed to cirrhosis, but this process is slow, taking as long as 20 years. Periodic exacerbations are marked by rises in alanine aminotransferase values. Histologically, there is little to distinguish HCV infections from other forms of viral hepatitis, except for the presence of lymphoid follicles within the portal tracts. There is also intense periportal infiltration with lymphocytes, and damage to the lining of the bile ducts. The 'ground glass' appearance of liver cells infected with HBV, due to accumulation of surface antigen, is of course absent.

HCC seems to be a direct consequence of the cirrhosis, rather than integration of nucleic acid sequences into the host-cell genome, as is the case with hepatitis B.

Research points to a direct effect of HCV polyproteins in causing hepatocyte apoptosis. There is up-regulation of pro-inflammatory, pro-fibrotic, and pro-apoptosis genes in cirrhosis patients. Obviously there is a paradox here, because apoptosis of virus-infected hepatocytes is an important part of the host defence to prevent spread and to eliminate infected hepatocytes.

Immune response

The cell-mediated response is more prominent than humoral immunity, but active proliferation of T-helper and cytotoxic lymphocytes seems insufficient either to clear the infection or to prevent re-infection. HCV can mutate *in vivo*, thus escaping immune surveillance. Indeed, the natural immune response to HCV seems unusually inefficient, which accounts for its long-term persistence.

Laboratory diagnosis

A number of commercial ELISA and immunoblot tests have been licensed for detecting antibodies to viral proteins.

Tests for genome by quantitative **RT PCR** provide valuable confirmatory evidence of infection and allow sensitivity to as few as 100 RNA genome copies/ml plasma.

12.8.3 Treatment of hepatitis C with antivirals

The clinical targets of antiviral therapy are to reduce cirrhosis of the liver and cancer of the liver whilst reduction of viraemia will reduce virus transmission from person to person. An important goal is reached when patients no longer have viraemia (undetectable HCV RNA by RT PCR) six months after treatment is stopped. With the interferon-based therapies only 40–50% of patients infected with genotype 1 achieve this SVR (sustained virological response), whereas 80% of those infected with the genotype 2 and 3 achieve this SVR.

Treatment for acute hepatitis C with interferon, the classic treatment, is best initiated early; a commonly used dosage is 6 mU of IFN-α three times weekly for 16–24 weeks alongside the nucleoside analogue ribavirin. Pegylated IFN-αs have longer half-lives and slower clearance.

The DAA, such as telaprevir and boceprevir, are inhibitors of hepatitis C NS3/4A protease and show beneficial effects in patients where the combination of pegylated interferon and ribavirin has failed. There is considerable variation amongst the virus genotypes as regards sensitivity to these antivirals. Although telaprevir inhibits genotype types 2 and 4 proteases there is still little effect against genotype 3.

Telaprevir and boceprevir are peptidomimetic inhibitors of NS3/4A protease; they bind tightly to the enzyme catalytic site and compete with the natural substrate. Thereby the cleavage of the viral polyprotein, and hence new virion production, is blocked. Drug-resistant mutations arise near the protease catalytic site and such mutants are cross-resistant to boceprevir and telaprevir.

The most recent scientific and clinical advances have come from drug combinations of asunaprevir (a protease inhibitor) and daclatasvir (NS5A polymerase inhibitor). Of particular note is that patients who are traditional drug 'non-responders', including those with cirrhosis, particular ethnic groups, persons of IL28B genotype, and those with a high body mass index, respond to these new 'second generation' anti HCV drugs. However, patients (5–15%) with virus with pre-existing mutations in the virus NS5 and NS5A proteins have a high chance of treatment failure.

Combinations of simeprevir (a protease inhibitor) and sofosbuvir (an NS5B nucleotide inhibitor) are under clinical investigation.

Finally, and still somewhat in the experimental stage, only six weeks of therapy with a triple drug combination of ledipasvir, sofosbuvir, and GS9451 (an NS3 protease inhibitor) appears to lead to SVR in non-cirrhotic patients infected with genotype 1 virus.

On a less positive note it should be added that therapy uptake in Europe in drug injectors is low and the cost of these new anti-HCV compounds is too high for the health budgets of many countries.

Liver transplantation has proved of short- or medium-term benefit in some cases of cirrhosis or HCC, but re-infection of the graft probably always occurs.

12.8.4 The use of passive antibody and vaccines

The development of a vaccine is hindered by the number of virus genotypes and by the inability to propagate the viruses readily *in vitro*; although some encouraging vaccine results have been obtained recently in the laboratory. It is clear that experimental recombinant forms of E1/E2 glycoproteins induce strong humoral and cellular responses in chimpanzees.

Passive immunotherapy of patients using polyclonal antibodies reduces HCV RNA levels, at least in the short term.

12.8.5 The GBV A–C flaviviruses now renamed as pegiviruses

The alphabetic designations of the various hepatitis viruses (A–C) had been reasonably simple until a confusing situation arose when yet another virus was isolated from the blood of a surgeon supposedly with acute hepatitis. It is perhaps unfortunate that it was termed GB, the patient's initials, and doubly unfortunate that the virus did not itself cause hepatitis. Soon afterwards, similar agents were detected in New World monkeys and initially provided with further alphabetic designations. But this situation has been clarified by renaming the virus as pegivirus.

The roles of all these agents in causing clinical disease has yet to be elucidated. The main thing to remember is that sequence analysis of their genomes shows them to be closely related to known **flaviviruses**. Around 750 million persons have HPgV viraemia and the virus is transmitted by blood, sex, and birth. Strangely, the virus may have some benefits since the life of HIV infected persons may be prolonged if they are co-infected.

 Reminders

- There are 10 genes in a single-stranded positive-sense RNA around 11 kb in size, capped at the 5′ end but having no poly(A) tail at the 3′ end.

- The virion is a spherical enveloped particle with an internal nucleoprotein of icosahedral structure. There is a capsid protein C, two envelope proteins (E and M), and seven non-structural proteins.

- Replication takes place in the cell cytoplasm on internal cellular membranes.

- Most flaviviruses are transmitted via ticks or mosquitoes.

- Mammals and birds are unusually the main reservoir but humans can take on this role.

- Members of this family cause very widespread and often serious disease in S. America, Africa, South-east Asia, and the USA.

- Yellow fever is well controlled by live attenuated vaccine; as are TBE and Japanese equine encephalitis, but using killed virions.

- Zika, an African virus, has achieved notoriety in S. America causing microcephaly in newborns.

- Hepatitis C, a blood-borne flavivirus, is mainly prevalent in **intravenous drug abusers**. There are seven major genotypes of hepatitis C and as many as 70 subtypes. The virus has a positive-sense ssRNA of 9.6kb in size. Replication is cytoplasmic and a single virus poly protein is cleaved to give structural and non-structural proteins.

- HCV infections tend to become **chronic**, with eventual cirrhosis and sometimes hepatocarcinoma. Chronicity may be due in part to the poor immune response and possibly to mutations in the viral genome.

- Directly acting antivirals (DAA) targeting the HCV virus protease such as telaprevir, boceprevir, and asunaprevir, as well as inhibitors of virus RNA-dependent RNA polymerase (daclatasvir), are now used as an alternative to these existing treatments but are expensive and still fail to inhibit all virus genotypes.

Further reading

Callaway, E. (2014). Hepatitis C drugs not reaching the poor. *Nature* **508**, 295–6.

Chivero, E.T. and Stapleton, J.J. (2015). Tropism of human pegivirus and host immodulation: insights into a highly successful viral infection. *Journal of General Virology* **96**, 1521–32.

Gane, E. (2014). Hepatitis C beware—the end is nigh. *Lancet* **384**, 1557–60.

Li, C. *et al*. (2014). Origin of hepatitis C virus genotype 3 in Africa estimated through an evolutionary analysis of the full length genomes of nine subtypes including the newly sequenced 3d and 3e. *Journal of General Virology* **95**, 1677–88.

Pawlotsky, J.M. (2011). Treatment failure and resistance with direct acting antiviral drugs against Hepatitis C virus. *Hepatology* **53**, 1742–51.

Pettersson, J.H. and Fiz-Palacios, O. (2014). Dating the origin of the genus *Flavivirus* in the light of Beringian biogeography. *Journal of General Virology* **95**, 1969–82.

Pybus, O.G. and Gray, R.R. (2013). The virus whose family expanded. *Nature* **498**, 310–11.

Roby, J.A. *et al*. (2015). Post translational regulation and modifications of flavivirus structural proteins. *Journal of General Virology* **96**, 1551–69.

Tan, T.Y, and Hann Chu, J.J. (2013). Dengue virus infected human monocytes trigger late activation of caspase-1 which mediates pro inflammatory IL-1B secretion and pyroptosis. *Journal of General Virology* **94**, 2215–20.

 ## Questions

1. How are flaviviruses like dengue and Yellow fever maintained in nature?

2. Compare West Nile virus and TBE as regards disease, vector, and geography.

Coronaviruses (including SARS CoV and MERS CoV)

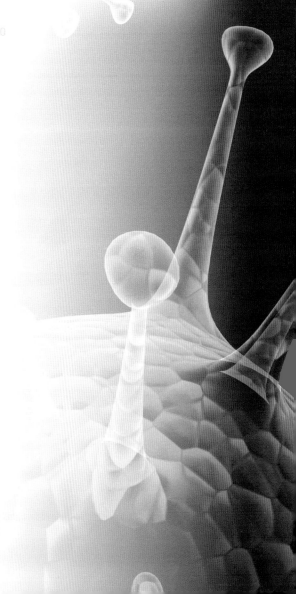

13.1 Introduction

Coronaviruses (Latin: *corona* = a crown; see Section 13.2.2) infect humans, birds, and other animals. The first two human strains identified infect the respiratory tract and are normally confined to the ciliary epithelium of the trachea, nasal mucosa, and alveolar cells of the lungs. A typical symptom is that of the 'common cold'. Some of the first isolates were obtained in the UK from volunteers at the Medical Research Council (MRC) Common Cold Research Unit, Salisbury. More recently, a third and fourth respiratory coronavirus were discovered (Fig. 13.1). Two other coronaviruses are the cause of Severe Acute Respiratory Syndrome (SARS) and Middle East Respiratory Syndrome (MERS). Coronaviruses have the largest known genomes of any RNA virus and a helical nucleocapsid more commonly possessed by negative-strand RNA viruses.

13.2 Properties of the viruses

13.2.1 Classification

The family *Coronaviridae* and sub family *Coronavirinae* belong to an order of viruses named *Nidovirales* and contain three genera, *Alpha-*, *Beta-*, and *Deltacoronavirus*. The first human coronaviruses to be isolated, 229E and OC43E, belong to group I alphacoronaviruses whilst the more recently isolated HKU1, and SARS CoV and MERS CoV, fall into the *Betacoronavirus* genus. The coronaviruses are classified as Baltimore Class IV (Fig. 13.2). Five serotypes of human coronavirus are known and around 15 more infect birds and other animals.

In 2002 and 2012 new coronaviruses were identified as the cause of SARS and MERS respectively. They emerged in South-east Asia (see Section 13.4) and the Middle East. The

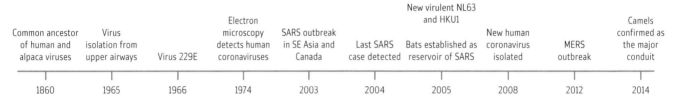

Common ancestor of human and alpaca viruses	Virus isolation from upper airways	Virus 229E	Electron microscopy detects human coronaviruses	SARS outbreak in SE Asia and Canada	Last SARS case detected	Bats established as reservoir of SARS	New virulent NL63 and HKU1 New human coronavirus isolated	MERS outbreak	Camels confirmed as the major conduit
1860	1965	1966	1974	2003	2004	2005	2008	2012	2014

Fig. 13.1 Timeline for coronaviruses.

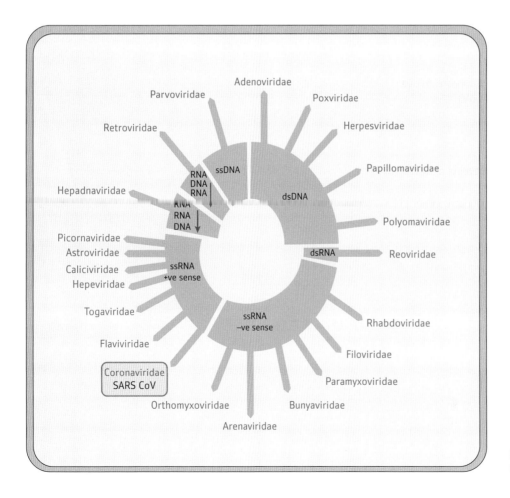

Fig. 13.2 Baltimore classification scheme and coronaviruses.

majority of new coronaviruses isolated are from bats, most of whom show no signs of illness; however, intermediate, possibly amplifier, animal species exist that are the more direct route of transmission to humans, for example the camel for MERS CoV.

13.2.2 Morphology

Coronaviruses are pleomorphic, ranging from 100 to 160 nm in diameter, and have club-shaped glycoproteins surface spikes about 20 nm in length (Fig. 13.3). These very large (200 kDa), heavily glycosylated spikes (S) give the virions the appearance of a crown, hence their name.

The RNA genome associates, unusually for a positive-strand RNA virus, with the virus N protein to form a long helical nucleocapsid known as the ribonucleoprotein (RNP). This is packaged within an enveloped virus particle. We have portrayed an internal RNP structure but some members of the family have a more stable structure resembling an icosahedron. In thin sections the envelope appears as inner and outer shells separated by a translucent space. Coronaviruses contain three major envelope proteins. The first, the matrix (M) protein, is a transmembrane glycoprotein. The second is the protein S that constitutes the surface glycoprotein and is responsible for eliciting neutralizing antibodies, receptor binding, membrane fusion,

and haemagglutinin (HA) activity, as well as pathogenicity. Genetic variation in this protein is detected particularly in the S1 region which is critical for virus neutralization. It is now considered that recombination involving regions of the S protein between animal, bird, and human coronaviruses allow this family to infect a range of animals and birds. The third protein has esterase activities, and so may have a role in virus exit, much like the neuraminidase (NA) of influenza virus. Some members of the family have such a protein with HA activity as well as (HE).

13.2.3 Genome

The genome is the largest of all the RNA viruses of humans (Fig. 13.4). It is positive-sense ssRNA, 27–32 kb in size, is 5′-capped, 3′-polyadenylated, and is infectious. The RNA genome is organized such that virus non-structural proteins are at the 5′ end of the RNA and the structural proteins are towards the 3′ end.

13.2.4 Replication

Virions initially attach to the cell plasma membrane through specific receptors. These have been identified for several coronaviruses; for example, human coronavirus uses the membrane-bound metalloproteinase, aminopeptidase N (APN), whereas OC43 simply binds to sialic acid groups on cell-surface proteins. SARS CoV uses the host-cell receptor angiotensin-converting enzyme 2 (ACE2) to gain entry into cells whereas MERS CoV uses the host receptor dipeptidyl peptidase 4 (DPP4). Uptake into cells is rapid and temperature-dependent, involving fusion with the plasma membrane or via endocytosis followed by a spike-mediated fusion in the endosome. Large multinucleated giant cells, syncytia, can be formed both in the laboratory and in an infected host.

Once released into the cytoplasm the virus positive-strand RNA is translated directly into two polypeptides (Fig. 13.5): ORF1a and ORF1b at the 5′ end of the genome. These are processed to form a replicase-transcriptase complex that possesses RNA polymerase activity. The RNA polymerase transcribes a full-length negative RNA strand, which acts as the template for transcription of multiple subgenomic virus mRNAs. Coronavirus mRNAs are unusual in that they all terminate at the common 3′ end of the genome, but start at various places from the 5′ end to produce a nested set of 3′ co-terminal transcripts.

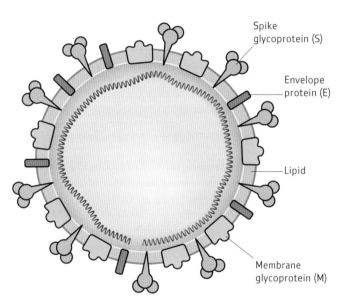

Fig. 13.3 Structural features of coronavirus.

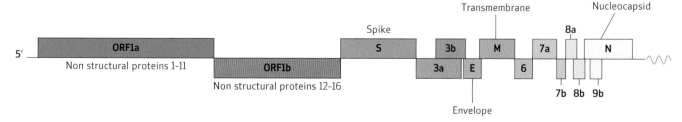

Fig. 13.4 Genome structure and encoded proteins of coronavirus (SARS). Cap (5′ end); ᐯᐯᐯ, poly(A) tail (3′ end).

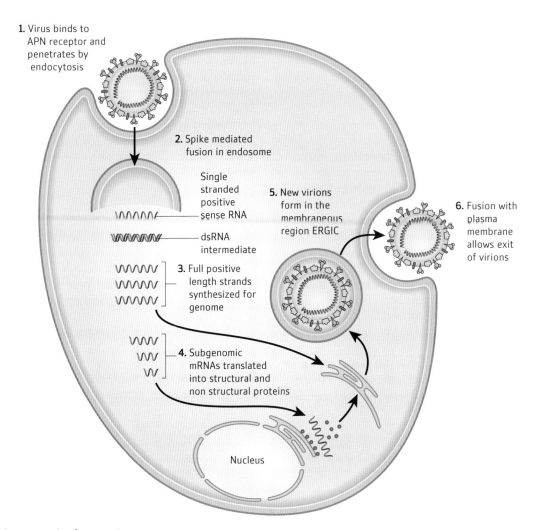

1. Virus binds to APN receptor and penetrates by endocytosis

2. Spike mediated fusion in endosome

Single stranded positive sense RNA

dsRNA intermediate

3. Full positive length strands synthesized for genome

4. Subgenomic mRNAs translated into structural and non structural proteins

5. New virions form in the membranous region ERGIC

6. Fusion with plasma membrane allows exit of virions

Nucleus

Fig. 13.5 Replication cycle of coronavirus.

Each of the eight mRNAs, except for the smallest, therefore encode for multiple proteins, with the longest one being, in effect, full-length coronavirus genome RNA and the others in descending order of size being S, E, M, and N. Generally, each subgenomic virus mRNA is the template for translation into one protein.

There are 16 non-structural proteins (1–16nsp), some of which have proteinase functions or are polymerases, including RNA-dependent RNA polymerase (nsp12) and endoribonuclease (nsp15).

Virus proteins that constitute the virus particle, namely N, M, and S, are produced in the infected cell and new virion assembly occurs initially in the cytoplasm on smooth-walled vesicles located between the ER and the Golgi known as ERGIC (endoplasmic reticulum Golgi intermediate compartment). There newly formed RNP interacts with the M protein from the ER, and M interacts with the S and other proteins to form the infectious virus which buds into the Golgi, thereby acquiring a lipid envelope. Envelope proteins are glycosylated in the Golgi. Virions are released by fusion of smooth-walled virion-containing vesicles with the plasma membrane.

As with other RNA viruses, the lack of proofreading functions in the virus RNA polymerase leads to a high rate of mutation in the new virus genomes. The very long genomes, together with the discontinuous RNA replication, can favour recombination leading to new genotypes with varying pathogenicity. There remains also the possibility of recombination between zoonotic coronaviruses and between human viruses. Recombination can allow coronaviruses to rapidly evolve and adapt to new ecological niches.

13.3 Clinical and pathological aspects of coronaviruses other than SARS and MERS

It is thought that 2–10% of common colds are caused by coronaviruses, typified by OC43 and 229E. Infection can precipitate wheezing in asthmatics and exacerbate chronic bronchitis in adults. In a typical case there is an incubation period of 3 days, followed by an unpleasant nasal discharge and malaise, lasting about a week. Patients excrete virus during this period. There is little or no fever, and coughs and sore throat are not common.

13.3.1 Pathogenesis

Replication is confined to the cells of the epithelium in the upper respiratory tract. Inflammation, oedema, and exudation occur in the tract for several days following destruction of cells by the virus.

13.3.2 Epidemiology

Coronavirus colds occur in the colder months of winter and early spring with sizeable outbreaks every 2–4 years. Antibody surveys show that most people have been infected at some time in their lives and it is thought that re-infections are quite common, either because of the **poor immune response**, or as a result of antigenic mutations, or both.

13.3.3 Laboratory diagnosis

Commercial diagnostic multiplex RT PCR kits able to detect a range of respiratory viruses are now widely used with samples from nasal swabs. Most routine clinical virology laboratories are not equipped to isolate coronaviruses, which replicate poorly in cell cultures and may even require organ cultures of human embryo trachea or nasal epithelium.

13.4 A newly emerging coronavirus: Severe Acute Respiratory Syndrome (SARS CoV)

13.4.1 Properties of the virus

The first cases of a new respiratory syndrome emerged in November 2002 in the Guangdong Province of China. By April 2003 the virus had spread to nearby Hong Kong and then worldwide to 32 countries by air travel, affecting 8460 individuals and resulting in 808 deaths. Unexpectedly, the virus was identified as a coronavirus (CoV), which emerged suddenly from a mammalian bat reservoir in China to infect humans, via the civet cat, which is used for exotic foods, or a rat upon which the civet preys (Fig. 13.6).

After a relatively long incubation patients excrete virus in the upper airways shortly before the first symptoms occur (Fig. 13.7). High temperature is characteristic. Sputum has a high virus load. Faecal–oral transmission is not uncommon and over half the patients have diarrhoea. Most patients recover.

13.4.2 Genome

Remarkably, the genome structure was established in a matter of weeks and was typical of other members of the family, with 11 ORFs coding for 23 proteins. Most of the NS proteins are encoded in the first half (5' end) of the genome (such as the proteases nsp1, nsp2, and the RNA polymerase nsp9), whereas the structural protein spike (S), membrane (M) envelope, and nucleocapsid proteins are positioned towards the 3' end (Fig. 13.4).

Unlike most other coronaviruses, the SARS virus possesses no HE protein. This genetic analysis immediately indicated that the virus was not a recombinant, with portions of human and perhaps of avian or animal coronaviruses, but a completely novel virus.

Nucleotide sequence comparison of isolates in Hong Kong, Taiwan, Vietnam, and Canada showed them to have a local genetic fingerprint, but they were all related to the original virus, indicating, fortunately, some genetic stability.

13.4.3 Replication

Genome expression commences with the translation of two large polyproteins, ppla and pplab, encoded by the viral replicase gene. These two polyproteins are processed by two viral proteases to the smaller functional components of the viral replicase complex. This latter enzyme then facilitates viral genome replication and transcription of a nested set of subgenomic

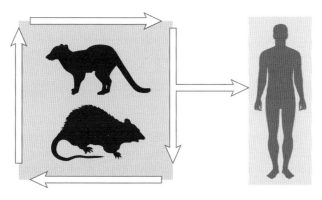

Fig. 13.6 Transfer of SARS from a civet cat or rat to infect humans.

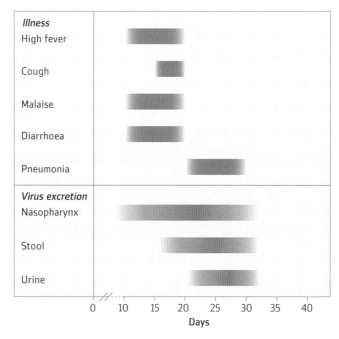

Fig. 13.7 Clinical and virological events during infection with SARS.

mRNAs, which code for the structural proteins S, E, M, and N, and also a number of so-called accessory proteins. The viral replicase is probably the most complex of enzymes in the RNA virus families. A striking observation is the unusually large number of genes encoding potential non-structural proteins interspersed with structural genes on the 3′ end.

13.4.4 Clinical and pathological events of SARS

Clinical

The clinical symptoms include high temperature (over 38 °C), dry cough, myalgia, and breathlessness. The majority of patients developed diarrhoea during the illness and most patients developed abnormal liver function tests. Asymptomatic and mild illness was uncommon, whilst children had less severe illness than adults. A lower respiratory tract infection developed in 20% of cases, needing hospitalization and intensive care. The overall fatality was 10%, reaching 50% in the over 65s or in persons with underlying illness. The incubation period could be as long as 10 days and virus excretion, at least in the throat, begins with early symptoms and peaks 7–8 days later, followed by a slow decline. The virus infects both upper respiratory tract and alveolar epithelial cells resulting in lung injury. Immunopathology may contribute to the seriousness of the disease with high levels of pro-inflammatory cytokines, such as IP10, MCP-1, TNF-α, and RANTES, expressed.

Epidemiology

Person-to-person transmission was relatively inefficient and necessitated close contact. A working hypothesis is that the preferred site of replication is the lower respiratory tract thus reducing the chance of virus in secretions from the nose and throat and hence reducing person-to-person spread by aerosol. Transfer of infection was from cough droplets, or via hand contamination, lift or door knobs, or shared drinking cups. High levels of personal hygiene, hand washing, and cleanliness of furniture and cooking utensils, together with rapid quarantine of suspect and real cases, were able to break the chain of transmission. The R_0 value for the virus (Chapter 6) was quite low, about 2, and this, together with the long incubation period during which the person was not infectious for others meant that the public health measures, such as quarantine of suspect cases and contacts, introduced to break the chains of transmission were successful.

New vaccines are being produced against SARS CoV and antiviral inhibitors developed, but the most effective prevention is good virological surveillance to detect human cases quickly, and to quarantine them to prevent person-to-person spread.

The most vulnerable sector in the community is the healthcare groupings—nurses and doctors—because an initial case is liable to arrive in Accident and Emergency as a 'community-acquired pneumonia' and to be treated with antibiotics as a bacterial infection. In retrospect the outbreak was an advance warning of both MERS and later Ebola, both crippling the hospitals and healthcare sector even in well developed countries. A

single case of SARS diagnosed at the Royal London Hospital was treated in intensive care and needed over 12 staff. A further six cases could easily have overwhelmed the hospital.

The virus is likely to persist in bats and mammals in Southeast Asia with the potential to flare up: occasionally, during these episodes, it could be carried to the rest of the world by persons incubating the disease.

13.5 A newly emerging coronavirus: Middle East Respiratory Syndrome (MERS CoV)

This new beta coronavirus was isolated in 2012 in the Middle East from a fatal case of pneumonia with renal failure. A cluster of cases was identified in hospitals in Saudi Arabia, followed by an eastwards spread to Jeddah. More recently cases, now totalling 1000, have been identified in Jordan, Qatar, Tunisia, and the UAE. Cases in the UK, France, the USA, Germany, Italy, and most recently South Korea have all had a link with the Middle East either by travel of the patients or a direct or indirect contact with a case from that region who has travelled during the incubation period. The S. Korea outbreak was hospital linked, as are half of the cases in the Middle East. In Korea a single infected visitor to Saudi Arabia caused a hospital-based outbreak where over 180 persons were infected and 40 died; 18 000 persons were placed in restrictive quarantine.

13.5.1 Properties of the virus and pathology

Studies of MERS CoV infected cells in the laboratory show pro-inflammatory cytokine induction of IL-1B, IL-6, and IL-8, but unlike SARS no induction of innate antiviral cytokines TNF-α, IFN-β and IP-10. Infection *in vitro* is also linked with attenuated innate antiviral response compared to SARS CoV. Drug combinations with IFN-α2b and ribavirin are current options for treatment.

MERS CoV genomics are classified into two clades, A and B, and both are genetically related to viruses recovered from bats and camels. The working hypothesis of human contagion is that the source of the virus was originally the bat but with the virus primarily spreading to humans from camels (Fig. 13.8). Although the camel is a conduit into humans the exact mode of transmission, for example unpasteurized milk, handling of uncooked meat, or via petting or close contact in camel farms, has not been ascertained.

13.5.2 Clinical aspects

A typical case recorded in Saudi Arabia was a 16 year old who had a fever, malaise, sore throat, cough, and wheezing, and these symptoms worsened for four days, at which point he was admitted to hospital with a temperature of 38.8 °C, watery diarrhoea, abdominal pain, myalgia, worsening cough,

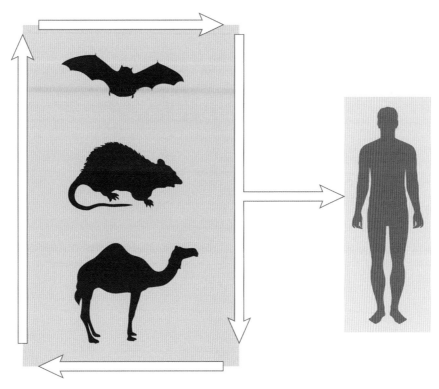

Fig. 13.8 Transfer of MERS from a bat or rat to a camel and thence to humans.

and headache. Chest X-ray showed bilateral hilar infiltrates which resolved over the next three days, when he was discharged. But a more serious illness was recorded in a 39-year-old patient who had the above symptoms upon admission but also had coalescing patchy densities at the base of the right lung. Within two days he was admitted to the Intensive Care Unit and he deteriorated and died. Approximately 40% of cases die, especially if they have underlying problems with smoking, asthma, immunotherapy, or obesity.

At present there has been documented person-to-person spread only in families and in hospitals. As with SARS CoV, the working hypothesis is of a deep-seated lung infection, thus obviating the chance of virus to spread in excretions from the upper airways and nose by cough and sneeze droplets.

 # Reminders

- **Coronaviruses** are **positive-strand RNA viruses** with a large 30 kb genome. Four classic serotypes are known. They cause **upper respiratory tract** infections (but see MERS CoV and SARS CoV). These viruses are difficult to grow in the laboratory but can spread readily and infect humans, causing up to 10% of common colds around the globe.

- Severe Acute Respiratory Syndrome Coronavirus (SARS CoV) is a relatively new and much more serious respiratory disease with high case fatality. It emerged quickly to humans in South-east Asia in 2003 from a reservoir in bats via civet cats in markets. The long incubation time of the disease enabled quarantine to be used to break the human to human transmission chain.

- A new virus, Middle East Respiratory Syndrome Coronavirus, (MERS CoV) emerged as a human disease in Saudi Arabia in 2013 and causes serious respiratory illness. Similarly to SARS CoV the virus is not easily transmitted between humans but once infected a person has a 40% chance of dying. Camels appear to be the main conduit into the human population.

- No universally effective vaccine or antiviral exists to counter the effects of these groups of viruses, although experiments are in progress with live attenuated adenovirus carrier vaccines and with specific antiviral drugs.

- These are potentially pandemic, newly emergent viruses and are taken very seriously by WHO and the global community to the extent of restricting air traffic from afflicted countries.

Further reading

Assiri, A., McGeer, A., Perl, T.M., Price, C.S., Al Rabeeah, A. *et al.* (2013). Hospital Outbreak of Middle East Respiratory Syndrome Coronavirus. *New England Journal of Medicine* **369**, 407–16.

Drosten, C., Gunther, S., Preiser, W., van der Werf, S., Brodt, H.R., and Becker, S. (2003). Identification of a novel coronavirus in patients with SARS. *New England Journal of Medicine* **348**, 1967–76.

Memish, Z.A., Zumla, A.I., Al-Hakeem, R.F., Al-Rabeeah, A.A., and Stephens, G.M. (2013). Family Cluster of Middle East Respiratory Syndrome Coronavirus Infections. *New England Journal of Medicine* **368**, 2487–94.

Questions

1. Explain the emergence of these viruses and how they cause high mortality. Could a vaccine be formulated?

2. Discuss the replication of the less virulent members of the family mentioning any unique features.

Group 2 Negative-sense single-stranded RNA viruses

Orthomyxoviruses: influenza A, B, C

14.1 Introduction

Influenza has long been with us (see Fig. 14.1); indeed, the name itself refers to the ancient belief that it was caused by a malign and supernatural influence.

In Florence during the time of the Renaissance, astrologers linked a curious juxtaposition of stars with an outbreak of infection in the city and attributed it to the 'influence' of the stars, hence, influenza. Known in the sixteenth and seventeenth century as 'the newe Acquayntance', influenza still causes major outbreaks of acute respiratory infection. It has, indeed, been described as 'the last great uncontrolled plague of mankind' and in this chapter we shall show how the property of causing epidemics and even pandemics is related directly to the ability of the causal viruses to undergo antigenic variation ('drift') or massive genetic reassortment ('shift'), and thus evade their hosts' immune defences. The

virus is a so-called emergent virus because originally, at the time of the Ice Age 10 000 years ago, it crossed the species barrier from birds to humans. It is still predominantly an avian virus spread silently around the world by migrating geese and ducks.

Meanwhile the human version has evolved. Massive problems can still occur when the avian and human viruses co-infect a human or pig and result in a brand new pandemic virus.

14.2 Properties of the viruses

14.2.1 Classification

Although the general public (and some doctors who should know better) refer to many incapacitating respiratory infections as 'flu', true influenza is caused by the small family of the *Orthomyxoviridae* (see Fig. 14.2).

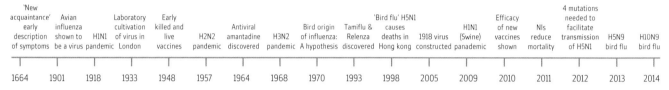

Fig. 14.1 Timeline for influenza virus.

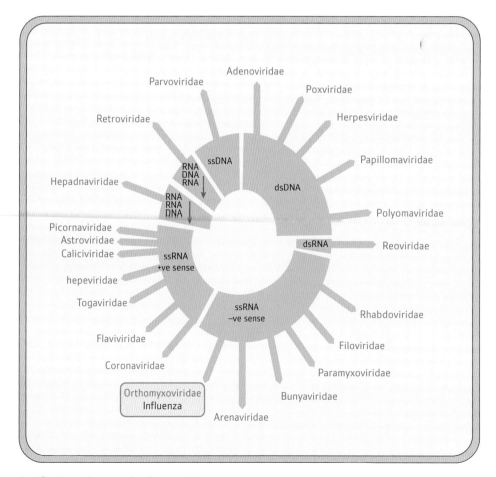

Fig. 14.2 Baltimore classification scheme and orthomyxoviruses.

Myxo derives from the Greek for mucus and refers to the ability of these viruses to attach to mucoproteins on the cell surface; *ortho* means true or regular, as in orthodox, and distinguishes these viruses from the paramyxoviruses (Chapter 9).

There are four genera, distinguished serologically on the basis of their matrix (M) and nucleoprotein (N) antigens. They are:

- influenza virus A;
- influenza virus B;
- influenza virus C;
- 'thogoto-like viruses', which do not infect humans and are not considered here.

Influenza C differs significantly from A and B, and is of much less importance in infections of humans; the descriptions that follow relate only to the A and B viruses.

Influenza A viruses have been designated on the basis of the antigenic relationships of the external spike haemagglutinin (HA) and neuraminidase (NA) proteins, H1–H18 and N1–N10. Of these, only viruses with H1, H2, H3, or H5, and N9, N1, or N2 are known to infect humans or to cause serious outbreaks. NAs are subdivided into two groups with N1, N4, N5, and N8 in group 1 and the remainder in group 2, with the exception of N10 from a bat virus which has a unique structure. HAs are similarly grouped with group I containing H1, H2, H5, H6, H8, H9, H11, H12, H13, and H16 whilst group II contains H3, H4, H7, H10, H14, and H15 subtypes. The avian viruses, H5 and H7, can infect humans, but do not appear to spread easily from person to person. You may also see, for example, in descriptions of the latest influenza vaccine, the designations of the individual strains from which it was prepared: these follow the pattern A/California/07/09 (H1N1), where A is for influenza A, followed by the place where it was isolated, the laboratory number, the year of isolation, and the H and N subtypes.

Type B strains are designated on the same system, but without H and N numbers as major changes in these antigens have so far not been observed. Recently B viruses of two distinct lineages, Victoria and Yamagata, co-circulated and now representatives of both are included in the vaccine.

14.2.2 Morphology

The virions are 100–200 nm in diameter, and are more or less spherical. The virus genome encodes ten virus proteins (Table 14.1). The lipid envelope is covered with about 500 projecting spikes, which can be seen clearly under the EM (Fig. 14.3). About 80% of them are the haemagglutinin **(HA)** antigen. The HA is a rod-shaped glycoprotein with a triangular cross-section (Fig. 14.4). It was first identified by its ability to agglutinate erythrocytes, hence its name, but it is now apparent that it also has important roles in the attachment and entry of virus to the cells of the host and in determining virulence. The HA spike has been crystallized and X-ray crystallography has revealed a trimeric complex with a stalk region attaching the HA to the

Table 14.1 Influenza genes and protein coding

Gene	Polypeptide	Functions
1	PB2	Part of RNA transcriptase: recognizes host cell capped mRNA
2	PB1	Part of RNA transcriptase: endonuclease activity
	PB1-F2	Induction of apoptosis
3	PA	Part of RNA transcriptase: role in transcription?
4	HA	Binds virus to cell via surface receptors. Fusion activity
5	NP	Associates with RNA to form nucleoprotein structure
6	NA	Enzyme active in virus release
7	M1	Major structural protein
	M2	Forms an ion channel through virus lipid membrane
8	NS1 (NEF)	Interferon antagonist
	NS2	Viral nuclear export protein

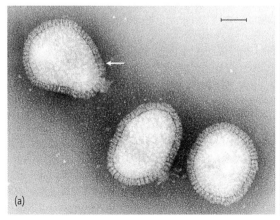

Fig. 14.3 Electron micrographs of influenza viruses. (a) Negatively stained virions. The surface HA and NA antigens are arrowed. Scale bar = 50 nm. (b) Scanning electron micrograph of virus particles budding from the cell surface. Many thousands are present on an infected cell. (Courtesy of Dr David Hockley.) Scale bar = 500 nm.

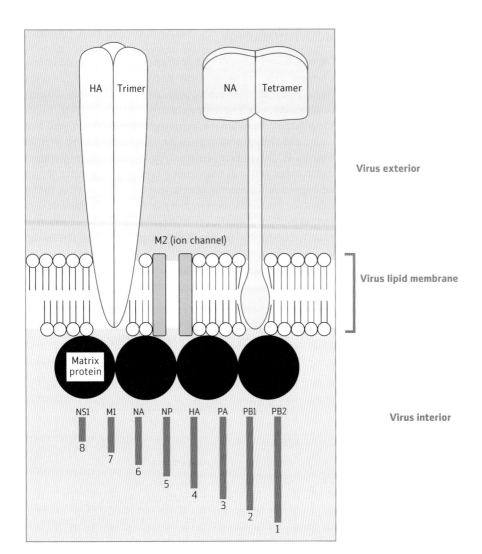

Fig. 14.4 The surface HA and NA antigens of influenza and internal M protein and genes.

virion and a distal global head. The head, made up of the HA1 polypeptide, has the important receptor binding site which is highly conserved and mutations here can alter the receptor specificity of HA. The stem region is made up from the HA2 polypeptide and anchors the HA to the virus itself. It has an important structure called the fusion peptide at the junction between HA1 and HA2 which, after massive low pH induced conformation change in the cellular acidic endosome, induces a structure which results in fusion of the endosome and viral membranes, allowing exit of the RNAs and their subsequent migration through the nuclear pores into the nucleus where RNA transcription and replication occur (see Section 14.2.4).

The mushroom-shaped neuraminidase (NA) can remove neuraminic (sialic) acid from receptor proteins. Its main function seems to be connected with release of new virus from cells. It is the target for the neuraminidase inhibitor drugs, oseltamivir and zanamivir. X-ray crystallography has delineated binding sites for these new drugs.

Penetrating the lipid membrane of the virus are molecules of M2 that form ion channels, allowing protons to enter the interior of the virus during replication in the cells. Inside the matrix shell (M1), which is an important stabilizer of the structure, are the **nucleoprotein** and an **RNA-dependent RNA polymerase** complex, which are essential for transcription of viral RNA to mRNA during replication and for genome replication (Section 14.2.4). There are considerable antigenic and genetic similarities between the M2, M1, and NPs of all influenza A viruses and it is possible to envisage a 'Universal Influenza Vaccine' composed of these proteins inducing CD4 and CD8 immune cells to block a wide range of, if not all, influenza A viruses.

14.2.3 Genome

Orthomyxoviruses are negative-sense single-stranded segmented RNA genome viruses classed as Baltimore class V. Segmented means the genome is not one contiguous nucleic acid, but is rather a collection of, in the case of influenza, eight discrete fragments of the negative-sense single-strand RNA genome, which together add up to approximately 13 kb in size. These viral RNAs are complexed with proteins (NP, PA, PB1, and PB2; Table 14.1) to form an RNP arranged in a helix.

14.2.4 Replication

Influenza viruses bind to neuraminic (sialic) acids on the cell surface via the virus HA to initiate infection. Species tropism is, to some extent, controlled by which type of sialic acid the different virus HAs can bind to. Viruses able to infect humans bind preferentially to glycoprotein whereby the neuraminic acid is covalently attached to the carbohydrate sugar by an α2,6 linkage, whereas avian influenza viruses prefer an α2,3 linkage. The initial attachment is low affinity but since multiple HA molecules are involved the strength of the interaction is very much increased.

The α2,3 linkage receptors are abundant in the intestinal tract of birds, thus explaining the enteric nature of 'bird influenza' in its natural host. However, α2,3-linked sialic acids are also present in the human lower respiratory tract. It follows then that whenever, albeit rarely, an avian influenza virus such as H5N1 and H7N9 can reach the human lower airways it can replicate and cause serious pneumonia. Efficient transmission from human to humans depends on respiratory droplets from the upper airways

and so paradoxically these avian viruses in humans cause serious disease and mortality but cannot easily spread from person to person. When the virus attaches to the cell surface, it triggers activation of RTK (receptor tyrosine kinases) cascades.

After attachment (Fig. 14.5) virions are taken into the cell by receptor mediated endocytosis involving clathrin. Many host cell proteins are involved in virus entry (Table 14.2). An alternative entry pathway is by macropinocytosis, which is essentially uptake through actin-dependent formation of large endocytotic vesicles called macropinosomes. Virus is ultimately transported to vacuoles called endosomes, where the acid pH induces a change in the configuration of the HA. This structural rearrangement of the HA brings a special set of catalytic amino acids, the 'fusion sequence', in contact with the lipid of the vacuole wall of the cells.

Simultaneously, protons pass along the M2 ion channel to the interior of the virion and cause the M1 protein to be released from the ribonucleoprotein (RNP) complex, which contains the virus RNA polymerase complex PB1, PB2, and PA

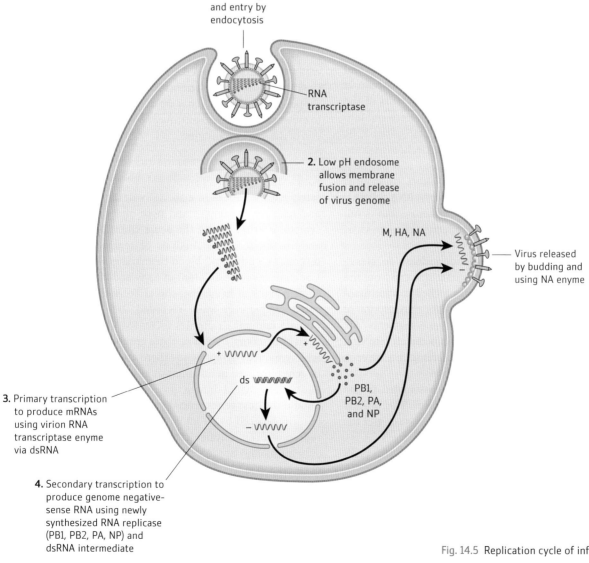

Fig. 14.5 Replication cycle of influenza virus.

Table 14.2 Cellular gene products involved in influenza virus intracellular replication

Cellular function	How do the cellular proteins help influenza?
Signalling	Virus replication and export of vRNAs
Actin	Virus movement and also transcription
Proteases	Aids cleavage of HA and later effects post transcriptase events
Nuclear and general trafficking	Cross membrane and glycoprotein movement as well as import of vRNA into the nucleus
Ubiquitination	Post-translational modification of virus proteins
Microtubules	Intracellular virus protein transport
Recycling pathway of endosomes	At entry and also nuclear export of vRNPs

together with the virus RNA and the NP protein. This 'released' RNP complex can enter the cell nucleus for transcription and replication. Unusually for an RNA virus influenza replicates in the cell nucleus and, as a result, the eight segments of the genome must be transported across the nuclear pore and into the nucleus. This is facilitated by the NP protein of the RNP. Once in the nucleus the virus negative-sense RNA is transcribed into viral mRNA, which is 5' capped and polyadenylated. The viral mRNA is therefore distinct from virus genome positive-sense RNA (complementary RNA, cRNA), which is produced by the same virus polymerase, but is full length and serves as a template for production of more virus negative-sense RNA genome segments.

Uniquely, influenza can use its endonuclease to cleave 10–13 nucleotides and the 5'-methyl guanosine cap from the normal nuclear mRNA of the cell. This 'snatched' cap and associated nucleotides is then used as a primer for transcription of each of the eight negative-sense mRNA gene segments. Six newly transcribed mRNAs are translated immediately into viral structural proteins, while two primary RNA transcripts are each spliced into two mRNAs coding for NS proteins.

Newly synthesized RNP complexes are assembled in the nucleus and then transported to the cytoplasm. This process involves the virus nuclear export protein (NEP/NS2) and the virus matrix protein (M1).

The new influenza virions are assembled at the host cell's plasma membrane and released by a process of budding in which both HA and NA are involved. The M1 protein, still associated with the RNP, interacts with the cytoplasmic tails of HA and NA, and is thought to act as the join between the virus RNP and the HA/NA with the M1 protein forming a layer

beneath the lipid envelope. The final virion also includes M2 in the lipid envelope. How a virus packages the correct eight segments of its genome is still not clearly understood. Two models have been proposed, the first, random incorporation is based on the fact that a virus particle randomly packages eight or perhaps more segments. A consequence of this model is that a large proportion of virions would be non-infectious as they would have the wrong complement of genome segments. A second model suggests that each segment has its own unique packaging signal thereby allowing the correct association of eight genome segments.

Finally, the new virus particle must bud from the cell. The viral NA has the important function in the active release of budding virus particle of cleaving sialic acid from viral and cellular glycoproteins, thus preventing virus aggregation and allowing individual virions to be released from the cell. Some cells die, but others may allow many viral replication cycles to occur without specific damage to themselves.

The manner of temporal regulation of virus replication is a hot research topic and many laboratories have, for example, recorded differential accumulation of certain of the eight RNAs. This could result from specific up-regulation and could reflect different rates of viral RNA synthesis or even differential stability of mRNAs. Messenger RNAs from NP and NS1 genes have been detected early in the virus replication cycle, in a matter of hours, whilst RNAs from HA, NA, and M1 have been detected later on. Control is also exerted at translation as host protein synthesis is suppressed whilst viral genes are selectively translated. NS1 acts to enhance certain virus mRNAs by binding to the 5' non-coding end of the mRNA transcripts.

Involvement of host genes in virus replication

For the first time scientists are able to infect young volunteers safely with influenza in a specially designed quarantine unit and take blood samples and nose and throat swabs before symptoms arise, during the acute clinical phase of infection and in the recovery. Analysis of temporal transcriptional response identified 42 blood biomarkers where gene expression patterns were different in early and later infection when clinical signs of headache, throat soreness, cough, and elevated temperature were recorded. Inflammatory cytokine ('innate immunity') profiles increased and preceded clinical signs by 36 hours. By contrast, in a group of infected volunteers who remained asymptomatic there was significant regulation of genes involved in inflammosome activities and antioxidant and cell-mediated innate immune response (Table 14.2). Should the host genes crucial to influenza replication be temporarily blocked by an inhibitor then such a molecule would represent an entirely new class of virus blockers. An integrated systems biology approach, using RNA interference to remove each human gene one at a time, detected nearly 300 host cofactors, even at the early stage of replication, and particularly proteins involved in phosphatase activity, kinase-regulated signalling, and ubiquitination that are required for influenza virus replication.

14.2.5 Genetic variation in influenza viruses

Influenza A viruses readily undergo gene 'swapping' or **reassortment** (Fig. 14.6), so that, in a cell infected simultaneously with two different parent viruses, the progeny virions may contain mixtures of each parent's genes. Add this property to the ability of influenza A virus to infect animals such as pigs and birds that often live in close association with humans, and we have a situation in which double infections with viruses of human and non-human origin may result at unpredictable intervals in the formation of new strains with genetic compositions differing from those in general circulation. A perfect example is the outbreak of swine influenza A (H1N1) virus in 2009, which has genes from pig influenza virus, avian, and human viruses. This reassortment of genes, which can result in **antigenic shift**, can, of course, also take place between two influenza A viruses of human origin, and all

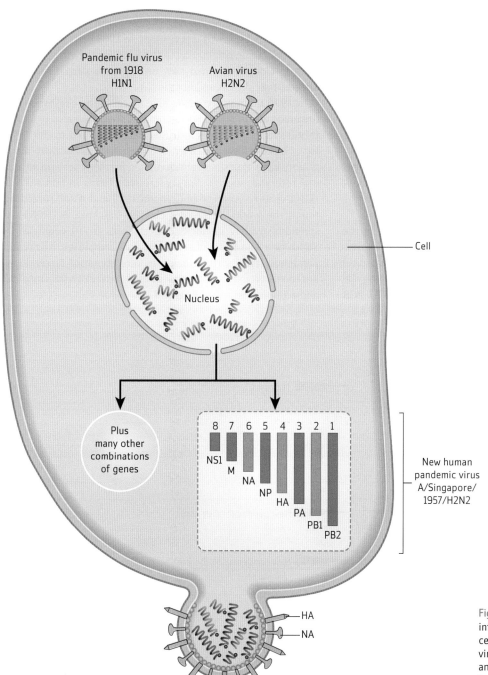

Fig. 14.6 Genetic reassortment with influenza A virus. Infection of a single cell by two different influenza A viruses could result in exchange of any of the eight genes of each parent to create a 'new virus' with genes from both parents.

types of antigenic shift can allow the emergence of new pandemic viruses. Reassortment cannot occur between influenza A and B viruses.

RNA viruses tend to have high mutation rates—more than 10 000 times higher than that of human DNA genome replication error rates—and this is true of all the influenza viruses. The viral RNA-dependent RNA polymerase is a low-fidelity enzyme, so transcription errors accumulate. Moreover, there are no proof-reading or corrective functions to this enzyme. These mutations give rise to changes in the viral polypeptides, such as HA, which, out of a total of 250 amino acids, undergoes two or three amino acid substitutions each year. Some of these mutations can change the binding sites for antibodies and this is known as **antigenic drift.**

Both influenza A and B are subject to antigenic drift, but only A viruses undergo antigenic shift and hence have the potential for causing pandemics.

14.2.6 Pandemic influenza

Figure 14.7 shows the relationship between antigenic drift, shift, and epidemics and pandemics of influenza A.

The great pandemic of Spanish influenza in 1918 was especially terrible in its effects in Europe; worldwide, it killed about 50–100 million people, far more than lost their lives in the whole period of the First World War.

It is not surprising therefore that the virus has been the object of intense research. From lung samples of victims stored in pathology museums like the Royal London Hospital (J. Oxford) and the Armed Forces Institute of Pathology in Washington, influenza genes have been amplified by RT PCR

and the entire virus reconstructed by reverse genetics. Some clinical material has been found in frozen victims of the pandemic from the Arctic. Surprisingly, the 1918 virus is not hugely virulent in animal models, which does suggest that the events of 1918, huge movements of young people, and semi-starvation, all contributed to the death toll. Of course this was in a pre-antibiotic era and many young people, rather than the aged, were most affected, succumbing to bacterial superinfection with pneumococcus and staphylococcus. Pre-existing immunity in older people to previous pandemics in the nineteenth century also helps explain why the young were most affected in 1918.

An influenza pandemic that occurred nearly 40 years later in 1957, when a strain differing completely in both HA and NA appeared in China, was less virulent. In 1968 there was another pandemic, again originating in the Far East; the virus, first isolated in Hong Kong, had now undergone a partial shift that affected only the HA and not the NA.

In 1976 there was considerable alarm in the USA when the dreaded swine-type H1N1 influenza A virus appeared in a military barracks. This was a drifted variant related to the 1918 virus; it could have arisen again by genetic reassortment, but it is more likely that the strain circulated for many years in an animal reservoir, such as pigs, before resurfacing in humans. An H1N1 virus re-emerged in China in 1977 and is still spreading round the world. At present chicken influenza A (H5N1) is spreading widely in ducks and chickens in 10 countries in South-east Asia and could re-assort, for example with the swine pandemic H1N1 virus. Obviously, dire scenarios are endless, but the fact remains that pandemics cause very serious medical and economic issues in the world.

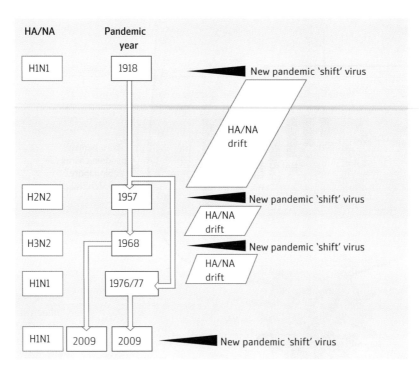

Fig. 14.7 Antigenic shift and drift.

14.2.7 The origin of pandemic or shifted strains of influenza A: the role of birds in H7N9 and H5N1 infection

Why do pandemic strains of influenza A seem to originate in southern China—or do they? The answer may lie in the close association of humans with domestic animals and birds. Migrating geese and ducks are the end source of all 18 antigenic subtypes of influenza and most often the infection is silent. The birds excrete virus from their respiratory tract and also in their droppings. Domestic birds become infected and the virus soon mutates to a virulent form, killing chickens by the tens of thousands. Largely unremarked except by farmers and veterinarians, influenza A viruses are constantly circulating in pigs, horses, and birds, including poultry; and it is a reasonable assumption that, in areas of very intensive small-scale farming such as in southern China, the chances of interchange of viruses between humans and other species—and, hence, of genetic reassortment—are considerable. Other factors that may be important are the very high human population density in that area, in which lives half the world's population, and, most significant of all, live chickens are sold in the markets where purchasers can inadvertently become infected.

Live bird markets in Hong Kong, or, indeed, in the USA or anywhere else in the world, give a particular opportunity for mutated avian influenza A virus to cross the species barrier to infect persons (Fig. 14.8). In the last few years there have been outbreaks of avian H5 and H9 viruses in Hong Kong and H7 in the Netherlands and also China. Pandemic viruses can arise anywhere that infected birds, pigs, and humans co-exist.

The emergence of swine influenza A (H1N1) pandemic virus in Mexico in 2009

By mid-March 2009 a surge in respiratory disease, COPD, and hospital pneumonias was recorded in Mexico. The causative virus was a mixed quadruple reassorted virus, with PB2 and PA genes from North American avian influenza, the HA, NP, and NS from the 1918 pandemic virus, NA and M from Eurasian and American swine influenza gene lineages, and PB1 from seasonal H3N2 viruses. Unlike previous 'swine' influenza H1N1 viruses, this one began rapid human-to-human transmission. Transferred by holiday makers in Mexico to the USA and to the UK and Spain, the virus soon began a widespread person-to-person community spread. Upon its arrival in the Southern hemisphere in June WHO declared a level 6 global pandemic. Case fatality was rather low and the over 60s had immunity from similar influenza A (H1N1) viruses encountered in the 1950s. Because of prior preparations for a potential pandemic of bird flu (H5N1), the global vaccine, antiviral, and infectious disease sectors of the world community reacted promptly and the pandemic was classed as the mildest within recorded history; however, despite being mild it is estimated that 280–580 000 people died around the world. There were at least two waves in the pandemic and not every country experienced the outbreak at the same time.

A new virus in China H7N9 in 2013

Unexpectedly, a new and potentially pandemic influenza A virus (H7N9) emerged in Shanghai in 2013 and caused fatal

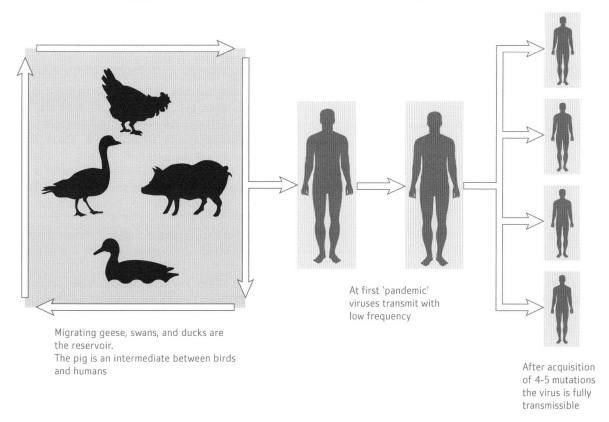

Migrating geese, swans, and ducks are the reservoir.
The pig is an intermediate between birds and humans

At first 'pandemic' viruses transmit with low frequency

After acquisition of 4-5 mutations the virus is fully transmissible

Fig. 14.8 Origin of pandemics—the eternal cycle.

pneumonia in local residents. The virus soon spread to other provinces across China and also to Taiwan. A new feature of this virus is that it causes no fatalities in domesticated chickens in live bird markets and yet is still able to spread to persons in contact with these birds, albeit rarely, and cause serious human disease. To date the virus has not acquired mutations to allow person-to-person spread. Case fatality is around 40%, especially if the antiviral neuraminidase inhibitor (NI) drugs such as Tamiflu are not administered or are given late after symptoms appear. Sequence analysis shows an avian origin and a complicated reassortment event with genes from wild birds (H7N9), a duck virus (H7N3), and a virus from bramblings (H9N2).

Scientists deliberately passaged bird influenza H5N1 in ferrets and mutated the virus. The mutant virus was able to spread by aerosol or droplet and has only four mutations at amino acid position Q2262, N224K, N158D and T318I. The former two amino acid changes are in the receptor binding site whilst the latter two are in the HA stalk.

14.2.8 Pandemic planning on a global scale by WHO

Interim guidance was issued by the World Health Organization in 2013 to help countries to revitalize and to update their existing plans. It cannot be overemphasized that such plans can help communities prepare for any infections emerging such as SARS CoV and MERS CoV and not just influenza. The old pandemic phases 1–6 are replaced by a continuum and each country now has freedom of action to declare their position and to act somewhat independently of 'the world'. These changes are brought about by experience of the last H1N1 pandemic in 2009. It was clear then that Mexico had a pandemic at phase 6 whilst most of Europe was still at phase 2/3. The new continuum is designated as follows: in inter pandemic, alert, pandemic, transition, and returning to inter pandemic. Actions will depend less on WHO announcing a 'global' pandemic and more on the 168 member states assessing the situation at national level. In other words national actions will be uncoupled from global phases. WHO has initiated a pandemic influenza preparedness (PIP) framework to encourage sharing the viruses so allowing more equitable access to vaccines and medicines.

14.3 Clinical and pathological aspects

Clinical onset is rapid after a short incubation time of 2–3 days and is preceded by virus excretion in the upper airways. Pneumonia is a secondary and serious complication, often being caused by streptococcus. Patients excrete virus for 1 day before the onset of symptoms. Virus excretion can be prolonged in children.

Medical case story   Influenza

You may come into medicine or science with a sense of invulnerability—that you are young, healthy, and will never develop any of the serious or lethal diseases that you see in your patients. Imagine then, if you will, that a new influenza pandemic is sweeping the globe. Nobody knows yet how infectious it is or how dangerous. Early mortality figures appear to be high. The government quietly prepares for a worst case scenario of one death per 1000. There is no vaccine.

This was the situation when the following scenario took place.

Your next patient is an 'immediate and necessary', and therefore temporarily registered at your practice. You know nothing about her except her name, Mrs K.

She shuffles down the stairs to your room in her pyjamas, flanked on one side by a man carrying a large plastic bowl, and behind her a younger woman. Mrs K has blood on her face.

She sits down on the seat and stares into the middle distance clutching her bowl. You introduce yourself, check her name, and the names of the man, who is her husband, and the other woman, who is her daughter and *is* a patient of yours.

'Who's going to tell me what's been happening?' you ask brightly. Mr K speaks up. He describes how 2 days before Mrs K had a sore throat, then a cough, then this morning she vomited, and then after that fainted in the shower and cut her head. He has an Australian accent. You know that the new variant of influenza is circulating at high levels in Australia.

Your heart sinks. You think, 'This is new-variant flu'.

You mentally review the Public Health UK guidance: pyrexia 38 °C or history of fever AND flu-like illness (two of the following symptoms: cough, sore throat, rhinorrhoea, limb/joint pain, headache, diarrhoea, vomiting).[1] In fact, you have quite a collection of print-outs about swine flu from the HPA, also from the PCT (primary care trust), the Department of Health, and WHO (World Health Organization). These have been updated regularly, sometimes daily, since the first cases of this outbreak were reported. The algorithm is also up on every GP surgery, pharmacy, and hospital door, and says; 'STOP. If you have swine flu or think you have swine flu do not come into the surgery.'

You take her temperature: it is 39.5 °C. 'Where are you from?' you ask just to confirm. 'Australia' says Mr K, 'The best address in the World!'

At this point Mrs K vomits into the bowl, which she then drops on the floor. Vomit comes sliding towards you. You leap out of the way, and then have to move back again to catch Mrs K, who slowly collapses forward, putting her head on her knees. 'Let's lay her flat on the floor,' you advise, but Mrs K prefers to stay put.

She then appears to improve. You are relieved to be able to take a history from her directly. You are aware that you have jumped to the conclusion of influenza and want to go over things again to make sure you are not missing other possibilities. What other diagnoses do you need to consider? What would you ask next?

She is 75, has no other illnesses, and is not on any medications. She is alert, orientated, and comfortable. She is not short of breath and a review of systems (ROS) reveals no other complaints: specifically no headache or diarrhoea. You listen to her chest, which is clear; her throat is also fine. She has no signs of meningism. You inspect the cut on her head, it is minor. Often the only sign to be found in influenza is the high temperature.

Other differentials apart from influenza, pneumonia, and tonsillitis, might include malaria (with a travel history), septicaemia, meningitis, and HIV seroconversion illness. However, at the end of your re-assessment you are still confident that new-variant influenza is the most likely diagnosis.

You explain your findings. All three look horrified. They reveal that Miss K is flying to Spain to be married in 5 days' time. She has a sore throat, as does her fiancé. She has a houseful of guests including her sister and her niece all over from Australia. Also her bridesmaid has a temperature.

At this point Miss K holds her head. 'I feel faint,' she says.

Miss K lies on the couch, Mrs K grips her chair, Mr K grimly sits tight. There is vomit on the floor. There is a quiet pause while you collect your thoughts.

'We need to get you home,' you say, 'Swine flu is very infectious and I have vulnerable patients in the surgery.'

You tell them that you will ring back later to discuss who else at home will need Tamiflu. You call them a cab. You advise Mrs K to stay at home, isolate herself and be very careful of hand hygiene. She should drink fluids slowly to avoid further vomiting and make sure she does not get dehydrated. She can take paracetamol if she needs it for her sore throat and any other aches and pains. You advise her to take a course of Tamiflu. You explain that she will probably feel ill for a week and then not back to normal for a further 2 weeks. She should not fly until she has been symptom-free for 48 h. By this you mean no temperature and no vomiting. If she is not able to drink, or if she has breathing difficulties, she should contact you again.

Miss K needs to empty the house and isolate in their rooms anyone who feels ill, such as the bridesmaid. Out of sympathy, but not clinical need, you feel it is reasonable to give the bride and groom a course of prophylactic Tamiflu. Everyone else can telephone the pandemic flu line.

After they have gone you get out your antiseptic floor cleaning wipes. You clean the floor, the surfaces, the door handles and banisters and change the paper on the bed. You clean your stethoscope and thermometer with alcohol wipes. You wash your hands twice. The virus can stay active on hard surfaces for 24 h; on soft surfaces it can stay active for 4 h. It can stay as an aerosol for 24 h, but the exact length of time depends on the particle sizes. Ultraviolet light reduces transmission.

The next patient is a 4-year-old girl. You are uncomfortably aware of the aerosol virus that is floating around the room.

After clinic the room is deep-cleaned.

Notes

[1] https://www.gov.uk/government/collections/seasonal-influenza-guidance-data-and-analysis (accessed 19 Feb. 2016).

★ Learning Points

- **Virological:** Influenza virus can cause a range of clinical illness ranging from subclinical to obvious, with the classic symptoms being high temperature, aches and pains, cough, sore throat, headache, and in some instances diarrhoea and vomiting. Incubation is 1–3 days and infection rate varies from 2–10 days.
- Diagnosis is by nasal swab using PCR, and is used for monitoring purposes at sentinel practices, rather than to assist clinical decisions. When influenza is circulating at above 200/100 000 this is counted as epidemic level and the accuracy of a clinical diagnosis will improve from 1/10 to 1/3: (that is, for every three people *thought* to have influenza, one *will* have influenza).

- Influenza can be prevented by immunization, but the vaccine has to be specific for the circulating virus and takes 6 months to mass produce. Influenza treated within 72 h by Tamiflu can reduce the duration of the illness, complication rate, and viral shedding.
- **Clinical:** It is sometimes helpful and not at all shameful to revisit the history or the examination. You may have forgotten an important question or part of the examination, or thought of a new differential diagnosis that needs excluding.
- **Personal:** Have you decided what you would do in the event of a pandemic?

14.3.1 Clinical features

After an incubation period of 2–3 days there is usually a very **abrupt onset,** with shivering, malaise, headache, and aching in the limbs and back. Characteristically, the patient is prostrated and has to take to bed. The temperature rises rapidly to about 39 °C. Influenza is **not** characterized by runny noses or sore throats at the beginning, as are common cold infections (Fig. 14.9).

Fortunately, influenza is usually short-lived in younger persons. In older people and the 'at-risk' group, however, recovery may take much longer, with persistent weakness and lassitude

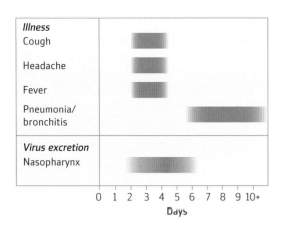

Fig. 14.9 Clinical and virological events during infection with influenza.

sometimes for 3–6 months. In general, the severity of influenza is proportional to age. There are no differences between influenza A and B as regards the clinical picture. There is one important caveat here: babies are as much at risk of complications as the elderly. This is a virus that strikes at both ends of the age spectrum.

Apart from the important and sometimes life-threatening secondary bacterial infection caused by *Streptococcus pneumoniae* there are few complications, but one rare condition, **Reye's syndrome**, is sometimes associated with influenza in children, often of the B type. The taking of aspirin has also been implicated in the causation of this syndrome, which involves encephalopathy with fatty degeneration of the liver and other viscera; it is often fatal (see also Chapter 30).

14.3.2 Pathogenesis

Infection is acquired by the respiratory route and is usually an infection of the upper respiratory tract. Virus multiplies in the epithelial cells in the nose and sinus passages, and destroys the cilia, which are an important element in the defence of the respiratory system.

There is increasing evidence of infection by small particles (<5 μm) which can stay airborne and spread widely. Large particles (>20 μm) travel less than 1 metre and also tend to settle on surfaces. These larger droplets can be blocked by simple masks. For care of patients healthcare workers (HCWs) have to be particularly mindful of equipment creating aerosol and small particles and wear protective fitted masks (N95 masks). Also recent studies in hospitals do raise concerns that small virus particles can travel 2 metres from the patient, projected by coughs and sneezes, and can infect other patients or HCWs.

Molecular pathology and the 'cytokine storm'

Influenza virus is recognized in the infected patient by host cell pattern recognition receptors (PRRs) such as RIG-1 and TLR-7 which note the presence of 5′ ssRNA with an uncapped phosphate. Nucleotide-binding oligomerization domain receptors,

(NOD-like receptors, NLRs) are also activated in cells. All together the PRRs start off many signalling cascades which give the first warnings to both the innate and adaptive immune arms of a viral invasion. This is now recognized as a two-edged sword, benefiting the patient but also putting the patient at risk of serious pathology caused by an immune over-reaction in a 'cytokine storm' (Chapter 5). As the infection progresses in the respiratory tissue a seemingly endless list of cytokines and chemokines are released (Table 14.3). One of them, IFN type I, is so effective that the virus has evolved a special protein (NS2) to block this pathway. Multiple immune cell types are recruited. Whilst 'cytokine storm' has been used to highlight the turmoil of the many cytokines released in response to infection, it is now becoming clearer that precise cytokine regulation is important in controlling infections and disease pathology. Alterations in this fine regulation can lead to different pathology outcomes, therefore the 'cytokine storm' might be better thought of as cytokine dysregulation. Infection with bird influenza H5N1, for example, may be exacerbated by high levels of cytokines IFN-γ, TNF-α, IL-1, IL-6, IL-12, and IL-18, and chemokines M1P-2 and M1P-α/β.

Viral infection of the lower respiratory tract, in the form of influenzal pneumonia, sometimes occurs, presenting as an overwhelming toxaemia with a high mortality. The virus replicates in epithelial cells of the alveoli, causing exudation into the air sacs and pneumonia. Pneumonia is, however, often due to **secondary infection with bacteria**. Of these, *Streptococcus pneumoniae* and *Staphylococcus aureus* are both the most frequent and the most dangerous, leading to respiratory distress, cyanosis, and collapse within 2–3 days of the onset of infection. More recently, research has shown that infection with influenza enhances the chance of stroke and heart disease.

It is clear that, for infectious diseases, changes in host genes in the form of mutations (single nucleotide polymorphisms, SNPs) can alter the function of genes. This contributes to the normal genetic diversity of humans. When such mutations occur in genes involved in control of virus infections or in genes required for viruses to replicate in cells then the course of infection can be changed. Sometimes the SNPs confer an advantage to the human, meaning the infection is mild or asymptomatic, but sometimes the SNP is deleterious, meaning the human is more susceptible to infection or to severe disease. Therefore, changes in the human genome can alter the pathogenesis to a virus, so that even when the virus is identical in two infected people, different pathogenesis can occur. Influenza has provided one of the best examples of this. For the virus to enter cells it has to exit the acidic endosome and the human protein IFITM3 efficiently blocks this from happening. Recent data indicates that some individuals have a gene IFITM3 with a SNP called rs12252, which can be either TT (due to 2 human chromosomes), which provides a degree of protection to serious influenza diseases, or CC, which is associated with a fivefold greater risk of severe and fatal influenza infection. Therefore human genetic variation can alter influenza pathogenesis.

14.3.3 Immune response

We have seen that protection against infection is mediated predominately by antibodies, the **anti-HA** being the most important in this respect. **Anti-NA antibody** prevents the release of newly formed virus from the host cells. Antibody of the IgA class is probably important in preventing infection, acting as it does at the mucous surfaces of the respiratory tract where the virus first attaches.

In addition to these humoral factors, CMI, in the form of cytotoxic T cells and alveolar macrophages, plays an important part in recovery. Cytotoxic T cells either CD8 or CD4 may have a wider response against all influenza A viruses, particularly if they are directed towards broadly cross-reacting epitopes of the type-specific NP or M proteins or stem regions of the HA. This is the subject of intense research to make a more universal influenza vaccine protecting against a wider range of antigenic HA variants and even novel pandemic viruses.

14.3.4 Laboratory diagnosis

In general practice, the diagnosis of influenza is made on the evidence of the characteristic clinical picture, backed up by the knowledge that an outbreak is in progress. In the UK, figures of influenza-like illness are published weekly by Public Health England (PHE).

The older technique of isolation of influenza virus in culture for further examination is now superseded by amplification of virus RNA by RT PCR in the laboratory. These results can be obtained in a few hours. Data from subsequent sequence analysis can show early signs of emergence of drug resistance by specific mutations in the NA gene.

14.4 Prevention and cure

14.4.1 Drugs against influenza

Three anti-NA drugs, which inhibit viruses from budding from the cell surface, are used in many countries. The anti-NA drugs (**oseltamivir, zanamivir**, and **peramivir**), so called neuraminidase inhibitors (NIs), have the big advantage, compared with amantadine, of inhibiting influenza B, as well as influenza A viruses.

Influenza A viruses, but not B or C, are inhibited by **amantadine**, a primary amine, and **rimantadine**, a methylated derivative (Chapter 31).

These two classes of anti-influenza drugs have to be given to the patient within 36 h of the onset of influenza symptoms, when they reduce virus excretion, alleviate symptoms, and reduce hospitalization, death, and the misuse of antibiotics. In the family environment, once an index case of influenza is identified, rapid distribution of antivirals to the rest of the family can reduce transmission by at least 80%. This is called post-exposure prophylaxis or PEP. Experience over the next few years will show how the drugs can be used with greatest benefit. At present and given the impact of influenza, both personally and in the community, they are underused and the most important reasons appear to be economic, rather than the needs of public health per se. Recent community studies show that the NIs reduce death by 50% in adults and pregnant women but the drugs need to be used within 2–3 days of the symptoms starting. One further point to note is that mutations in the NA around the enzyme active site can confer drug resistance but fortunately many of these mutants are concomitantly crippled and do not spread in the community.

Human monoclonal antibodies used as 'passive antibodies' are the subject of intense research. A novel feature is to select an epitope which is conserved on the HA where attachment of monoclonal antibody would inhibit the conformational change required for membrane fusion and virus uncoating. Such a place is the stalk of the HA and these antibodies might have a wide spectrum to inhibit influenza A viruses across the 18 subtypes.

14.4.2 Vaccines

At present, immunization, rather than chemoprophylaxis, remains the method of choice for preventing both influenza A and B, although antivirals are gaining momentum to be used in therapy. Even so, vaccine use poses a particular problem: every time a new strain of influenza A or B appears, the rapid production of large quantities of vaccine virus with the required antigenic characteristics, together with the need for routine tests of safety and efficacy, limits the amount of vaccine available.

Up until recently most national authorities immunized about 10% of the population annually: these are **the 'at-risk' group**, who have a much increased chance of serious clinical complications after an attack of influenza, often caused by superinfecting bacteria such as pneumococci. There are two main categories of at-risk persons and it is most important that these individuals are vaccinated each year:

- The elderly (>65 years in the UK or >60 years in the EU), or debilitated, and those of any age with chronic heart, respiratory (asthmatics and chronic bronchitics), renal, or endocrine (diabetes) disease.
- People in closed institutions, such as residential homes for the elderly, in which attack rates may be high.
- Pregnant women.
- Asthmatics and persons with COPD.

However, there are more thoughts now to immunize all sections of the community and especially children. It is also considered sensible to immunize groups in community service, such as healthcare staff, nurses, and doctors, and police, who may need protection against wholesale sickness at times of major epidemics. Canada, particularly, has pressed ahead to immunize nurses, doctors, and carers of the 'at risk' who could, unimmunized, carry the virus to their patients or from the hospital to home. In the USA vaccine is now recommended for persons of all ages in the community.

Inactivated vaccines are prepared from appropriate strains of influenza A and B grown in the chick embryo allantoic cavity

(a)

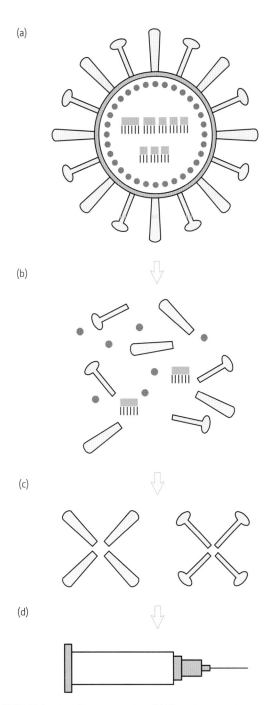

(b)

(c)

(d)

Fig. 14.10 Subunit influenza vaccine. (a) Virus is grown in eggs, or mammalian cells purified, and concentrated. (b) Viral lipid is dissolved by detergent, releasing proteins and nucleic acid, which is removed. (c) HA and NA subunits are purified by ultracentrifugation. (d) Standardized quantities of HA and NA are incorporated into the final product (approximately 15 μg per dose).

so-called 'split' vaccines that have been extracted with detergent to reduce the side-effects of whole-virus vaccines. Apart from local erythema and soreness, sometimes with fever, these vaccines are generally very safe. **Live attenuated vaccines** are made by reassorting genes of viruses possessing the required HA and NA antigens with various laboratory-derived mutants selected previously for inability to grow at 37 °C or for ability to grow only at low temperatures, for example, 25 °C (cold-adapted mutants). The reassortants have mutations in at least six of the eight genes of the virus and are of diminished virulence for humans. These vaccines are now considered of particular relevance in children aged 2–17 years, both to protect them and to break the chain of onward infection in the family to the grandparents.

The most recently developed cold-adapted vaccine, FluMist, is targeted in the USA and now the EU at children from the age of two years. The concept here is to break the normal chain of infection within the family, from the child to the parents and grandparents. There is some evidence that in this situation the virus-induced mortality of the grandparents is reduced during the yearly influenza season. An essentially similar vaccine developed in Russia is used also in India and South-east Asia.

Table 14.3 Multiple pattern recognition receptors sense infection of a cell and trigger innate immunity

The receptors	Examples
Toll-like receptors	TLR3
	TLR7
Cytosolic receptors	RIG-1 (retinoic acid inducible gene)
Nucleotide binding domain proteins	NLR
Leucine-rich repeat proteins	NLR P3
Influenza dsRNA receptors	TLR3
Influenza ssRNA receptor	TLR7

Cytokines and chemokines produced	
IFN-α	
Chemokine ligand	(CCL)—2
	(CCL)—5
IFN-γ	

Immune cell recruitment
Activated macrophages
Activated monocytes
Activated natural killer cells
(Expression of CD69 activation marker on above cells)

(Chapter 30); the infected fluids are harvested, purified by ultracentrifugation, and inactivated with formalin. Two more modern vaccines are produced using cell culture techniques with Vero (monkey kidney) or MDCK (dog kidney) cells growing in large fermenters very much like a brewery, whilst a cloned HA vaccine has now been licensed. Most of the vaccines are either subunit preparations containing purified HA and NA (Fig. 14.10) or

Reminders

- Influenza is primarily a short-lived infection of the upper respiratory tract; its severity is most directly related to extremes of age, although younger people with underlying medical problems are also at risk. Pathology is caused both by virus replication and also elements of a cytokine storm. Complications include secondary bacterial infection and, rarely, **Reye's syndrome**. Each year hundreds of thousands of deaths worldwide are attributed to influenza. Essentially it is a disease of childhood which spreads onwards to parents and grandparents.

- Influenza is caused by the **orthomyxoviruses**, which contain segmented ssRNA genomes of negative polarity, 13 kb in size. They have **helical** symmetry.

- There are four genera, of which only two, the A and B viruses, are important in infections of humans. Influenza viruses are also prevalent in a number of mammalian and avian hosts, and some can be transmitted between species.

- The **HA** and **NA** antigens are important in the infection of cells, and the corresponding antibodies to these spike- and mushroom-shaped proteins play a major role in preventing infection.

- The **genome has eight segments**, allowing reassortment of genes (**shift**). Shifted strains appear at long intervals and cause worldwide epidemics (pandemics) at long intervals. Virologists are planning for a new pandemic.

- By contrast, antigenic **drift**, due to minor mutations, affects both A and B viruses, is slowly progressive, and causes more frequent, but localized outbreaks.

- The laboratory diagnosis is made rapidly by RT PCR of nasopharyngeal washings.

- Two anti-NA drugs (**oseltamivir** and **zanamivir**), have prophylactic and therapeutic activity, and are now being used in clinics throughout the world. These drugs must be used as soon as an infection is apparent and certainly within 36 h of the first symptoms. These drugs are stockpiled in preparation for a global outbreak (pandemic) and the stockpiles were opened for the A/Swine/H1N1 pandemic of 2009/10. They can prevent 80% of illness in the family or work group, so-called post-exposure prophylaxis or PEP when given early, or reduce the duration of illness by 1–2 days when administered within 36 h of appearance of symptoms and reduce deaths by 50% or more.

- **The main method of control is at present by means of inactivated virus and live attenuated vaccines. The important components of the killed vaccine are the HA and NA proteins**, which are reformulated at intervals to match the prevalent strains. Influenza vaccines are up to 80% effective in preventing illness and reducing deaths by respiratory disease, stroke, and heart attacks in the elderly. Immunization of children is now viewed as a key public health measure to protect them and to reduce onward spread to grandparents.

- The swine influenza H1N1 pandemic of 2009–10 will go down in history as a relatively mild outbreak attacking younger persons and causing enhanced mortality in the obese and diabetics alongside pregnant women. However, the virus could still mutate in the future to allow diffusion into the over 65-year cohorts where it would be able to cause fatalities.

Further reading

Dawood, F.S., Iuliano, A.D., Reed, C. *et al.* (2012). Estimated global mortality associated with the first 12 months of 2009 pandemic influenza A H1N1 virus circulation: a modelling study. *The Lancet Infectious Diseases* **12**, 687–95.

Edingert, T.O., Pohlt, M.O., Stertz, S. (2014). Entry of influenza A virus: host factors and antiviral targets. *Journal of General Virology* **95**, 263–77.

Gao, R., Cao, B., Hu, Y. *et al.* (2013). Human infection with a novel avian-origin influenza A (H7N9) virus. *N Eng J Med* **368**, 1888–97.

Huang, Y, Zaas A.K., Rao, A. *et al.* (2011). Temporal dynamics of host molecular responses differentiate symptomatic and asymptomatic influenza A infection. *PLOS Genetics* **7**: e 1002234.

Imai, M., Watanabe, T., Hatta, M. *et al.* (2012). Experimental adaptation of an influenza H5 HA confers respiratory droplet transmission to a reassortant H5 HA/H1N1 virus in ferrets. *Nature* **486**, 420–8.

Killingley, B., Enstone, J., Booy, R. Oxford, J. S. *et al.* (2011). Potential role of human challenge studies for investigation of influenza transmission. *The Lancet Infectious Diseases* **11**, 879–86.

Li, Q., Sun, X., Li X. *et al.* (2012). Structural and functional characterization of neuraminidase-like molecule N10 derived from bat influenza A virus. *Proc. Natl Acad. Sci USA* **109**, 18897–902.

Novel Swine-Origin Influenza A (H1N1) Virus Investigation Team, Dawood, F.S., Jain, S., Finelli, L., Shaw, M.W., Lindstrom, S., Garten, R.J. (2009). Emergence of a novel swine-origin influenza A (H1N1) virus in humans. *N Eng J Med* **360**, 2605–15.

Oxford, J. S. (2002). The so-called Great Spanish Influenza Pandemic of 1918 may have originated in France in 1916. *Philos Trans R Soc London* **356**: 1857–1859.

Teijaro, J.R., Walsh, K.B., Rice, S. *et al.* (2014). Mapping the innate signalling for cytokine storm during influenza virus infection. *Proc. Natl Acad. Sci. USA* **111**: 3799–3804.

Questions

1. What advance preparations would you make to combat the second pandemic of our century? Where and how is the new virus likely to arise?

2. Compare and contrast live attenuated and killed influenza vaccines. Would you recommend large scale immunization in epidemic years?

3. Outline the stages of replication of virus paying attention to intervention with new drugs.

Arenaviruses: Lassa and haemorrhagic fevers

15.1 Introduction

This family of viruses causes widespread haemorrhagic fevers in tropical countries especially in Africa and South America. There is an animal reservoir and periodically the viruses jump the species barrier to infect humans, often agricultural workers.

15.2 Properties of the viruses

15.2.1 Classification

The family *Arenaviridae* (Fig. 15.1) contains a number of viruses (Table 15.1) with a distinctive 'sandy' appearance (Latin *arena* meaning sand) when virus sections are examined by EM.

15.2.2 Morphology

The virions are somewhat pleomorphic and range in diameter from 50 to 300 nm (Fig. 15.2). Glycoprotein spikes (GP1 and GP2) project through the virus envelope and on their inner side contact the internally situated nucleoprotein (Fig. 15.3). The nucleoprotein (NP) has a rather unusual structure of two helical closed circles. The 'sandy' appearance is caused by the inclusion in the virion of cellular ribosomal material during virus budding and of course viral RNA and NP.

15.2.3 Genome

Arenaviruses are unusual negative-stranded RNA viruses with two ambisense single-stranded RNA segments, L(ong) and S(hort), meaning that protein-coding open reading frames are present in both directions 5′ to 3′ and 3′ to 5′ of the virus RNA segments (Fig. 15.4). The genome segments have complementary termini and therefore can form pan-handle structures by base pairing. The S RNA encodes for the virus glycoprotein in the 5′ to 3′ direction and the nucleoprotein on the opposite strand in a 3′ to 5′ direction. The L RNA encodes for the virus Z protein in the 5′ to 3′ direction and the RNA-dependent RNA polymerase on the opposite strand in a 3′ to 5′ direction. In both S and L RNAs an intergenic region separates the two ORFs, which fold into stable hairpins and may allow the release of the RNA polymerase once the gene has been transcribed. Similarly to influenza, the mRNA derives its 5′ cap by 'cap snatching' short cellular heterogeneous RNAs.

15.2.4 Replication

The arenaviruses have a broad host range and tissue tropism, and consistent with this is cell binding and entry through a common cell surface receptor namely OCDG. The virus is internalized by uncoated vesicles and following low pH mediated fusion of virus and cellular membranes the virus NP is

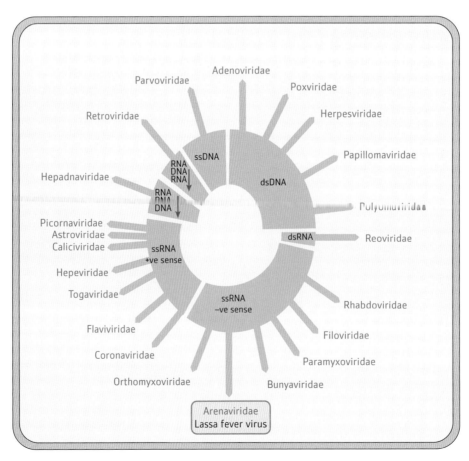

Fig. 15.1 Baltimore classification scheme and arenaviruses.

Table 15.1 Arenaviruses that infect humans

Virus	Geographical distribution	Disease
Junin	Argentina	Argentinian haemorrhagic fever
Machupo	Bolivia	Bolivian haemorrhagic fever
Lassa	West and Central Africa	Lassa fever
Lymphocytic choriomeningitis	Worldwide	Lymphocytic choriomeningitis

deposited into the cytoplasm where virus RNA synthesis is initiated (Fig. 15.5). The infecting RNA serves as a template for both transcription and replication. A subgenomic mRNA coding for the N (nucleocapsid) protein is transcribed by the virus-associated polymerase from the 3′ minus-sense half of the S segment. At the same time a subgenomic L (transcriptase) mRNA is transcribed from the 3′ minus-sense L segment and translation of both these mRNAs proceeds prior to replication of the viral genome. Only at this stage can the mRNA for the G spike protein and another protein called Z be transcribed from

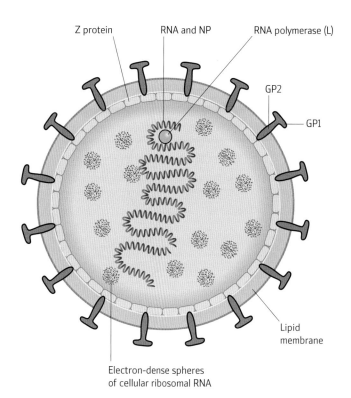

Fig. 15.3 Structural features of the arena (Lassa) virion.

the other end of the S and L segments. It is thought that mRNA transcription is favoured at first due to low levels of NP, preventing the RNA polymerase from passing the internal genomic hairpins and thereby preventing full length genome copying.

15.3 Clinical and pathological aspects

15.3.1 Clinical features of arenavirus haemorrhagic fevers

In endemic areas, subclinical infections are frequent. Clinically apparent disease may be severe; the fatality rates in hospitalized patients with Lassa fever range from 15% to 25%. The incubation period is commonly 1–2 weeks and the first signs are non-specific, including fever, headache, and sore throat (Fig. 15.6). A **rash** on the face and neck and a worsening of the patient's general condition usually signals the next stage.

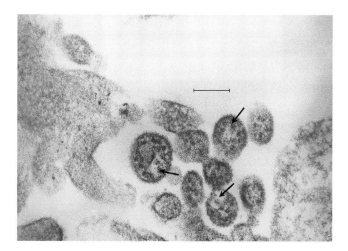

Fig. 15.2 Electron micrograph of an arenavirus (Lassa). This is a thin section of virus in an infected liver. The granules (arrowed) are host-cell ribosomes that have become incorporated in the virions. (Courtesy of Drs David Ellis and Colin Howard.) Scale bar = 100 nm.

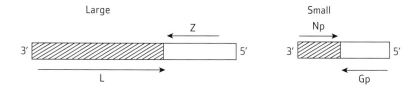

Fig. 15.4 Genome structure of arenavirus.

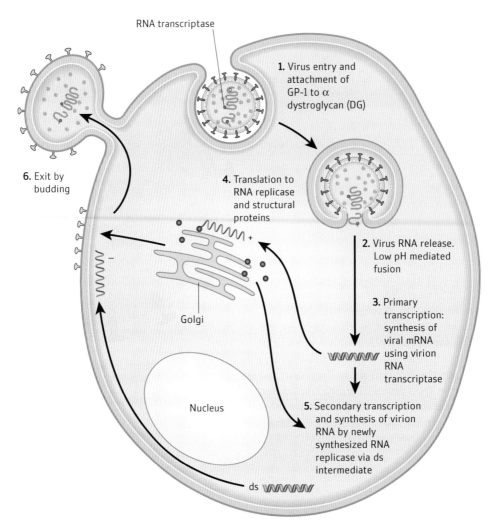

Fig. 15.5 Replication cycle of an arenavirus (Lassa).

In the second week of the illness there may be **gastrointestinal and urogenital tract bleeding** and a **shock syndrome**. Even if the patient survives, the convalescence is prolonged; **neurological sequelae** are a prominent feature, especially in Lassa fever. Argentinian and Bolivian haemorrhagic fevers are similar clinically.

Note: Lassa fever should be suspected in any febrile patient from endemic areas, particularly rural Sierra Leone, Nigeria, Liberia, and Guinea, and is the most often imported haemorrhagic fever in the EU and USA.

By contrast with the haemorrhagic fevers, lymphocytic choriomeningitis (LCM) is a comparatively mild infection, beginning with headache, fever, and malaise. The illness usually resolves after this stage, but a small proportion of patients develop **meningitis** or **choriomeningitis**, which again usually resolves without sequelae; deaths are rare.

15.3.2 Pathology of haemorrhagic fevers

Extrapolation of data from experimental infections of primates suggests that viral replication in humans occurs in the

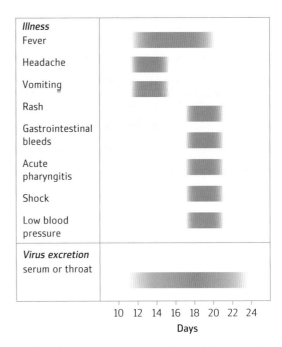

Fig. 15.6 Clinical and virological events during infection with Lassa.

hilar lymph nodes and lungs following droplet infection. The subsequent **viraemia** causes wide dissemination to other organs including **liver, spleen, heart**, and **meninges**. **Bronchopneumonia**, either primary viral or secondary bacterial, is a common finding. Large amounts of viral antigen are detected in autopsy samples of spleen, bone marrow, and viscera.

Exactly how these arenaviruses cause haemorrhagic fevers is an open question. Extensive macrophage infection causes release of TNF-α and other cytokines, and platelet-activating factor. Immune complexes, complement activation, or disseminated intravascular coagulation are not thought to play any role in the pathogenesis.

15.3.3 Epidemiology

The distribution of arenaviruses (Figs. 15.7, 15.8) is determined by that of the host rodents. Persistent infection of the rodent reservoir, often without overt signs of disease, maintains the virus in nature and humans are only incidental hosts. The natural reservoir of Lassa and Old World adenoviruses is the domestic rodent *Mastomys* which breeds in houses and shops. In South America, with the exception of Tacaribe, which is isolated from fruit bats, the reservoir is in cricetid rodents breeding in open grasslands and forest.

In rodents, infection may be transmitted vertically *in utero* or via milk; transmission via saliva and urine also occurs. Humans

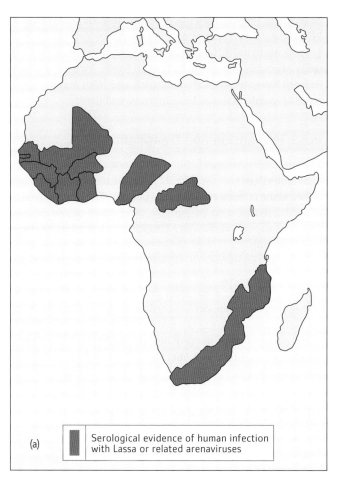

(a)

| | Serological evidence of human infection with Lassa or related arenaviruses |

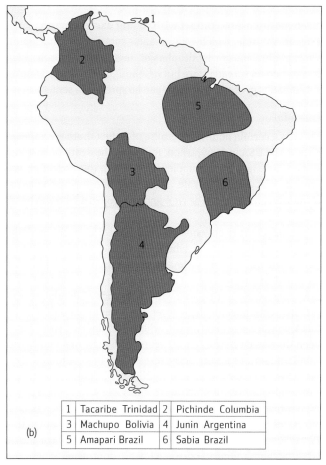

(b)

1	Tacaribe Trinidad	2	Pichinde Columbia
3	Machupo Bolivia	4	Junin Argentina
5	Amapari Brazil	6	Sabia Brazil

Fig. 15.7 (a) Prevalence in Africa of human infection with Lassa or related arenaviruses, as shown by serological tests. (b) Prevalence in South America of arenaviruses. Only the Machupo and Junin viruses are known to infect humans.

Argentinian (Junin) fever	Trinidad (Tacaribe) virus	Bolivian fever (Machupo)	Lassa fever (Nigeria)	Lassa fever (Sierra Leone)	Brazilian fever (Sabia)	Venezuelan fever (Guanasito)	Whitewater Aroyo (Mexico and USA)	Continuing outbreaks in S. America and W. Africa
1943	1956	1959	1969	1972	1990	1991	2001	2014

Fig. 15.8 Timeline for arenavirus.

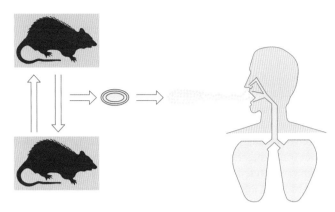

Fig. 15.9 Animal reservoir of Arena virus and transmission to humans via inhalation of dried urine.

catch the disease by **contact with rodent excreta**, particularly (Fig. 15.9) urine, which contaminates surfaces and may enter via skin abrasions or by droplets or, less likely, aerosols. Person-to-person spread is rare but it is thought that droplet/aerosol spread is possible. The virus can be spread via semen.

Lassa fever

This has a certain notoriety because of its first dramatic appearance in the USA in 1969, when, following the death of hospital staff in Nigeria, clinical samples were sent to the world-famous arbovirus laboratory in Yale where they infected and killed a technician and laid low the well-known virologist who was head of the unit.

The infection is endemic in rural West Africa, particularly Nigeria and Sierra Leone, where subclinical infections are frequent, but where there have also been a number of outbreaks with high mortality rates. There are more than 100 000 new infections yearly in this area with thousands of deaths. It was the only viral haemorrhagic fever, until the Ebola outbreak in 2014, to have reached the UK, where 10 imported cases have been diagnosed in the last 20 years. Although this number is small, the serious nature of the disease dictates continued vigilance when dealing with febrile illnesses in patients arriving from endemic areas within the previous 3 or 4 weeks. Rarely, it may spread from person to person, possibly via skin cuts and abrasions. In the UK there is a dedicated special Lassa unit at the Royal Free Hospital which was used to treat the healthcare workers and military personnel who contracted Ebola in W. Africa in the 2014 outbreak.

Argentinian haemorrhagic fever

This is caused by the Junin arenavirus. Those predominantly infected are male field workers who come into contact with the excreta of chronically infected wild rodents. Thousands of cases may occur when the maize is harvested in January–April; the case fatality rate is 10–20%.

Bolivian haemorrhagic fever

This is due to another arenavirus, Machupo, which again is transmitted from rodent excreta, although person-to-person

spread is not unknown. Outbreaks have occurred both in town and country dwellers, but the incidence has greatly diminished in recent years.

Lymphocytic choriomeningitis (LCM)

LCM is widespread throughout the world in house mice, who are chronically infected without clinical signs. Contact with mice or their excreta brings the virus to humans as a 'dead end' infection, with no further spread to others.

15.3.4 Laboratory diagnosis

Remember that only high-security category 4 laboratories should handle blood and urine from patients with suspected haemorrhagic fevers. Consensus oligonucleotide primers for New World arenaviruses are now available and **diagnosis by PCR** is the method of choice.

Virus and viral RNA can be recovered from the blood and urine of acutely ill patients for several weeks. Virus may also be isolated from throat swabs and urine and from autopsy specimens of lymphoid tissue, bone marrow, and liver.

15.3.5 Prevention and treatment

The specific treatment of choice of Lassa fever is the nucleoside analogue **ribavirin** (Chapter 31) administered intravenously, either alone or with convalescent plasma. The drug should be administered as soon as possible after onset of the disease. Bacterial superinfections are common and need to be monitored carefully.

As for all the haemorrhagic fevers, supportive nursing care is often life-saving, and comprises careful maintenance of fluid and electrolyte balance, protein replacement therapy, and appropriate support for cardiac complications. Isolation and barrier-nursing experience is vital to prevent spread to other patients or to nursing and doctoring staff. Control of rodents is an important method of controlling arenavirus infections, but may be impracticable in rural areas.

15.4 Risk categories of patients

Where haemorrhagic fevers are endemic or causing an epidemic, the possibility of such infections will be well to the fore and the diagnosis of moderate or severe illnesses will probably be fairly obvious. This may not be true of travellers from such areas who fall ill after arriving in a non-endemic country. **The importance of taking a careful history from anyone developing a febrile illness within a month of arrival from an endemic area cannot be overstressed**. In many cases the illness is due to malaria, which should always be tested for, but even so, the possibility of a double infection should not be overlooked. Figure 15.10 summarizes the procedures for dealing with such situations. It is most important that the patient be assigned to the appropriate risk category to avoid over- or under-reacting to the possibility of a haemorrhagic fever.

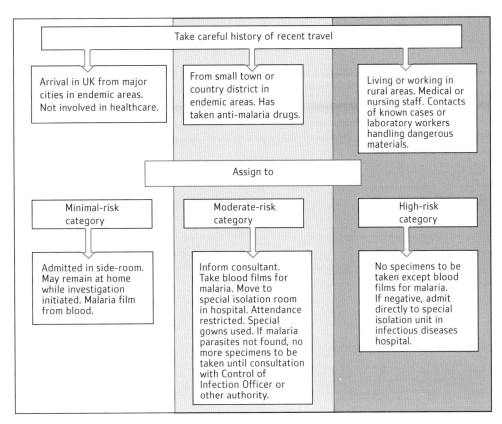

Fig. 15.10 Summary of procedures for dealing with suspected haemorrhagic fevers.

 # Reminders

- These are spherical enveloped viruses with a sandy appearance internally.
- The arenaviruses Lassa, Junin, and Machupo (family *Arenaviridae*) have a 'sandy' appearance due to incorporation of cellular ribosomes in the virion. Their genomes are segmented, and are ambisense with positive-stranded RNA at the 5' end and negative-stranded RNA at the 3' end.
- The **arenaviruses** cause persistent infections in rodents and are transmitted to humans by **contact with rodent excreta** either as an aerosol or via skin lesions or by ingestion of contaminated food or water. They are prevalent in Latin America and parts of Africa. LCM virus occurs worldwide.
- In humans, infections are often inapparent. **Lassa, Junin**, and **Machupo** arenaviruses may cause **severe haemorrhagic fevers** with high fatality rates; **LCM** virus may cause 'aseptic' **meningitis** or rarely, meningoencephalitis.

- Lassa infections are **pantropic**, causing lesions in many organs, notably the liver and spleen. Prostration, diarrhoea, rashes, vomiting, and bleeding from the **respiratory, gastrointestinal**, and **urogenital tracts** are features of the haemorrhagic forms of these illnesses.
- Prompt treatment with **convalescent plasma** and **intravenous ribavirin** may be of some value in Lassa, but supportive care with careful maintenance of fluid and electrolytes to prevent shock are key elements for survival. Adherence to infection control procedures for actual and suspect cases is important.
- The diagnosis of these illnesses is based on the **history and circumstances of travel in endemic areas**, and the exclusion of malaria and other tropical fevers. Laboratory investigations must be carried out in specialist category 4 laboratories. PCR is used now for rapid diagnosis, particularly in remote communities, but only by trained staff.

Further reading

Ambrosia, A. *et al.* (2011). Argentine haemorrhagic fever vaccines. *Hum Vaccine* **7**, 694–700.

Bowen, M.D., Rollin, P.E., Ksiazek, T.G., Hustad, H.L., Bausch, D.G., Demby, A.H. *et al.* (2000). Genetic diversity among Lassa virus strains. *Journal of Virology* **74**, 6992–7004.

Ishii, A. *et al.* (2012). Molecular surveillance and phylogenetic analysis of old world arenaviruses in Zambia. *Journal of General Virology* **93**, 2247–51.

Questions

1. Describe the symptoms of Lassa. How are patients infected and how is the virus controlled in the community?

2. Describe the genome of Lassa and summarize replication strategies of the family.

3. What are the routes of infection of arenaviruses? How are these infections best controlled?

Bunyaviruses: Hanta, phlebo, and nairo

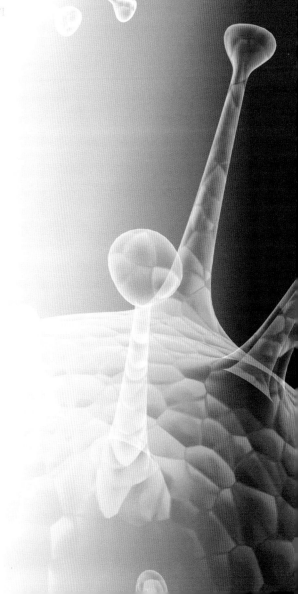

16

16.1 Introduction

The family (Fig. 16.1) takes its name from the prototype Bunyamwera virus, which, like many arboviruses, is named after the place where it was first isolated, namely in Uganda. This family of some 350 viruses was first established in 1975 (Fig. 16.2).

There are five genera, of which four are arboviruses pathogenic for humans and other animals, namely *Orthobunyavirus*, *Phlebovirus*, *Nairovirus*, and *Hantavirus* (Table 16.1).

Laboratory studies of these viruses are carried out at high containment level 4 (the same containment level where anthrax and Ebola are studied) to obviate aerosol infection of staff. Some of these haemorrhagic fevers are classified as emerging human pathogens. They are spread via mosquitos, sand flies, and ticks with the exception of hantaviruses, which are spread through aerosols of urine of mice and rats.

These viruses amplify by persistent infection in mosquitoes, midges, sand flies, and ticks, and spread by vertical transmission in the egg or venereally. Reassortment of the three genes of bunyaviruses can occur, most likely in their insect vectors such as mosquitoes, and, given the vast number of these viruses, the potential for new viruses to emerge is very considerable.

16.2 Properties of the viruses

16.2.1 Morphology

Bunyavirus morphology is that of an enveloped virus, with spherical virions of 90–100 nm in diameter. These viruses contain no matrix proteins. Cryo electron tomography has given insights into the structure of the bunyavirus virion and shows

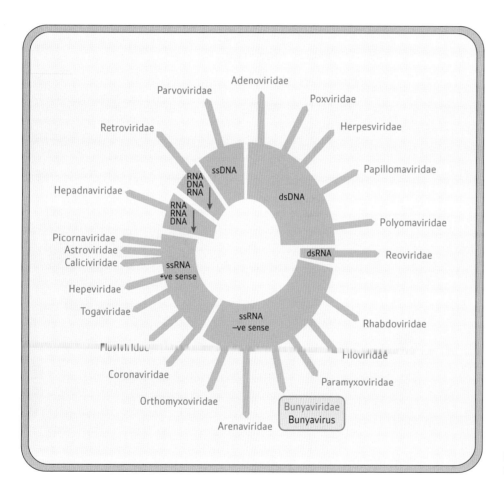

Fig. 16.1 Baltimore classification scheme and bunyaviruses.

Chinese records	Phlebotomus fever described in Mediterranean basin	Sandflies transmit Phleboviruses	Clinical description of virus	Rift Valley Fever isolated in Kenya	Nairovirus Crimean-Congo haemorrhagic fever isolated in Crimea	Phlebovirus Uukuniemi isolated in Finland	Virion of HFRS isolated	Sin Nombre causes HPS	Bunyaviridae virus factories in Golgi discovered	Hantavirus transmission via rodent litter	Kidney cells of voles are sensitive to Hantavirus growth
960	1820	1900	1913	1930	1944	1960	1976	1993	2008	2010	2012

Fig. 16.2 Timeline for family *Bunyaviridae*.

Table 16.1 Classification of bunyaviruses

Family	Genus	Approximate number of viruses	Representative diseases	Vector	Distribution
Bunyaviruses	Orthobunyaviruses	170	California encephalitis, Oropouche fever, La Crosse	Mosquito	America S. America N. America
	Phlebovirus	50	Sand fly fever, Rift Valley fever	Sand fly Mosquito	Europe, Africa, Asia
	Nairovirus	35	Crimea-Congo haemorrhagic fever	Tick	E. Europe, Africa, Asia
	Hantavirus		Hantaan, haemorrhagic fever with renal syndrome (HFRS), Puumala, mild HFRS	Bank vole	W. Europe
			Sin Nombre with pulmonary syndrome	Deer mouse	N and S America

that the external spike glycoproteins, called Gn and Gc, can form ordered structures such as the icosahedral lattice of the phleboviruses. The interior of the virion is however, unorganized (Fig. 16.3). Hantavirus virions, however, show a non-icosahedral Gp structure with a lattice of tetragonal spike complexes with the basal portion of the Gp spikes extending to the centre where they are in proximity to the RNP.

16.2.2 Genome

The bunyavirus genome is negative-sense single-stranded RNA in three segments, large (L), medium (M), and small (S), with sizes of approximately 7, 4, and 2 kb, respectively (Fig. 16.4).

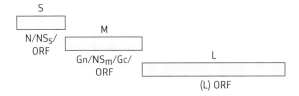

The labelled proteins are Gn, Gc (envelope spikes), N (nucleocapsid), L (polymerase), NS_s, NS_m (non-structural proteins).

Fig. 16.4 Genome structure and encoded proteins of bunyavirus. Labelled proteins are G1 and G2 (envelope spikes), N (nucleocapsid), L (polymerase transcription protein), NS (non-structural protein), o (RNA transcriptase).

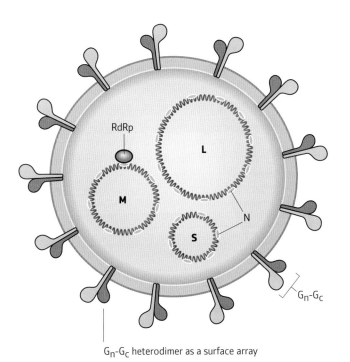

Nucleocapsid (N) wrapping the three RNA segments (S, M, and L) in a circular configuration

G_n-G_c heterodimer as a surface array

Fig. 16.3 Structural features of the bunyavirus virion.

Each segment has an open reading frame and is flanked at the 5′ and 3′ ends by complementary terminal nucleotide sequences. Since each genome segment has complementary sequences at the 5′ and 3′ ends, base pairing results in the formation of pan-handle-like structures. Total genome size varies amongst the sub-groupings of the family and for example genome size reaches 19kb for nairoviruses with a very large L segment.

16.2.3 Replication

The viruses attach to a cellular receptor and enter a cell by endocytosis. For example, hantavirus attaches to integrins and phlebovirus attaches to DC-SIGN of dendritic cells (Fig. 16.5). A low pH is needed for successful infection. Replication is entirely cytoplasmic. Unusually for a negative-strand RNA virus, repli-cation and virus assembly occur in tube-like factories around the Golgi complex. Primary transcription of the virus genome segments is initiated by the virus L protein. Similarly to influenza virus, 12–15 nucleotides are cleaved from the 5′ end of cellular mRNA molecules. These 'snatched caps' are then used as primers for transcription of each of the negative-sense RNA gene segments into subgenomic mRNAs. The viral M segment encodes a polyprotein that is cleaved into Gn and Gc proteins by host-cell proteases. The Gn and Gc heterodimer glycoprotein spikes mediate virus assembly as well as virus attachment to cells. The small segment encodes the nucleocapsid (N) and a NS protein in overlapping open reading frames. Different start codons are used and are encoded in different ORFs, the NSs being located within the N ORF. The NS protein is an important virulence factor that antagonizes the interferon response.

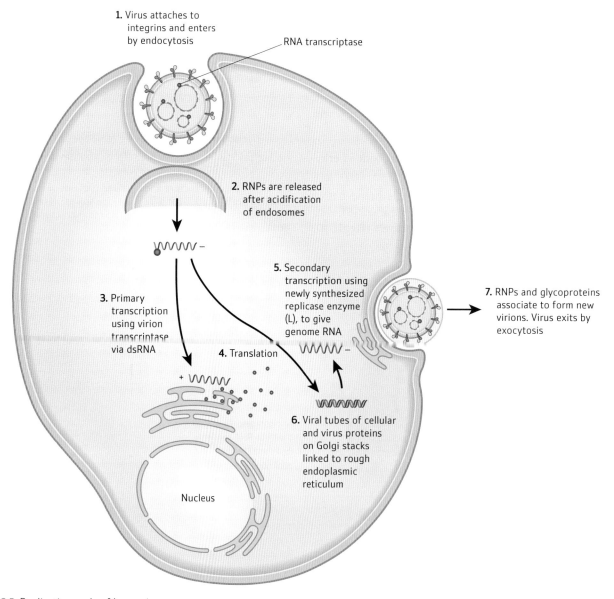

1. Virus attaches to integrins and enters by endocytosis

RNA transcriptase

2. RNPs are released after acidification of endosomes

3. Primary transcription using virion transcriptase via dsRNA

4. Translation

5. Secondary transcription using newly synthesized replicase enzyme (L), to give genome RNA

6. Viral tubes of cellular and virus proteins on Golgi stacks linked to rough endoplasmic reticulum

7. RNPs and glycoproteins associate to form new virions. Virus exits by exocytosis

Nucleus

Fig. 16.5 Replication cycle of bunyavirus.

The viral N protein has RNA binding properties and in some hantaviruses facilitates preferential attachment of ribosomes and capped virus transcripts. The switch from primary transcription to genome replication requires the L protein. Genomes and anti-genomes are encapsidated by the N protein to form biologically active RNPs. Viruses bud into secretory vesicles and are trafficked to the plasma membrane.

Unlike other negative-strand viruses, bunyaviruses assemble and mature at the Golgi. The virus glycoproteins accumulate in the Golgi apparatus, associate with nucleocapsids, and are eventually released by budding into the Golgi. Virions are transported within secretory vesicles and are released from the cells when these vesicles fuse with the plasma membrane.

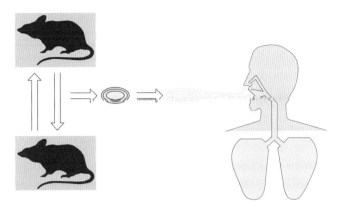

Fig. 16.7 Animal reservoir of hantavirus and transmission to humans via inhalation of dried urine

16.3 Clinical and pathological aspects

16.3.1 Haemorrhagic fever with renal syndrome (HFRS)

HFRS (haemorrhagic fever with renal syndrome) ranges in severity from mild to a severe life-threatening infection with a mortality of $\approx 5\%$ and is currently a notable public health problem causing, for example, 100 000 cases per year, mainly in China (Fig. 16.6). Viruses causing the syndrome are Hantaan, Dobrava, Seoul, and Puumala. The viruses are spread by contact with rodent excretion (Fig. 16.7). The incubation period varies widely, but is usually about 3 weeks. The onset is marked by **malaise, fever, backache**, and **abdominal pain**, and is typically followed by a **hypotensive phase** 5 days later. There may be thrombocytopenia, with petechiae on the face and trunk, and severe bleeding in the gastrointestinal tract and CNS. **Renal function becomes impaired** by the ninth day, with oliguria, proteinuria, and elevated blood urea and creatinine. About half the deaths occur at this time and are due to renal failure, pulmonary oedema, or shock. With good management, the mortality rate is about 5%, but may be higher. Convalescence is often prolonged.

Nephropathia epidemica is caused by the Puumala strain of hantavirus; it is prevalent in Scandinavia and western Europe, and resembles HFRS, but is less severe, with a mortality of >1%.

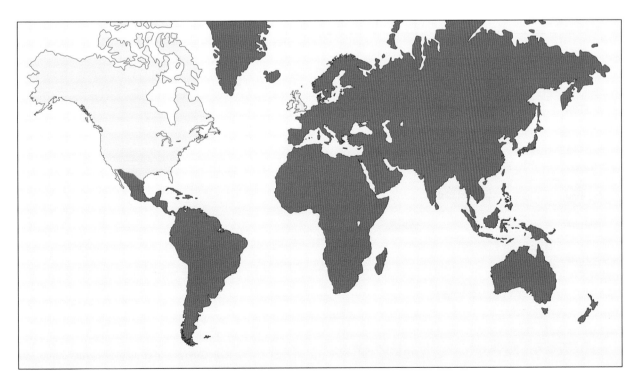

Fig. 16.6 Geographical distribution of hantavirus.

16.3.2 Hantavirus pulmonary syndrome (HPS)

In 1993, there was an outbreak of severe acute respiratory illness in the south-western states of the USA, with a mortality rate of 50%. The virus is called Sin Nombre and the disease itself hantavirus pulmonary syndrome (HPS).

The onset is sudden, starting with cough and muscle pains, followed by dyspnoea, tachycardia, pulmonary oedema, pleural effusion, and hypotension. Death is due to respiratory failure.

Immune pathology

In hantavirus infections there is increased expression of a wide range of T-cell inflammatory cytokines, including TNF-α, IL-1, IL-5, IL-6, IL-10, IL-15, and IFN-γ and -β. This cytokine dysregulation causes pathological changes which may be mediated by direct interaction of viral and host cell proteins. Epithelial cells, monocytes, and macrophages are the main target cells and spread the virus within the body. Lung endothelium and kidney tubular cells are involved in HPS and HFRS, respectively.

16.3.3 Bunyamwera virus

This infection, the prototype virus of the family, is widely distributed in Africa. Clinical findings are typically fever, maculopapular rash, vertigo, neck stiffness, and loss of visual acuity (Fig. 16.8). Another member of the genus, namely California Encephalitis Virus, is found across western parts of the USA and Canada. Usually 100 cases are diagnosed yearly, with these figures increasing tenfold in an epidemic year. The vector of this virus, and the closely related La Crosse which can be prevalent in the midwest states of Iowa and Ohio, is *Aedes triseriatus*, a mosquito breeding mainly in tree holes or vehicle tyres. Not unexpectedly most infections occur in forest workers or persons on holiday. There is a sudden onset of fever, headache, lethargy, nausea, and vomiting. More severe cases may have seizures and life-threatening convulsions.

16.3.4 Phleboviruses

A typical phlebovirus, Rift Valley fever, is epidemic in sub-Saharan Africa and may be caught from insect vectors or from freshly killed animal carcasses. Following an incubation period of 2–6 days most persons have a benign febrile illness with fever, myalgia, and malaise. In 1% of patients gastrointestinal bleeding, jaundice, and shock develop and mortality in this group may exceed 50%.

16.3.5 Nairoviruses

A typical nairovirus, Crimean-Congo haemorrhagic fever, follows the geographical range of the host tick which includes Africa, Eastern Europe, the Middle East, and Asia. After an incubation period of 3–6 days the patient has fever, chills, myalgia, headache, and nausea and is often flushed and vomiting. The haemorrhagic features develop about 3–6 days later with blood in the urine and a range of rashes with somnolence and dizziness. Later shock from internal haemorrhage leads to coma and death. The case fatality is up to 50%.

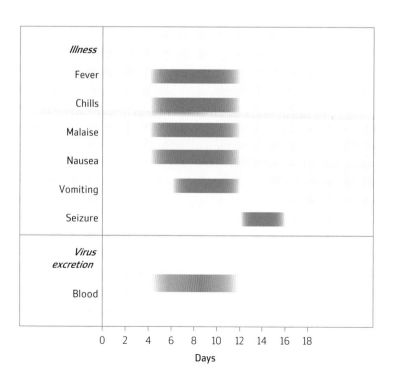

Fig. 16.8 Clinical and virological events during infection with Bunyamwera virus.

Reminders

- The virions are spheres with an envelope and projecting spikes.
- Bunyavirus genomes are segmented **negative-sense ssRNA**; the three unique genome segments (S, M, and L) are wrapped as a helical ribonucleoprotein complex.
- Virus replication is cytoplasmic in tube-like factories with virus assembly at the Golgi.
- Viral mRNA synthesis primed by capped oligonucleotides from host cell mRNAs.
- Bunyavirus infections are **zoonoses**, with reservoirs mostly in **small mammals** and **birds**. With the exception of hantavirus, which is spread from rodent urine, they are transmitted by the bites of blood-sucking arthropods, i.e.

mosquitoes, sandflies, and **ticks**. The illnesses they cause are often **biphasic**, an initial **viraemia** being followed some days later by infection of the target organs. Many infections are **inapparent**, but some are more severe, resulting predominantly in:

- – fever/rashes/arthritis/myositis; *or*
- – meningitis/encephalitis/encephalomyelitis; *or*
- – haemorrhagic fevers.

- There are no vaccines or antiviral drugs and prevention and control depend upon anti-mosquito nets, chemicals, and avoiding skin exposure at dusk. Hantavirus infections are prevented by controlling household penetration by small mammals.

Further reading

Hepojoki, J. *et al.* (2012). Hantavirus structure—molecular interactions behind the scene. *Journal of General Virology* **93**, 1631–44.

Swaneppel, R. (2000). Bunyaviridae. In Zuckerman, A.J., Banatvala, J.E., and Pattison, J.R. (eds), *Principles and Practice of Clinical Virology*, 4th Edition, pp515–49. John Wiley, Chichester.

Walter, C. and Barr, J.N. (2014). Recent advances in the molecular and cellular biology of bunyamwera. *Journal of General Virology* **92**, 2467–84.

Questions

1. Describe the clinical features of hanta virus infection.
2. Write short notes on:
   a. Bunyaviruses.
   b. Cytoplasmic replication of members of the family.
   c. The reasons for the absence of vaccines and antivirals.

17 Paramyxoviruses: measles, RSV, mumps, parainfluenza, metapneumovirus, and the zoonotic henipaviruses

17.1 Introduction

The *Paramyxoviridae* cause a variety of diseases, predominantly involving the respiratory tract, in humans, birds, and other animals (Fig. 17.1)

In humans, they include measles, respiratory infections caused by Respiratory Syncytial Virus (RSV), parainfluenza viruses, and metapneumovirus, and the more innocuous salivary gland infection of mumps. Although few in number, these viruses are responsible for half the cases of croup, bronchiolitis, and pneumonia in infants.

The paramyxoviruses are worldwide in their distribution and cause respiratory disease in all age groups, but predominantly in children. They are also responsible for a number of economically important infections in domestic and farm animals. The henipaviruses Nipah and Hendra are pathogenic zoonotic members of the family and are also examples of emerging viruses.

The viruses are transmitted by airborne droplets or hand contact, and spread is rapid among children in institutions.

17.2 Properties of the viruses

17.2.1 Classification

Members of the family *Paramyxoviridae* are enveloped, negative-stranded **ssRNA** viruses (Fig. 17.2), 150–200 nm in diameter, with a nucleocapsid of helical symmetry. Two subfamilies are recognized: the *Paramyxovirinae*, containing the genera *Paramyxovirus*, *Rubulavirus*, *Avulavirus*, *Respirovirus*, *Henipavirus*, and *Morbillivirus*; and the *Pneumovirinae*, with two genera, *Pneumovirus* and *Metapneumovirus* (Table 17.1). The morbilliviruses differ from those in the other genera in not possessing a NA.

17.2.2 Morphology and structural proteins

Because of the flexibility and fragility of the lipoprotein envelope, the viruses often appear distorted or disrupted in the EM, with the nucleoprotein spewing out from inside the virion (Fig. 17.3). The structural polypeptides of these viruses are

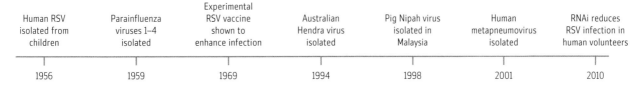

Human RSV isolated from children	Parainfluenza viruses 1–4 isolated	Experimental RSV vaccine shown to enhance infection	Australian Hendra virus isolated	Pig Nipah virus isolated in Malaysia	Human metapneumovirus isolated	RNAi reduces RSV infection in human volunteers
1956	1959	1969	1994	1998	2001	2010

Fig. 17.1 Timeline for paramyxovirus.

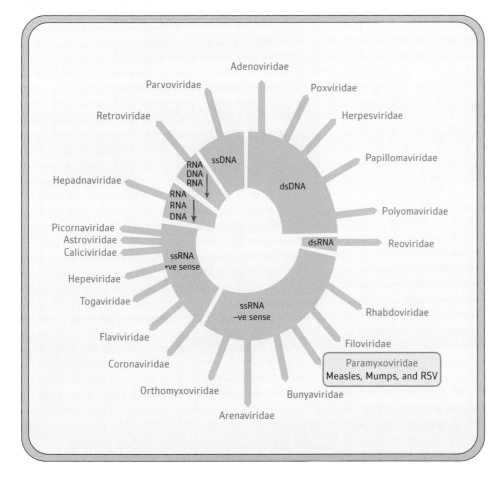

Fig. 17.2 Baltimore classification scheme and paramyxoviruses.

Table 17.1 The paramyxoviruses and their diseases

Subfamily	Genus	Representative viruses	Diseases of humans	Economically important diseases of domestic animals
Paramyxovirinae				
	Paramyxovirus	Human parainfluenza viruses types 1, 3	URTI, notably croup (type 1); pneumonia, bronchiolitis (type 3)	URTI in various species
	Henipavirus	Nipah	Neurological disease	CNS in pigs
		Hendra	Respiratory effects	URTI in horses
	Rubulavirus	Human parainfluenza viruses types 2, 4	URTI, including croup (type 2)	URTI in various species, including Newcastle disease in poultry
		Mumps	Mumps	None
	Morbillivirus	Measles virus	Measles	Canine distemper
				Rinderpest in cattle
	Respirovirus	Parainfluenza virus types 1, 3	URTI	Bovine parainfluenza
Pneumovirinae				
	Pneumovirus	Human respiratory syncytial virus	URTI, notably bronchiolitis	Respiratory infections of cattle and poultry
	Meta-pneumovirus	Human metapneumovirus	URTI	None

URTI, upper respiratory tract infection.

diverse. The rubulaviruses, respiroviruses, and avulaviruses encode a sialic acid binding haemagglutinin-neuraminidase (HN) protein embedded in the virus envelope which allows virus entry via the plasma membrane. Viruses in the morbillivirus and henipavirus genera encode spike proteins, H and G respectively. These are structurally related to HN, but bind to cell-surface protein receptors rather than sialic acid. Viruses of the *Pneumovirinae*, which includes RSV and MPV, encode a different attachment glycoprotein (G), with no structural relationship to those in the *Paramyxovirinae* subfamily. Receptor binding of HN, H, or G allows the triggering of the virion membrane F (fusion) glycoproteins, which form the surface spikes and mediate membrane fusion. The **F** or fusion protein, formed by proteolytic cleavage of a larger precursor polypeptide, is important, as it also mediates the fusion of infected cells that then form the giant cell syncytia so characteristic of infections with this group of viruses. Within the virions is the M (matrix) protein (Fig. 17.4 and Table 17.2), which, together with three other proteins and the virus RNA, form the nucleoprotein core of the virion, and the RNA-dependent RNA polymerase possessed by all negative-stranded viruses (Table 17.2). There are two NS proteins.

17.2.3 Genome

The genome of the *Paramyxovirus* is an **ssRNA** molecule of negative sense and is some 15–19 kb in length containing the 10 genes coding for the 11 known virus-specific proteins (Fig. 17.5). These viruses are classified by Baltimore as class V

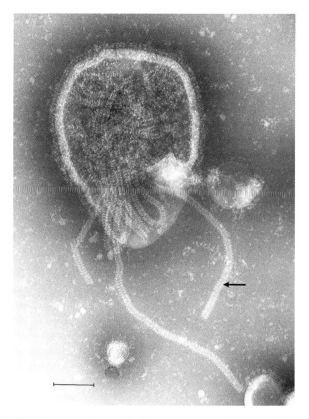

Fig. 17.3 Electron micrograph of a paramyxovirus. The virion has ruptured, spilling out coils of ribonucleoprotein (arrowed). (Courtesy of Dr David Hockley.) Scale bar = 100 nm.

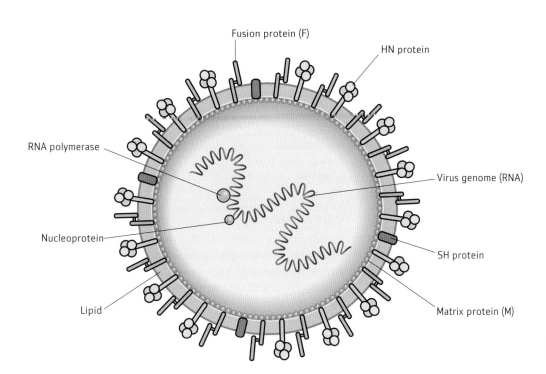

Fig. 17.4 Structural features of paramyxoviruses.

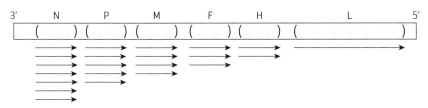

mRNAs are produced by a stop-start mechanism with the polymerase falling off the template at the intergenic boundaries. The chances of successful re-initiation are reduced and a gradient is established with less abundant production of the L protein compared to N.

Fig. 17.5 A 'generalized' paramyxovirus genome structure illustrating slippage. Labelled proteins are N (nucleocapsid), P (polymerase), M (matrix), F (fusion), HN (HA–NA), and L (large polymerase).

Table 17.2 Proteins of paramyxoviruses

Protein	Function
Nucleoprotein (N)	Nucleocapsid protein binding to RNA
Phosphoprotein (P)	Polymerase co-factor
M2 ORF 1 protein (M2-1)	Transcription elongation/anti-termination factor
Large (L)	Polymerase component with catalytic domains
Matrix (M) Fusion (F)	Lines inner surface of envelope. Transmembrane envelope surface protein. Mediates fusion in viral penetration and syncytium formation: F0 precursor activated by cleavage into disulphide-linked F1, F2
Glycoprotein (SH)	Transmembrane envelope surface protein. Attachment protein
Small hydrophobic (SH)	Transmembrane envelope surface protein. Function unknown
M2 ORF2 protein (M2-2)	Status as structural/non-structural unknown; low abundance; shifts balance of RNA synthesis from transcription to replication, gene deletion virus is viable but attenuated
Non-structural 1 (NS1)	Non-structural protein; interferes with interferon-κ/β induction
Non-structural 2 (NS2)	Non-structural protein; interferes with interferon-κ/β signalling

and importantly the genome is non-segmented (the genome is just one piece of ssRNA).

The genomes of paramyxovirus encode up to 10 genes. The beginning and end of each gene contain transcriptional control regions that are also copied into the mRNA. Between each gene are intergenic regions, which are exactly three nucleotides for morbilliviruses like measles, but can be variable in length for other paramyxoviruses. The gene order for the negative sense genome can be generalized as 3′–N-P-M-F-[HN or H or G]—L-5′. Additional genes are present in different genera and overlapping open reading frames in the *P* gene also increase the protein-coding capacity of the genome. For example, in measles the gene codes for three proteins (P, V, and C) by a curious dual mechanism of internal initiation of translation and insertion of a non-templated guanosine residue into mRNA to shift the reading frame. This guanosine is inserted into the mRNA during transcription by a so-called 'stuttering' of the polymerase. The large P protein is translated from an mRNA copy of the complete gene. The smaller C protein is read, following initiation from an internal initiation codon, in a different reading frame. The V (NS1) protein has an extra guanosine inserted into the mRNA during transcription, making this frame distinct from that utilized to translate the C protein.

17.2.4 Replication

All aspects of paramyxovirus replication occur in the cytoplasm and the replication cycle is completed in 14–30 h in cell culture. The strategy of replication is that of a typical negative-stranded RNA virus (Chapter 3). For the respiroviruses and rubulaviruses sialic acid acts as the cell-surface receptor. The receptors for other paramyxoviruses are not well characterized; however, the measles virus uses the cell-surface receptor CD150SLAM and/ or CD46 to infect cells. The virion enters the cell by mediating fusion of its lipid envelope with the external plasma membrane of the cell and this vital event of catalysed 'fusion from without' (Step 1, Fig. 17.6) is mediated by the F protein. The nucleocapsid containing the N protein and virus genome is released immediately into the cytoplasm. (Step 2, Fig. 17.6)

The paramyxoviruses carefully control the amount and timing of expression of each virus gene during the replication cycle (Step 3, Fig. 17.6). This is known as primary transcription. Later the increase in negative sense genome production allows the secondary transcription phase (Step 5, Fig. 17.6) to take place, leading to the rapid production of high levels of viral RNA.

In primary transcription only the intact negative-sense viral genome is transcribed by the vRNAP using the RNA-dependent RNA polymerase enzyme carried into the cell by the virion to give 6–10 subgenomic, generally monocistronic (i.e. only one protein is encoded by a single mRNA species) mRNAs (Fig. 17.6). The vRNAP binds at the 3′ end and moves along the genome beginning with the first gene N. After the gene has been transcribed, the vRNAP enters the terminator region, stops transcribing, and follows one of two alternative pathways. First, the vRNAP may fall off the gene and, therefore, must bind again at

the 3′ end and start transcription again. Second, the vRNAP can move along the gene until it reaches the next transcription signal where it starts to transcribe the second gene, namely P. Therefore genes at the 3′ end are transcribed to higher levels than those at the 5′ end of the genome. Production of viral proteins inhibits viral transcription and helps the switch from genome transcription to replication (Step 5, Fig. 17.6). Shortly afterwards the newly synthesized negative RNA strands interact with N protein and the vRNAP, and these nucleocapsid structures then combine with viral M protein. The viral spikes already inserted into the plasma membrane of the cell interact with the M protein. The viruses are released from the infected cell by budding from the plasma membrane (Step 6, Fig. 17.6). Because of the tendency of infected cells to fuse, they may, under certain conditions, release hardly any new viruses into the extracellular environment; instead, the viruses may spread from cell to cell, behaving as typical 'creepers' (Chapter 4).

17.3 Clinical and pathological aspects

17.3.1 Measles

'Measles' derives from a Middle English word, *maseles*, and so has been with us for several centuries (Fig. 17.7). Its Latin name, *morbilli*, is a diminutive of morbus, a disease, and thus signifies a minor illness. In temperate countries measles is a comparatively mild infection and serious complications are rare; however, in some tropical areas it is still a killer of children on a large scale.

Typically of the family, the virus genome codes for six proteins, three of which participate in the formation of the viral envelope, namely the matrix (M), haemagglutinin (H), and fusion (F) proteins. Both H and F are responsible for host cell attachment and fusion.

Other encoded proteins are N (nucleoprotein), P/V/C (phosphoprotein and non-structural proteins), HN (haemagglutinin-neuraminidase proteins), and L (large polymerase protein) (Fig.17.8).

Clinical features

The period from infection to appearance of the rash is fairly constant at about12 days, but may be less by a few days (Fig. 17.9). Before the rash there is a **prodromal stage lasting 2 or 3 days**, with running eyes and nose, cough, and moderate fever. At this time, careful examination of the buccal mucous membrane adjacent to the molar teeth may reveal **Koplik spots**, which resemble grains of salt just beneath the mucosa. They may be many or few in number, but when present are pathognomic of measles.

The **rash** first appears on the face and spreads to the trunk and limbs within the next 2 days. It is a dull-red, blotchy, **maculopapular** eruption, usually characteristic enough to have its own designation, 'morbilliform'. If severe, it may have a livid purple (purpuric) appearance resulting from bleeding into the skin from small haemorrhages called petechiae.

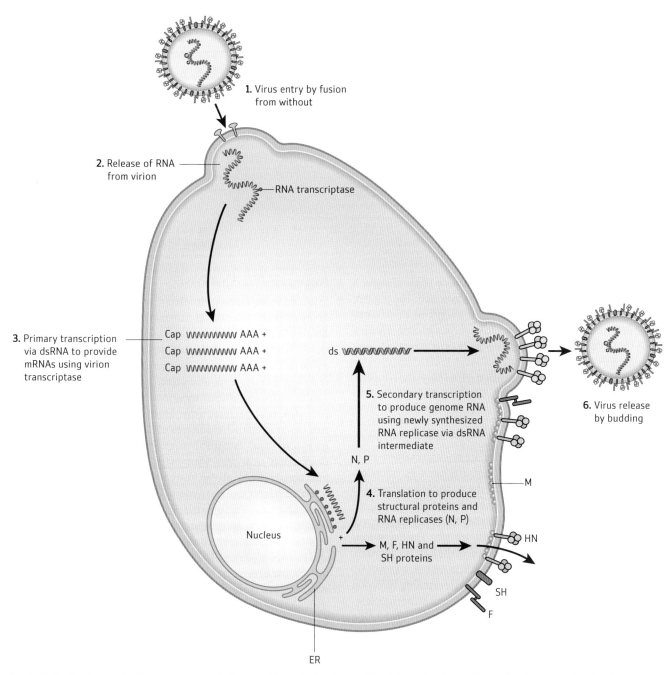

Fig. 17.6 Replication cycle of paramyxovirus in the cell. Step 1: fusion from without; Step 2: release of negative RNA strands; Step 3: primary transcription; Step 4: translation; Step 5: secondary transcription and genome replication; Step 6: virus release.

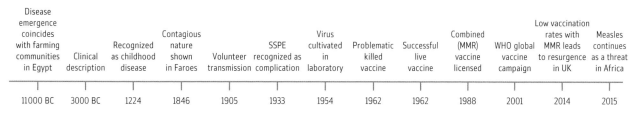

Disease emergence coincides with farming communities in Egypt	Clinical description	Recognized as childhood disease	Contagious nature shown in Faroes	Volunteer transmission	SSPE recognized as complication	Virus cultivated in laboratory	Problematic killed vaccine	Successful live vaccine	Combined (MMR) vaccine licensed	WHO global vaccine campaign	Low vaccination rates with MMR leads to resurgence in UK	Measles continues as a threat in Africa
11000 BC	3000 BC	1224	1846	1905	1933	1954	1962	1962	1988	2001	2014	2015

Fig. 17.7 Timeline for measles.

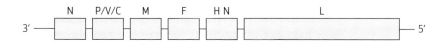

Fig. 17.8 Genome structure of measles.

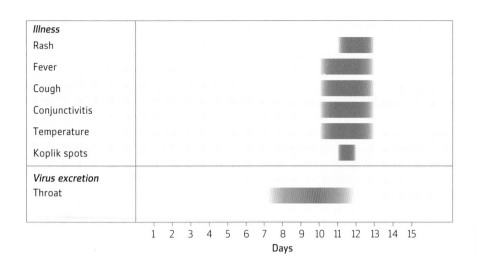

Fig. 17.9 Clinical and virological events during infection with measles.

Although the rash is one of the most obvious physical signs, it is but one manifestation of a generalized infection. The temperature rises sharply to about 40 °C, and in every case there is evidence of **bronchitis** and **pneumonitis**, with cough and 'crackles' in the chest. Occasionally, diarrhoea in the early stages indicates inflammatory lesions in the gut.

Within 2–3 days of onset, the rash starts to fade, the temperature subsides, and the child feels better; recovery is usually uneventful.

Complications

Along with the fever and rash there are widespread lesions of various body systems. Although they are listed as complications, it is arguable whether most of them should be referred to as such, as they are all part of the generalized infection and are thus present in most cases; whether they surface in the form of signs and symptoms depends on their severity. The exceptions are those in which the immune response of the host is abnormal. They include **post-infection encephalitis** (about 1 per 1000 cases), and giant-cell pneumonia, a life-threatening infection seen occasionally in immunodeficient children. **Subacute sclerosing panencephalitis** is particularly insidious, becoming clinically apparent only years after the initial measles infection, perhaps in 1 per 1 000 000 cases.

Table 17.3 Complications of measles

Site	Complication
Respiratory system	Croup (in prodromal stage); bronchitis; giant-cell pneumonia (in immunocompromised patients)
Eye	Conjunctivitis; corneal ulceration (rare)
Ear	Otitis media; possible secondary bacterial infection
Gut	Enteritis with diarrhoea
Central nervous system	Febrile convulsions (acute phase); post-exposure encephalitis (rare); subacute sclerosing panencephalitis (very rare)

Pathogenesis and pathology

Infection is acquired via droplets that enter the respiratory tract or eye. Measles is one of the most highly infectious viruses known. The pathogenesis is that of an acute generalized infection (Chapter 4), the secondary viraemic phase corresponding to the height of the fever. After a minor burst of replication in the cells lining in the respiratory tract, the virus multiplies in local lymph nodes and infects mononuclear cells, particularly lymphocytes, which release virus into the blood whence the virus reaches most epithelial surfaces of the body. Cells of the respiratory tract and conjunctiva are particularly susceptible to destruction and, coincident with this phase, 10 days post-infection there is an abrupt onset of symptoms of a chest infection.

The rash is not due to a direct CPE of virus on cells, but to a reaction by cytotoxic T cells against viral antigen appearing in the skin cells. In addition, antigen–antibody complexes form on the capillary endothelium with consequent cell damage, vasodilatation, and leakage of plasma. The rash is thus a clear sign that a satisfactory immune response is in progress, and that recovery is on the way; conversely, its failure to appear (e.g. in immunodeficient patients) is a bad prognostic sign. In these patients giant-cell pneumonia may occur, often many weeks after the acute infection.

During the prodromal and acute phases, virus is shed in body fluids, including respiratory secretions; it replicates in lymphocytes, causing lymphopenia and a transient immune deficiency. During the acute phase, virus-infected giant cells containing up to 100 nuclei have been found in the pharynx and tonsils, skin, respiratory epithelium, lymph nodes, and Peyer's patches. The virus is widespread in the skin, but disappears quickly with the onset of the rash and the appearance of circulating antibody.

Immune response

During the acute phase, replication of virus within monocytes and other white cells depresses cell-mediated responses to other antigens, although cytotoxic T lymphocytes specifically directed against measles M protein in infected cells are important in the recovery process and also provide protection against subsequent infection. This immunosuppression explains the high mortality before MMR vaccine: and can also explain super-infection by opportunistic infections. The effect of the immune deficiency has recently been uncovered in a study that showed MMR vaccination has an additional positive effect on reducing other, different childhood infections in young children, the corollary being that measles infection is associated with a subsequent increase in other childhood infections.

Epidemiology

Measles virus is highly infectious and has a worldwide distribution. It becomes endemic only in countries with populations large enough to provide a continuing supply of susceptible children; in small, isolated communities it dies out until a fresh importation causes a major epidemic (see Chapter 6). Because of vigorous vaccination campaigns, large epidemics are now unknown in most European countries and the USA. In developing countries, on the other hand, measles still has a high incidence in infants less than 2 years old and is much more severe, with unusual clinical features, such as blindness, and a high case fatality rate of 3–6% (see 'Measles in developing countries' later in this section). The World Health Assembly plans to eradicate measles and, with this target in view, plans to reduce measles mortality by 95% (compared to 2000) and to reduce incidence to less than five cases per million by 2020. All six WHO regions have agreed targets.

Laboratory diagnosis

In day-to-day practice, laboratory confirmation of the clinical diagnosis is not needed because the rash is so typical. In atypical cases, usually seen in hospitals, RT PCR is the method of choice.

Prophylaxis

Modern attenuated vaccines are very effective and, ideally, are given in combination with mumps and rubella vaccines (MMR) at 13–15 months of age (see Chapter 30), by which time there is no risk of neutralization by maternal antibodies. The preparation of vaccine is simplified by the fact that there is only one virus serotype and no evidence of marked antigen variation. In the USA measles is nearly eradicated by achieving vaccination rates approaching 100%. As part of the campaign, in the USA children were not admitted to state schools without a certificate of vaccination. Measles deaths were about 400 per annum in 1960, dropping to fewer than 10 per annum in 2014, and measles encephalitis around 300. In the UK notifications of measles have dropped from 200 000 to 1144 in 2014. The economic saving in medical care and hospital beds is therefore enormous. However, isolated outbreaks still occur in ethnic minority groups and, in some cases, university students. To maintain high levels of immunity, children are now given a booster vaccine at age 3–5 years (11–12 years in the USA).

There has however, been a setback in the vaccination programme in the UK, where flawed data were published attempting to show that the vaccine virus was associated with autism. In fact, there is no sound evidence of such an association, but meanwhile parental confidence in the MMR vaccines was seriously undermined and has only recently returned.

Passive protection with normal immunoglobulin, in a dose sufficient to modify but not completely prevent measles, is valuable for protecting debilitated or immunodeficient children who have been exposed to infection.

Measles in developing countries

Over 1 million malnourished children die each year of measles in the developing world. Two host factors seem to be of primary importance in explaining why measles is much more severe in developing countries than in the USA and Europe:

* on average, children acquire the infection at a younger age;
* these children are more likely to be poorly nourished.

In these children, measles differs considerably from the relatively benign illness described above, and more nearly resembles the disease as it existed in Europe 100 years ago. The rash is more severe and exfoliates extensively, exposing large areas of skin to bacterial invasion. Inflammation in the mouth (stomatitis) interferes with eating and drinking. Vitamin A deficiency and measles infection combine to cause corneal ulceration and blindness. The virus is cleared less rapidly than in well-nourished people, which results in:

* more damage to lymphocytes with depression of CMI and increased susceptibility to bacterial infections;
* prolongation of the period of virus shedding by a week or more.

Immunosuppression induced by measles allows invasion by other viruses and bacteria. Bronchopneumonia and persistent diarrhoea with protein-losing enteropathy often result in death, mortality rates being of the order of 10%.

Measles vaccine in developing countries

Inoculation of infants at 9 months of age is a top priority of the WHO. The heat lability of measles vaccine complicates its distribution in tropical climates, and maintaining the 'cold chain' is essential. The freeze-dried vaccine has to be held at between 2 and 8 °C until minutes before injection and, unless this temperature range is maintained, the vaccine rapidly loses its potency. The other problem in providing effective immunization for children in developing countries is reaching them during the brief interval between the loss of maternal antibodies and the acquisition of natural disease (the so-called 'window of opportunity'). It was thought that this could be solved by the use of more concentrated ('high-dose') vaccine, which is not swamped by maternal antibody and is effective in babies less than 9 months old. However, field trials showed that suppression of T-cell immunity to other diseases could occur and so the new strategy is to use the standard Schwartz strain of virus at 9 months.

Measles in developed countries

There have been large-scale outbreaks of measles in France, Bulgaria, and Ukraine together with significant problems in Wales (UK). In 2007, 6936 cases in Europe were reported to WHO but these cases increased to 110 000 in 2012. Most cases were in the general population, although some emerged in religious groups. Political and public complacency are important factors in the suboptimal vaccination in the WHO European region, as well as residual fears about the MMR vaccine.

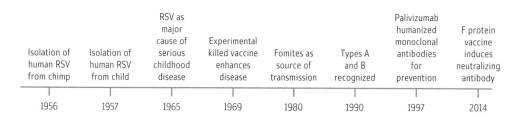

Fig. 17.10 Timeline for RSV.

17.3.2 Respiratory syncytial virus infections

This respiratory syncytial virus, first isolated in 1956 (Fig. 17.10) and called RSV for short, is highly contagious and causes sharp outbreaks of respiratory disease throughout the world, particularly in infants. The burden on children is huge. In the USA annual hospitalization rates reach 17 per 1000 children under 6 months of age and overall this virus causes 20% of childhood hospitalization in the winter period. It is spread by aerosol and hand contact. It may cause death in the young (although deaths are rare in some developed countries such as the USA) and also the elderly and immunocompromised. In the northern hemisphere the yearly outbreak usually occurs in December and in the corresponding winter months in the Southern Hemisphere.

Children excrete virus in the upper airways shortly before symptoms appear (Fig. 17.11). Wheeze and bronchitis are characteristic. RT PCR would detect RNA in the throat for in excess of 16 days and the respiratory symptoms are also prolonged. The disease can be equally serious in the elderly.

Clinical aspects

The illness often starts like a common cold, but within 24 h a baby in particular may be acutely ill with cyanosis and respiratory distress (Fig. 17.11). Typically, there is bronchiolitis, with

or without involvement of the lung parenchyma causing pneumonitis. There is some evidence that RSV infection in infancy may cause long-term respiratory problems. Severe illness carries a significant risk of death at both extremes of age, and in the elderly the infection may be confused with influenza, which is also prevalent at the same time of year. Infection and re-infection are common in the first years of life.

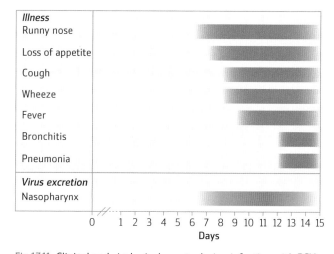

Fig 17.11 Clinical and virological events during infection with RSV.

Medical case story   Respiratory syncytial virus

Colour can be very important to a doctor, to be able to recognize, for example, the white, glassy look of anaemia, the neat-edged red rash of cellulitis, the grey sweat-sheened face of a patient with sepsis, and the subtle blue-mauve lips of cyanosis. Indeed, there are even courses to help colour-blind doctors cope.

One afternoon you call your next patient and check the notes as you wait. It is a 4-month-old baby girl, a second child. Her mother brought her in 2 days ago because of a cough, and came back again yesterday, the child still with a cough, to a different colleague.

This multiple presentation immediately alerts you to a potential hospital referral. You look at the notes again. They are unremarkable: the other doctors have noted the feeding pattern and have checked for respiratory distress. They have both diagnosed a viral cough, and have advised the mother to monitor the baby's breathing and feeding, and to come back if there is further concern.

As the mother enters the room with the baby on one arm, you are alert for signs of respiratory distress, but the baby looks fine. She is not breathing fast, but is relaxed and alert with pink cheeks.

The mother immediately says, 'The other doctors told me to come back if I was worried'; she is obviously anxious that she is taking up too many appointments. You reassure her, you are always keen to see babies with respiratory symptoms promptly. Then you say, 'But she looks fine, doesn't she?'

You go through the story again with the mother: 4 days of coughing, temperatures (but not very high), feeding fine, but less than usual. 'But she is coughing a lot,' says the mum. 'She coughs and coughs, and doesn't seem to be able to catch her breath'.

You undress the baby on the couch, gently pulling off her top and vest. You know it is important to check for signs of respiratory distress with the baby completely undressed. It's surprising how well their clothes can disguise fast breathing.

However, this little baby has no tracheal tug, no intercostal recession, no abdominal breathing, and, most importantly, is not breathing fast. Her respiratory rate is 36 breaths/min and her chest sounds clear.

The baby does not like you listening to her chest: she feels chilly. She pouts and looks for her mother, and then draws in breath and starts to cry. This sets off a coughing fit. She coughs and coughs. You stop talking to the mother to watch. She goes an alarmingly deep red, which clashes catastrophically with her orangey-red hair.

'This is what has been happening,' says the mother. 'She's better when she's upright.'

She picks up the baby who draws breath at last, but who then cough-coughs again, then a third time, and then a fourth. Fortunately, it is a well-lit room, because you are sure her lips have just gone blue.

Then the baby settles: her face goes back to the pale pink colour and her lips to a nice raspberry shade, so that you wonder if you saw anything at all.

You think carefully while the mother dresses the baby. Something is not right. She has presented three times and you think you have seen cyanosis. Could this be whooping cough? Or could this be an unusual presentation of respiratory syncytial virus (RSV)? You mentally slap yourself for your earlier silly remark about the baby looking fine. That will teach you to make glib diagnoses before properly assessing. You make a decision.

'I see exactly what you mean now,' you say to the mother. 'I think we had better get you to go straight up to the hospital.'

The mother looks serious as you discuss why. You then ask her to wait outside whilst you sort out a referral letter and speak to the hospital.

The letter is important: it has to grab the attention of the on-call doctors. '*Third presentation*' and '*cyanosed lips*' need to feature in the first line. It takes time to get through to the hospital, and then for the paediatric registrar to answer his bleep. Fortunately, it is rare for you to refer someone to hospital (less than one patient a week, or one in 90 consultations), although to the hospital it can feel like GPs are sending up every patient they see.

You explain the situation to the paediatric registrar, and he says, 'We are incredibly busy here. Are you sure the patient needs to come up? If she does come here we may not have any beds available'. He adds, 'It sounds like she's feeding fine and managing ok'.

You pause for thought and then you say, 'Look, I don't see cyanosed lips every day and I've never seen it before in a baby of this age.' This baby is going to the hospital whether he accepts her or not. He accepts the patient.

One week later you ring the mother to find out what happened. She tells you that the paediatric team found that, during the baby's coughing fits, her saturations dipped to 77% on air. A respiratory swab revealed RSV. The baby was admitted for 4 days and nursed in an oxygen-filled head box, but is now well and back at home.

The mother is very grateful to you for taking her concerns seriously and referring her. You reflect on this unusual presentation of RSV bronchiolitis.

★ Learning Points

- **Virological:** RSV is spread easily by respiratory secretions on hands and on toys and utensils. It has an incubation of 5–7 days and it continues to be infectious for up to 2–3 weeks after symptoms have settled. Infection with RSV does *not* always confer immunity. Symptoms include runny nose and cough before progressing—typically in babies and infants—to wheeze and respiratory distress, clinically known as bronchiolitis. Treatment is supportive with oxygen and nasogastric tube feeding. There is no vaccine.
- **Clinical:** The signs of respiratory distress include abdominal breathing, intercostal recession, and tracheal tug, and these are all important signs to teach to parents. However, the most important of all—and the one that is the most objective and most easily measured—is the respiratory rate. This is a key

sign to teach to parents. They should go to the hospital if their infant's respiratory rate is, or is approaching, 60 breaths/min (easy to remember: one breath every second).
- **Personal:** The feeling or instinct of a doctor about a patient is a widely accepted risk factor for serious illness. For example, one of the identifiers for a high risk of serious illness, in the NICE guideline: *Feverish illness in children* is, simply: '*Appears ill to a healthcare professional*.'[1]

Notes

[1] NICE Clinical Guideline 47 (2007). Feverish Illness in Children. NICE, London.

Pathogenesis and pathology

The incubation period is about 5 days followed by a mild prodromal period. Later in the disease progression the virus can cause lower respiratory tract infection. There is a necrotizing bronchiolitis in which partial blocking of the bronchioles leads to the collapse of areas of lung. Peribronchial infiltration may spread to give widespread interstitial pneumonitis. Inapparent infections also occur. The host inflammatory response called a 'cytokine storm' (Chapter 5), induced early in infection, is an important contribution to clinical disease and persists after viral replication has finished.

RSV has been much investigated recently and features of the replication and functions of virion and non-structural proteins (Fig. 17.12) discovered.

The two non-structural proteins of RSV (NS1 and NS2) inhibit activation of interferon regulatory factor 3 and also mediate inhibition of cytokine production. The virus G protein regulates type 1 interferon as well as production of cytokines and chemokines. Finally, expression of virus G protein can down-modulate host micro RNA expression and thus can abrogate, or at least reduce, host antiviral response.

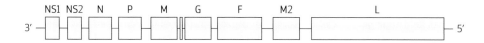

Fig. 17.12 Genome structure of RSV.

Immune response

As infants as young as 6 weeks are often infected, maternal antibody does not seem to provide protection for very long after birth. RSV induces both antibody- and cell-mediated responses; impairment of the latter in immunocompromised patients may lead to **persistent infection**. In later life, **re-infections** are comparatively frequent, suggesting that first infections do not always induce long-term immunity. Antigenic variation between strains of RSV may also contribute to re-infections.

Epidemiology

Respiratory syncytial viruses fall into two serological groups, A and B, which can be distinguished by specific antisera and by nucleotide sequencing of the G and F genes (Fig. 17.12).

RSV infections are transmitted by inhalation of **respiratory secretions** as small droplets and by **contamination of hands** with **fomites** such as bedding. In temperate climates, they occur annually in epidemic form during the winter months, but in the tropics, the incidence may be highest during the summer months or rainy season. Babies aged from 6 weeks to 6 months are predominantly affected; indeed, RSV is the most important respiratory pathogen in young infants.

During epidemic periods, spread within hospitals, crèches, and day nurseries (**nosocomial infections**) often takes place, facilitated by close personal contacts and the liability of those infected to shed virus for up to 3 weeks after the acute phase. Adults may also acquire RSV, particularly the elderly, in whom it may exacerbate existing bronchitis and cause an increase in deaths, easily misdiagnosed as influenza.

Re-infections are common, although they may often be subclinical. These subclinical infections maintain a huge reservoir of infective virus. In a unique study at the Antarctic base at the South Pole, parainfluenza viruses were isolated continually, although the personnel were completely cut off from the outside world for several months during the winter.

Laboratory diagnosis

Detection of viral RNA by RT PCR is the most commonly used method nowadays. The technique of clinical sampling has to be precise and the sample must reach the laboratory quickly. The virus can be isolated in a specialized laboratory in continuous lines of human cells (e.g. HeLa), in which CPEs with formation of syncytia appear after 2–10 days. **Indirect immunofluorescence** detects viral antigen in cells from a nasopharyngeal washing, which gives an immediate result, or detection of viral antigen in nasal fluid may be done using enzyme immunoassay kits.

Prevention

Attempts to develop attenuated vaccines have so far met with little success; the use of recombinant DNA technology for making RSV vaccine is now being studied.

In the absence of a vaccine, limitation of spread within paediatric units, nurseries, and the like depends on good hygienic practice, such as hand washing, covering of the mouth when coughing or sneezing, and careful disposal of paper handkerchiefs.

Treatment

The nucleoside analogue ribavirin (Chapter 31), given by continuous aerosol inhalation to babies with serious RSV infections, may abrogate infection but not commonly enough to make it a routine treatment.

High-titre human immunoglobulin or humanized monoclonal antibodies, such as palivizumab are efficacious in high-risk infants, but the treatment is expensive and is used only in emergencies. These antibodies target the F protein of RSV.

17.3.3 Mumps

The **name probably** originates from an old word meaning 'to mope', an apt description for the miserable child afflicted by this common illness. Mumps was one of the first infections to be recognized and was described by Hippocrates as early as the fifth century BC (Fig. 17.13). Robert Hamilton in *An Account of a Distemper by the Common People of England Vulgarly Called the Mumps*, noted in 1790 that 'The catastrophe was dreadful; for

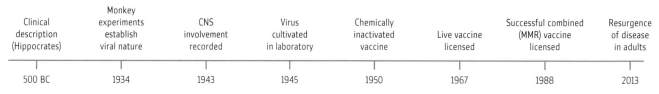

Clinical description (Hippocrates)	Monkey experiments establish viral nature	CNS involvement recorded	Virus cultivated in laboratory	Chemically inactivated vaccine	Live vaccine licensed	Successful combined (MMR) vaccine licensed	Resurgence of disease in adults
500 BC	1934	1943	1945	1950	1967	1988	2013

Fig. 17.13 Timeline for mumps.

the swelled testicles subsided suddenly the next day, the patient was seized with a most frantic delirium, the nervous system was shattered with strong convulsions, and he died raving mad the third day after'. Fear of orchitis and consequent sterility, in reality greatly exaggerated, explains the continued interest in immunization against what is otherwise a comparatively benign disease.

After a long incubation of 16 days or so virus is excreted from the nasopharynx shortly before the first clinical signs are apparent (Fig. 17.14) and later can also be recovered from the urine and much more rarely from the CSF. Meningitis is rare.

Clinical aspects

The onset after a long incubation time of 16 days or so is marked by malaise and fever followed within 24 h by a painful enlargement of one or both parotid glands; the other salivary glands are less often affected. In most cases, the swelling subsides within a few days and recovery is uneventful. Inapparent infections are common. (Fig. 17.14.)

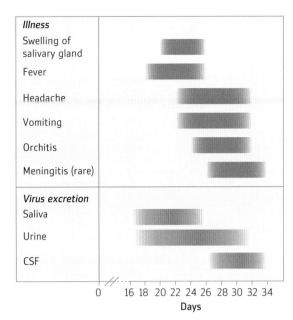

Fig. 17.14 Clinical and virological events during infection with mumps.

Medical case story   Mumps

You are a GP registrar in your first month of practice. Mr F is a new patient. He is 30. He walks into the room with a confident stride, but a doleful expression. His face is swollen on one side.

You stare at him with great interest, and then you recognize the swelling is over the parotid glands. It is more on one side than the other and he looks a little hamster-like. This is the first time you have ever seen a mumps patient.

You have read in the free GP magazine *Pulse* that there is a mumps epidemic amongst young adults this autumn.

Mr F felt unwell for 2 days before his parotid gland swelled up. He still has a temperature and his face feels sore. He is able to drink without a problem and is taking paracetamol. Three weeks ago he was visiting his younger brother at university, who has since been diagnosed with mumps.

You advise him that he has mumps and that this is a viral infection. You say that it is self-limiting (it will get better by itself). He should expect his condition to improve over the next week. In the meantime, he can continue to take paracetamol for pain/tenderness, and he should continue to drink water, even if he doesn't feel like eating. You say he should come back if he is not able to drink or if he has any other concern.

He asks you if there are any complications and you reassure him.

As soon as he leaves the room you reach for your book on infectious diseases. You are aware of a big hole in your knowledge of mumps.

You find out from a patient leaflet[1] that the incubation period for mumps is 2–3 weeks, which fits with Mr F's contact history. He would have been infectious for 5 days before the temperature

came on and for 5 days after. As it happens, you were right about the illness lasting about a week.

Your case study book reads; 'complications: meningitis 5–10%, encephalitis 0.02%, orchitis: 40% post pubertal males, 5% post-pubertal females, risk of sterility small'.[2]

You light-heartedly remark to the practice manager about your mumps patient, and she says, 'Have you notified it?'

You find out that mumps is a notifiable disease in this country, because it is an illness that we vaccinate against. You ask the practice manager how to notify and to whom, and she hands you the appropriate form to send to Public Health England (PHE).

You are curious, so you use the internet (look up the PHE website)[3] and find the following interesting quote about young adults with mumps:

Dr Mary Ramsay who monitors cases of mumps for the Agency said; 'This age group (born between 1981–1989) were too old to be routinely offered MMR vaccine when it was first introduced in 1988, so many will have not received any mumps containing vaccine ...'

... Many of the young adults in this cohort are now at university. The large number of people in semi-closed institutions allows the disease to pass more easily from one person to another. As the susceptible group is quite large we do not expect the number of cases of mumps to reduce in the near future ...

... Along with the Department of Health, our advice is that school leavers and other young adults who have not received MMR or only received one dose, should ensure that they take up the offer of MMR vaccination.

You reflect that adults born before 1981 would have caught mumps as a child. However, as herd immunity increased with the introduction of the MMR, the circulation of mumps would have decreased, leaving those born after 1981 vulnerable as adults.

You ought to ring Mr F to offer further advice and to see if he has any housemates who might need vaccination.

Notes

[1] Leaflet, *Mumps*, 2009. Available at: http://patient.info/health/mumps-leaflet (accessed 19 Feb. 2016).

[2] Humphreys, H. and Irving, W. (2004). *Problem-orientated Clinical Microbiology and Infection*, 2nd Edition. Oxford University Press, Oxford.

[3] http://www.infectioncontroltoday.com/news/2004/11/mumps-cases-among-young-adults-continue-to-increa.aspx (Accessed 19 Feb. 2016—recognized as an old reference but still relevant for this cohort.)

★ Learning Points

- **Virological:** The mumps virus is a paramyxovirus that is spread by saliva and respiratory secretions. The incubation period is 2–3 weeks and infectivity is from 5 days before onset until 5 days after. It causes swelling of one or both parotid glands. Complications include orchitis, meningitis, and encephalitis; the risk of sterility is low. Mumps can be prevented by vaccination.

- **Clinical:** The presentation of mumps is so specific that it has been described and recognized since 500 BC.
- **Personal:** What do you think of the way this doctor handled the question on complications?

Complications

Orchitis

Orchitis develops in about 20% of males who contract mumps after the age of puberty; it may develop in the absence of preceding parotitis. Typically, there is pain and swelling of one or both testicles 4–5 days after the onset of parotitis. The pain is often severe enough to demand strong analgesics. There is often an accompanying general reaction with high temperature and headache. Symptoms tend to subside after a week. Although some degree of testicular atrophy follows in about 30% of cases, sterility following mumps orchitis is rare.

Inflammation of the **ovary** (oophoritis) and **pancreas** has been reported, but these complications do not seem to have serious long-term effects.

Central nervous system

The incidence of 'aseptic' **meningitis** is higher after mumps than after any other acute viral infection of childhood. Rates of 0.3–8.0% have been reported in the USA. This complication almost always resolves without after-effects (sequelae). Post-infection **encephalitis** is, however, more serious and carries an appreciable mortality. These syndromes are described in Chapter 30.

Some degree of **deafness** is a residual complication in a small percentage of cases.

Pathogenesis

The infection is spread in saliva and secretions from the respiratory tract, and is acquired by the **respiratory route**, either by aerosol or hand contact. The incubation period is 16 days (Fig. 17.14). Viraemia during the acute phase is followed by generalized spread of virus to various organs, including the parotid gland. Virus is shed for several days before and after the first symptoms, not only from the respiratory tract, but also in the urine.

Immune response

The appearance of specific IgM, IgA, and IgG antibodies follows the sequence usual for these acute viral infections. Cell-mediated immunity (CMI) is probably important in the recovery process.

Epidemiology

Nucleotide sequence analysis of the F and SH genes has identified 12 (A to N, excluding E and M) genotypes of mumps with genotypes A and G dominating. Recent outbreaks in young adults may indicate the increasing importance of variants alongside waning immunity from the vaccine. Mumps has a worldwide distribution and nowadays mainly affects those aged less than 15 years. In temperate climates, sporadic cases occur all year round, although the incidence is highest in winter, but there is no seasonal variation in tropical countries. Mumps is highly infectious; outbreaks in institutions are common and there have been a number of epidemics in military recruits in barracks. Formerly, it was more common than measles and was an important disabling virus for the military.

Laboratory diagnosis

RT PCR is now used to detect virus RNA in saliva, blood, and urine, whilst virus genotyping of patient isolates can identify transmission pathways. Virus can be isolated in specialized laboratories in various cell lines and in monkey kidney cells and identified by haemadsorption or haemagglutination inhibition (Chapter 29).

Prevention

Live mumps vaccines are prepared most commonly from the Jeryl Lynn attenuated strain. There is an appreciable incidence of post-vaccination meningitis with the Urabe strain, which has now been abandoned in the UK. In the UK and USA, mumps vaccine is combined with measles and rubella vaccines (MMR). Where uptake rates are high, there have been very substantial reductions in the number of cases of mumps and its complications; immunity can last for 20 years. The incidence of mumps has increased in the UK, to a total of 7628 in 2009, and also in the Netherlands. There has been no corresponding increase in rubella cases.

17.3.4 Parainfluenza viruses types 1–4

Parainfluenza viruses cause up to one-third of all respiratory tract infections and nearly one-half of respiratory infections in pre-school children and infants (Table 17.1). Much less is known about these viruses, especially their immune pathology, than with other members of the family (RSV, measles, and mumps) and this relative dearth of knowledge needs to be corrected. In contrast, the gene structure has been well studied by molecular virologists (Fig. 17.15).Types 1 and 2 are most often associated with laryngotracheobronchitis (croup) (Fig. 17.16), boys being affected more often than girls; type 3 usually causes infection of the lower respiratory tract (e.g. bronchiolitis and pneumonia).

What little is known of the pathogenesis and immune response suggests similarities with those of RSV infections. The epidemiology is also broadly similar to that of other respiratory infections due to paramyxoviruses. The methods of laboratory diagnosis are similar to those for RSV infections—in particular, the use of RT PCR in targeting the M and L genes (Fig. 17.16) or rarely nowadays by indirect immunofluorescence for detecting antigen in nasopharyngeal washings. Virus is detected in throat washing samples just before the first clinical signs (Fig. 17.17).

Prevention

Inactivated vaccines for use in humans have so far proved unsuccessful. Nevertheless, the efficacy of an attenuated vaccine for the related and economically important virus infection of poultry, Newcastle disease, suggests that similar vaccines may eventually be prepared for use in humans.

(a) Parainfluenza types 2 and 4

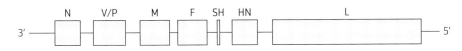

(b) Parainfluenza types 1 and 3

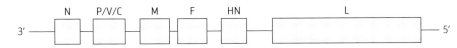

Fig. 17.15 Genome structure of parainfluenza. (a) Parainfluenza types 2 and 4 (b) Parainfluenza types 1 and 3.

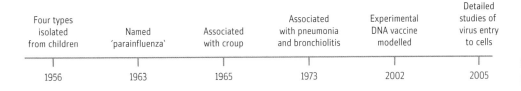

Fig. 17.16 Timeline for parainfluenza types 1–4.

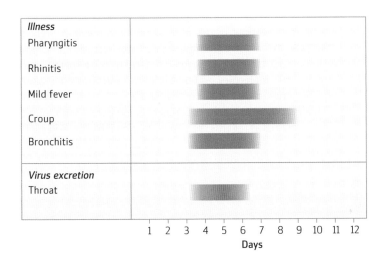

Fig. 17.17 Clinical and virological events during infection with parainfluenza types 1–4.

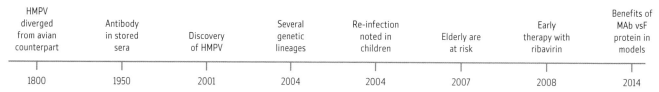

Fig. 17.18 Timeline for human metapneumovirus (HMPV).

17.3.5 Human metapneumovirus (HMPV)

The virus was discovered in 2001 and was cultivated with difficulty with a requirement for trypsin in the culture medium and a lengthy incubation in cell culture. This virus has now been shown to cause acute respiratory illness in children including severe bronchiolitis (Fig. 17.18) and pneumonia (Fig. 17.19) in many countries of the world, and also in the elderly and immunocompromised. It causes peak numbers of infections in the winter months, and in the USA 12% of paediatric hospital visits are associated with HMPV. By five years of age virtually every child has been infected. Not unexpectedly, infection with metapneumovirus can exacerbate asthma as with RSV, and re-infections are not uncommon. Four major lineages of the human virus (A1, A2, B1, and B2) have been described along with antigenic differences in the G spike protein. There are at present no vaccines or antiviral drugs against these viruses. Genetic analysis (Fig. 17.20) indicates an avian origin, probably moving to humans 200 years ago.

The difficult task of differential diagnosis

To diagnose HMPV unambiguously from other parainfluenza virus infections, and from adenovirus and influenza, is not an easy task. Moreover, in older children acute bronchospastic disease from inhalation of allergens may complicate the clinical diagnosis.

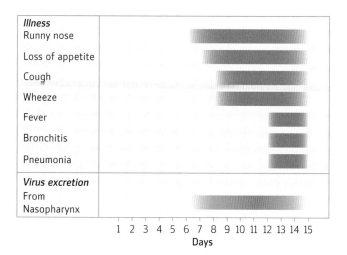

Fig. 17.19 Clinical and virological events during infection with metapneumovirus.

Diagnosis of HMPV in infants is usually made by analysis of symptoms (Fig. 17.19), the time of the year, and epidemiological information, such as virus spread in the community.

Research laboratories have developed RT PCR kits to distinguish these important infections using the L and M genes (Fig. 17.20) using clinical throat samples.

Vaccines and therapies

Experimental animal model studies have shown that protective immunity can be achieved using the F protein as a basis to a vaccine. Also monoclonal antibodies to F induce a prophylactic effect. In contrast the G protein neither induces neutralizing antibody nor protection. The problem of vaccine efficacy is of course confounded by the natural history of the disease where re-infections are the norm rather than the exception, particularly in infants.

The successful clinical use of palivizumab and motavizumab in RSV infections has encouraged research for a comparable passive antibody for HMPV. The first monoclonal antibody produced against the F protein has reduced virus titre and abrogated serious pathology in animal models. Standard immunoglobulins can inhibit HMPV in vitro, and RNA interference has shown clinical effect against RSV, which has encouraged researchers to select an Si RNA targeted against the G protein of HMPV, which is required for virus growth in vivo.

Ribavirin is used clinically at the present on the basis of in vitro inhibition of TNF-α, interferon gamma and interleukin 10. This combination of effects on virus transcription alongside a down-regulation of T-cell mediated cell damage is a good starting point for the discovery of further inhibitors, perhaps, as with influenza, targeting host cell gene products vital for virus replication.

Replication and genetics of human metapneumovirus

The virus attaches via the hydrophobic N terminus region of the G protein, which binds to heparin sulphate molecules of the cell. Fusion from without is mediated by the F protein and, unlike other paramyxoviruses, requires exogenous activation before fusion can start. Fusion is triggered by binding to one of the integrin group of molecules. As far as is known the virus then takes on typical replication events of the family.

However there are some differences compared to the 'typical' paramyxovirus as regards the non-coding regions between each ORF amounting to 23 to 209 nucleotides, with signals for gene starting and gene ending as well as intergenic regions.

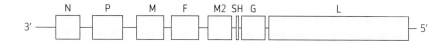

Fig. 17.20 Genome structure of human metapneumovirus (HMPV).

17.3.6 The zoonotic henipaviruses, Hendra and Nipah

The first member of the *Henipavirus* genus was detected in a suburb of Brisbane (Hendra) in 1994 when the virus was isolated from horses ill with a respiratory infection (Fig. 17.21). Fourteen of the 21 horses and a trainer died.

A few years later a second virus of the genus, Nipah, was recovered from pig farmers ill with encephalitis in Malaysia (Fig. 17.21). The infection spread in the pigs and 1 million were killed as a precaution. Within two months, 265 human cases were notified, with over a hundred deaths. The virus is not only found in Malaysia: the case fatality reaches 75% in Bangladesh where yearly outbreaks happen.

Genomes of henipaviruses and virus replication

The RNA genome of Hendra is 15% larger than other paramyxoviruses whilst the Nipah genome is larger still (Fig. 17.22). As with other paramyxoviruses, all the Hendra genes except the

L gene have untranslated regions at the 3′ end, and between the 6 genes N, P, M, F, HN, and L are stop and start signals. The Hendra virus and Nipah genomes are conserved because sequence analysis of isolates from flying foxes, equines, pigs, and humans are virtually identical, although the Nipah virus is diverse in the flying foxes in South-east Asia

Reverse genetic experiments have demonstrated that the N, P, and L proteins are required for viral RNA replication, as with other members of the paramyxovirus family. The cellular protease cathepsin L cleaves the virus F protein in an endosomal vacuole to give a form which can actively fuse with the plasma membrane, thus allowing entry of RNA into the cellular cytoplasm.

The primary sites of replication of Nipah virus are endothelial cells and virus is recovered from urine and the trachea and nasopharynx early in human disease (Fig. 17.23). Few post-mortem samples have been made available, but animal modelling in hamsters, cats, and guinea pigs show vascular pathology in brain, lung, liver, kidney, and heart, with the brain being the focus of pathology.

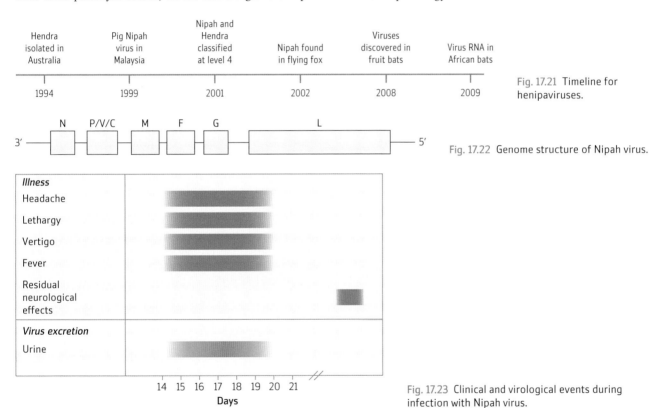

Fig. 17.21 Timeline for henipaviruses.

Fig. 17.22 Genome structure of Nipah virus.

Fig. 17.23 Clinical and virological events during infection with Nipah virus.

 ## Reminders

- The family *Paramyxoviridae* contains two subfamilies: the *Paramyxovirinae*, containing the genera *Paramyxovirus*, *Rubulavirus*, *Morbillivirus*, *Respirovirus* and *Henipavirus*, and the *Pneumovirinae*, with two genera, *Pneumovirus* and

Metapneumovirus. The viruses are about 200 nm in diameter, of helical symmetry.

- The ssRNA genome of negative polarity is 15 kb in length with 6–10 genes. The genome is transcribed into 6–10 subgenomic monocistronic mRNAs encoding glycoprotein spikes (G, H, or HN), fusion protein (F), and transcription related proteins (N, P, and L).

- All the viruses code for a fusion protein (F), which causes adjacent infected cells to fuse and form multinucleate giant cells (syncytia).

- Morbilliviruses cause measles in humans and other generalized infections, including canine distemper, in animals. Measles is an acute febrile illness of childhood associated with a characteristic maculopapular rash. Severe forms of measles with high mortality rates are seen in some developing countries and are associated with malnutrition.

- Active immunization with live attenuated measles vaccine, alone or in combination with live mumps and live rubella vaccines (MMR), is highly effective.

- RSV affects the lower respiratory tract in infants and the elderly; it may cause necrotizing bronchiolitis and pneumonitis. In the absence of an effective vaccine, prevention in institutions depends on good hygiene.

- Mumps is a generalized infection of childhood, in which the salivary glands, especially the parotids, are attacked. It is a comparatively benign illness. In 20% of infected adolescent and adult males orchitis develops, but rarely results in sterility.

- The zoonotic viruses Hendra and Nipah, emerging from bats, both cause encephalitis in persons in South-east Asia with some respiratory involvement particularly with Hendra. Case fatality can be high.

- In some countries bats are widely hunted as bush meat. Indeed a co-evolutionary relationship may be present between chiroptera and paramyxoviruses.

- Parainfluenza viruses types 1–4 are a major cause of respiratory tract infections, especially in infants and children. Types 1 and 2 are particularly liable to cause croup, and type 3, bronchiolitis. All four serotypes also cause upper respiratory infections. The pathogenesis, epidemiology, and methods of diagnosis are similar to those of RSV.

- Metapneumovirus was discovered in 2001, and causes acute respiratory disease in children and also in the elderly and immunocompromised.

⟫ Further reading

Baker, K.S., Todd, S., Marsh, G., Fernandez-Loras, A., Suu-Ire, R., Wood, J.L.N. *et al.* (2012). Co-circulation of diverse paramyxoviruses in an urban African fruit bat population. *Journal of General Virology* **93**, 850–6.

Gouma, S., Sane, J., Gijselaar, D., Cremer, J., Hahné, S. *et al.* (2014). Two major mumps genotype G variants dominated recent mumps outbreak in the Netherlands (2009–2012). *Journal of General Virology* **95**, 1074–82.

Haas, L.E.M., Thijsen, S.F.T., van Elden, L., and Heemstra, K.A. (2013). Human metapneumonia in adults. *Viruses* **5**, 87–110.

Strebel, P.M., Papania, M., Finkelkorn, A., and Halsey, N. (2012) Measles vaccines. In Plotkin, S., Orenstein, W.A., and Offit, P.(eds), *Vaccines*, 6th Edition, pp 352–87. Saunders, Philadelphia.

World Health Organization (2013a). Measles and rubella elimination 2015. Package for accelerated action: 2013–2015.

World Health Organization (2013b). The guide to tailoring immunization programmes. Increasing coverage of infant and child vaccination in the WHO European Region. http://www.euro.who.int/__data/assets/pdf_file/0003/187347/The-Guide-to-Tailoring-Immunization-Programmes-TIP.pdf (Accessed 19 Feb. 2016).

? Questions

1. Describe the constituent viruses of the MMR vaccine. Is one virus more important clinically or prevalent than the other two?

2. Why do viruses emerge? What are the genetic features of the two emergent paramyxoviruses?

3. Is it possible to eradicate measles by using live vaccine?

4. Are any of the paramyxovirus diseases treatable?

5. Describe the essential replication steps of the paramyxoviruses. You could focus on one member of the family.

Filoviruses: zoonotic Marburg, and Ebola

18

18.1 Introduction

Most of the viruses described in this chapter (Table 18.1) have been isolated or characterized in the last four decades, although there is no reason to think that they have not existed and evolved, like other viruses, over many thousands or even millions of years. They cause haemorrhagic fevers in tropical countries and their common property is existence in an animal reservoir in which they persist quietly until disturbed by human intrusion. They are sometimes known as **exotic viruses**; some are named after the town or area where an outbreak was first investigated. An example is the filovirus 'Marburg', named after the town in Germany where seven persons died of what was, at that time, a new and unrecorded disease. In this outbreak the common link between the infected persons was the handling of monkeys or monkey tissues. The monkeys had been imported from Africa to provide kidney tissue for preparing poliomyelitis vaccine. Nearly 10 years later an outbreak of an even more lethal haemorrhagic disease caused by another filovirus was described in Zaïre and the Sudan—'Ebola disease', named in this case after a river in Zaïre.

The filoviruses are so dangerous that only high-security category 4 laboratories handle them. Hence, five decades after their discovery, they have still not been as thoroughly studied as some viruses that were isolated much later.

Since the mid-1990s the frequency of outbreaks in Africa of Ebola has been increasing, most likely because of human movement, farming, and deforestation. The R_0 of these viruses is usually just above or just less than one, which, coupled with a long incubation time, no aerosol spread, and a long generation time, makes these viruses controllable by the classic methods of contact tracing and quarantine. Until recently Ebola has only been a problem in remote villages on the edge of jungle areas but in 2014 Ebola reached two large cities in West Africa and infected persons who subsequently travelled to Europe, the UK, and the USA. But Ebola is extremely unlikely to become rooted in any country outside Africa where the public health infrastructure, including safe water and sewage systems, and the hospital system is fragile and even in some areas non-existent. It is to be hoped that developed world countries will now aid in the reconstruction of those West African nations.

Table 18.1 Filoviruses

Family	Genus	Important viruses	Geographical distribution
Filoviridae	Marburg	Marburg	West and Central Africa
	Ebola	There are several 'species', termed Sudan, Zaire, Reston, and Bundibugyo (Uganda)	West and Central Africa

18.2 Properties of the viruses

18.2.1 Classification

The family *Filoviridae* (Fig. 18.1) is composed of extremely pleomorphic viruses: its name derives from the Latin *filum*, a thread, which refers to their morphology (see Fig. 18.2). The family belongs to the order *Mononegavirales*. Marburg and Ebola viruses can be distinguished from each other by the size of their genomes and their different protein composition; they also differ serologically.

18.2.2 Morphology

These viruses have an extraordinary filamentous morphology and are sometimes longer than common bacteria, up to 1400 nm, often with branched, circular, and bizarre-shaped forms (Fig. 18.2). They have lipid envelopes, beneath which a nucleocapsid structure containing RNA can be visualized by EM. The nucleocapsids have helical symmetry. The virion surface is covered by (PVP24 and G) spikes 10 nm in length (Fig. 18.3) which penetrate the envelope to the matrix (VP40) protein.

18.2.3 Genome

The genome of filoviruses is negative-sense single-stranded RNA, 19 kb in size (Fig. 18.4). It is the largest genome of all the viruses in this order. It has seven ORFs coding for the seven or eight known structural proteins. There are stop and start signals at the boundaries of each gene similar to those of the paramyxoviruses, rabies, and measles. There is a start site at the 3' genome end and genes are terminated with a transcription stop site. Termination of transcription occurs at a series of 5–6 Us where 'stuttering' (repeated copying) by the viral RNA polymerase results in the addition of long poly(A) tails to the transcripts. An unusual feature of the filovirus genome is the presence of gene overlaps in the 5' and 3' non-coding regions of certain ORFs. The genome is organized with structural proteins at the beginning of the negative-sense genome and the non-structural RNA-dependent RNA polymerase (the L protein) at the opposite end. However, the exact gene order is complex as a component of the polymerase complex, the VP35 protein, is encoded within the structural protein region.

18.2.4 Replication

The filoviruses have a broad host range and tissue tropism (Fig. 18.5). The GP proteins of the virus bind to a range of receptors on cells including an asialoglycoprotein receptor, folate receptor, integrins, and DC-SIGN (dendritic cell specific intercellular adhesion molecule grabbing non integrin) (Fig. 18.5). They infect and enter cells by low-pH mediated endocytosis, and transcription and genomic RNA replication take place in the cytoplasm. Intracellular entry receptors have also been identified for Ebola virus, such as Niemann–Pick disease, type C1 (NPC1) protein, a membrane protein that mediates intracellular cholesterol trafficking in

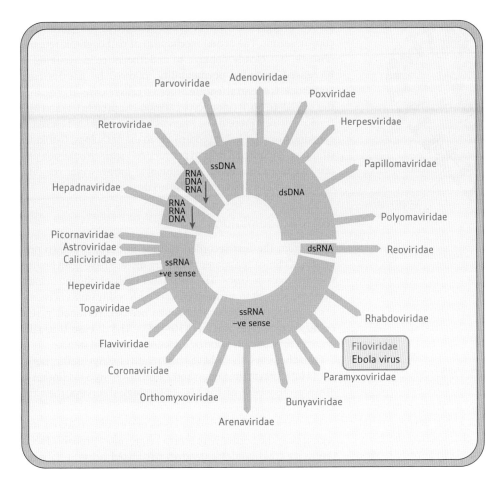

Fig. 18.1 Baltimore classification scheme and filoviruses.

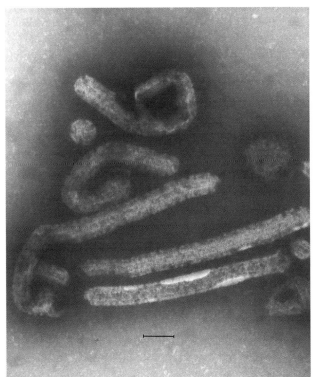

Fig. 18.2 Electron micrograph of a filovirus (Marburg). The virions are highly pleomorphic. (Courtesy of Dr Anne M. Field.) Scale bar = 100 nm.

mammals, but which also serves as a critical factor for host cell entry, binding to Ebola virus GP in the endosomes. Transcription is the initial event and starts at the 3′ end of the genome with each virus ORF producing a monocistronic mRNA by the polymerase starting and stopping at conserved initiation and termination signals in the genome. Translation leads to accumulation of NP and VP35, which trigger a switch to production of full length antigenomes, which in turn act as templates for genome synthesis. During cellular infection the viral glycoprotein (GP) spikes insert into the plasma membrane and nucleocapsids (NP) accumulate in the cytoplasm. The two come together at the inner cell surface just before budding and virus release. Other viral proteins are the small secreted sG, VP40, L protein and VP35 (an RNA-dependent RNA polymerase), NP and VP30 (nuclear proteins), and finally a VP24 of unknown function. The cycle is complete in 12 h.

18.3 Clinical and pathological aspects

18.3.1 Clinical features

A filovirus may be suspected in an acute febrile patient from an endemic area (Central and West Africa, 10° North and South of the equator), but can be misdiagnosed with malaria, typhoid, or cholera. It must be emphasized though that a mild, almost subclinical, febrile disease is not uncommon in these Ebola patients.

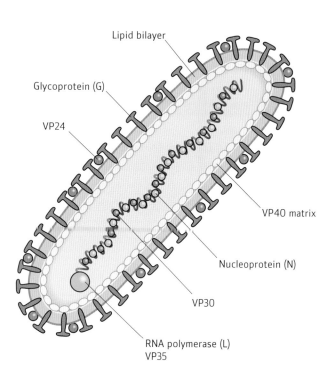

Fig. 18.3 Structural features of the filovirus virion.

The illnesses caused by Ebola viruses have an abrupt onset after an incubation period of 3–16 days (Fig. 18.6). Severe frontal headache, high fever, and back pains characterize the early phase. The patient is rapidly prostrated and can have diarrhoea and vomiting lasting about a week; conjunctivitis and pharyngitis are usually present. A transient non-itching **maculopapular rash** with peeling skin may appear after 5–7 days. At this time with certain species of Ebola, but not the one responsible for the 2014 outbreak, **severe bleeding** starts in the lungs, nose, gums, gastrointestinal tract, and conjunctiva in a proportion of patients, preceded and accompanied by **thrombocytopenia**. Death is caused by severe shock and blood loss, resulting in low blood volume, and usually occurs between days 7 and 16. The mortality is very variable, ranging from 25% to 90%. Good doctoring and nursing care with intravenous fluids to combat shock can reduce mortality, as can transfusion of antibodies from recovered patients and passive administration of synthetic recombinant antibodies.

18.3.2 Pathology

Ebola is a so-called pantropic virus: this virus infects and causes lesions in many organs, but especially the liver and spleen, which become enlarged and dark in colour. In both these organs severe degeneration and necrosis occur. Large quantities of virus are

Fig. 18.4 Genome structure and encoded proteins of filovirus. Labelled proteins include NP (nucleoprotein), GP (glycoprotein spike), L (polymerase), and VP40 (matrix).

present in organs and may be even visualized by electron microscopy in serum. The actual mechanism of pathogenesis remains obscure, but clearly damage to endothelial cells, resulting in **increased vascular permeability** followed by haemorrhage and shock, is a central feature. Infected cells release large quantities of TNF-α, MCP (monocyte chemotactic protein)-1, and MIP (major intrinsic protein)-α, whereas IFN-α and -β are blocked. A cytokine storm is likely to contribute to the pathology. Cytokines stimulate endothelial cells to produce cell-surface adhesion and pro-coagulant molecules. Disseminated intravascular coagulation and platelet dysfunction also occur.

18.3.3 Epidemiology

Apart from the first recorded episode in Europe, known outbreaks of clinically apparent Marburg disease have been limited to individuals in Africa. The virus most likely originated in bats. Most cases result from direct contact with bodily fluids of cadavers or patients, or via contaminated needles. Fortunately, secondary attack rates are low (10–15%). The disease is usually brought to wider attention by infection of hospital staff.

Over the years, there have been relatively large outbreaks of Ebola infection in the Sudan and Zaïre, involving a maximum of 400 persons where, on the basis of serological surveys, the virus seems to be endemic.

The 2014–2015 outbreak of Ebola originated in Guinea in a region near the border with Sierra Leone, most probably facilitating its cross-border spread to urbanized areas in Liberia, Sierra Leone, and Guinea. The epidemiology was more complex than previously encountered in rural settings with the population density and people movements associated with cities exacerbating the infection. The outbreak was the largest recorded to date with over 11 000 deaths and over 28 000 infected persons.

In spite of public unease, Ebola is unlikely to cause widespread disease in the EU, USA, and Australia, for example, where hygiene is high and is joined with safe water and sewage systems. These are mainly diseases of impoverished and fractured nations.

The outbreak in a primate facility in Reston, USA, led to identification of a third virus, Ebola-Reston. The virus came from imported crab-eating macaques from the Philippines. In this outbreak in a monkey colony in the USA, many macaques died, but in the four employees who were serologically diagnosed as infected there were no clinical signs. This is an encouraging observation and may indicate that certain Ebola strains are non-pathogenic for humans.

The Ebola virus can spread among chimpanzees and gorillas in the wild in Central Africa and Côte d'Ivoire. In some West African countries as much as 80% of food protein comes from 'bush meat', and collecting bats and primates for food is an 'at risk' occupation (Fig. 18.7). But classic methods of quarantine and contact tracing prevented an outbreak in Nigeria in 2014 after an infected person arrived by plane from Sierra Leone. Apart from direct contact with animals, very close contact with an infected patient is a prerequisite for infection, and high-level

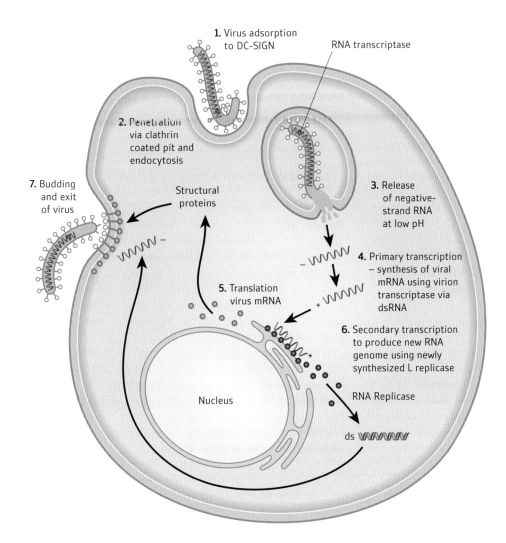

1. Virus adsorption to DC-SIGN

RNA transcriptase

2. Penetration via clathrin coated pit and endocytosis

7. Budding and exit of virus

Structural proteins

3. Release of negative-strand RNA at low pH

4. Primary transcription – synthesis of viral mRNA using virion transcriptase via dsRNA

5. Translation virus mRNA

6. Secondary transcription to produce new RNA genome using newly synthesized L replicase

RNA Replicase

Nucleus

ds

Fig. 18.5 Replication cycle of a filovirus.

isolation and barrier-nursing methods reduce transmission to HCW. Up to the present initial diagnosis will be on clinical grounds but some local laboratories in West Africa now have diagnostic capability, using RT PCR. The recent international aid following the 2014 Ebola outbreak in West Africa will

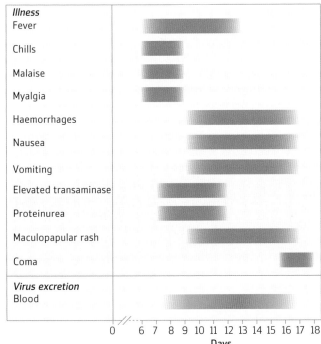

Illness	Days
Fever	
Chills	
Malaise	
Myalgia	
Haemorrhages	
Nausea	
Vomiting	
Elevated transaminase	
Proteinurea	
Maculopapular rash	
Coma	
Virus excretion Blood	

0 6 7 8 9 10 11 12 13 14 15 16 17 18
Days

Fig. 18.6 Clinical and virological events during infection with filovirus.

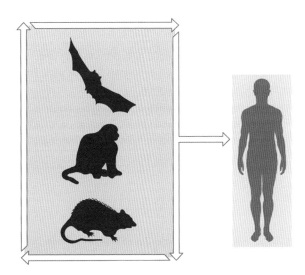

Fig. 18.7 Transmission of filovirus Ebola from an, as yet, unidentified reservoir, but most likely a fruit bat.

strengthen this important early laboratory diagnosis and help to prevent future outbreaks.

18.3.4 Laboratory diagnosis

RT PCR is used to detect and quantify viral RNA. The clinical sample can be treated with guanidinium-containing buffer in the high-security laboratory to destroy virus infectivity and to allow safe handling in the molecular laboratory. Clinical samples are blood, vomit, and diarrhoea. Rapid sequencing can also now be carried out in laboratories in West Africa. However only certain national laboratories designated category 4 have the experience and high-level containment facilities to propagate these viruses in quantity in Vero cells. There are only a dozen or so of these facilities in the world, including four in the USA, one in Canada, three in mainland Europe, one in S. Africa, and three in the UK. A significant rise in **antibody**, as measured by ELISA, is also diagnostic.

18.3.5 Prevention and treatment

Only experimental vaccines and chemotherapeutic agents are available. In the 2014–2015 outbreak in West Africa transfusion antibodies from recovered patients were investigated for therapy, but not in controlled trials. An alternative immune therapy has resulted from the use of human antibody genes which are cloned into an agrobacterium which is then used to infect tobacco mosaic plants.

Two filovirus genes have been cloned into a chimpanzee adenovirus for an experimental live vaccine. As an alternative the Ebola virus GP has been cloned into the animal virus vesicular stomatitis virus (VSV), a rhabdovirus. This type of vaccine offers protection in a primate model; the first volunteers have been immunized in field studies and the vaccine has shown some good protective efficacy.

18.4 Risk categories of patients

In Africa transmission is by direct contact, close droplet infection, or contact with body fluids. Fortunately the R_0 value of the virus is very low. **Healthcare staff are particularly at risk**, as most or all infections involve close contact with sick patients. In hospitals with adequate facilities, strict barrier nursing, training with personal protective equipment (PPE), and correct handling of bodies after death, the attack rate in staff is low or non-existent, but where they are not, transmission rates to staff of 80% have been recorded.

Traditional funerals are a source of infection as friends and relatives wash the bodies of Ebola victims and themselves become infected. On an optimistic note filoviruses are easily destroyed by hot water, soaps, detergents, and disinfectants, and a combination of hygiene, new vaccines, change of burial practices, and organization of the health structures in West Africa will abrogate future outbreaks.

 Reminders

- The filoviruses Marburg and Ebola (family *Filoviridae*) are **enveloped** viruses with a bizarre **filamentous morphology**.

- They have ssRNA genomes of negative polarity with seven ORFs. Replication is cytoplasmic and virus is released by budding.

- The filoviruses cause zoonotic infections in **bats and monkeys**, and may infect other animals such as pigs. They spread to hunters of 'bush meat'.

- Filovirus infections are **pantropic**, causing lesions in many organs, notably the liver and spleen. Prostration, diarrhoea, rashes, vomiting, and bleeding from the **respiratory,**

gastrointestinal, and **urogenital tracts** are features of the haemorrhagic forms of these illnesses.

- Prompt treatment with **convalescent plasma** is used experimentally for Ebola fever, but supportive care with careful maintenance of fluid and electrolytes to prevent shock are key elements for survival. For Ebola safe burial practices are key to preventing spread, alongside training of HCW and the use of rapid diagnostics to allow segregation of Ebola cases away from suspect cases.

- The diagnosis of these illnesses is based on the **history and circumstances of travel in endemic areas**, and the exclusion of malaria and other tropical fevers. RT PCR is used now for rapid diagnosis, particularly in remote communities.

 Further reading

Baize, S. *et al.* (2014). Emergence of Zaire Ebola virus disease in Guinea—preliminary report. *New England Journal of Medicine* **371**, 1418–25.

Carroll, M.W. *et al.* (2015). Temporal and spatial analysis of the 2014-2015 Ebola virus outbreak in West Africa. *Nature* **524**, 97–104.

Marzi, A. and Feldman, H. (2014). Ebola vaccines: an overview of current approaches. *Exp. Rev. Vaccines* **13**, 521–31.

Tong, G.Y. *et al.* (2015). Genetic diversity and evolutionary dynamics of Ebola virus in Sierra Leone. *Nature* **524**, 93–6.

Questions

1. Describe the symptoms of Ebola. How are patients infected and how is the virus controlled in the community?

2. Discuss the genome of Ebola and summarize replication strategies of the family.

3. Could a vaccine be developed against Ebola, and how would it be used?

19 Rabies: zoonotic rabies

19.1 Introduction

This devastating disease of animals, comparatively rarely transmitted to humans, has been recognized since the dawn of history and references appear in the Babylonian Eshnunna Code before 2300 BC. The infection spread to Europe and then to the world by natural migration and also colonial activity. Celsius first described hydrophobia in AD 100 and recommended cautery of animal bites with a hot iron; this remained the treatment of rabid animal bites until 1884, when Louis Pasteur introduced his famous human rabies vaccine. With very rare exceptions—with, in fact, only four persons documented as survivors—the disease is fatal in humans and many people still die each year, especially in the developing countries. There are 24 000 deaths in Africa alone, almost all contracting rabies from dogs. The burden in Asia is larger. It is fair to say that this is a neglected enzootic disease, presenting a serious public health threat in developing countries.

Increasingly travellers worldwide are entering areas where rabies is rife in the animal population. With the exception of the UK and Norway some parts of Europe are not 'rabies free'. In the Southern hemisphere rabies free areas are the islands of New Zealand and Australia.

19.2 Properties of the virus

19.2.1 Classification

The order Mononegavirales consists of four virus families that are all phylogenetically related. These are the *Bornaviridae*, *Filoviridae*, *Paramyxoviridae*, and the subject of this chapter, *Rhabdoviridae* (Fig. 19.1). Rabies virus is a Lyssavirus (Greek: *lyssa* = madness), which is one of the six genera in the *Rhabdoviridae* (Greek: *rhabdos* = a rod), a family of characteristically **bullet-shaped RNA** viruses (Fig. 19.2), that contains over 150 animal, fish, insect, and plant viruses. Rabies is one of fifteen viruses within the *Lyssaviruses* that infect vertebrates (Table 19.1) and these are divided into three groups. The group including Ikoma virus is antigenically distinct from the other members of the groups, therefore, the vaccine gives protection to all viruses except Ikoma.

Vesicular stomatitis virus belongs to the genus *Vesiculovirus* within the *Rhabdoviridae*, and affects horses and cattle; it may cause a mild febrile illness in humans exposed to it. Other genera in the family affect only insects or plants.

There are seven genotypes of rabies with genotypes 1, 2, and 3 circulating in Africa. The African type 3 genotype is restricted

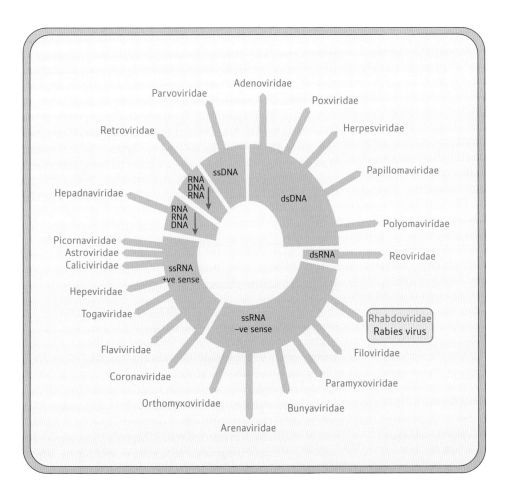

Fig. 19.1 Baltimore classification scheme and lyssavirus.

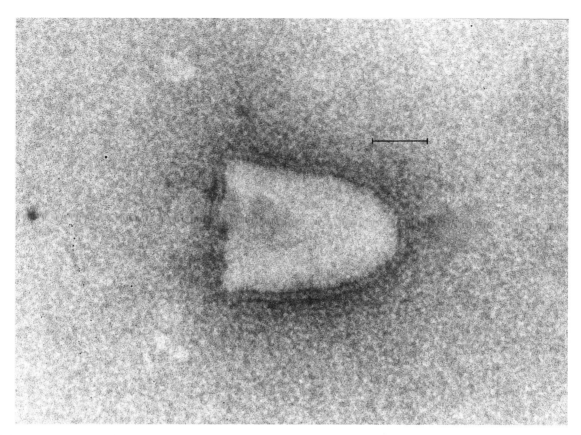

Fig. 19.2 Electron micrograph of rabies virus. Note the characteristic bullet shape and the fringe of glycoprotein spikes on the surface. (Courtesy of Dr David Hockley.) Scale bar = 50 nm.

to South Africa and is adapted to mongooses. The virus was introduced into Africa <200 years ago by colonization and urbanization, taking 100 years to occupy the whole region.

19.2.2 Morphology

The rabies virion consists of a **helical** nucleocapsid contained in a bullet-shaped lipoprotein envelope (Fig. 19.3) about 180 nm in length. Visibly protruding from the lipid envelope are

Table 19.1 Some members of the family *Rhabdoviridae*

Genus	Members
Lyssavirus	Rabies
	Lagos bat virus
	Australian bat lyssavirus
	Mokola
	Duvenhage
	European bat virus 1; virus 2
	Ikoma bat virus, antigenically separate from the above 7 viruses
Vesiculovirus	Vesicular stomatitis virus and other viruses infecting vertebrates and invertebrates

approximately 200 glycoprotein (G) spikes of the virus, responsible for viral attachment to cellular receptors and subsequent fusion activity. G also has HA activity and has important antigenic sites, which are neutralized by specific antibody.

The M or matrix protein is the major structural protein of the virus, lying internally beneath the lipid membrane; it may contact the end of the G spike, helping to stabilize the structure of the virion (Fig. 19.2). The nucleoprotein (N) encapsidates and protects the RNA from degradation by RNAase enzymes. Closely attached to the RNA and N protein in the virus particle is the L or large protein, which functions as the virus RNA dependent RNA polymerase and also has 5′ cap methylase, 3′ poly(A) polymerase, and protein kinase activities.

19.2.3 Genome structure

Rhabdoviruses contain a single-stranded RNA genome of negative polarity and 11–12 kb in size (Fig. 19.4).

There is a leader region of 50 nucleotides at the 3′ end of the genome and a 60 nucleotide untranslated region (UTR) at the 5′ end of the virus RNA. The genome lacks a 5′ cap or a 3′ polyadenylated tail and, therefore, cannot function as mRNAs. Viruses in this class contain a virion-associated RNA dependent RNA polymerase (the L protein), which is responsible for the production of viral mRNAs in infected cells (Chapter 3). The gene order from the 3′ end of the genome is N, P, M, G, and L gene, where the structural proteins occur first, an order that is very similar to

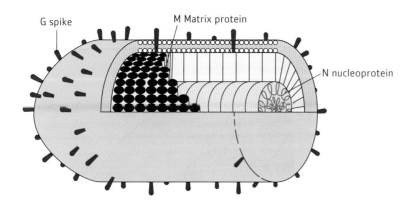

Fig. 19.3 Structural features of rabies virus. Glycoprotein (G) spikes protrude through the lipid of the bullet-shaped virion. The M protein subunits are represented as short cylinders. The nucleoprotein (N) is closely associated with the RNA.

the paramyxoviruses. There is a conserved polyadenylation signal at the end of each gene and short intergenic regions between the five genes.

19.2.4 Replication

Rabies virus attaches via the G spike and several neuronal cell proteins have the ability to bind G including the nicotinic acetylcholine receptor, the neural adhesion molecule CD56, and the low affinity nerve growth factor receptor p75^{NTR} (Fig. 19.5). The virus enters a susceptible cell by viropexis, in much the same way as influenza virus. A coated pit is formed which then becomes an endosome, with virus fusion and exit occurring from the acidic endosome. Uncoating and release of the virus genome RNA occurs in the cytoplasm and replication of the viral RNA genome also occurs in the cytoplasm. The first step in virus replication is primary transcription where the infecting genome is used to produce virus mRNAs. The replication of the virus genome to produce progeny genomes requires the synthesis of virus proteins from these primary transcripts. At that stage production of the positive sense antigenome from the parental infecting genome can occur. This then serves as the template for the production of many more virus genomes and as the source for the production of virus mRNAs as secondary transcripts. Therefore the essential difference between primary and secondary transcripts is their abundance, with many more secondary transcripts being produced. Five monocistronic virion mRNAs are synthesized and this process is essentially a stop start mechanism along the virus genome, with new transcripts initiated at 3′ UUGUC 5′ sequences. Transcripts terminate at 3′ AUAC-UUUUUUU 5′ sequences and polyadenylation may occur by RNA polymerase slippage or 'stuttering' at each inter-gene stretch of seven U residues. The enzyme then moves on to the next gene initiation site. There is higher molar abundance of gene products from the 3′ end of the genome. In contrast, full-length complementary copies of genomic RNA are thought to be synthesized by complete read through of intergenic regions by a modified RNA polymerase enzyme, which consists of L and P proteins and, therefore, explains the dependence on newly synthesized virus proteins for the initiation of genome replication.

Minus strands of genomic single-stranded RNA associate with the N protein to form nucleocapsids, which together with the M protein trigger budding at areas of the plasma membrane, where G protein is already inserted. The G protein is produced through transport and maturation within the ER and Golgi. The virus is released by budding from the plasma membrane of the infected cell. The cell may not die, but continues to act as a source of budding viruses.

19.3 Clinical and pathological aspects

19.3.1 Clinical features

Humans

Ninety-nine percent of rabies is acquired from the bite of an infected animal, but simple licking of abraded skin may also transmit the virus; the infection has also been acquired from aerosols in bats' caves.

Human-to-human transmission of rabies via infected transplant donors is an unusual mode of acquisition and depends on the unfortunately timed transfer of tissue from a donor who is incubating the disease. There have been several such episodes involving transplants of corneal tissue and of lung, liver, and kidney.

The **incubation period** (Fig. 19.6) in humans varies from 10 days to a year or more, but is on average 1–3 months, the time depending on the quantity of virus deposited and—because the virus has to reach the brain via the peripheral nerves—on the distance of the bite from the head. As quantity of virus is so important multiple bites may transmit the disease more readily than single bites.

The onset of disease is usually insidious, with a 1–10-day prodromal period of malaise, fever and headache, and hypersalivation. There may also be psychological disturbances including anxiety and aggression; indeed, one case in the UK was at first misdiagnosed as acute schizophrenia. Pain and tingling around the area of the bite, sometimes accompanied by small jerky

Fig. 19.4 Genome structure of rabies virus.

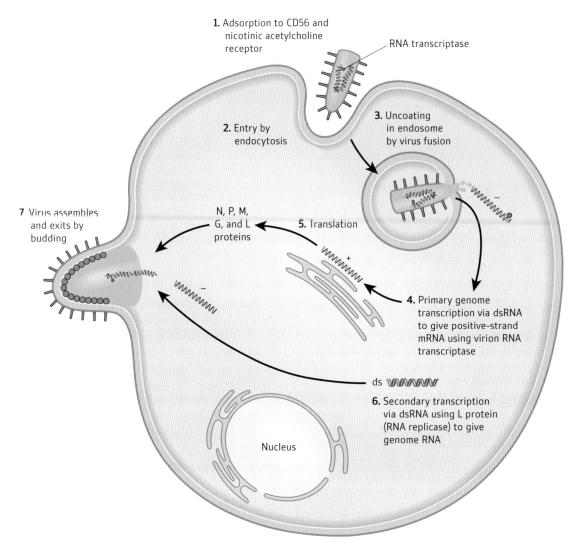

Fig. 19.5 Replication cycle of rabies virus.

movements, are particularly suggestive of incipient rabies. The subsequent course may take one of two forms.

Furious rabies

The patient passes into the 'stage of excitement', with anxious and apprehensive expression, fast pulse, and rapid breathing (Fig. 19.7). The physical signs are protean, their nature depending on the areas of brain affected. Cranial nerve and other paralyses are frequent, and there may be greatly increased activity of the autonomic nervous system and hyperpyrexia. The classical sign, present in most cases, is **hydrophobia**; this is particularly distressing, as the patient needs to drink, but any attempt to do so, or even the sight of water, elicits violent spasms of the respiratory and other muscles, accompanied by a feeling of extreme terror. Periods of lucidity alternate with impaired consciousness; after a week or so, the patient dies in coma with generalized paralysis and cardiovascular collapse. About one-fifth of infected patients present with this form of the disease.

Paralytic ('dumb') rabies

The course is less dramatic. An illness lasting as long as a month is characterized by ascending paralysis; hydrophobia is not a prominent feature. In these cases, the spinal cord and medulla are affected more than the brain. They are particularly associated with bites from vampire bats, rather than dogs. As with furious rabies, death is inevitable.

Animals

Both forms of the disease occur in dogs and cats (Fig. 19.8), 'dumb rabies' predominating. The incubation period in dogs can be as long as 8 months. The first sign is usually a change in behaviours. It should be remembered that the classical image of a rabid dog running amok and biting all and sundry is not always true; there may be intervals in which both dogs and cats become abnormally friendly, and make repeated attempts to lick those near them. A small proportion of dogs may recover from rabies. Most rabid cats enter a furious phase, scratching and biting without provocation.

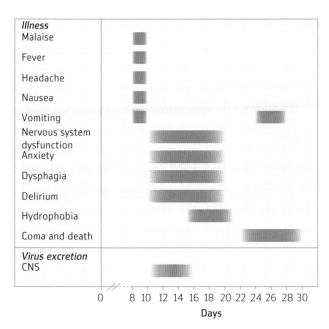

Illness	
Malaise	
Fever	
Headache	
Nausea	
Vomiting	
Nervous system dysfunction Anxiety	
Dysphagia	
Delirium	
Hydrophobia	
Coma and death	
Virus excretion CNS	

0 8 10 12 14 16 18 20 22 24 26 28 30
Days

Fig. 19.6 Clinical and virological events during infection with rabies. Very rapid onset is illustrated here, with a patient being bitten near the CNS, perhaps the face. More serious symptoms appear later in this biphasic disease. Hydrophobia aids the clinical diagnosis but is not always apparent. Virus is present in the brain and in some additional tissues but can only be transferred by a transplantation (cornea). Thus the human patient is normally a dead end host.

19.3.2 Pathogenesis

Following an animal bite, rabies virus reaches the CNS in humans by way of **peripheral nerves** and is a classical example of centripetal spread of a virus followed by centrifugal spread from the CNS. The virus first replicates in epithelial or striated muscle cells at the site of the bite or in the mucosal cells of the respiratory tract, and gains access to the peripheral nervous system via the **neuromuscular spindles**. In another major site of

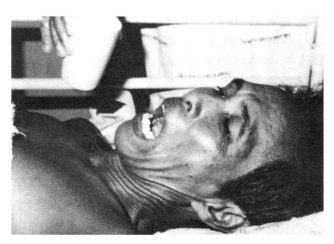

Fig. 19.7 Patient with rabies. [Photograph courtesy of Professor C. Kaplan.]

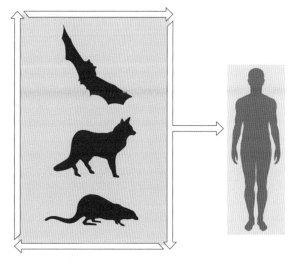

Fig. 19.8 Cycling of rabies to humans

neuronal invasion the virus binds specifically to the cholinesterase-positive binding sites at neuromuscular junctions. The rate of centripetal progress of the virus along the axons of the peripheral nerve has been estimated experimentally in mice as 3 mm per day. Once the virus has replicated in the spinal cord and throughout the CNS it may spread centrifugally along the neuronal axons of the peripheral nerves to other tissues, including the salivary glands and hair-bearing tissues. In persons infected by aerosols, e.g. in bat-infested caves or in laboratory accidents, the virus probably reaches the CNS via nerves supplying the conjunctiva or the upper respiratory tract, including the olfactory nerves. The unfortunate victim almost invariably dies of encephalitis. Strangely, there is comparatively little pathological evidence of neuronal necrosis, but the virus may interfere with neuronal transmission. Therefore, the precise pathology still remains a mystery. Myocarditis is often present and the characteristic cytoplasmic inclusions (Negri bodies) are detectable in the hearts of some patients.

19.3.3 Epizoology

Rabies vectors are mainly carnivores (Fig. 19.8), but the virus may transmit to 'cul de sac' animals such as ourselves. The virus has two epidemiological forms: urban rabies mainly with domestic dogs as reservoirs and transmitters; and sylvatic rabies with various wildlife species fulfilling these roles. Rabies virus is capable of infecting all warm-blooded animals. The reservoirs of infection vary according to the geographical area; dogs and cats are the most important sources of human infection, which is nowadays most frequent in developing countries. The main sylvatic reservoirs are wolves in eastern Europe, the red fox in western Europe, mongooses and vampire bats in the Caribbean, skunks and racoons in the USA and Canada, and vampire bats in Latin America.

Except for New Zealand, Norway, Australia, and the UK, rabies is present in every continent. The UK has, however, remained rabies-free because of its strict quarantine regulations, which are, of course, relatively easy to enforce in an island. Elaborate precautions have been taken to prevent wild fauna getting to the UK from continental Europe via the Channel Tunnel.

However, with the introduction of highly effective rabies vaccines, combined with confirmation of vaccination in the form of an implanted microchip, and serological evidence of vaccination, the need for maintaining quarantine in the UK has now vanished, and pets with documentation of rabies vaccination can move freely within the EU.

About 700 rabies deaths are officially reported each year worldwide, but these probably represent only a fraction of the total number, which is estimated at 100 000, mainly in the developing world. In contrast only one or two cases per year are reported in the USA. Over 40% of human cases are in children aged 5–14 years; most cases are male, presumably because of greater contact with animals.

The epizootiology of the rabies-related viruses is not clear; but Duvenhage virus, carried by bats, has been recovered from cases of rabies in South Africa, Finland, and Russia.

19.3.4 Laboratory diagnosis of rabies

Fortunately, virus isolation, which requires a high-security level 4 laboratory, is rarely, if ever, needed for diagnostic purposes. Most if not all cases are diagnosed by detecting virus RNA by RT PCR. However, if necessary, samples of brain tissue, saliva, CSF, or urine may be injected intracerebrally into newborn mice. The central feature of classical laboratory diagnosis is **demonstration of rabies antigen by immunofluorescence**, and avidin-biotin immunochemistry in cells obtained from **corneal impressions** or **hair-bearing skin**. This method may also be used on brain smears obtained post mortem from humans or animals. A less sensitive, but useful post-mortem technique is the search of brain smears for the eosinophilic **cytoplasmic inclusions** known as **Negri bodies** named after the Italian physician who first discovered them. The tissue is taken from the Ammon's horn region of the hippocampus and stained with Mann's stain. These diagnostic procedures for rabies or for virus isolation may be undertaken only in specialized, high-security laboratories.

19.3.5 Prophylaxis in humans using vaccines

The original Pasteur rabies vaccine

Pasteur used a vaccine isolated from the brain of a rabid cow (Fig. 19.9). Rabies viruses isolated from the wild are called 'street viruses'. After passage in the laboratory their virulence is reduced and stabilized. This virus was passaged 90 times

intracerebrally in rabbits, by which time the incubation period had become shorter and **fixed** at 6–7 days. The rabbit spinal cords were then air dried. In 1885, a 9-year-old boy, Joseph Meister, was admitted to hospital with severe bite wounds from a presumed rabid dog. Pasteur gave 13 injections of rabies-infected cord suspensions. and Joseph Meister survived. A year later Pasteur reported the result of treatment of 350 cases; only one person in this group had developed rabies, a child bitten nearly 4 weeks before treatment started. Contemporary figures show that 50% of those bitten should have developed rabies. Within a decade there were Pasteur Institutes around the world and, by 1898, 20 000 persons had been treated, with a mortality of only 0.5%.

Modern rabies vaccines

A major advance in rabies vaccines was cultivation of the virus in **human diploid cells (HDC)**, such as WI-38, and inactivation by β-propiolactone. This method provided a potent vaccine that is considerably less reactogenic than those preceding it. **Human diploid cell strain (HDCS) vaccine** has become the vaccine of choice for prophylactic and therapeutic use. The vaccine is given prophylactically to veterinary surgeons, animal handlers, or others at risk from rabies, including travellers to endemic countries, in three doses 1 month apart with a booster at 2 years.

Some Asian countries have developed rabies vaccines from virus-infected hamster kidney cells, which are less costly to produce than those made in WI-38 cells.

If a person is unfortunate enough to be bitten by a suspected rabid animal, the wound should be thoroughly washed with soap and water, alcohol, iodine, or a **quaternary ammonium compound** (to which rabies virus is particularly susceptible). Good wound care remains the cornerstone of rabies prevention and is thought to reduce the risk of rabies by 90%. Appropriate **anti-tetanus treatment** should be given to those not immunized against this infection within the past 3 years. The management then follows the scheme in Table 19.2. A full course of HDCS vaccine consists of six doses given intramuscularly on days 0, 3, 7, 14, 30, and 90 after exposure. This modern vaccine is free of unpleasant side-effects.

Rabies is one of the few diseases in which vaccine is effective when administered during the incubation period. Over 1.5 million people are still immunized yearly throughout the world, often with the Semple-type vaccines made from phenol inactivated virus in brains of rabbits, sheep, and goats, which are given as 21 daily injections subcutaneously. Side-effects, particularly encephalomyelopathies, are not uncommon with such vaccines. They are due to an allergic response to myelin, which is present as an impurity in the vaccine, but these vaccines are so

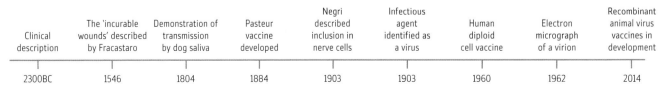

Fig. 19.9 Timeline for rabies

Table 19.2 Treatment of people in contact with animals with suspected or actual rabies

Nature of contact	Status of animal*	Treatment of patient†
Indirect contact only	Appears healthy or has signs suggesting rabies	Not needed
Licks to skin	Appears healthy or has signs suggesting rabies	
	(a) Under observation for at least 10 days after the contact	Start vaccine immediately. Stop only if animal is normal 10 days after the contact
	(b) Escaped	Full course of vaccine immediately
	(c) Killed	Start vaccine immediately. Stop only if laboratory tests on animal for rabies are negative
Bites	Schedules as for 'Licks to skin' plus human rabies immunoglobulin (HRIG), 20 international units/kg body weight, of which half is injected around the bite(s) and half is given intramuscularly	

*Note that the risk of contracting rabies from a wild animal is significantly greater than from domestic dogs and cats.

†Regardless of any pre-exposure prophylaxis, persons exposed to a significant risk of contracting rabies should be given at least two doses of HDCS rabies vaccine.

much cheaper than the cell culture preparations that the risk of side-effects is judged to be the lesser evil.

19.3.6 Principles of rabies control in animals

Domestic animals

The removal of stray animals and vaccination of all domestic dogs and cats are essential features of control programmes in rabies-endemic areas. As a result of these procedures, rabies in dogs has decreased dramatically in the USA. Immunization of cats is now being encouraged.

Any domestic animal that is bitten or scratched by a bat or by a wild carnivorous mammal is regarded in the USA as having been exposed to a rabid animal, and unvaccinated dogs and cats are destroyed immediately or quarantined for 6 months. Vaccinated animals are revaccinated and confined for 90 days.

Wildlife

The control of rabies in wildlife is a difficult or, some might say, impossible task. In the USA, continuous trapping or poisoning as a means of rabies control is not recommended, but limited control is maintained in high-contact areas, such as picnic or camping grounds in National Parks. Similarly, bats are eliminated from houses.

In Europe, attempts are being made to control rabies in foxes by a unique immunizing technique. Chicken heads are impregnated with **live attenuated rabies vaccine and tetracycline**, and dropped by helicopter into remote mountainous areas. The foxes eat the heads and become infected, and hence are vaccinated with the attenuated rabies strain; concomitantly, the tetracycline is deposited in their bones. A simple fluorescence test on the bone tissue of captured foxes can detect tetracycline and indicate whether they have been immunized. By this method, rabies has been drastically reduced in parts of Switzerland, and many other European countries are now trying the system, but unvaccinated foxes often move into areas where it was previously considered that all foxes had been vaccinated. Tourists within the EU are now allowed to import pet dogs and cats to the UK so long as they have a history of rabies vaccination and a certificate and/or microchip as proof.

 # Reminders

- The genus *Lyssavirus* belongs to the family *Rhabdoviridae*, and contains rabies and other viruses that infect vertebrates.
- The viruses are **bullet-shaped** and glycoprotein spikes project from the lipid envelope. The genome is **ssRNA of negative polarity** and 11–12 kb in size.

- A series of five monocistronic mRNAs are transcribed by the virion associated RNA transcriptase. RNA transcriptase 'stuttering' at intergene poly U residues results in addition of poly(A) tails to the preceding mRNA and initiation of transcription at the next gene. Full-length copies of genomic RNA are synthesized by read-through of intergene regions by the newly synthesized viral RNA replicase.

- The natural reservoir of rabies is the wild animal population and the animal involved varies in different continents. In Europe, **domestic dogs** are an important source of infection for humans, whereas the main reservoir in the wild is **foxes**. Some countries, including Australia and New Zealand, impose a strict **quarantine**, but the advent of effective **vaccines** and techniques for confirming that immunization has been performed are making this method of control unnecessary in the UK and pets that have been vaccinated can have entry.

- Following a bite, virus replicates in muscle and moves along **peripheral nerves** to the CNS. Centrifugal spread along peripheral nerves to other tissues follows. The **incubation period** varies from 10 days to a year or more with an average of **1–3 months**, the time depending upon the distance of the bite from the CNS.

- The onset of clinical disease is insidious with malaise, fever, and headache. The 'stage of excitement' is characterized by local paralysis, swallowing difficulties, and **hydrophobia**. The patient dies within 1 week of cardiovascular collapse and coma. The illness may also take a predominantly paralytic form ('dumb rabies').

- Rabies may be prevented by immunization even after infection. The safest vaccine is the **HDCS** virus inactivated with β-propiolactone. At least six doses of vaccine are given by deep intramuscular injection, in combination with **human rabies immunoglobulin (HRIG)** as soon as possible after a bite. The vaccine is also used prophylactically, as one or two doses, to protect persons, such as veterinary surgeons, at special risk and travellers to areas in Africa, India, and South-east Asia where the virus is endemic.

Further reading

Horton, D.L. *et al.* (2014). Antigenic and genetic characterisation of a divergent African virus Ikoma lysavirus. *Journal of General Virology* **95**, 1025–32.

Questions

1. Discuss the natural history of rabies and comment on the different host reservoirs.
2. Outline the replication strategy of rabies in a cell.
3. Describe features of vaccines against animal and human rabies.

Group 3 Double-stranded RNA viruses

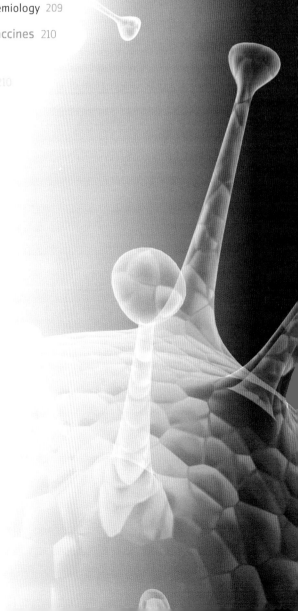

20

Reoviruses: diarrhoea-causing rotaviruses

20.1 Introduction

These reo (respiratory enteric orphan) viruses were identified in the 1950s and were noted to cause very minimal or no disease. Scientifically they became noteworthy as the first viruses discovered to have a double-stranded RNA genome. Later a genus of the family, namely rotavirus, was shown to be an important cause of gastroenteritis. Rotavirus is now recognized as the most important cause of gastroenteritis worldwide in children under the age of five. The virus causes over 500 000 deaths annually, the majority in developing countries of Asia and Africa.

20.2 Properties of the viruses

20.2.1 Classification

The viruses of this chapter are members of the family *Reoviridae* (Fig. 20.1), 'reo' standing for 'respiratory enteric orphan' because although they could be identified in the respiratory tract and gut, the virions were at first thought not to be associated with any disease, hence 'orphan'. *Rotavirus* is a medically important genus within the family.

20.2.2 Morphology

The shape of these viruses suggested their name (Latin *rota* meaning wheel). The capsid has a double shell and is about 100 nm in diameter; some smaller, single-shelled particles may

also be seen by electron microscopy. The virus particle comprises three layers (Fig. 20.2), an inner capsid, an intermediate capsid, and an outer layer. Eight proteins make up the virion but admittedly the nomenclature is the most complicated of any human virus! For this reason we have presented a simplified version. Moving from the inside towards the outside, the icosahedral core is composed of a structural protein, an RNA-dependent RNA polymerase (replicase), and an RNA binding protein. The inner icosahedral capsid is also the group-specific antigen. The outer icosahedral capsid is the neutralization antigen. Twelve spikes of the HA project outward at the icosahedral apices (Fig. 20.3). These have a cell attachment protein.

Rotaviruses are traditionally classified serologically into eight distinct groups (A to H), dependent upon the composition of the inner capsid protein, with groups A, B, and C found in both animals and humans, but groups D to H only found in animals. There is extensive antigenic variation amongst these individual groups.

20.2.3 Genome

The *Reoviridae* differ from all other RNA viruses in that their genome is 11 segments of double-stranded RNA. The 11 segments total 18 kbp in size and encode for structural and NS proteins. Except for gene 11, each genome segment only encodes for one protein. Six segments encode for virus structural proteins, and five segments encode for six non-structural proteins. The 11 segments of reoviruses can reassort their genomes in the same way as influenza viruses (see Chapter 14). This occurs

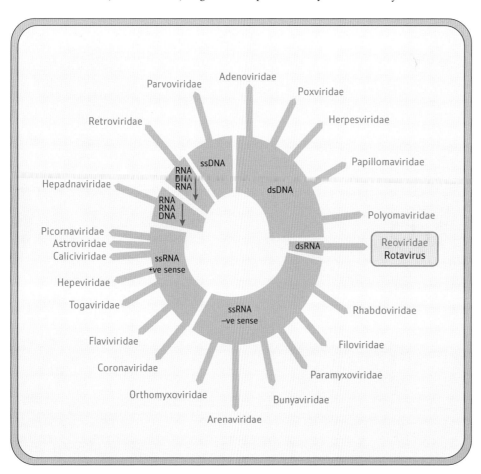

Fig. 20.1 Baltimore classification scheme and rotaviruses.

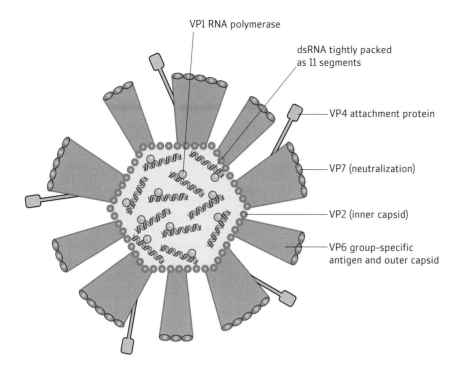

VP1 RNA polymerase

dsRNA tightly packed
as 11 segments

VP4 attachment protein

VP7 (neutralization)

VP2 (inner capsid)

VP6 group-specific
antigen and outer capsid

Fig. 20.2 Structural features of rotavirus.

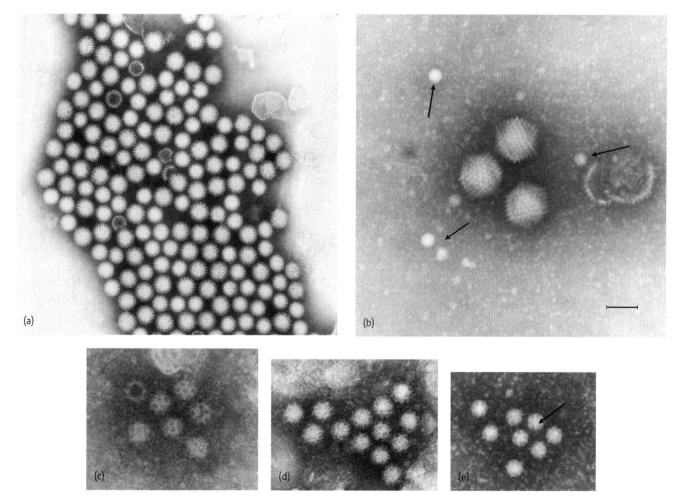

Fig. 20.3 Electron micrographs of rotavirus compared to other gastroenteritis viruses found in stools of patients with diarrhoea. (a) Rotavirus
(courtesy of Dr June Almeida). (b) Adenovirus from a child with diarrhoea. Note also the very small adeno-associated parvoviruses (arrowed).
(c) Caliciviruses from faeces of a child with diarrhoea. Note the cup-like depressions on the surface. (d) 'Small round structured virus'. (e)
Astroviruses. The six-pointed star pattern of the capsomeres is arrowed. (Parts (b) and (e) courtesy of Dr Ian Chrystie.) Scale bar = 50 nm.

when two genetically distinct reoviruses infect one cell and the resulting 22 genome segments end up in different 11 segment combinations in new virions. It is now recognized that reassortment can occur between human and animal strains from pigs, for example, and is a major process of genetic evolution.

20.2.4 Replication

The cell type naturally infected by rotaviruses is the enterocyte of the small intestine. Viruses bind to sialic acid receptors and also to junction adhesion molecule A (JAM-A) at tight junctions of cells via the (spike) cellular attachment protein, and enter cells by clathrin-mediated endocytosis. Additional receptors could be required, such as integrins, post attachment. The endocytosed virus is released into the cytoplasm and undergoes partial degradation by cellular proteases to give sub virion particles (Fig. 20.4). Internalized and disrupted virions act as factory sites in the cell cytoplasm. The factories contain all the machinery to make and cap mRNAs. By 8 hours post infection these can be

seen by electron microscopy. The virion-associated RNA-dependent RNA polymerase transcribes positive-stranded mRNAs from the negative RNA strand of the genome segments (initially for early viral enzymes) that then exit the factory sites via the hollow apices of the apical spike protein. The virus mRNAs contain a 5′ cap-like cellular mRNA, but lack a polyadenylated tail. Negative RNA strands are then produced from the mRNA molecules and, as a necessity of transcription, dsRNA forms are made which remain within the core. The dsRNAs can then serve as templates for multiple rounds of transcription of late mRNAs. Late mRNAs encode mainly virus structural proteins that trigger the self-assembly process. The translation of reovirus mRNAs is closely regulated. There are nucleotide sequences in the 3′ non-translated regions of reovirus mRNAs that effect translation efficiency. As virus protein synthesis increases host cell protein synthesis decreases. A unique feature of rotavirus assembly is that subviral particles that assemble in the cytoplasm bud through the endoplasmic reticulum (ER) and, therefore, briefly acquire a lipid envelope. This is lost as the virus traffics through the ER, being

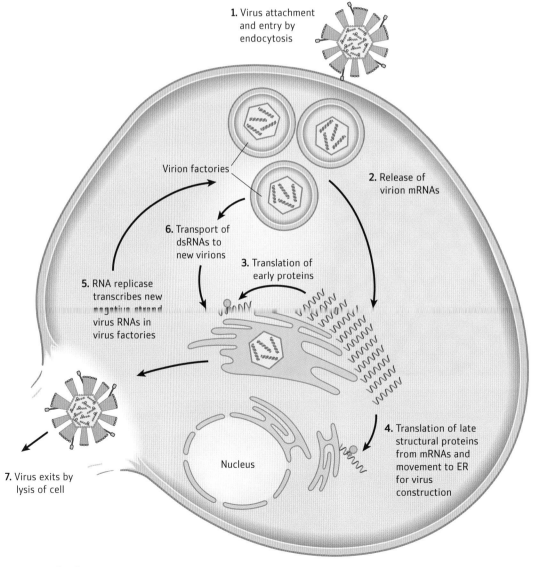

Fig. 20.4 Replication cycle of rotavirus.

replaced by a final layer of virus-encoded protein. Double-layered particles accumulate and form pseudocrystalline aggregates. The infected cells lyse and release the accumulated new virions. However, not all cell types are killed by rotavirus replication, and indeed rotaviruses can establish a persistent infection in cell culture and in immune-deficient patients.

20.2.5 Laboratory diagnosis

Rather rapid agglutination tests are carried out directly from diarrhoea, there being vast numbers of virions in these clinical samples. The same diarrhoea sample is subjected to RT PCR or more instant commercial diagnostic kits.

Unlike other reoviruses, rotaviruses can, if treated with trypsin, be propagated in primary monkey kidney cell cultures, but this method is not used for routine diagnosis.

20.3 Clinical and pathological aspects

20.3.1 Clinical features

Rotaviruses primarily infect the young of many species, including humans. **Babies under 2 years of age** are the main victims, but there may also be outbreaks in the elderly, particularly those in institutions.

The incubation period can be as short as 2 days; the characteristic syndrome comprises **vomiting**, **diarrhoea**, and **fever**, but silent infections also occur (Fig. 20.5). Virus may continue to be excreted for a week after diarrhoea has stopped. The disease symptoms are in part caused by one virus protein, NSP4, which has been identified as an enterotoxin and can lead to diarrhoea in mice that mimics the disease caused by rotavirus infection. **Dehydration** must be dealt with promptly; it should be no problem in countries with adequate facilities, but causes untold numbers of infant deaths in developing countries.

20.3.2 Pathogenesis

Observations on infected animals show that rotaviruses attack the **columnar epithelium** at the apices of the villi of the **duodenum** and **upper ileum**. There are some data available using clinical samples from patients but for obvious reasons this is limited. Virus infection activates programmed cell death (apoptosis) probably via caspases, which are cellular proteases. Regeneration from the bases of the villi is normally rapid after the acute attack.

20.3.3 Immune response

Following infection, antibodies are demonstrable in the serum; tests for specific IgG are useful in epidemiological studies, which show that most adults have been infected at some time or another. IgA antibody produced in the gut is probably an important factor in the immune response, and there is also some cross-protection among the various serotypes.

20.3.4 Epidemiology

We tend to associate diarrhoea and vomiting (D&V) with the summer months, but in developed countries rotavirus outbreaks typically occur in the winter. Spread is probably by the **faecal–oral route**; but the seasonal incidence suggests that respiratory infection cannot be ruled out. Outbreaks are common in institutions such as crèches and hospital-acquired infections are also fairly frequent. Unfortunately, the virus is relatively resistant to chemical disinfectants and spreads readily when hygiene is inadequate. In developing countries, infections with rotaviruses, along with other viruses and bacteria, are responsible for large numbers of infant deaths every year and at all times of the year.

Transmission to infants from asymptomatic health care staff or relatives can occur; conversely, there is evidence that babies can infect adults in close contact with them. **Chronic rotavirus**

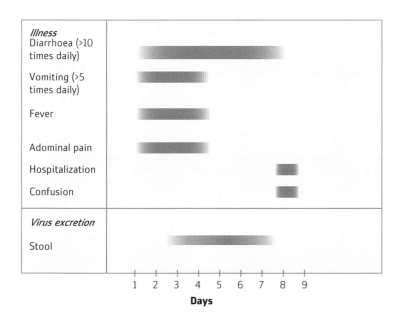

Fig. 20.5 Clinical and virological events during infection with rotavirus.

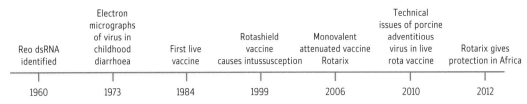

Fig. 20.6 Timeline for rotavirus

infection can be troublesome in patients **with primary immunodeficiencies**.

There are eight genotypes (A–H), identified by antigenic differences in the capsid antigen VP6, but most human rotaviruses are members of group A, including those which cause life-threatening diarrhoea in children. Most of the 100 million childhood infections and half a million deaths in children yearly are caused by group A viruses. Group A viruses are further subdivided into four antigenically distinct subgroups. However, genetic analysis has identified a further 27 genotypes. Classification schemes are obviously in a state of flux and have yet to be authorized by the International Committee on Taxonomy of Viruses (ICTV).

The segmentation of the rotavirus genome permits reassortant strains to be made in the laboratory. As these viruses are widespread in nature and infect most, if not all, domestic and farm animals and birds, there is at least a theoretical possibility that infection of humans with mammalian or avian strains might take place, with the emergence of 'shifted' strains analogous to those of influenza A (Chapter 14); but, so far, such hybrids have not been reported.

20.4 Rotavirus vaccines

Two live attenuated vaccines, Rotateq and Rotarix, were developed between 1984 and 2005 (Fig. 20.6) and offer a major benefit. Rotateq is based on a mixture of five reassortants in a bovine genetic 'backbone' whereas Rotarix is monovalent and derived from a human strain. In addition a more recent live vaccine derived from a calf rotavirus has been licensed in China. Clinical trials of the first two vaccines in Europe, Asia, Latin America, and the USA showed an efficacy of 85–98% against severe rotavirus gastroenteritis. In Africa the vaccines are less efficacious against the more diverse range of genotypes found there.

Reminders

- Rotaviruses have a complex double shell.
- Possess a double-stranded and segmented RNA genome of 18kbp.
- mRNAs are synthesized and capped inside intact virus cores (factories) and exude into the cytoplasm.
- Rotaviruses are important causes of infantile gastroenteritis and diarrhoea.

- Both infants and the elderly in institutions are at risk. There is year-round infection in developing countries whilst in developed countries infection is seasonal, especially in the winter.
- Three licensed live attenuated vaccines are giving significant protection both in developed and developing countries.
- Hand and surface hygiene are important in preventing and controlling outbreaks.

Further reading

Holloway G. and Coulson, B. (2013). Innate cellular responses to rotavirus infection. *Journal of General Virology* **94**, 1151–1160.

WHO (2013). WHO position paper. Rotavirus vaccines, *Weekly Epidemiological Record* **No 5**, 49–65.

Questions

1. Describe the morphology of rotavirus. How does the virus cause diarrhoea?

2. Outline the medical impact of reoviruses.

3. Describe the replication cycle of rotavirus.

Group 4 Double-stranded DNA viruses

21

Polyomaviruses

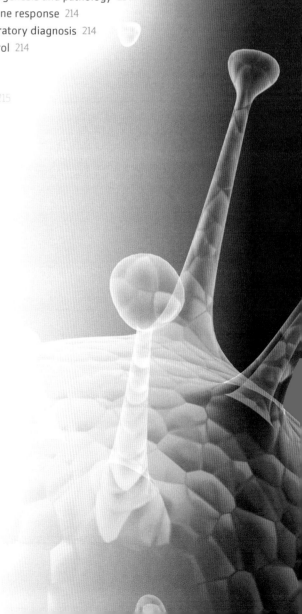

21.1 Introduction

Polyoma means what it sounds like—'many tumours'. This family of viruses (Fig. 21.1) has been the object of intense study to investigate how cells become transformed into tumour cells and also other basic parameters of cellular RNA and DNA synthesis. Like papillomaviruses, the viruses of this family are widespread in the community, but unlike papillomaviruses their disease impact is very small.

21.2 Properties of the viruses

21.2.1 Morphology

Polyomaviruses are small non-enveloped viruses 45 nm in diameter with an icosahedral structure of 72 capsomeres as pentamers of the protein VP1.

21.2.2 Genome

The polyomavirus genome is a circular double-stranded DNA of 5.3kb (Fig. 21.2). Cellular histones are attached and together with the viral DNA form a mini-chromosome in the infected cell.

The polyomavirus genomes have compact regulatory regions and have a number of overlapping genes to compensate for the small size of the genome, namely only 5.3kb coding for 6–8 proteins. The three regions of the genome are the non-coding region (NCCR), containing the origin of replication (ori), and the early and late coding regions. The small size of the genome justified separating the polyomavirus and the papillomavirus families.

21.2.3 Replication

Polyomaviruses bind to sialic acid-containing glycoproteins and gangliosides, and enter cells by endocytosis of specialized vesicles called caveolae. Intracellular fusion releases virions from the endosomes to the perinuclear regions of the cell and virus DNA enters the nucleus. Polyomaviruses are completely dependent upon inducing the S phase of cells as are parvoviruses. There are essentially two stages of viral replication. Firstly, the mRNAs of the early viral proteins are transcribed from one DNA strand in a counterclockwise direction. The early proteins are the Large, Medium, and Small T antigens. The three different messenger RNAs each code for a single virus protein and are made from each gene transcript by RNA splicing. The large T antigen binds to and inactivates cellular p53 and Rb proteins, and also regulates polyomavirus gene expression by binding to three sites on the viral genome. Binding to one site stops attachment

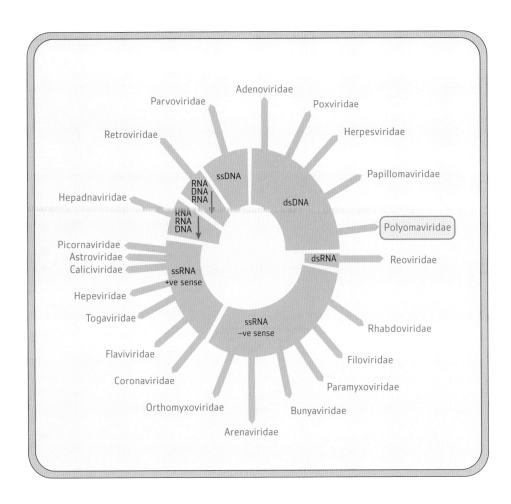

Fig 21.1 Baltimore classification scheme and the polyomaviruses.

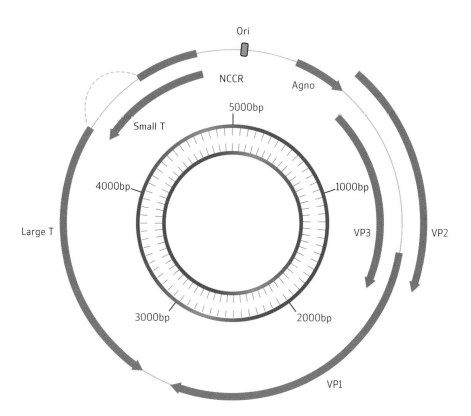

Fig. 21.2 Genome structure and encoded proteins of human polyomavirus. Large T and small T are illustrated on the counterclockwise portion of the genome. The three late mRNAs are formed by alternative splicing and code for structural proteins VP1, VP2, and VP3. The agno protein gene is not always present.

of RNA polymerase and thereby halts early transcription. In addition, the large T antigen starts viral DNA replication, which begins with the displacement of viral DNA histones and signals the start of late gene expression. The small T antigen also helps to initiate the S phase of the cell. The mRNAs for late viral proteins are transcribed from the opposite genome DNA strand in a clockwise direction. Differential splicing occurs for late transcripts producing VP1, VP2, and VP3 proteins. All this activity centres upon the viral mini-chromosome in the cell nucleus.

The new viruses begin to assemble in the nucleus and are released by lysis of the cell.

Very high numbers of virions are produced in urine, faeces, and saliva. Indeed the presence of polyomaviruses is used as a marker of sewage contamination of recreational water!

21.3 Clinical and pathological aspects

21.3.1 Clinical features

Three polyomaviruses cause clinical disease in humans, but only rarely. However antibodies to these viruses are present in a high proportion of sera suggesting widespread community infection. Most often they cause subclinical infections only reactivating during immunosuppression.

JC virus and progressive multifocal leucoencephalopathy

First described in 1958, this syndrome is characterized by hemiparesis, disturbances of speech and vision, dementia, and a relentless progression to death within a few months. It is note-

worthy that progressive multifocal leucoencephalopathy (PML), which is quite rare, often occurs in **elderly** patients already suffering from disease of the **reticuloendothelial** system; the first three to be described suffered from chronic lymphatic leukaemia or Hodgkin's disease. The lesions in the brain comprise multiple areas of **demyelination, abnormal oligodendrocytes**, and, in the later stages, pronounced astrocytosis. The name of the syndrome derives from these clinical and pathological features.

PML was a comparatively rare disease until it came to greater prominence as a complication of AIDS, or in severely immunocompromised patients.

BK virus: another persistent agent

In the same year as the detection of JC virus, a similar agent, termed BK, was isolated from the urine of an immunosuppressed renal transplant patient in London (Fig. 21.3). Although its physical characteristics are very similar to those of JC virus, its pathogenicity for humans is quite different.

BK is ubiquitous in the human population, infecting children asymptomatically and then persisting in the kidney. There are four subtypes. Subtype I is global, whilst subtype II is prevalent in Asia and Europe. Subtypes III and IV are rare, but are found worldwide. As with other members of the family a dramatic rise in overt infection has accompanied the HIV epidemic.

MC and newer polyomaviruses revealed by DNA sequencing

The more widespread use of DNA sequencing methods have uncovered eight more polyoma viruses including two which cause disease in elderly and immunosuppressed patients. Not

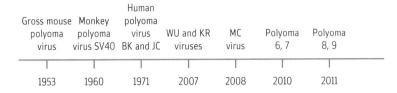

Gross mouse polyoma virus	Monkey polyoma virus SV40	Human polyoma virus BK and JC	WU and KR viruses	MC virus	Polyoma 6, 7	Polyoma 8, 9
1953	1960	1971	2007	2008	2010	2011

Fig. 21.3 Timeline for polyomaviruses.

all of these viruses are pathogenic. The KI and WO polyomaviruses are detectable in respiratory tract samples from persons with acute disease. MC polyomavirus is found in Merkel cell sarcomas, a rare aggressive cutaneous tumour of neuroendocrine origin of elderly and immunosuppressed patients. The incidence of the cancer is around 0.2 per 100 000 and has an association with face skin and sun exposure. The MC genome is integrated in the human genome in Merkel cell cancers. Human polyomaviruses 6, 7, and 9 are found in healthy human skin whilst polyomavirus 10 was recovered from a stool sample and later from a wart.

It is possible that these polyomaviruses arose from cross-species infection as emergent viruses, perhaps from orangutans. We await further discoveries with the daunting thought that the number of polyomaviruses could even reach that of papillomaviruses.

21.3.2 Epidemiology

Polyomaviruses are widespread in the community and virus DNA is detectable even on door handles, presumably originating from saliva. Not unexpectedly, the prevalence of BK antibody, for example, in the general population is high; people seem to become infected with polyomaviruses early in life without suffering obvious ill effects. Healthy people do not excrete BK virus unless they become immunocompromised or pregnant, in both of which cases the virus is liable to reactivate. Even when it does, however, overt disease is rare, the main exception being small numbers of cases of ureteric stenosis in renal transplant recipients, and of acute haemorrhagic urethritis or cystitis.

21.3.3 Pathogenesis and pathology

Primary infections are asymptomatic and little is known about the portal of entry. The virus does not move easily between communities, but around 2% of the population have JC virus in their bloodstream. The presence of virus in the urinary tract gives a clue to at least one site where BK virus may lurk after the primary infection, but apart from this, little is known of its pathogenicity.

Oncogenesis and polyomaviruses

The early virus genes act as oncogenes and induce the start of the DNA synthesis (S) phase of the cell cycle. Study of the virus T antigens helped unravel important cellular regulatory pathways which are changed at cell transformation. A key discovery was that oncogenesis would only happen in cells which did not support full growth of the virus—the cells were non permissive. Only a small fraction of the non-permissive cells are transformed; in such cells the early viral genes are integrated randomly into cell DNA and the cells continue to express T antigens. The studies observed that when key cellular signalling pathways which can increase cell growth are changed these cells evolve into tumours.

21.3.4 Immune response

By contrast with papillomaviruses, polyoma agents induce a good **antibody response**, but little is known about the role of CMI. In view of the persistent nature of these infections, it seems that immune mechanisms serve only to keep the viruses at bay, without actually eradicating them.

21.3.5 Laboratory diagnosis

Diagnosis of PML can be confirmed at autopsy by the characteristic histological appearances. Virions, often in crystalline arrays, can be found by EM in ultrathin sections of oligodendrocyte nuclei.

Diagnosis of BK may be made by staining exfoliated cells in the urine for viral antigen by immunofluorescence or immunoperoxidase methods. Of course, molecular diagnostics using PCR are important aids to diagnosis.

21.3.6 Control

Limited success has been claimed for cytarabine in the treatment of PML, but for practical purposes the prognosis is dire. Other than this, there is no treatment for these rare polyomavirus infections.

 Reminders

- Polyomaviruses are widespread in humans and animals, causing **persistent, but silent infections of the urinary tract**. They are, however, oncogenic if injected into newborn animals or into other species.

- The genome is circular double-stranded DNA of 5.3kb. The DNA is packaged as a mini-chromosome alongside cellular histones.

- The three best characterized human polyomaviruses which cause clinical disease are JC, BK, and MC: **JC virus** causes **PML**, a fatal demyelinating disease of the CNS, usually seen in later life and particularly affecting those with disease of the reticuloendothelial system.

- **BK virus** also reactivates in **immunocompromised patients**, but rarely causes overt disease.

- MC causes Merkel cell sarcoma.

- An additional eight polyomaviruses have been uncovered in human skin, stools, and respiratory secretions, but to date only one is associated with disease, namely MC virus in Merkel cell sarcoma.

Further reading

Feltkaup *et al.* (2013). From Stockholm to Malawi: recent developments in studying human polyomaviruses. *Journal of General Virology* **94**, 482–96.

Questions

1. How do polyomaviruses infect people and what is their strategy of cellular replication and their 'lifestyle'?

22 Papillomaviruses

22.1 Introduction

Up to the Third Edition of our book the papilloma and polyomaviruses were placed in the same family. However, distinctions in molecular replication and the organization of the virus genomes (Fig. 22.1) have now separated them.

The -oma component of the name immediately alerts you to the fact that these agents have something to do with neoplasia (Greek *oma* = tumour) and, indeed, the papillomaviruses are now recognized in relation to cancers of the genital tract (Fig. 22.2).

22.2 Properties of the viruses

Human papilloma virus (HPV) is the most common sexually transmitted infection in the USA, with millions of new cases each year. Importantly, oncogenic HPV types are responsible for 25–35% of oral cancers, 90% of anal cancers, and 40% of penile cancers in the USA. Both of these clinical problems are of worldwide concern but fortunately there is now a vaccine to prevent cancers induced by these viruses.

22.2.1 Morphology

Papillomaviruses are non-enveloped small icosahedral viruses (Fig. 22.3) that are 52–55 nm in diameter.

22.2.2 Genome

The papillomavirus genome is a double-stranded circular DNA that is 8 kbp in length and encodes for 9–10 ORFs depending on the virus (Fig. 22.4). The genome is efficiently used with overlapping coding regions.

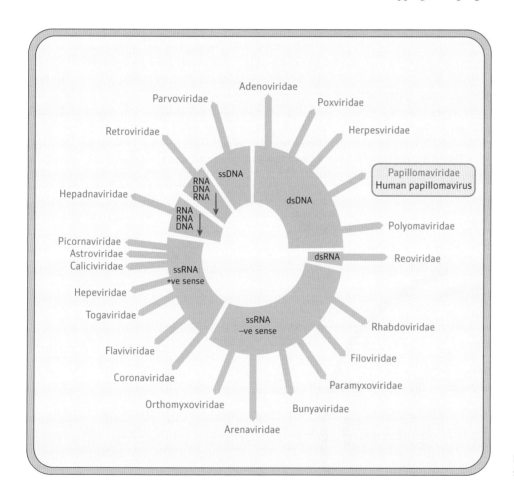

Fig. 22.1 Baltimore classification scheme and papillomaviruses.

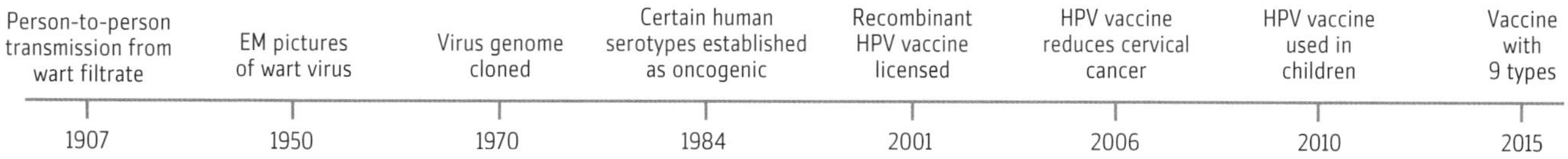

Person-to-person transmission from wart filtrate	EM pictures of wart virus	Virus genome cloned	Certain human serotypes established as oncogenic	Recombinant HPV vaccine licensed	HPV vaccine reduces cervical cancer	HPV vaccine used in children	Vaccine with 9 types
1907	1950	1970	1984	2001	2006	2010	2015

Fig. 22.2 Timeline for papillomaviruses.

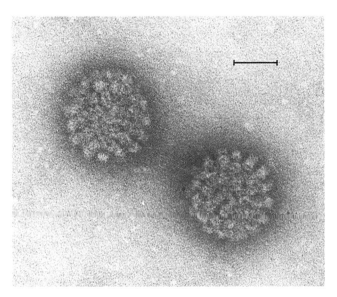

Fig. 22.3 Electron microscope of papillomaviruses. Scale bar = 50 nm.

22.2.3 Replication

Papillomaviruses have a strong tropism for squamous epithelial cells. Replication of the virus is intrinsically linked to the differentiation state of the epithelial cell (Fig. 22.5). As the basal epithelial cells of squamous epithelium are the only dividing cells during keratinocyte differentiation, papillomaviruses must infect these cells at this time to initiate an infection. The virus binds to heparin sulphate proteoglycans and the surface located α_6-integrin via the virus L1 protein. Following virus entry by endocytosis the viral DNA enters the nucleus and maintains itself as a free circular mini-chromosome. The viral DNA genome is transcribed from virus promoters to produce a complex array of differentially spliced transcripts. The early ORFs code for viral non-structural proteins. Early E1 (helicase and ATPase) and E2 proteins are involved in early viral DNA replication. E2 may also repress the activity of virus promoters by binding closely to them. E5 is a small protein and together with E6 and E7 is oncogenic, most likely as a by-product of interacting with cellular growth factor receptors. However, only E6 and E7 are required for cellular immortalization for high-risk oncogenic human papillomaviruses (cervical cancer).

E6 and E7 proteins, like SV40 Large T antigen and adenovirus E1B, target cellular p53 and retinoblastoma protein (Rb), and thus interfere with these cellular proteins' ability to negatively regulate the cell cycle, thereby initiating the cell cycle. Therefore, virus proteins E6 and E7 create the cellular environment conducive for virus genome replication, but also highlight the potential for this aspect of host–virus interaction to be oncogenic.

Late virus DNA replication, capsid protein synthesis (L1 and L2 proteins), and virion assembly occur only in differentiated keratinocytes. Multiple splicing patterns ensure that there are many more proteins than ORFs in the genome. In addition, a

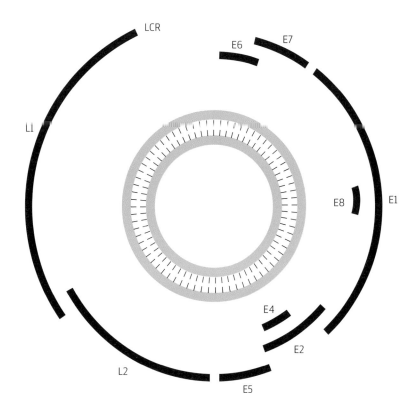

Fig. 22.4 Genome structure and encoded late proteins of papillomavirus. Labelled late proteins are L1 and L2 (capsid proteins), early proteins are E6, E7 (oncogenic proteins), E4 (virus release), E5 (stimulator of cell maturation), and E1, E2 (DNA replication proteins). LCR is the long control region.

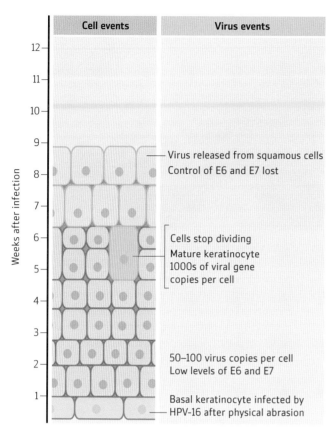

Fig. 22.5 Infection of squamous epithelial cells with papillomavirus leading to warts.

long central region of the genome (1 kb in length) called the Upstream Regulatory Region, which lies between the early and late ORFs, is thought to determine human papillomavirus tissue specificity.

22.3 Clinical and pathological aspects

For the first few of weeks of an infection in a patient virus DNA copy number is stabilized at 50–100 per cell. During this early infection of cells oncogenic E6 and E7 proteins are expressed at low levels. However, cessation of division of the host cells and the start of maturation into the mature keratinocyte induces the virus to increase gene copies to 1000s per cell. In the top layers of the epithelium virus genomes are encapsulated and emerge as infectious virions.

This entire step-by-step infection progress takes 12 weeks or so in a patient. There is no blood-borne viraemic phase and the cells continue to proliferate during infection.

22.3.1 Clinical features

Human papilloma viruses, when analysed by nucleotide identity, fall into five clades designated A–E. They cause disease only in skin and mucous membranes, where they give rise to warty lesions. These are usually benign, but some may become malignant; there are distinct—but not invariable—associations between the type of HPV, the anatomical site involved, and the potential to cause malignant lesions. Table 22.1 gives examples of lesions caused by papillomaviruses and the types of virus most commonly associated with them; this list is not exhaustive and may vary somewhat according to the authority.

Predominantly benign lesions

Cutaneous warts

The virus is transmitted from infected skin, either by direct contact or through fomites, and enters its new host through abrasions. Swimming pools and changing rooms are fertile sources of infection, and skin warts are most liable to affect the younger age groups.

The common wart (**verruca vulgaris**, Fig. 22.6) has a characteristically roughened surface; the excrescences are usually a few millimetres in diameter and may occur in quite large numbers anywhere on the skin, but especially on the hands, knees,

Table 22.1 Lesions caused by human papillomaviruses (HPV)

Type of lesion	Site	HPV types*
Predominantly benign lesions		
Common warts	Skin, various sites	2,4
Plantar and palmar warts	Hands, feet	1,2,4
Butchers' warts	Hands	7
Flat warts	Skin, various sites	3
Genital warts (condylomata acuminate)	Cervix, various sites	6,11
Juvenile laryngeal papilloma	Larynx	6,11
Malignant or potentially malignant lesions		
Flat warts	Skin	10
Bowenoid papulosis	Vulva, penis	16
Pre-malignant and malignant intraepithelial	Cervix, penis	Many types, including 6, 11, **16**, **18**, **31**, 39–45, 51–56
Neoplasia		
Carcinoma	Cervix, penis	**16**, **18**, **31**, 33, 35, 39, 45, 51, 52, 56, 58, 59, 68, 73, 82
Papilloma/carcinoma	Larynx	16
Epidermodysplasia verruciformis	Skin, various sites	Many types including 5, 8, 9, 10, 12, **14**, 15, 17, 19–29

*HPV types particularly liable to cause malignancies are printed in bold.

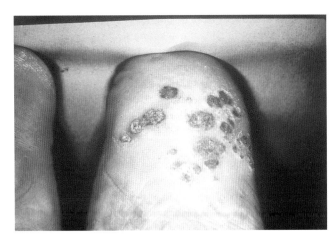

Fig. 22.6 Plantar warts.

Fig. 22.7 Penile warts (condylomata acuminata).

and feet. They usually cause no inconvenience, apart from the cosmetic aspect, but plantar warts on the soles of the feet are pressed inward by the weight of the body; they may be painful and call for more urgent treatment than those in other sites. **Flat warts (verrucae planae)**, as their name implies, are flatter and smoother than common warts; they also predominantly affect children. 'Butchers' warts', caused by HPV-7, are an occupational hazard. Why this HPV type affects butchers—and only butchers—is a mystery; there is certainly no relationship between it and any of the bovine papillomaviruses.

Genital warts

It is at this point, as we work down through Table 22.1, that we arrive at the somewhat blurred interface between the benign and malignant lesions caused by papillomaviruses.

As might be expected, genital warts are acquired by sexual contact: they are, in fact, one of the most common sexually transmitted diseases, and often occur in association with others, e.g. gonorrhoea or chlamydial infection. It follows that their highest incidence is the late teens and early adulthood. In the USA, where good figures are available, 20 million persons are considered to be infected and the prevalence in the population is about 1%, but could be 10 times higher in more sexually active groups.

Ordinary skin warts sometimes occur on the penis, but the most frequently seen lesions, both in men and women, are **condylomata acuminata** (condyloma acuminatum, if, as rarely happens, there is only one of them). This Greco-Latin name means 'pointed lump', but sounds nicer. They are fleshy, moist, and vascular, and may grow much larger than the common skin warts. By contrast with the latter, they may be pointed or filiform.

Condylomata acuminata may be seen at the following sites:

- In men:
  - on the penis, affecting the area around the glans and prepuce more than the shaft (Fig. 22.7);
  - within the urethral meatus and urethra itself;
  - around the anus and within the rectum, particularly in homosexuals; in those practising receptive anal intercourse this site is more frequently affected than the

penis. It is unusual for condylomata acuminata to become malignant, but they occasionally progress to squamous cell carcinoma. This is particularly true of the comparatively rare giant condyloma, associated with HPV6 or HPV11, which forms huge masses of tissue on the penis or in the anoperineal area.

- In women:
  - on the vulva;
  - occasionally, in the vagina;
  - on the cervix. Here, the typical lesion is the flat, intraepithelial type, rather than the fleshy variety seen on the external genitalia. It is difficult to distinguish clinically between this lesion, caused by a papillomavirus, and other forms of cervical dysplasia;
  - around the anus and on the perineum.
- In children these lesions have been observed on the external genitalia of both male and female infants and children. Although they may be indicators of sexual abuse, it is dangerous to jump to this conclusion, as the infection may be transmitted from the mother during delivery or even by close, but innocent, contact within the family.

Malignant or potentially malignant lesions

Bowenoid papulosis

This syndrome manifests as multiple papules on the penis or vulva; it is usually seen in young people and, although usually benign, may become malignant. It takes its name from Bowen's disease, which occurs in older people and is not associated with HPV, but with which there are histological resemblances.

Premalignant intraepithelial dysplasia

Irregularities in the histological pattern of the epithelium ('atypia') may occur on the penis, vulva (vulvar intra-epithelial neoplasia, VIN), vagina (VAIN), or cervix (CIN). They are staged according to the degree of dysplasia and the most severe form (CIN 3 in the case of cervical lesions), which involves all layers of the stratified epithelium, has a high chance of progression to metastasizing carcinoma. It is now known that many of these lesions are associated with papillomaviruses. **HPV types16** and **18** in particular are heavily implicated in the causation of carcinoma both of the **cervix** and of the **penis**.

Laryngeal infections with human papillomavirus

It is probably no accident that the types of HPV causing warts in the **juvenile larynx** are the same as those associated with genital warts, namely **HPV6** and **11**; they may well be transmitted during delivery and establish a persistent infection. Although they tend to recur after treatment, these papillomas do not become malignant in children; this is not, however, true of the **adult** variety, associated with **HPV16**, which may develop carcinomatous changes.

Epidermodysplasia verruciformis

This is a rare autosomal recessive disease associated with defects in T-cell function and numbers, but not with other varieties of immune deficiency. The multiple flat lesions may persist for years; they are associated with a wide range of HPV types and may become malignant, especially in areas of skin exposed to sunlight.

Immunosuppressed patients

Allograft recipients receiving immunosuppressive treatment are liable to develop squamous cell carcinomas of the skin; HPV5 has been demonstrated in the lesions of at least one such patient.

22.3.2 Pathogenesis and pathology

The papillomaviruses are something of a curiosity in that they can multiply only in **proliferating stratified squamous epithelium**, which cannot be grown as conventional cell cultures; the point of attack is often where one type of cell changes to another, e.g. the junction of the columnar epithelium of the cervical canal with the stratified squamous epithelium of the outer cervix.

As we noted in Section 22.3.1, the virus establishes a chronic infection of the stratified squamous epithelial cells and the HPV DNA is integrated via non-homologous recombination, stabilizing the high expression of E6/E7 and producing more severe lesions.

A characteristic feature of the benign skin warts is **hyperkeratosis**, i.e. a massive proliferation of the keratinized layers of the dermis (Fig. 22.8). Particularly in plantar warts, there are deep extensions of the hypertrophic epidermis. Large, pale, vacuolated cells are present in the granular layer; these are more pronounced in plane warts than in verruca vulgaris. There are many eosinophilic cytoplasmic inclusions that are not of viral origin, but which consist of abnormally large keratohyalin granules. There are, however, **basophilic inclusions in the nuclei** of the epidermal cells in which typical papillomavirus virions can be seen by EM. HPV stimulates cell proliferation in the stratum spinosum by early expression of the immortality functions of E6 and E7 proteins. Late gene expression of L1 and L2 proteins occurs in the uppermost stratum corneum where virions are assembled and released.

The histology of condylomata acuminata is quite similar, except that hyperkeratosis is not a significant feature. The presence of vacuolated cells in cervical scrapings (Fig. 22.8) is diagnostically useful; they are known as **koilocytes** ('empty cells') because of their appearance in stained preparations, but the term is inappropriate as the apparently empty vacuoles contain enormous numbers of virions.

At this point, the histopathology of papillomavirus lesions merges into that of precancerous and cancerous changes.

The DNA of high-risk HPVs can be separated from the low-risk HPVs by their ability to transform human fibroblasts and their transforming virus proteins E5, E6, and possibly E7. The long interval between the first infection with HPV and the cancer suggests the contribution of environment and host factors, and/or that a hit-and-run mechanism may operate.

Cancer and HPV

Cervical cancer has been recognized as a sexually transmitted disease and caused by certain HPV. The Dutch virologist Harold Zur Hausen was awarded the Nobel Prize for his discovery of the oncogenic properties of HPV. It is the second most common malignancy among women worldwide with 0.5 million new cases diagnosed each year. About 80% of cervical cancer occurs in developing countries. In the USA 12 000 cases are diagnosed yearly and the rate increases two-fold in African Americans. Although 40 HPV types are found in the genital tract, only a subset is regularly found in genital cancer and these (16, 18, 31, 33, 35, 39, 45, 51, 52, 56, 58, 59, 68, 73, and 82) are designated as high risk, with four sero types (16, 18, 31, and 45) making up 80% of HPV positive cancers. In most areas of the world type 16 is found to be the most common type. Oncogenic HPV types 16 and 18 are estimated to cause around 30% of oral cancers, 90% of anal cancers, and 40% of penile cancers in the USA, and by extrapolation elsewhere. Laboratory cell lines from cervical cancers have portions of the HPV genome integrated into their genome and a classic example is the HeLa cell. Originally derived from an American, Henrietta Lacks, this widely used cell line expresses E6 and E7 proteins.

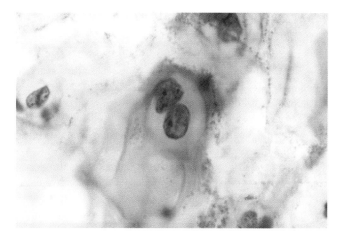

Fig. 22.8 Koilocyte in cervical scraping. The double nucleus, which is characteristic, is surrounded by an area of unstained cytoplasm. (Courtesy of Dr David Oriel.)

22.3.3 Immune response

The failure to culture papillomaviruses except in very low yielding cultures has greatly hindered our ability to unravel the immune reactions to infection. This is a pity, because the little that we do know poses tantalizing suggestions that the immunology of papillomavirus infections may have some unique characteristics. Why, for example, may cutaneous warts persist for years and then suddenly disappear for no obvious reason? What is the relationship between the immune response and oncogenesis? What are the respective roles of CMI and antibody? We know from immunofluorescence studies that both IgM and IgG antibodies may appear, particularly in those with regressing warts; but not all those with antibody have warts and not all warty individuals have antibody. On the other hand, cell-mediated immunity may be equally or more important, as regressing lesions contain many macrophages and lymphocytes, of both T-suppressor (CD8+) and T-helper (CD4+) subclasses. Circumstantial evidence of the importance of the immune responses is provided by patients with epidermodysplasia verruciformis, who have an inborn defect of immunity, and by the high prevalence of warts in immunosuppressed patients.

There is an association between histocompatibility (HLA) type and the development of cervical carcinoma. The mechanisms by which papilloma and other viruses give rise to malignant tumours are dealt with in Chapter 6.

22.3.4 Laboratory diagnosis

The presence of warty lesions is usually so obvious to the eye, aided when necessary by a magnifying glass, colposcope, or urethroscope, that laboratory methods are largely redundant except for determining, if possible, whether a given lesion is benign or malignant. At colposcopy, lesions in the vagina (or rectum) are rendered more visible by painting the area with 5% acetic acid. The histological appearances are characteristic, and Papanicolaou staining of smears may reveal characteristic features, notably the presence of koilocytes. For typing HPVs within lesions, recourse must be had to real time PCR.

22.3.5 Control with vaccines

In spite of the success of the prophylactic anti-cancer HPV vaccine there remains a large burden of infection in the community because the vaccines contain only a handful of HPV types respectively. In the US, where the quadrivalent vaccine was licensed in 2006, 90% efficacy for the prevention of external genital warts and 70% reduction in cervical cancers caused by types 6, 11, 16, and 18 has been recorded. The vaccine is also licensed and recommended for men aged 9–26 years of age. New vaccines have 9 HPVs.

A key discovery for vaccine development was that the L1 virus protein, which contains the virus neutralization epitope following the cloning of its gene into a cell line, can self-assemble into virus-like particles (VLPs), essentially DNA-free empty capsids. These VLPs are highly immunogenic, producing high levels of virus neutralizing antibody. Almost complete protection of vaccinated women against persistent infection and dysplasia from virus types 16 and 18 has been shown over 4 or more years post-vaccination. The cross-immunity also extends to other virus types such as 45. Since the L2 protein contains more cross-reactive epitopes, a vaccine based on this protein may protect against an even broader number of HPV serotypes.

Medical case story The case for HPV vaccination

Much modern consultation work is now done over the telephone, particularly where an examination is not needed, as here.

You receive a message from the receptionists on your computer attached to a 12-year-old girl's notes, 'Please ring patient Katey Jennings's mother to discuss vaccine,' it says. After surgery and a quick break for a cup of tea, you begin your telephone calls at about midday.

'Hello, this is Dr X from X surgery,' you flick into the notes, 'phoning to speak to Katey Jennings's mother'.

'Oh thank you so much for calling. It's about this cervical vaccine. They've started giving it at the school, and what with that girl dying straight after having it last week, I don't know whether Katey should have it or not'.[1]

'I think they have found that the girl didn't die from the vaccination, haven't they? Didn't they find she had an invasive heart and lung cancer? They did a post-mortem'.

'Oh I know, but I'm still so worried about it. After all it's so young to be having it, isn't it? It's not as if she's going to have sex for years'.

'Mmm, I suppose. It's not that we think anyone's necessarily having sex at her age, it's just that as soon as a teenager does have sex we more or less assume they have caught human papillomavirus

(HPV). It's so common, you see, that more or less everyone catches it. Then most people clear it, but a few don't, and those are the people who then get cell changes on their cervix'.

'But cervical cancer's not that common, is it?'

'About 3000 people a year get it in this country, and about 1000 a year die from it. It used to be much more common, but now that everyone gets their three-yearly cervical screening, fewer people get cancer. Cervical screening in the UK started in 1980 and probably cut the mortality rate by about half. Perhaps, one day, everyone will be vaccinated and then gradually we will be able to phase cervical screening out'.

'Do you think vaccinating against these warts will really stop cervical cancer? Is it like a cancer vaccine?'

'Well in a way, yes. You know of course that if you smoke you are more likely to get lung cancer? Well, the link between HPV and cervical cancer is even stronger than that. But it won't stop all cervical cancer, just the cancers associated with HPV 16 and 18. These two account for about 70% of the cervical cancers'.

'OK. But do you think it is safe? All those other children at the school where that girl died said they were feeling faint and dizzy and things'.

'Dizziness and headaches are among the listed side-effects.[2] The vaccine has been tried on 20 000 women. In fact, in the last year they gave it to over half a million girls in the UK. I am sure it is safe. It contains only synthetic virus, no actual virus at all. It also contains adjuvants to help generate the immune response, but these have been used for many years in other vaccines. The government doesn't rush out these vaccination campaigns unless it is absolutely certain they are going to be effective, because they cost so much money. It has to be absolutely sure it's worthwhile and safe.'

'Would you use it?'

'Yes. Yes I would. I don't have teenagers but I have two small daughters, and when they are big enough I will definitely have them vaccinated.'

'Oh. It's so difficult to decide isn't it? I'm just so worried about it. It's so new'.

'Well one nice thing is that there is less pressure on you to get your daughter vaccinated. We don't need herd immunity like we do with the baby vaccines. If someone chooses not to get vaccinated it affects no one but herself.' (In fact children under the age of 16 are allowed to exercise Gillick Competence and decide to have it against their parents' will. They would not be allowed to decline it.)

'She seems so young ... Do we know how long it will last?'

'No, not as yet. It's been followed up for 5 years so far. The prediction based on the immune response is that it will last at least 20 years. But don't worry if Katey doesn't have the vaccination now. She can choose to have it herself later in life and still benefit—not as much as if she were to have it now, but she will still benefit. They have tested the vaccine in women up to the age of 45.'

'Yes. Well thank you so much. You've been very helpful; I've been really worried about it all. I'll have another think about it.'

'If you want more information there is a good NHS website at immunisation.nhs.uk.'[3]

'Ok, I'll have a look. Thank you so much.'

'Thank you. Bye now.'

'Good bye'.

Later you are discussing HPV vaccines during a practice meeting. 'I've had a phone call about that,' says one of your colleagues.

'I've had two,' says the partner, 'I think it's because they've just started to bring it out in schools this year. It should have been the other one: the 4-component vaccine not the 2. The Government's just being cheap' she adds.

'It's not cheap, it's cost effective. They have to make sensible decisions about spending money,' argues your colleague.

'It's not either,' you speak up. It happens you went to a lecture on this topic by a consultant gynaecologist who had been involved in the decision to use Cervarix. 'They outlined exactly what they wanted in the vaccine before they researched it. It had to be effective for HPV 16 and 18—the cervical cancer ones— and it had to produce a good immunological response. The 4-component vaccine protects against four different HPVs, but the government is not in the business of preventing warts: the idea is to prevent cervical cancer.

'That fits with the story I heard,' agrees your colleague. The partner looks wry but says nothing.

'I wonder how much it is costing. Will they have a catch-up campaign?' you wonder.

Following your discussion with the partner you are annoyed enough to go and look up more information about it. You try the *Drug and Therapeutic Bulletin*[4] first because you know it will compare the two vaccines head-to-head.

You are completely wrong about the reasons for the vaccine choice.

It turns out that both vaccines had equally good efficacy (99% of patients generated a response), and equally good levels of antibodies generated—over twice that of a natural infection. Nobody knows exactly what the government paid per vaccine. Now there is a 9-component vaccine with types 6, 11, 16, 18, 31, 33, 45, 52, and 59.

Notes

[1] BBC News. (2009). Schoolgirl dies after cancer jab. Available at: www.bbc.co.uk/1/hi/health/8279855.stm (accessed 20 Feb. 2016). https://www.gov.uk/government/uploads/system/uploads/attachment_data/file/345955/8874-HPV-the-Facts-05.pdf (accessed 20 Feb. 2016).

[2] Medicines and Health care products Regulatory Agency. (2009). Suspected adverse reaction analysis Available at: www.mhra.gov.uk (accessed 29 October 2009).

[3] http://www.nhs.uk/conditions/vaccinations/pages/vaccination-schedule-age-checklist.aspx

[4] BMJ Group. (2008). Vaccination against human papillomavirus. *Drug Ther Bull* **46**, 89–93.

★ Learning Points

- **Virological:** There are over 80 human papillomaviruses, of which the ones causing the majority of cervical cancer are 16 and 18. These two types are spread very readily during sexual contact, generally causing no symptoms, and are usually cleared by the immune system within a few months. If the virus is not cleared then it can lead to cervical cancer, usually taking about 10 years.

- **Clinical:** The cervical screening programme will continue, as the vaccination programme will not reach or protect everyone for at least another generation.

- **Personal:** Which vaccine would you have purchased?

Specific antivirals

A number of key virus proteins such as E6 and E7 have been targeted with anti-sense RNA ribozymes and peptide aptamers with, it has to be admitted, mixed success.

Other measures

Warts, especially on the feet, are readily acquired in **swimming pools and changing rooms**, where adequate hygienic measures should be maintained. Measures for preventing genital warts are similar to those for other sexually transmitted diseases.

Skin warts may need treatment on cosmetic grounds, or, in the case of plantar warts, because of pain and disability on walking. They may be removed by treatment with **podophyllin**, extracted from the roots of the American mandrake, or by **freezing** with liquid nitrogen. There are a variety of methods of dealing with cervical lesions, including **freezing**, **electrodiathermy**, and **cone biopsy**. Recurrent laryngeal warts may be removed surgically, but must never be treated with irradiation, as this induces malignant change. Injection of **IFN** directly into the lesions has met with limited success, but by no means in all patients so treated.

Reminders

- The human papillomavirus (HPV) genome is double-stranded circular DNA, 8kbp in length, coding for 9–10 ORFs.
- HPV replicates only in dividing (S phase) basal epithelial cells.
- Virus proteins E6 and E7 are oncogenic and block cellular pathways of cellular p53 and Rb proteins which normally suppress tumours.
- Many of the 100 viruses in the family cause warts in **skin** and **mucous membranes**, which are usually benign, but may become malignant, especially in the genital tract. Certain types of HPV, notably 16 and 18, are strongly implicated in

the causation of cancer of the genital tract, notably **carcinoma of the cervix**.

- HPV are particularly liable to cause malignant disease in **immunocompromised patients**.
- HPV infection is now the most common **sexually transmitted disease** in the UK.
- There is also a tetravalent vaccine containing HPV6, 16, 11, and 18. A newer vaccine has types 6, 11, 16, 18, 31, 33, 45, 52, and 59. Both these vaccines are used for cancer prophylaxis in both women and men.

Further reading

Stanley, M.A. (2012). Genital human papilloma virus infections: current and prospective therapies. *Journal of General Virology* **93**, 681–91.

Questions

1. Describe the range of diseases caused by HPV. Which viruses are targeted by the HPV vaccine?

Herpesviruses: herpetic lesions, zoster, cancer, and encephalitis

23.1 Introduction

A herpetologist is not a virologist who specializes in herpes infections; he or she is, in fact, an expert on reptiles (Greek, *herpeton*). The name herpes (meaning to creep or latent) seems to have become attached long ago to this group of viruses because of a rather fanciful idea of the creeping nature of the lesions caused by some of them. The name is, however, appropriate in a more modern context, because these viruses are excellent examples of 'creepers' (Chapter 4).

The herpesviruses form a large and most important group of infective agents, and affect many species both of warm- and cold-blooded animals. In humans they cause a wide range of syndromes, varying from trivial mucocutaneous lesions to life-threatening infections including certain cancers (Fig. 23.1). To give an idea of the size of the problems, 80–90% of the population are infected, most unknowingly, with human herpesvirus 4 (HHV-4), otherwise known as Epstein–Barr virus.

An important property of all herpesviruses is their ability to cause latent infections where the virus can persist in an infected cell without expressing many of the protein-encoding genes of the virus. Latent viruses can subsequently become reactivated. They are an ancient family of viruses, predating the advent of agriculture 10 000 years ago, when virus success did not necessarily depend on rapid person-to-person spread. After a primary infection, herpesvirus persists for the lifetime of the host, reactivating from time to time to allow transmission to new uninfected hosts.

From a scientific viewpoint these large genome viruses have a correspondingly large budget, allowing a more timely regulated virus-host interaction, establishing more often a long-term relationship, fatal infections being the exception rather than the rule. Molecular virologists still place the mechanisms of herpes latency and reactivation under intense research scrutiny.

23.2 Properties of herpesviruses

23.2.1 Classification

There are currently over 130 known species of herpesviruses. The order *Herpesvirales* is divided into three families and here we are interested in the family *Herpesviridae* (Fig. 23.2) (Table 23.1).

The nine human herpes viruses are divided into three subfamilies, *Alpha-*, *Beta-*, and *Gammaherpesvirinae* (Table 23.1), with these subfamilies having 4 genera in each.

23.2.2 Morphology

The herpesviruses are large enveloped viruses around 200 nm in diameter and with double-stranded DNA genomes (Fig. 23.3).

The virions are **icosahedral**, built up of 162 tubular capsomeres surrounding a core of DNA. The core, in the form of a torus, is surrounded by an amorphous mass called a tegument, and outermost there is a large trilaminar **envelope** derived from altered cell membranes (Figs 23.3 and 23.4); unenveloped particles may also be present. The tegument is layered and has a selection of 14 or more proteins, which on virus entry to a new cell can manipulate the intracellular environment, whilst also stimulating expression of virus genes. There are even some virus mRNAs in the tegument that can be translated very rapidly upon infection (Fig. 23.4). Around 10 different viral glycoproteins are embedded in the envelope as short protrusions, probably in total exceeding 1000 per virus. The capsid, which encloses the DNA, has six proteins and the major protein VP5 makes up the 162 capsomeres. They are arranged as 150 hexamers with 12 pentamers (VP26) at the vertices of the icosahedron.

23.2.3 Genome

The virus genome is double-stranded DNA and is linear in the virus particle. The genome is infectious. The genome size varies according to the subfamily, from 125 to 250 kbp. The genome organization of the subfamilies differs remarkably; those of HHV-1 and HHV-2 resemble each other most closely. However, it is obvious from the fact that many core genes are conserved across all herpesviruses that the α, β, and γ herpesviruses have evolved from a common ancestor. All nine human herpesviruses have had their genomes completely sequenced and annotated.

Herpes simplex is considered the prototype virus of the *Herpesviridae*. It has a 152 kbp genome, containing over 80 genes. Each gene is expressed from its own promoter, although there are many overlapping reading frames due to the fact that both strands of the herpesvirus genome encode for genes. The HSV-1 genome has two covalently joined sections termed the

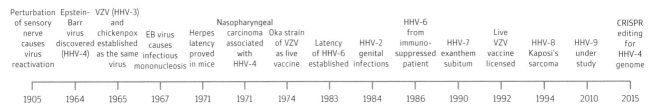

Fig. 23.1 Timeline for herpesviruses.

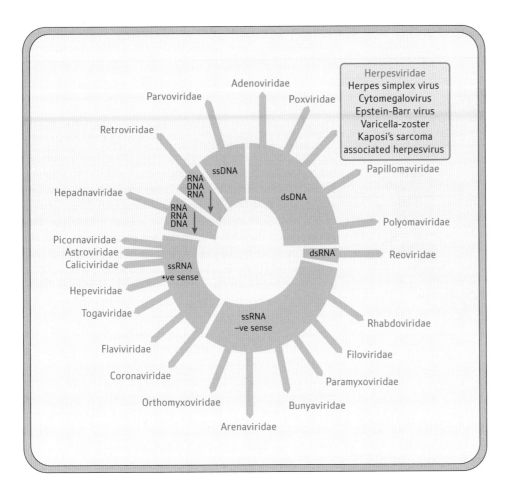

Fig. 23.2 Baltimore classification scheme and herpesviruses.

Table 23.1 Members of the family *Herpesviridae* that infect humans

Subfamily	Colloquial name and abbreviation	Numerical designation
Alphaherpesvirinae	Herpes simplex type 1 (HSV-1)	HHV-1*
	Herpes simplex type 2 (HSV-2)	(HHV-2)
	Varicella-zoster (VZV)	HHV-3
	Herpesvirus B (or herpes simiae virus)	
Betaherpesvirinae	Cytomegalovirus (CMV)	HHV-5
	Human herpesvirus 6	(HHV-6)
	Human herpesvirus 7	(HHV-7)
Gammaherpesvirinae	Epstein–Barr virus (EBV)	HHV-4
	Kaposi's sarcoma associated herpesvirus (KSHV)	HHV-8

*HHV = human herpesvirus

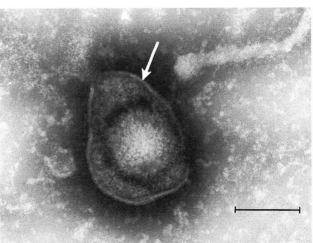

Fig. 23.3 Electron micrograph of HSV-1. The baggy appearance of the outer envelope (arrowed) is characteristic, but is an artefact arising during preparation of a specimen. (Courtesy of Dr David Hockley.) Scale bar = 100 nm. The envelope has numerous (1000 or so) short protrusions of glycoproteins.

unique long and **unique short** regions (Fig. 23.5). Each is bounded by inverted repeats, which allow structural rearrangements of the unique regions (U_L and U_S) and hence **the HSV-1 genome exists as a mixture of four isomers**, which

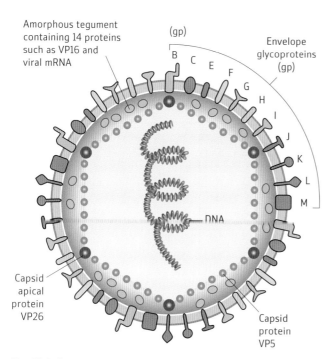

Fig. 23.4 Structural features of the herpes virion.

are functionally equivalent to each other. The unique short (U_S) genome region codes for 14 viral proteins while the unique long (U_L) codes for 65 proteins. All herpes genomes have repeated sequences and this introduces another variation in genome size in the family.

Members of the same subfamily share most of the genes, which have similar layouts. For example, there are only five genes in HHV-3 (VZV) that do not have counterparts in HHV-1 (HSV-1). Up to 80% of the total herpesvirus proteins have homologues in a different herpesvirus, whereas 20% appear to be unique to particular genomes, sometimes existing as multiple copies. The function of many of the gene products of different herpesviruses is still not known. Between 9% and 30% of genes in a given herpesvirus have clear homology to host-cell proteins, suggesting the viruses have captured these genes from the host as they are beneficial for the viruses' life cycle. Finally, some genes present in the prototype virus HSV-1 are missing in other viruses. Thus, the gamma herpesviruses do not have a thymi-

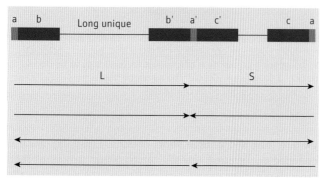

Fig. 23.5 Genome structure of herpesvirus.

dine kinase (TK) gene. There is also genome variation between the subfamilies, for example, in the genes encoding the virion surface glycoproteins.

23.2.4 Replication

The docking proteins of the most studied member of the family, HSV-1, are the glycoproteins gC and gB, which attach to cellular heparin sulphate proteoglycans (Fig. 23.6). Afterwards gD binds to two additional receptors, nectin 1 and HVEM, a tumour necrosis factor receptor family. Other viral glycoproteins (gB, gL, and gH) then become involved in a complex fusion process of the viral lipid envelope with the plasma membrane. This releases both the virus capsid and tegument proteins into the cytoplasm. These proteins are transported to a nuclear pore along the microtubule network, at which point the linear DNA is released and enters the nucleus of the cell where most events of transcription, viral DNA replication, and capsid assembly occur. The virus DNA is circularized in the nucleus by direct ligation of the DNA ends and by recombination between the sequences at the ends. Following infection of the nucleus, a complex cascade of virus gene expression is initiated. For the alphaherpesviruses, the virus enters what is known as the lytic cycle, leading to the production of new virus particles followed by latency events. However, the beta- and gammaherpesviruses tend to enter a latent state of infection more rapidly whereby the virus genome is maintained in the cell nucleus without the expression of the virus genes necessary to make new virus particles. What controls the choice of latent or lytic gene expression, or how latent viruses are reactivated at a later point, is not fully understood.

During the lytic cycle of all herpesviruses, there is an initial burst of transcription activity from the immediate early viral genes followed by the initiation of viral DNA replication (Fig. 23.6), and then intermediate (or early) and late gene expression. Cellular RNA polymerase II carries out gene transcription. Not unexpectedly, the whole process is carefully controlled. In HHV-1 transcription the immediate early 'α' mRNAs are stimulated in the first hours by a viral tegument protein (VP16) and once these mRNAs are transferred to the cytoplasm, translation into α proteins commences. The α proteins are, on the whole, *trans*-acting regulatory proteins, which are very important and which control the expression of early or β genes expressed 4–8 hours post infection.

The optimal expression of virus α genes is reliant upon tegument proteins, particularly VP16, which is complexed with two or more cellular proteins. The complex binds to viral DNA at a pre-existing sequence upstream from the promoter of α genes and transactivates them. At this point in the life cycle viral mRNAs are transcribed in abundance. Transcription and translation of cellular mRNAs are blocked by other herpes proteins. Eventually another tegument protein (virus host shutoff protein, Vhs) degrades α gene mRNAs. At this point the finesse of the whole complicated but well-rehearsed scheme comes into play: viral mRNA is suppressed soon after the start of DNA synthesis, by binding to VP22 and VP16.

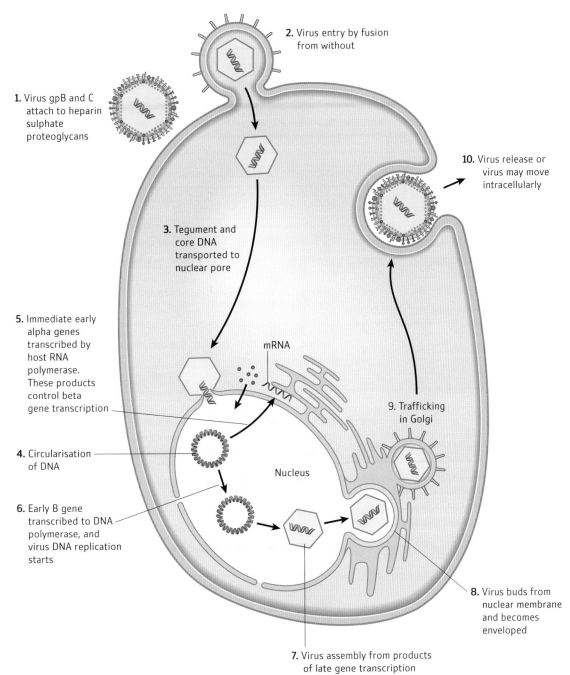

1. Virus gpB and C attach to heparin sulphate proteoglycans

2. Virus entry by fusion from without

3. Tegument and core DNA transported to nuclear pore

5. Immediate early alpha genes transcribed by host RNA polymerase. These products control beta gene transcription

4. Circularisation of DNA

6. Early B gene transcribed to DNA polymerase, and virus DNA replication starts

mRNA

Nucleus

10. Virus release or virus may move intracellularly

9. Trafficking in Golgi

8. Virus buds from nuclear membrane and becomes enveloped

7. Virus assembly from products of late gene transcription

Fig. 23.6 Replication cycle of herpesvirus.

Replication of viral DNA is under the control of β genes, which code for viral TK, DNA polymerase, and helicase. They start by directing the viral DNA polymerase to start DNA synthesis. This is essentially a classic rolling stone mechanism. One of the strands of parental DNA is nicked, allowing one of the DNA strands to extend itself. Viral DNA replication itself triggers viral γ genes and these genes code for viral structural proteins. Transcription of late genes coding for structural proteins is totally dependent upon virus DNA synthesis.

VP5, an important capsid protein, accumulates around a scaffold structure in the cell nucleus made up of other late, mainly structural, proteins, VP21, VP22, and VP24. Other late virus proteins force the entry of viral DNA into the capsids, and at the correct DNA length, since cleavage stops further addition of DNA. The scaffolding is then dismantled and the DNA is safely sealed into the virus.

The next stage of viral replication is the subject of some controversy. The new virions may bud through into the space between the inner and outer nuclear membranes and, in doing so, acquire an envelope with tegument proteins. It is probable, but not certain, that the virions are transported to the Golgi membranes via vesicles budding from the outer nucleus membrane. Some viruses may not exit but move to adjacent cells presumably by fusion events at the plasma membrane.

The host cell, of course, activates several hundred genes to counteract the above viral attack, but many of these host mRNAs are degraded by Vhs. An excellent example is the blockage of host interferon production and concomitant presentation of virus peptides on the cell surface. Finally, herpes infection blocks the host response of induced apoptosis.

23.2.5 How herpes viruses establish latency

Herpes viruses such as HHV-1 can infect nerve endings and undergo retrograde transport along axons to cell bodies at the dorsal root neurons, where virus DNA enters the nucleus, circularizes, and becomes dormant. We have noted in Section 23.2.4 above that during normal lytic infection virus alpha genes are activated by a complex of viral VP16 and cellular transcription proteins, but little VP16 reaches the nucleus of the nerve and, moreover, the local cells have proteins which actually prevent the formation of the transcription complex; thus, the end result is no or very late virus gene transcription. There is no DNA replication either because the α, β, γ viral genes are themselves blocked.

23.3 Clinical aspects of HHV-1 and HHV-2

HHV-1 and HHV-2 have many features in common, but can be distinguished by laboratory tests such as genome sequencing. There is no sharp demarcation between the clinical syndromes caused by the two viruses. All of them may be caused by either virus, but as a general rule HHV-1 primarily affects the upper part of the body whereas HHV-2 is more usually, but not exclusively, the cause of genital infections. Infection can be clustered in families and schools but herpesvirus infections are never epidemic.

23.3.1 Primary oropharyngeal infection

The primary or first infection with HHV-1 is often acquired in infancy or early childhood from close contact with an older person, often as the result of a kiss. HHV DNA is detected in the blood of one third of patients at this stage, suggesting a systemic seeding. Neonates can exhibit clinical figures of disseminated infection. In most instances the infection is asymptomatic, but it may present as **acute gingivostomatitis** (Fig. 23.7): vesicles on the gums and oral mucosa break down to form ulcers, there is a sore throat and swollen lymph glands, and a high temperature.

In later life, recurrences and reactivation manifest around the mouth, commonly at the lip margin (Fig. 23.8), as a cluster of vesicles, which are heralded some hours previously by an itching sensation. They are milder, more localized, and of shorter duration than the primary or initial infection; the lesions, which at first contain a clear watery fluid, crust over and heal within a few days. With the passage of

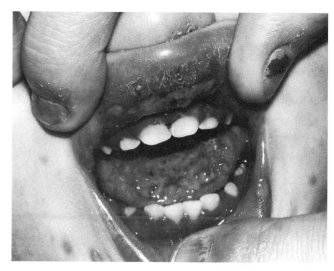

Fig. 23.7 Herpetic gingivostomatitis in a child. Note the vesicles around the mouth and on the tongue and the intensely inflamed gingivae. (Courtesy of the late Dr W. Marshall.)

years, recurrences tend to become less frequent and often cease altogether.

23.3.2 Dermal infections

Primary infections of the skin are not uncommon in healthcare staff whose hands come into contact with oral secretions.

Sportsmen, e.g. wrestlers, may acquire skin lesions ('herpes gladiatorum') from contact with another competitor, or from contaminated mats or other articles.

People with eczema are particularly liable to acquire skin infection with HSV, an unpleasant and occasionally fatal condition known as **eczema herpeticum**. Overall incidence is 2.4 per 100 000 people per year in the USA.

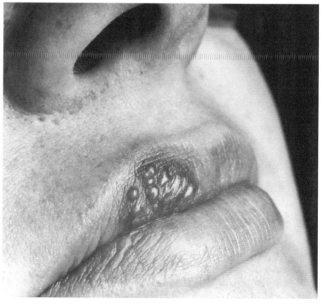

Fig. 23.8 Recurrent herpes simplex.

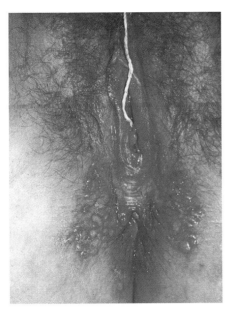

Fig. 23.9 Genital herpes (photography courtesy of the late Dr David Oriel).

23.3.3 Genital infections

Because most people have been infected with HSV-1 early in life in the oropharynx, genital infection is more often acquired as an initial episode, rather than as a true primary infection. In the male it presents as a crop of vesicles on the penis; lesions may also occur within the meatus causing dysuria. **Herpetic proctitis** is sometimes seen in homosexuals. In the female, the lesions are on the **labia, vulva**, and **perineum**, sometimes extending to the inner surface of the thighs (Fig. 23.9). **Cervicitis** with vesicular lesions also occurs. **Inguinal lymphadenopathy** is pronounced in both sexes and there is often some fever and malaise. Not infrequently, particularly in male homosexuals, there is also an attack of **meningitis**. Recurrences of infections with HHV-2 are likely to be more frequent and severe than those due to HHV-1. They are the cause of much sexual disability and mental distress.

In children, herpetic vulvovaginitis may indicate sexual abuse, but, needless to say, other causes of infection, such as auto-inoculation from an oral lesion, must also be considered.

Medical case story   Genital herpes

Complaints are an upsetting and difficult part of anybody's job. They are also difficult to predict. The consultation you felt you made a complete hash of will likely enough be the one that results in a thank-you note, whereas the person you bend over backwards for will be the one sending in a long detailed description of your deficiencies. Sometimes having an unpleasant diagnosis communicated gets misinterpreted as an unpleasant communication technique, as perhaps it does here.

Your next patient, Miss T, enters your room. She is a 52-year-old African American lady with a long history of chronic fatigue. She is a petite, rather thin lady who looks upset. She tells you that she is worried about some sore patches she has recently had on her vulva.

What possible diagnoses should you think of at this point? You think first of thrush, ordinary spots, and lichen sclerosis. VIN (vulval intracytoplasmic neoplasia) is less likely, but still a possibility.

What questions would you ask to narrow down your list? You need to ask about the time course, and also about itchiness, pain, and discharge.

She says the patches started 5 days ago as a tingling feeling and then became painful. Initially, they looked like blisters, but now they are just sore patches. They are not itchy and there is no discharge.

You tentatively ask if she has a partner. She says she has recently had sex for the first time ever with her very first partner. You now think of sexually transmitted infections (STD). Genital herpes is the only thing that fits.

However there are other causes of sexually transmitted genital ulcers.

You say 'May I have a look?' Miss T agrees, but is clearly jumpy and nervous.

During the examination her nervousness makes you nervous. You give her clear instructions to settle her down. You ask her to lie on her back with her feet near her bottom and to let her legs flop to each side. You gently part the labia and see a cluster of ulcers looking inflamed and sore. There are no intact blisters.

You explain that it looks very much like genital herpes. Miss T is keen for you to swab the blisters to make absolutely sure, but unfortunately you do not have a viral swab to hand. You excuse yourself for a moment, and check your swab drawer and then your colleague's drawers, and you even run upstairs to the storeroom. However, there are no viral swabs left in the practice.

The patient is not pleased to be left behind the curtain while you pop in and out. Although there is no doubt in your mind that this is genital herpes, Miss T does not see it this way. She is deeply unhappy with both you and your diagnosis.

You exit the curtained area to allow Miss T to dress. Then you ask her to sit down again.

You start carefully by asking if she has heard of genital herpes, and what she knows about it. She has an idea that you can get it again and that it will infect all future partners.

You say it is possible to get it again, although many people only ever get it once. Some people do have other episodes, although not usually as bad as the first. A few people have regular episodes (for example, once a month) and they often benefit from taking regular aciclovir. You explain that treatment with aciclovir (five times a day for 5 days) is helpful for each episode in the first 72 h, but that it would not be beneficial for her to take it this time as we are 5 days in.

'Does this mean my partner has been unfaithful?' asks Miss T. This is always a heart-stopping moment. You swallow and open your mouth to answer when she says, 'Actually, I know he has been'.

You explain that he could have caught it a long time ago, but still be having occasional outbreaks of blisters that shed virus. In fact, up to a third of adults have had herpes simplex virus 2 (HSV-2) infection, of which at least three quarters will have had a sub-clinical infection without any genital ulcers at all.

You can tell that Miss T is not really listening or taking in information well. You offer her a leaflet on genital herpes and suggest she contact you again should she need more information. You also suggest painkillers for the ulcers and also a lignocaine gel for topical use. You expect them to heal up within the next 2–3 weeks. If they were to get worse, rather than better, she should call you.

She leaves quietly, holding the leaflet. After she has gone you sit for a moment on your own, as the awkwardness and misery of the consultation empties from the room.

Two days later the nurse practitioner remarks to you that Miss T had rung back, unhappy that no viral swab had been taken and making a complaint verbally that the doctor who saw her was unsympathetic. The nurse practitioner says that she advised Miss T to go to a walk-in sexual health clinic for a full screen, because the presence of one infection means she may have others.

In fact, you felt sympathetic toward this patient. You also make a mental learning point to offer full sexual health screening to similar patients in the future.

Differential	Aetiology
Genital herpes	Herpes simplex virus
Syphilis	*Treponema pallidum* spirochaete
Chancroid	*Haemophilus ducrevi* bacteria
Granuloma inguinale	*Donovania granulomatis* (papules on genitals and groin)
Lymphogranuloma venereum	*Chlamydia trachomatis* (ulcer plus lymph nodes which can rupture)

★ Learning Points

- **Virological:** Genital herpes is caused by herpes simplex virus-2. The primary attack is the most severe and sometimes the only episode a patient has, causing pyrexia, malaise, lymphadenopathy, and widespread genital ulcers, usually lasting 2–3 weeks. Some patients have recurrent attacks. Complications include urinary retention and pelvic inflammatory disease. Treatment is with aciclovir. There is no vaccine.
- **Clinical:** If you have identified one STD in a patient it is important that they be screened for everything else, particularly chlamydia, gonorrhoea, HIV, hepatitis B and C, and syphilis.

This is often best done at an STD clinic where they can also do contact tracing.
- **Personal:** The acute upset from a complaint can last several days and is not something you forget easily. If the complaint is a written one, your work place has an obligation to write an acknowledgement within 7 days, and you are obliged to answer the complaint within 14 days. The mission is to apologize, to explain, and not to escalate the situation. Also it is helpful to get some good out of a bad situation by identifying any learning points, as was done here.

23.3.4 Ophthalmic infections

Herpetic keratoconjunctivitis is a comparatively frequent infection, nearly always caused by HHV-1. It is often characterized by **dendritic ulcers**, so called because of their branching appearance. The corneal stroma may also be affected (disciform keratitis), and **iridocyclitis** is a serious complication. Recurrent attacks of keratitis or keratoconjunctivitis, if untreated, may also cause considerable damage to the eye, with corneal scarring and loss of vision. Nearly 300 000 cases are diagnosed each year in the USA. Recurrence is not uncommon.

23.3.5 Meningitis and encephalitis

We mentioned earlier (in Section 23.3.3) that meningitis occasionally complicates genital tract infections. By itself, genital herpes is a benign condition. In contrast, HHV-1 encephalitis is a serious, life-threatening illness with a high rate of neurological sequelae after recovery from the acute infection. It may occur at any age and is usually caused by HHV-1; there may be concurrent signs of herpes elsewhere, but more often than not it strikes without warning. The onset is often insidious with malaise and fever lasting a few days, followed by headache and changes in behaviour. Clouding of consciousness proceeding to coma is a bad prognostic sign. Untreated, the mortality is about 70% with a high rate of neurological sequelae, including mental defect, in survivors. In the USA herpes encephalitis is seen in 50 people per year.

Medical case story Herpes simplex encephalitis: view from the hospital ward

By the time Paul gets to you he already has a diagnosis of herpes simplex encephalitis and he is not well. It's not that he is unstable, but he is so profoundly affected by amnesia, so that he cannot remember his own name, which for a man of 33 is totally devastating.

Paul had been vaguely unwell in the week leading up to admission, with non-specific viral-sounding symptoms including a temperature, general lethargy, and some diarrhoea. He had gone to work at the building site for the first few days but then took to his bed, which was unusual for this fit and young man. His wife Susan didn't know what to make of it; both her and their toddler had experienced similar mild viral symptoms (it was the winter season after all) but had carried on just the same. She had reassured herself that he had the 'man flu', but kept a close eye on him all the same.

On the Friday when she came home from work he was worse. She had seen him get up to stagger to the bathroom and nearly collapse. On the Saturday he started to act strangely, and by the evening when she checked on him again he was confused and crying out for help, at times unmanageable. Susan called the ambulance, and by the next morning he was in the Intensive Treatment Unit (ITU) having been sedated for a CT head scan and a lumbar puncture.

'Everything just happened so quickly,' Susan says to you, when Paul is stepped down to your ward from ITU. 'The ambulance took him straight to A&E because they said his temperature and heart rate were high and blood pressure low. It took three doctors and a nurse to hold him down to take blood tests and things because he was so delirious. The doctor looking after him said quite early on that she thought he might have infection of the brain because she said it was so unusual for people of his age to get that confused unless they are taking drugs.'

Looking back though his admission notes you see that his initial examination did not reveal any positive findings other than the patient's agitation. His GCS was 14 and he was confused, but he was speaking, his eyes were open spontaneously, and, when more settled with minor sedation, he was able to follow simple commands. The ITU team had been called early in the proceedings because he was initially haemodynamically unstable with a blood pressure of 85/50 and heart rate of 140 in the context of a temperature of 39.8 °C.

Given that the A&E team strongly suspected encephalitis or meningitis with delirium he needed a CT head scan and a diagnostic lumbar puncture without delay, which could only safely be done under sedation given his cognitive state. The ITU team sedated and intubated him and took him via the scanner up to ITU for a lumbar puncture and initiation of intravenous antimicrobial treatment including aciclovir, ceftriaxone, and amoxicillin to cover for possible viral (aciclovir) and bacterial (ceftriaxone and amoxicillin) infection. You are impressed to see that this was all done within an hour and a half.

Paul remained less than 48 hours in ITU as his blood pressure and heart rate improved with fluids alone, and he was extubated soon after the lumbar puncture. The preliminary results from the lumbar puncture confirmed a viral-looking picture; the opening pressure was normal and the CSF clear, there were 15 cells per microlitre, predominately lymphocytes, and the protein/glucose ratio was normal. Positive PCR results for HSV-1 came back within 24 hours. An MRI added further weight to the diagnosis of herpes simplex encephalitis by demonstrating bitemporal high signal intensity, which is a classic radiological feature.

Susan is relieved that Paul is no longer in ITU because she feels it is a frightening environment, and coming to your ward means he's already on the road to recovery. She is right; in many ways he is better than he was and he is on the correct treatment for herpes simplex encephalitis, but you know he will need a lot of neurocognitive rehabilitation.

'He will get better won't he? I mean he'll get his memory back and everything, right?' Susan asks. You understand why she might be worried because Paul is the main earner in the family and Susan already has a toddler to look after let alone a dependent husband.

You have to admit to Susan that you don't really know the answer. You excuse yourself and go to the shelf in the doctors' room to pull down a giant textbook of clinical microbiology. Leafing through the thin pages with dense writing and horrible pictures of rashes you will come to the chapter on herpes simplex encephalitis. It seems old-fashioned to read a textbook these days, but actually you find it very helpful and it saves time trawling through unvalidated information on the internet.

But then you read with a sinking heart that full recovery from herpes simplex encephalitis is worse than you thought; only around 40% of patients pull through without permanent neurological deficits. Even something like a mild personality change can have an impact on family life. Would Susan get back the man she knew?

Paul remains on IV aciclovir for 21 days and during that time you begin to notice small improvements in his condition. He sleeps less, he begins to eat, and starts being able to hold a bit of a conversation. Susan is by his side most days and brings in photos of the family to jog his memory as well as drawings from their toddler. Some of his building colleagues come to visit, and although you see that Paul seems to know them and enjoy their company, he cannot tell you who they are or what they do. Every day when you go in to see him you repeat the same short cognitive screening test and at the beginning he rarely gets any answers correct but you see a flicker of recognition that he is remembering much more and he surprises you one morning when he greets you and addresses you by your name!

Quite early on in his admission you have involved a consultant from the neurological rehabilitation team who you know will be key to the ongoing management of any cognitive dysfunction Paul may have. It's good that Paul can remember your name and can tell you where he is, but he will need more intense and specialist input from the neuro-rehab team who can teach him how to live again. He'll need to know how to get from A to B, to remember a shopping list, to wire a plug, to cook a meal and remember to turn the oven off afterwards.

With regret you hand over the reins of Paul's care to the neuro-rehab consultant; it would have been great to have been involved in his journey from acute admission to discharge home but you know he needs more of a specialist approach than you can provide. When you ask after Paul some weeks down the line you are pleased to hear that he responded well to neuro-rehab and was discharged back to his family with very little long-term neurological damage.

★ Learning Points

- Herpes simplex encephalitis is mainly caused by the HSV-1 variety. It typically affects adults, with a male preponderance, and presents with features of sepsis including a high fever, and most characteristically confusion or reduced consciousness. There can also be behavioural changes, speech disturbance, and seizure activity.
- The importance of early recognition and prompt initiation of IV antivirals cannot be emphasized enough. Diagnosis is usually made by obtaining a thorough collateral history, and taking a lumbar puncture to analyse the opening pressure, cell count, protein/glucose ratio, and most usefully a PCR test for HSV. Brain imaging such as a CT scan in the early stages is not often useful, except in the situation of diagnostic uncertainty, in order to rule out other causes. However a CT or MRI scan a couple of days following diagnosis typically shows temporal or frontal involvement. The presence of haemorrhage carries a poor prognosis.

- Treatment with IV aciclovir has revolutionized the outcome for herpes encephalitis. Before such a treatment was available the mortality rate was as high as 70%. Now, the expected mortality rate at 6 months is around 20%. However, in two thirds of those who survive neurological symptoms can persist, ranging from mild to severe. Very few patients go on to fully recover.
- Neuro-rehabilitation is key to the treatment of patients after the initial acute phase of the infection. Specialists can thoroughly assess the deficits present and can advise on suitable rehabilitation techniques, although recovery may take time.

Reference

Solomon, T., Hart, I., and Beeching, N. (2007). Viral encephalitis: a clinician's guide. *Practical Neurology* **7**, 288–305.

23.3.6 Infections of the newborn

These are almost always caused by HHV-2 during passage through an infected birth canal.

23.3.7 Pathogenesis

Skin and mucous membranes are the portals of entry in which the virus also multiplies, causing lysis of cells and formation of **vesicles**. In mucous membranes these soon rupture to form shallow ulcers, but in skin they remain intact for several days before crusting over and healing. The clear fluid within the vesicles contains large numbers of virions.

Local replication is followed by virus spread to the regional lymph nodes, and then by viraemia to the CNS, causing meningitis or even encephalitis. Such events are, however, comparatively rare; the nervous system is involved by a more subtle mechanism that is central to the pathogenesis of HSV infection, viz. latency.

Soon after replication is under way in the skin or a mucous membrane, virions travel to the **root ganglia** via the sensory nerves supplying the area (Fig. 23.10). Nerve cells are most unusual in their anatomy and have a long cylinder shape (the axon) which can be many metres long, terminating in a 'cell body'. Many of these aggregate into ganglia. The herpes virions travel along the axon to the ganglia where the infection becomes latent. Primary infections of the orofacial and genital areas involve, respectively, the **trigeminal** and **lumbosacral** dorsal root ganglia.

Once HHV-1 has established itself in a ganglion *in vivo*, **reactivations** are liable to take place at intervals of weeks or months thereafter. Such attacks may be triggered by a variety of stimuli. The popular terms 'cold sores' or, in the USA, 'fever blisters' illustrate their association with colds and other febrile illnesses; sunlight, menstruation, and therapeutic irradiation are other well-known factors, as is surgical interference with the trigeminal nerve. The host cell is stimulated to produce infective virions that pass down the sensory axons to replicate once again in dermal cells, with the production of vesicles at or near the site of the original infection. It may be that those nerve cells in which virus is reactivated become lysed in the process; if so, it is plausible that the gradual destruction of the whole pool of latently infected cells, possibly in combination with the immune response, accounts for the fading away of recurrences.

23.3.8 Treatment

The classic drug aciclovir (ACV) remains the drug of choice for treating and, in some situations, preventing HHV-1 infections. Intravenous, oral, and topical preparations are available. The prodrugs most widely used are valaciclovir and famciclovir, giving improved pharmacokinetics compared with the parent drugs.

The mode of action is described in Chapter 31. In clinical practice, the following general points are important.

As with all antiviral therapy, ACV must be given early to be fully effective. It should not be used indiscriminately for trivial infections. In the eye, topical applications are relatively effective. ACV may, however, be indicated for infections in immunodeficient patients that would not otherwise need treatment. ACV-resistant mutants are rare, and in general are isolated only from patients with immunodeficiencies.

ACV is excreted in the urine and the dosage must be reduced in patients with impaired renal function.

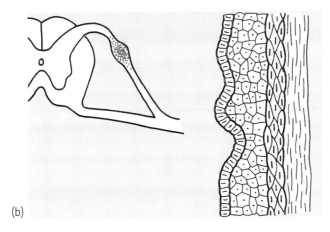

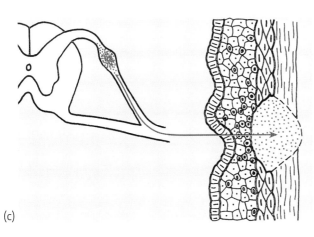

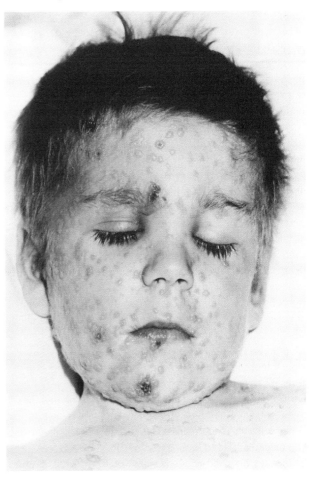

23.4 Clinical aspects of chicken pox/ varicella-zoster (HHV-3) virus

In contrast to the alphaherpesviruses there is only one varicella-zoster virus, named after the two main syndromes that it causes: **varicella**, or chickenpox, and **herpes zoster**, popularly known as shingles (or hives in the USA). There are similarities with HHV-1 infections in terms of lesions, pathogenesis, and pathology.

23.4.1 Clinical aspects

Chickenpox

Chickenpox is a common childhood infection. The virus is highly contagious, causing epidemics in winter and spring in temperate climates. The **incubation period** is about 2 weeks, but may vary by several days each side of this figure. In children, the rash is usually the first clinical sign, associated with a mild feverish illness. Adults usually suffer much more constitutional upset. The rash is characteristically **centripetal**, i.e. more pronounced on the trunk than on the limbs. It first appears as flat **macules**, which rapidly become raised into **papules**; these are succeeded by **vesicles**, which finally form **crusts** that are shed from the skin (Fig. 23.11). The rash appears in successive crops, so that all stages of the

Fig. 23.10 Latency and recurrence in herpes simplex infections. From the primary lesion, the virus (a) travels up the sensory nerves to the dorsal root ganglion, and (b) becomes latent; (c) when reactivated, it returns to the skin by the same route and gives rise to a recurrent lesion.

Fig. 23.11 Chickenpox rash. The lesions are mostly at the vesicular stage; the darker ones have started to form crusts. (Courtesy of the late Dr W. Marshall.)

eruption can be seen at the same time. Recovery is usually uneventful but infection of cells of the dorsal root ganglia leads to virus latency. Reactivation later leads to shingles and zoster.

The complications most to be feared are **varicella pneumonitis** and generalized varicella, which again mainly affect immunocompromised patients. **Leukaemic children** are at particular risk of pneumonitis, which is often fatal. The term 'generalized varicella' is tautologous, but is used to imply clinically apparent involvement of the viscera, joints, and CNS. These complications approach 11 000 per year in the USA with case fatality of 8 per 100 000.

Herpes zoster and shingles: the reactivation phase of chickenpox

'Zoster' is derived from the Latin word for a belt or girdle and refers to the characteristic distribution of the rash when a thoracic dermatome is involved. The attack is heralded by hyperaesthesia and sometimes by pain in the affected area, followed within a day or so by a crop of typical herpetic vesicles that eventually crust over and heal in the usual way. The highest incidence is in the elderly and immunosuppressed, and the risk is around 3.4 cases per 1000 per year. Lifelong risk approaches 50% for a person living 85 years. Disease burden in the USA is 800 000 cases a year.

Complications

- Post-herpetic pain in the affected area is frequent, particularly in the elderly. It usually ceases within a month, but may persist for years. If severe, its resistance to treatment can make life a misery for a minority of patients.

- Ophthalmic zoster is a potentially serious complication when, as often happens, the ophthalmic branch of the trigeminal nerve is affected. Both superficial and deep structures of the eye may be involved.

- Generalized zoster is similar to generalized varicella, but is distinguished clinically by the presence of vesicles distributed along a dermatome in addition to the more general rash. Encephalitis is a rare complication of herpes zoster. Reactivation in an immunosuppressed person can lead to virus dissemination in lungs, liver, CNS, and viscera.

Medical case story Chickenpox seen in the hospital

Although it is hoped that by the end of your training you will have set eyes on every relevant rash and sign that a patient can have, this is not always the case. The signs you will have seen will be partly a matter of chance and partly a matter of location. For example, a child with chickenpox would rarely present to the hospital; it is a mild illness that can be identified quickly and easily in the community. In fact, they do not even present that often to a community doctor, because lay people are very adept at identifying chickenpox for themselves.

Dr D has completed over 9 years of training without ever having seen a case.

He has just started as a registrar in neurosurgery. His first child is 9 months old, and is looked after by a child-minder when he and his wife are both at work.

He comes home one evening to find his daughter happily playing with the child-minder. As he lifts the baby up for a kiss he notices that his daughter has a spot on her arm. It looks like a normal spot: he doesn't think anything of it.

The baby needs her nappy changed so Dr D undresses her and then notices a spot on her pudenda. He looks up at the child-minder because it hadn't been there this morning when he handed the baby over. He feels concerned.

'Ooh, it looks like chickenpox,' says his child-minder.

'Does it?' says Dr D doubtfully.

'Oh yes', she says, 'They can get spots anywhere: on their head, on their bottom.' They turn the baby over. There is one on her back more like a blister. 'Yeah, sometimes they look like that: like a drop of water, to start with,' she adds.

Dr D takes his child-minder's word for it. After all, she has bought up three children of her own and looked after at least two families before his. She might have seen as many as eight cases at every stage of the illness.

Over the next few days, the baby gets several crops of spots. After the last ones have crusted over, she is no longer infectious.

The child-minder informs the playgroup that she normally takes the baby to and they don't go again until the baby is no longer infectious. 'Although I can't see the point' she says to Dr D, 'because they've all got to catch it, haven't they?'

Dr D quietly looks up a patient leaflet about chickenpox and finds himself privately agreeing. The incubation time is about 2 weeks (up to three) and the infectious period starts 2 days before the rash, so the baby probably transmitted it to quite a few children before the spots even appeared.

The baby is not that ill while she has the rash: she has only a mild fever and a slight cough. In fact, she looks more cross than ill. Her spots are itchy and look uncomfortable. Dr D keeps the spots covered up to avoid her scratching them. He puts aqueous cream with calamine on the spots, but it doesn't seem to do much. One of the spots gets quite red and angry-looking, but then it settles of its own accord. It leaves a small white scar.

Dr D and his wife take alternate days off work until the baby is well again.

Dr D reflects that, in America, they have been using the varicella vaccine in children since 1995. Since 2005 they have combined it with the MMR to make MMRV vaccine. The varicella component is live, attenuated virus and is given twice: once at 13 months and once pre-school, just like the MMR. At present no one is sure how long it protects for or what implications it has for shingles.

He wonders whether the UK will introduce the varicella vaccine. He finds this issue is under discussion in major journals. It has been introduced in Germany, but not in the rest of Europe. There is an analysis of how cost-effective it would be and how medically effective.

Common complications of chickenpox include serious skin infections (just like the mild version his own little girl had experienced), pneumonia, and encephalitis. The rate of chickenpox infection is approximately 500 per 100 000 people per year.[1] In each group of 500, about six people get a complication that needs a hospital admission.

The rate of death from chickenpox pre-vaccination in America was 0.04 per 100 000. In America they are very proud to have reduced their death rate by a seemingly massive 66%–0.014 per 100 000. Should they be very proud? What is the actual number of lives saved, assuming the population of America is 280 million?[2]

At present in the UK, vaccinations are targeted at vulnerable people, such as women who are planning pregnancy, but who are not immune to chickenpox. To protect vulnerable people who cannot have the vaccine, such as neonates or the immunocompromised, the herd immunity would have to be 95%. It might seem simple to combine the vaccine with the MMR as they have elsewhere. However, if the public were not confident with the vaccination, vaccination rates would fall and herd immunity

would not be achieved. Indeed, the current low levels of MMR vaccination (65% in some parts of the country) are such that herd immunity would not be achieved in any case.

Dr D translates the American figures to the UK and estimates that, at present, 24 people die of chickenpox a year. If varicella vaccine was introduced and had 95% coverage, then 16 lives might be saved. Eight people would still die of it. When he works out a back of an envelope cost, assuming £100 per vaccination per child per year, it comes to 4 million pounds per life saved. Is this cost-effective? How much is the National Institute of Health and Clinical Excellence (NICE) prepared to pay per life year?[3]

Dr D concludes that it is a fascinating and contentious issue. Should he privately vaccinate any future children of his own?

Notes

[1] Centres for Disease Control and Prevention (USA). Available at: cdc.gov/vaccine/varicella (accessed 1 June 2009).
[2] The actual number of lives saved using the above figures would be 72. The death rate from chickenpox would have fallen from 112 per year to 39.
[3] Vaccination costs approximately £27 per dose and administration of the vaccine is calculated at £20 per visit. The birth rate per year in the UK is about 700 000. NICE is prepared to pay £30 000 per life year. In 2000, 16 people died of chickenpox in the UK.

★ Learning Points

- **Clinical:** Decisions regarding the cost-effectiveness of a treatment are a fascinating and important part of a healthcare system. Healthcare rationing should be transparent and open so that doctors and the general public alike understand the necessity of it. A price can be and is put on a life.
- **Personal:** Lay people, patients, and supporting staff can all be very effective clinical teachers.

Medical case story Zoster seen in general practice

Until relatively recently, pregnant women were felt to be rather irrational and were sometimes kept in the house throughout their 'confinement'. Today though, they are expected to work as normal right up until 37 weeks, and are even asked to do night shifts if their work demands it. In truth, the reality lies somewhere in between. There is no doubt that people who are pregnant or postnatal have specific medical issues, and may interact with you in a very different way from their usual selves.

You have a telephone call from a 33-year-old Mrs P who sounds panicky. 'I've just had a baby and three days ago I got this rash on my bottom and the midwife thinks it's because of an allergic …' At this point you interrupt her. If she has a rash and is immediately postnatal you want to look at her. 'I can book you in for

4.40 today,' you say. She sounds relieved, although still flustered as she thanks you for the appointment.

Later that afternoon Mrs P laboriously walks into your room accompanied by her mother. The baby has been left with the father. You find her mother is a calming influence because Mrs P is very postnatal and emotionally labile. She looks postnatal too; plumper than usual, paler than usual, and not concentrating well. She is in a haze.

She had the baby 3 days ago. It was a vaginal birth with no complications. The baby did not latch well at first, but is now starting to feed. It is having eye contact when awake, and is passing urine and stool. You mentally put the baby aside.

Mrs P has been well since the birth with no high temperatures, no abdominal pain, normal amounts of bleeding, and her breasts feel ok. She is drinking and eating well. She has felt tired, but puts this down to the birth and lack of sleep.

Which common postnatal diagnoses are you trying to exclude with this history?

You now ask her about the rash. She says it came at about the same time as the birth, and that it feels sore, rather than itchy. She has not had anything like it before.

You ask to examine her and draw the curtains. You carefully explain that you would like her to lie on her left side facing the wall with her knees drawn up to her chest. It is better to give clear instructions about intimate examinations, rather than let a patient try to guess what to do. The position you have asked her to lie in is called the *left lateral position*.

What differentials should you think of as you prepare to look at the rash?

When she is ready you ask if you may come in. You see the rash immediately. As you look at the rash, your differential diagnoses of allergy and impetigo drop away. It is on one side of her buttock in the distribution of the 2nd sacral nerve. There are several groups of square-looking blisters, each 2 mm across, scattered in groups of 20 over an area of 6 cm^2.

'Oh!' you blurt out, 'I believe you have shingles'.

Mrs P sits up on the bed still undressed and looks blankly at you for quite a few seconds before her face crumples and she bursts into tears. 'Oh my God!' she sobs, 'The baby! The baby is going to catch it! What shall I do? Has he caught it already?'

'No, no it's alright,' you say, acutely aware of your clumsy communication just now. 'The baby has your immune system for the first 3 months of his life, and can't catch it'. You put your hand on her shoulder sympathetically and feel a need for the mother who is outside the curtain. 'Let's get you dressed,' you offer, and draw the curtain back as soon as she is ready, still crying.

'Listen to the doctor,' the mother agrees, 'It's fine'.

You explain that the baby has the mother's immune system for 3–4 months. At this point the mother's immunoglobulin fades away as the baby's own immune system becomes more and more active. Breast feeding supplements this a little, principally against any gastroenteritis infections. The baby's most vulnerable time is at this juncture. 'That would be the least good time for the baby to catch chickenpox,' you add, 'although most babies are still fine'.

Her shingles has come from her reactivated chickenpox virus (varicella). People can't catch shingles from her, but they can catch chickenpox from her if they have never had it.

She should keep the rash covered to reduce viral shedding. She should warn any pregnant ladies she comes into contact with. (If they have had chickenpox then they are not at risk). You prescribe her 5 days of aciclovir five times a day, as you are within 72 h of the outbreak. You also advise paracetamol and ibuprofen if she needs analgesia, and emphasize that both are safe during breastfeeding.

She wipes her continuing tears with the tissue you have given her. 'But what about the baby? Should I stay away from him?' Her mother is helping to remember your advice and reiterate. 'Your baby can't catch it from you,' you say again.

You advise her that the rash will go over the next 5–7 days and that she should ring you if she thinks of any questions after she gets home. She leaves clutching further tissues, a leaflet on shingles, and the prescription you have given her, supported by her mother.

★ **Learning Points**

- **Clinical:** It is well known that patients don't remember all the advice given to them in a consultation, and sometimes less than half of it. In the above description, the many repetitions of the advice given are omitted for the sake of the reader. It is considered good practice to summarize advice given, to ask for the patient to repeat it back to you, and to supplement this with written advice in the form of a leaflet or hand-written advice by the doctor.
- The common postnatal diagnoses you attempted to exclude at the beginning of the history are endometritis, retained products of conception, and mastitis.

- **Personal:** In this scenario the diagnosis was blurted out as enthusiasm for the rash overtook concern for the patient. The great advantage of doing daily clinics is that you can do it better next time. The next time a postnatal patient comes in with a worrying infection you will be able to talk to them once they are dressed and less vulnerable, and handle it in a more delicate manner. The next time a patient with a herpes zoster rash comes in you could check to see if they know about shingles before introducing the topic.

23.4.2 Pathogenesis

Varicella is acquired by the **respiratory route**. Dissemination of virus in the bloodstream is followed by a rash, at first macular, but rapidly developing into papules followed by vesicles similar to those caused by HHV-1 and, like them, containing many virions. The lesions become crusted and separate as scabs within 10 days of onset; contrary to early ideas, the scabs are virtually without infectivity.

In some patients the virus becomes **latent** in a dorsal root ganglion. This event is analogous to what happens in HHV-1 infections, but as one might expect in such a generalized infection, VZV may affect any sensory nerve, those most commonly involved being the thoracic, trigeminal, cervical, and lumbosacral; it is rare for more than one of these to be involved. Unlike HHV-1, VZV infects only humans, so that research on latent infections is very limited. Viral DNA can be detected in the sensory ganglia of cadavers, and this technique has yielded some information on the distribution of latent virus.

In humans, VZV can reactivate, causing an attack of herpes zoster with a characteristic distribution of vesicular lesions along the affected dermatome. Almost invariably, there is only one such episode in a lifetime. Even less is known about the trigger mechanism than in the case of HHV-1 infections, but it is noteworthy that many attacks of zoster occur in the elderly, in whom immunity is waning, and in **immunocompromised patients**.

Epidemiology

Varicella can be readily acquired from patients with an active zoster infection. This source is sufficient to produce epidemics each year. This can present a problem in wards containing immunodeficient patients who are liable to develop zoster if they have previously had chickenpox, or who are unduly susceptible to VZV infection if they have not. Furthermore, spread from patients to non-immune nursing staff occurs readily. Management of such situations includes adequate isolation of patients with VZV infections, who ideally should be attended only by staff with antibody to the virus.

23.4.3 Prevention and treatment

Active immunization

Varicella vaccines are lyophilized live attenuated virus. They can be given at the same time as MMR, but in a different arm. If the vaccine cannot be given simultaneously then VZV should be delayed for 1 month to avoid the MMR interfering with the varicella vaccine take. The VZV vaccine can be used as a post-exposure therapy for healthcare workers. For pre-exposure it is recommended in the UK for healthcare workers (persons with patient contact, such as cleaners, ambulance staff, and receptionists, as well as medical and nursing staff), laboratory staff, and carers of immunocompromised patients, and for persons over the age of 70. It is not currently recommended for routine use in children in the UK.

Passive immunization

Although there are no really well-controlled trials, **immunoglobulin** prepared from the sera of blood donors with high titres of antibody to VZV seems to be of some use in preventing or modifying severe disease in immunodeficient patients who come into contact with chickenpox or herpes zoster. This preparation, **zoster immune globulin (ZIG)**, must be given as soon as possible, and certainly within 4 days of contact. ZIG is recommended for VZV antibody-negative pregnant contacts at any stage of pregnancy providing it is given within 10 days of exposure to a case. It is of no value for treating established infection.

Antivirals

Valaciclovir and famciclovir are used to treat severe VZV infections. For intravenous use, twice the usual dose of aciclovir may be given, i.e. 10 mg/kg every 8 h.

ACV in high dosage can be given during the prodromal stage in an attempt to modify the severity of the infection, but seems to have little or no effect in reducing the incidence of post-herpetic pain.

Herpesvirus B

This alphaherpesvirus is not usually considered as an HHV as it normally infects only macaque monkeys, hence its alternative name, *Herpesvirus simiae*. Humans exposed to saliva, e.g. by being bitten, are liable to contract severe **encephalomyelitis**, which is almost always fatal. There have been several such cases in laboratory staff handling apparently normal monkeys.

23.5 Clinical aspects of cytomegalovirus (HHV-5)

The name means 'large cell virus' and derives from the swollen cells containing large intranuclear inclusions that characterize these infections (Fig. 23.12). Most infections are subclinical, whilst acute infections are often opportunistic in the immunocompromised and in AIDS patients.

At first encounter, CMV infections are confusing because they vary so much in terms of age groups affected, mode of acquisition, and clinical presentation. A good aid to memory is to classify them by the ages predominantly affected, as this factor strongly influences the type of disease likely to be encountered. CMV can be a problem even before the cradle and certainly to the grave.

CMV is transmissible to the **foetus** via the placenta, and is an important cause of neonatal morbidity and mortality. Newborns have a range of symptoms including microcephaly, lethargy, and seizures. Congenital CMV is an important medical and public health problem.

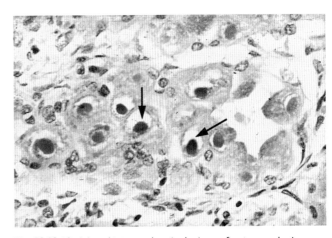

Fig. 23.12 'Owls' eye' intranuclear inclusions of cytomegalovirus (arrowed).

Normal infants can acquire infection from **colostrum** or **breast milk**; maternal antibody does not seem to confer protection. Such infections are usually asymptomatic.

Young children readily become infected, again without overt disease, when they enter crèches or play schools, in which the environment is liable to become contaminated by virus shed in **urine** and **saliva**.

The next wave of infection occurs at **adolescence and early adulthood**, when infection is spread by **kissing and sexual intercourse**. At this age, CMV sometimes causes a syndrome like the infectious mononucleosis resulting from infection with EBV. In all these groups CMV can cause a febrile illness with splenomegaly, impaired liver function—sometimes with jaundice—and the appearance of abnormal lymphocytes in the blood.

23.5.1 Transplant recipients

The potential for CMV complications must be considered for every patient having solid organ tranplantations and occurs usually 4–8 weeks after the operation. The source of infection can be:

* exogenous, deriving from the donor's tissues or from blood transfusions given in support of surgery;
* endogenous, i.e. reactivation of an existing infection in the recipient.

Both types of infection, and particularly the second, are facilitated by the associated immunosuppressive treatment. Primary infections are the most dangerous and may result in glomerulonephritis with rejection of a transplanted kidney or in CMV infection of the lungs. This is an interstitial pneumonitis with oedema and pronounced cellular infiltration. The mortality is about 20% in bone marrow recipients, but only 1–2% in kidney transplant patients. Reactivations are less serious than primary infections and may result in nothing worse than a febrile illness.

Although rare, cases of CMV encephalitis have been reported in apparently immunocompetent adults.

23.5.2 Epidemiology

Throughout life, there is **intermittent shedding of virus in body fluids**, including saliva, urine, cervical secretion, semen, and breast milk, giving ample opportunities for transmission; transfused blood may also be a source of infection. In most people, the infection is silent, but in overt disease almost any organ is liable to be damaged. As with other herpesviruses, **reactivations** are an important feature of pathogenesis, and to complicate matters further, **re-infections** with other strains are also possible.

As might be expected, the opportunities for acquiring CMV are greatest under conditions of poverty, overcrowding, and poor hygiene. In one sense, this is an advantage, because infection is acquired early in life when it is usually symptomless. Conversely, the average age of acquisition is greater in communities with high living standards and the outcomes, in terms of the more severe primary infections, are correspondingly worse; this is particularly true of first infections during pregnancy.

23.5.3 Prevention and treatment

Because CMV lacks the TK enzyme possessed by HSV and VZV, aciclovir and the corresponding prodrugs are inactive. Ganciclovir, a similar compound, however, is phosphorylated by a different enzyme and is used with some success in treating immunosuppressed patients, including those with AIDS. Viraemia can be reduced by the drug as well as virus shedding. Unfortunately, it is relatively toxic and its bioavailability is poor, so that a satisfactory treatment for CMV must await further advances in chemotherapy (see also Chapter 31). Valganciclovir can be used in these serious infections, whilst foscarnet and cidofovir are second line antiviral drugs. The first two antivirals are commonly used in transplant medicine as a pre-emptive approach in seronegative recipients from seropositive donors. Drug resistance can occur.

23.6 Clinical aspects of HHV-6 and -7

The comparatively recent identification of these viruses supports the speculation that, despite all the diagnostic advances in the last few decades, there are still others waiting to be discovered.

HHV-6 was first isolated in 1986 from blood leucocytes of six patients suffering from lymphomas or acute lymphocytic leukaemia. It induces cytopathic changes with inclusion bodies in human B lymphocytes and is morphologically similar to other herpesviruses, but varies in several biological characteristics. HHV-6 was at first thought to be associated with AIDS, but was later established as the cause of a febrile illness of children, **exanthem subitum**, or **roseola**; on the basis of its distinctive genome structure and other characteristics it was assigned to a new genus, *Roseolavirus*, within the subfamily *Betaherpesvirinae*. There are two groups, A and B, defined by differences in serological reactions and genetic composition. HHV-6B causes exanthema subitum, whilst HHV-6A has no known association with disease.

23.6.1 Clinical features

Both names of this illness refer to the **rash**, and 'subitum' reminds us of its sudden onset. These designations are not really appropriate, as the rash appears in only 30% of cases, although it is, of course, the feature causing most parental alarm. It usually follows the febrile stage, and is macular or maculopapular, most evident on the trunk. The cervical lymph nodes, and sometimes the spleen, are enlarged.

Convulsions are not uncommon soon after onset, suggesting some involvement of the CNS; nevertheless, complete recovery is the rule. Treatment is symptomatic. The infant develops a sudden fever for a few days followed by a rash on the trunk and face, which spreads to the lower extremities.

In normal adults, infection with HHV-6 may cause an illness resembling infectious mononucleosis. In immunocompromised patients, however, the illness may be severe, involving various organs, including the bone marrow.

Medical case story Roseola infantum

Roseola infantum (sometimes also known by a third name, erythema subitum) often has only a very brief mention in textbooks and is usually not especially memorable in these descriptions. This means you may not realize what high temperatures it can cause and how strange an illness it is, insofar as it can only be diagnosed once it has resolved.

Dr C is a haematology registrar on maternity leave. One morning she notices that her 4-month-old baby is breathing much faster than usual. She checks his temperature by touching his forehead: it is normal. She then looks at capillary refill by pressing the baby's finger tips. It is prolonged (at 3 s) and his limbs look mottled. He is awake and has good eye contact, is not crying, and still feeds well.

Dr C does not keep any medical equipment at home (no thermometer, auroscope, or stethoscope), precisely because she does not want to be in the position of having to examine her own family. However, this means she cannot proceed with an examination. She is stuck.

Dr C is very worried about her baby and has no clue what is wrong with him. She needs another doctor to assess him. How urgent is it? Has she time to get dressed? Should she call an ambulance? Does she remember the way to the hospital?

At the accident and emergency department (A&E), she queues up and registers the baby's details with the receptionist. She is seen by the triage nurse 55 min after first noticing the baby's fast breathing. The triage nurse notes the respiratory rate and the capillary refill time and takes the baby's temperature, which is 37.5 °C.

They are taken through to majors. A friendly A&E doctor assesses the baby. He arranges a chest X-ray and checks the baby's oxygen saturations, both of which are fine. He goes off to refer them to the paediatric team.

Then Dr C notices marks on the baby's bottom, which were not there an hour ago; bright red bands along the creases of his bottom. She feels absolutely frantic, finally giving in to the possibility of meningitis. She pleads with the nurse to call back the SHO.

The SHO comes back and calms down Dr C. He realizes the baby's temperature might be up. He measures it again and finds it is 38.5 °C. He prescribes paracetemol suspension and suggests that Dr C should feed the baby, who is crying.

By the time the paediatric doctor arrives, the baby's temperature is down, the rash has faded, and the capillary refill and respiratory rate are back to normal. Dr C looks up and recognizes the doctor as a colleague from a medical job 2 years before. She has no confidence in him. When he announces that her baby probably has coryzal symptoms (has a cold), she nods, and returns home.

Over the next week the baby continues to spike temperatures of 39.5 °C every 3–4 h—very high for a 4-month-old. Throughout, though, the baby feeds well and has no other symptoms: no cold/cough/vomiting/diarrhoea/rash.

Seven days after the initial episode, Dr C is changing the baby's nappy one night when she notices a rash. She rapidly undresses him and finds a macular red rash all over the baby's trunk. It looks like measles.

She hurries downstairs and empties her desk to find her microbiology case studies book.[1] Amongst the differentials of measles she finds roseola infantum. It says: 'causative agent; primary infection with human herpes virus types 6 or 7, age; 6 months to 2 years old, clinical features; 3–4 days pyrexia without localizing signs, with morbilliform rash appearing as fever subsides'. She then recalls her friend's child had it 10 days before.

The rash heralds the recovery of the baby, whose temperature settles.

Dr C writes to the A&E department to thank the A&E SHO for his kindness and to tell him the diagnosis.

Notes

[1] Humphries, H., and Irving, W. (2004). *Problem-orientated Clinical Microbiology and Infection*, 2nd edn. Churchill Livingstone, London.

★ Learning Points

- **Clinical:** The decision to seek help is a big hurdle that every patient you see will have had to make. Having made that decision there will be a delay approaching an hour before coming face to face with a doctor, no matter how close to the hospital a patient lives or how urgent the situation.

- **Personal:** When you are a patient, it is not uncommon to run into doctors that you have previously studied or worked with. It is an uncomfortable situation, and it will certainly affect your levels of trust—one way or the other.

23.6.2 Epidemiology

The virus initiates infection in the respiratory tract, possibly the tonsils. The virus is widely disseminated and is shed in body fluids, including blood.

The prevalence of antibody in infants is biphasic: most are seropositive at birth, having acquired it transplacentally. Most of this maternal antibody disappears by 6 months of age, after which the seroprevalence rate increases as a result of silent infections, so that by the second year of life nearly all children in developed countries are again seropositive.

Human herpes virus 7 was first isolated from CD4+ T cells taken from a healthy person. Like HHV-6, it is widespread and most people have antibody by late childhood. It too may cause an illness resembling infectious mononucleosis, but the results of primary infections are not as clear as those caused by HHV-6. Some cases of exanthem subitum have been associated with HHV-7, and a possible association with pityriasis rosea, a transient inflammatory rash, has also been reported.

The pathogenesis of both HHV-6 and -7 is undefined at present and possible vertical transmission needs to be studied. The practical and medical problems from an integrated genome are questions for further research.

23.7 Clinical aspects of Epstein–Barr virus (HHV-4)

EBV is in some respects the most sinister herpesvirus, for its association with malignant disease is now well established. In 1958, Burkitt described a tumour in African children that occurred in areas with a high prevalence of malaria. He thought that it might be caused by an infectious agent spread by mosquitoes. The mosquito theory was wrong, but 6 years later Epstein and his colleagues discovered a herpesvirus in cultures of the tumour cells. In fact malaria co-infection weakens the immune system, which may allow symptoms to flourish. In 1966, American workers showed the association of this virus both with infectious mononucleosis (glandular fever) and with another form of cancer, nasopharyngeal carcinoma, occurring mostly in southern China. In addition to its association with these three syndromes, EBV causes B-cell lymphomas in immunodeficient patients (Table 23.2).

There are two types of the virus, with differing biological characteristics. There are additional strains based on sequence differences in the latent genes. Worldwide, type A is predominant with evidence of infection of around 90% of Caucasians and Asians. Type B appears to have a particularly high prevalence in equatorial Africa and in immunocompromised patients, including those with AIDS. However, there is as yet no firm evidence that the two types differ in terms of pathogenicity or oncogenicity. Both types may co-exist in the same person.

The clinical contrasts between infectious mononucleosis and Burkitt's lymphoma are large, and both Burkitt and the virologist Epstein must have been very surprised to find herpes virus in cells from the tumour.

23.7.1 Infectious mononucleosis

This clinical picture is seen worldwide and most infections are subclinical.

EBV is shed intermittently by a substantial proportion of the population, many of whom have no symptoms. Infectious virus is generated in the **pharynx**, probably in lymphoid tissue, and appears in the saliva. Thus, like CMV mononucleosis, glandular fever is a 'kissing disease', and the peak incidence in the developed world is in **adolescents and young adults**. In contrast, in the developing world, children become infected in families before 3 years of age. The incubation period is a month or more. There is fever, **pharyngitis**, and **enlargement of the lymph nodes**, first in the neck and later elsewhere. In most patients the spleen is palpable and there is some **liver dysfunction**, occasionally with frank jaundice. There may be a transient macular rash; it is a peculiarity of the disease that patients given **ampicillin** develop a more severe rash due to the formation by transformed B cells of antibody to this antibiotic.

The symptoms parallel a CD8 T cell lymphocytosis, suggesting a special pathology with symptoms caused by pro-inflammatory cytokines such as IL-1 and TNF-κ. Not unexpectedly, the antiherpes drug aciclovir has little effect on disease progression.

Complications are relatively uncommon. They include Guillain–Barré syndrome or other signs of CNS involvement and, rarely, rupture of the spleen. Complete recovery within 3 weeks is the rule, but convalescence is sometimes lengthy, with prolonged lassitude and loss of well-being.

In poor communities, EBV is acquired early in life, when it causes mainly inapparent infections. The initial infection confers lifelong immunity to subsequent encounters. In developed countries the infection is acquired later in life, with a peak incidence at 16–18 years.

Virus isolation, depending as it does on the transformation of lymphocyte cultures, is impracticable; diagnosis in the routine laboratory is based on the haematological findings, on serological tests, and, not unexpectedly, by quantitative genome analysis using PCR.

The feature that gives infectious mononucleosis its name is the **raised leucocyte count** (20×10^9/l, 50% of which are lymphocytes). Up to 20% of the lymphocytes are atypical in appearance, with bulky cytoplasm and irregular nuclei. These are not, as you might suppose, infected B cells, but **T cells** reacting against viral antigens expressed on the B cells and capable of killing the latter.

Table 23.2 Epstein–Barr virus infections

Syndrome	Age group mainly affected	Remarks
Infectious mononucleosis	Adolescents, young adults	Worldwide distribution
Burkitt's lymphoma (BL)	Children 4–12 years	Endemic in sub-Saharan Africa and New Guinea
Nasopharyngeal carcinoma	Adults 20–50 years	Endemic in southern China
B-cell lymphoma	Children and adults	Occurs in some primary and acquired immunodeficiencies

Medical case story Epstein–Barr virus

There are illnesses that can be diagnosed with one glance at a rash or one key phrase in a history. There are illnesses that you *think* you can diagnose with one glance and one phrase, such as glandular fever. Diagnosing from a classic presentation (fever, pharyngitis, and lymphadenopathy) can be exciting, but it doesn't always work that way. Sometimes a presentation just doesn't follow the textbook description.

One afternoon your receptionist asks you to see an extra patient, in between appointments. She is a 26-year-old woman called Fern who has just walked in. The receptionist says she is 'literally shaking in the waiting area. I don't know what's going on but she looks really ill to me'.

Your receptionist is very experienced and has picked up patients with serious illness before and fast-tracked them through the booking system. You don't quibble. Instead you say 'Thank you. I'll see her next'.

Fern has stopped shaking by the time she sees you. She was completely well up until that morning. 'Then I had violent trembling which lasted about half an hour, and I had one just now again too,' she says tremulously. When she demonstrates the shaking you know she is describing a rigour. But when you look for an infection source, you can't find one.

You run through a review of systems: headache/throat/ear/cough/abdominal pain/dysuria/vaginal discharge. You check her temperature, ears, and throat. You dip a urine sample to exclude pyelonephritis: this often presents with rigours. The urine sample is negative.

You tentatively conclude a flu-like illness and ask the patient to come back if she gets worse, gets a new symptom, or isn't getting better after a week.

Three days later she re-presents still unwell. She feels tired, has aches, pains, and a tummy ache, and has been told she looks pale.

When you examine her abdomen she has lower abdominal tenderness. You consider pelvic inflammatory disease (PID). A sexual history reveals two unprotected encounters with British men in the last 6 months. You ask for her occupation and about travel. She is a gymnast, and has recently travelled in France and Germany.

You do a bimanual examination, which reveals cervical tenderness, so you take swabs and decide to treat her for PID. You look up in the British National Formulary (BNF) the recommended combination of antibiotics to cover both anaerobic and aerobic bacteria, and issue her with a 10-day course of ofloxacin and metronidazole.

Nine days later (12 days of illness) she is back again. Her temperature and abdominal pain have improved, but she has a sore throat. Her temperature is 37.3 °C and she has gigantic tonsils with the largest beds of pus you have ever seen. She has tonsillitis, but she is still on antibiotics. The cervical swabs are back and they are normal.

What is your usual differential of a sore throat? What about in this case?

Viral	Epstein–Barr Virus (EBV), cytomegalovirus (CMV), adenovirus, enterovirus, HIV
Bacterial	Streptococcal group A (can cause scarlet fever), diphtheria, mycoplasma, *Chlamydia*

You suspect glandular fever. You send bloods to the laboratory for monospot and HIV, and also for CMV, full blood count (FBC), and liver function tests (LFT). She had an HIV test 4 months ago, which was negative.

This time you arrange a review for 4 days' time.

However, she re-presents sooner than that, 2 days later, on your day off and is seen by your colleague. This time she has a diffuse macular rash on her torso and is feeling ill and weak. Your colleague refers her to the on-call infectious diseases team who admit her overnight.

The monospot test comes back as positive. She does have glandular fever.

The infectious diseases team conclude the rash was caused by her glandular fever plus the antibiotic and she is discharged.

You see her again for the fifth and final time 10 days after her overnight admission. You know her very well now, and have memorized her name and moods, and you know what contests she has had to withdraw from. She had an unprepossessing start as an extra patient presenting so early into an illness that there was nothing to see, but she turns out to have been your first case of Epstein–Barr virus.

'One thing I'm not sure about is whether I'll get it again' she says. 'I feel a lot better already, but I've looked on the internet and lots of people who have had glandular fever say they get relapses. Like, if they are run down, or stressed they say it comes back again for a few days, although not as bad'.

'As I understand it,' you say, 'you can only get it once. There are illnesses you can get again like CMV, genital herpes, shingles, but glandular fever is not one of them'.

★ **Learning Points**

- The term *infectious mononucleosis* means that there is a high level of atypical mononuclear cells, which are in fact T cells. This is a non-specific test result and can be caused by EBV certainly, but also by CMV, HIV, toxoplasmosis, and group A streptococcus.

- **Clinical:** Amoxicillin reacts with some of the antibodies made by the transformed B cells. Therefore, if amoxicillin is given to someone with a tonsillitis caused by EBV, they get a rash. For this reason tonsillitis (usually a bacterial infection) is always treated with penicillin, rather than amoxicillin, just in case it is glandular fever.
- Although this patient wasn't given amoxicillin she did develop a macular rash, possibly due to the antibiotics she

was given. Alternatively, it could simply have been part of the illness.
- **Personal:** Sometimes it is not possible or even helpful to diagnose and treat a patient all at once. An illness may have to evolve to be identifiable or it may not be possible to do certain examinations immediately. Each consultation should include advice on when a review would be necessary; this is sometimes known as a 'safety net'.

23.7.2 Burkitt's lymphoma

This highly malignant neoplasm (see Fig. 4.1(c)) occurs mainly in **African children** living in the belt where malaria is hyperendemic, roughly speaking between the Tropics of Capricorn and Cancer (Fig. 23.13(a)). The peak incidence is in children 6–7 years old. There is another focus in **Papua New Guinea** and

sporadic cases occur elsewhere. It presents as a tumour, usually of the jaw, less often of the orbit and other sites. Untreated, it is nearly always fatal within a few months, but is very responsive to **cyclophosphamide**, which, if given early enough, may effect a cure. There is limited clinical benefit in using the herpes antiviral aciclovir and its prodrugs.

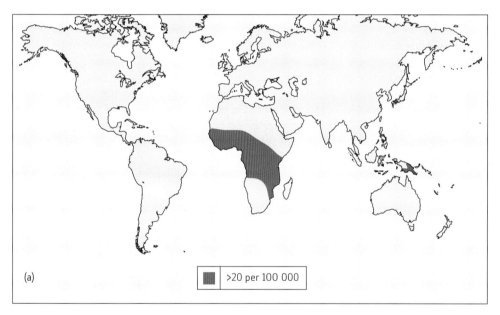

(a)   ▮ >20 per 100 000

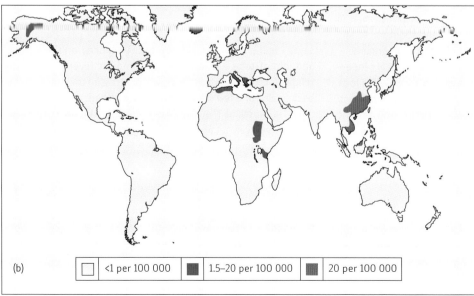

(b)   ▢ <1 per 100 000   ▮ 1.5–20 per 100 000   ▮ 20 per 100 000

Fig. 23.13 Geographical distribution of (a) Burkitt's lymphoma (b) nasopharyngeal carcinoma. (Redrawn, with permission, from Zuckerman, A.J., Banatvala, J.E., and Pattison, J.R. (1987), *Principles and Practice of Clinical Virology*. John Wiley & Sons, Chichester.)

EBV genome in circular episomal form can be demonstrated in the tumour cells of most cases of African and New Guinea BL, but not usually in those of the sporadic cases seen elsewhere. All children with African BL possess antibodies to the virus, and there is thus a strong association between the virus infection and the tumour. Falciparum malaria is an important co-factor; its similar geographical distribution is striking, and there is some evidence that its ability to cause immunosuppression interferes with the control by cytotoxic T cells of tumour development; but this is not as yet proven.

Patients in the earlier stages of AIDS are also liable to develop BL.

23.7.3 Nasopharyngeal carcinoma

People in **southern China** or who originate from there, and also Inuit and some populations in Northern and Eastern Africa, are subject to an undifferentiated and invasive form of nasopharyngeal cancer, usually presenting as enlarged cervical lymph nodes to which the tumour has metastasized. It is the most common form of cancer in that area, with an incidence of 0.1% of the population (Fig. 23.13(b)). It mainly affects people **20–50 years old**, males preponderating. Epithelial cells rather than lymphocytes are involved, and the epidemiology is quite different. Not only is the age of onset much later, but there is no association with malaria. Other co-factors appear to operate, one being an **association with certain HLA haplotypes**. It has also been suggested that dietary habits may play a part, in particular a high consumption of **nitrosamines** in salted fish foods. These tumours are relatively inaccessible to surgery or chemotherapy. Even after irradiation, the prognosis is poor.

23.7.4 B-cell lymphoma

Probably as a result of failure of T-cell control, those with primary or acquired immune deficiencies, including organ graft recipients and AIDS patients, are liable to develop B-cell lymphomas associated with EBV infection.

23.7.5 Oral hairy leucoplakia

EBV is often detectable in these lesions, which appear as white roughened patches on the buccal mucosa and the sides of the tongue, and are seen in male homosexuals who are HIV positive, with or without signs of AIDS.

23.7.6 Other syndromes

EBV has been linked with T-cell lymphomas and Hodgkin's disease, but causal relationships are not fully established.

23.7.7 Antiviral therapy

Like CMV, EBV does not code for thymidine kinase, and although the antiviral aciclovir in high doses has been claimed to have some therapeutic value in EBV-related neoplasms, and in oral hairy leucoplakia, the effect is only palliative.

23.7.8 Pathogenesis

EBV induces more than 80 virus-specified antigens in B lymphocytes, some of which are encoded by 'latent genes', i.e. are expressed in cells latently infected. The first of these to appear are viral nuclear antigens (Epstein–Barr (virus) nuclear antigen (**EBNA**)), which, as their name implies, are found in the nuclei of infected cells. Other important antigens are:

- LYDMA (lymphocyte-detected membrane antigen), the target for cytotoxic T cells and actually an operationally described antigen that results from the CTL recognition of the different EBV latent proteins.
- EA (early antigen), involved in viral DNA replication.
- VCA (viral capsid antigen), a structural protein complex.

In addition, there are several glycoproteins, mostly associated with the cell membrane and hence with infectivity. The major antigen of this sort is gp340/220, which induces neutralizing antibody.

23.7.9 Immortalization of host cells

In vivo, the full lytic cycle of virus replication takes place in differentiating squamous epithelium, from which the virus is shed, and in B cells as they terminally differentiate into antibody-producing plasma cells. When B cells are exposed to EBV *in vitro*, however, a proportion of them develop a latent infection, characterized by circularization of the viral genome. These cells can multiply indefinitely, a condition known as **immortalization**, and give rise to a **lymphoid cell line**. This process has obvious implications for the oncogenic potential of the virus. Similar events take place *in vivo* so that, once infected, a person carries B cells containing EBV genome throughout life. There are six EBNAs, three of which are involved in maintaining the immortalized state. A few cells in such lines possess a plasma cell antigen, PC1, which, together with the presence of VCA, marks those cells that are due to undergo a lytic infection, an activity triggered by cell differentiation.

23.8 Clinical aspects of Kaposi's sarcoma associated herpesvirus (HHV-8, KSHV)

The KSHV genome is typical of the gammaherpesvirus subfamily, along with EBV. These two viruses share life-cycle features, but have different latency properties. KSHV has many coding regions that are not conserved in other human herpesviruses, especially genes that are related to host genes. Like EBV, KSHV genome directs the production of small non-coding RNAs, some with unknown functions, and others which encode for virus mRNAs.

KSHV is found across the globe and nucleotide differences in the genome can pinpoint the source of virus origin of isolates. Although KSHV is necessary for KS development, other co-factors are likely to be required. Thus in AIDS-related KS, HIV is the co-factor through the destruction of the human immune systems leading to a lack of immune control of KSHV. An indolent form of KS used to be the main clinical presentation, mostly in elderly men of Jewish or Italian origin, but a much more aggressive form is now prevalent, mainly in the USA and Africa, where it is a characteristic feature of **AIDS**. The lesions occur mainly in the **skin**, **lymph nodes**, and **gastrointestinal tract**.

KSHV is not found in the population at the same level throughout the world. USA and Northern Europe has low prevalence. In these countries transmission occurs efficiently in male homosexuals and is correlated with the number of partners. Heterosexual spread to women is rather rare. The virus is detected most easily in saliva, rather than sexual secretions. Blood trans-

mission is inefficient. In higher prevalence countries, such as Africa and the Mediterranean, infection begins in childhood both vertically and horizontally.

There is evidence linking HHV-8 with other malignancies such as effusion lymphoma and multicentric Castleman's disease.

The tumour described by Kaposi in the late nineteenth century appears to arise from **endothelial cells** at multiple sites in the dermis. KS lesions are often infiltrated by many cell types. KS lesions can be locally and systemically invasive, and may require chemo- or radiotherapy for control. The spindle cells are still diploid. KSHV DNA is present in all KS tumours. Most spindle cells are latently infected, whilst 1–2% show markers of lytic herpes infection.

Medical case story Kaposi's sarcoma

You have been in Malawi now for 2 months volunteering as a medical student in a large hospital. You are working on the male ward in the General Medical Department where you and a Malawi-trained senior doctor look after one bay of patients. Since your grasp of the local language is getting better, you no longer need as much help with translation. There are 20 beds in your bay, but often there are 10 more patients using mattresses on the floor between the beds. Your job is to ensure that your patients receive the investigations and interventions needed, to chase up results, and to document any new developments. Most of your patients are in the 20–40 years age range and most have HIV.

Over by the window in the far corner of the bay lies Anthony. He is a very frail young man who can speak a little English and smiles sweetly when you go to see him. Most of the time he sleeps and is too weak to sit up for long. His hair is thin and tufty, and his palms are very pale. On his arms you can see dark purple smudges that are raised, hard, and painful, and there are a few flecks on his cheeks too.

He has been on the ward now for a day. On admission he complained of a cough with sputum tinged with fresh blood, and pain when swallowing, which had started 3 weeks earlier. He had generalized body pains and weakness, and looks very thin and emaciated. He has been receiving standard first-line antiretroviral therapy (ART) since he was diagnosed with HIV 2 years ago, but has steadily lost weight. The standard first-line ART in Malawi at that time was stavudine, lamivudine, and nevirapine (2009).

You reflect that, whilst Anthony was lucky enough to have access to HIV treatment, it may not have been the best regimen for him as he says he never really improved. Unfortunately, it was not routinely possible to test patients for HIV drug-resistance, due to limited resources in Malawi, although facilities to measure CD4 counts are available in the main treatment centres. Not as many alternative drugs can be offered to patients here in comparison with the UK.

When you examine him, he has a temperature of 38.5 °C, a blood pressure of 90/40 (using an adult cuff), a respiratory rate

of 30, and a heart rate of 110. You notice that he has a white coating all over his oral mucosa, plus some more purple smudges on his hard palate. You make the following brief notes of significant findings:

General examination

- Thin and weak
- Generally pale

Airways

- Clear

Breathing

- Tachypnoeic with diffuse crackles in both lung fields
- No notable findings on percussion or resonance

Circulatory

- Tachycardic with a thready but regular pulse
- No additional heart sounds

Development

- Cranial nerves 2–12 intact
- No peripheral neurological signs

Everything else

- Multiple purple lesions on limbs and torso
- Oral candidiasis
- Oral purple lesions
- Pale conjunctiva, but no jaundiced sclerae
- Gastrointestinal system—soft but tender abdomen with no organomegaly

In summary, Anthony is clearly weak and septic with compromised respiration and a tender abdomen. He has raised, non-blanching purple lesions all over his skin and buccal mucosa, and signs of severe anaemia. He also has oral and possibly oesophageal candidiasis.

Anthony's chest X-ray shows 'diffuse insterstitial shadowing', and a full blood count indicates that he has a microcytic anaemia with a haemoglobin level of 3.3 (based on a European standard range for men of 13–18 g/dL).

You compile a problem list, which strongly suggests WHO stage IV HIV/AIDS[1]:

- Sepsis secondary to pneumonia.
- HIV with ART treatment failure.
- Kaposi's sarcoma, possibly disseminated.
- Anaemia.
- Malnutrition.
- Oral/oesophageal candidiasis.

Following local guidelines, Anthony is given benzylpenicillin and chloramphenicol IV to cover for pneumonia, and is also transfused with two units of blood in an attempt to improve his haemoglobin levels, whilst receiving iron-replacement medication. The presence of the purple lesions indicates Kaposi's sarcoma (KS) and you suspect that he also has lesions disseminated throughout his organs. This could explain the abdominal tenderness, and possibly the findings on the X-ray. In order to confirm this diagnosis, Anthony undergoes a bronchoscopy, which clearly shows bloody patches of KS in the bronchi.

Anthony is referred to the palliative care team who can treat patients with KS using vincristine chemotherapy. The palliative care team are always busy in this hospital and in the community, where they help patients through painful illnesses, and support the families and carers. They come to sit by Anthony's bed and quietly explain to him about the purple spots and what the possible treatment options are. He seems to understand and asks some questions; when you go to see him afterwards he looks somehow content. However, chemotherapy for KS can be very toxic and you are not so sure that he will be well enough to cope, especially with such severe anaemia and sepsis, neither of which are resolving fast enough with treatment.

During the night, Anthony passes away before any more can be done for him, and when you come again in the morning, his bed has been taken by the patient from the neighbouring mattress on the floor. You wonder if you did enough to help him, or if you could have done any more. Anthony's illness was so severe that even with the best of intentions you think he would not have lived.

Notes

[1] World Health Organization. (2010). *Antiretroviral Therapy for HIV Infection in Adults and Adolescents. Recommendations for a Public Health Approach.* Available at: http://www.who.int/hiv/pub/arv/adult/en/index.html (accessed 27 July 2010).

★ Learning Points

- **Clinical:** Diagnosis of KS is usually clinical, and purple patches in the buccal mucosa are a useful sign especially in patients with dark skin. Early stages of the disease can be treated using local excision and liquid nitrogen therapy, but in the case of disseminated KS, chemotherapy is the best treatment.

Effective treatment with ART has reduced the incidence of KS in HIV patients.

- **Personal:** Dealing with the pain and suffering experienced by your patients can be very difficult for a clinician, especially a young medical student.

Reminders

- Herpes virions have four concentric layers: an outer envelope, a tegument, an icosahedral nucleocapsid, and an inner core. They are 120–200nm in diameter.

- Viral replication is entirely nuclear and is strictly controlled with α, β, and γ genes coding for corresponding proteins. The α products are *trans*-acting regulatory proteins, whereas β proteins are enzymes such as DNA polymerases; the γ proteins are structural.

- Following primary infection the herpesviruses establish a lifelong persistent infection of the host by establishing a latent infection. Reactivation from latency allows production of new virus particles and transmission to new uninfected host.

- HHV-1 and HHV-2 cause skin lesions and more serious tissue infections and encephalitis. HHV-3 infection results in chicken pox followed by reactivation as shingles and zoster many years later. Epstein–Barr (HHV-4) virus is oncogenic for African children and adults in China. Cytomegalovirus (HHV-5) causes foetal mortality and in adulthood can give clinical problems in immunosuppressed transplant patients. HHV-6 and HHV-7 cause febrile rash disease of children. HHV-8 with co-factors is linked with inducing Kaposi's sarcoma.

- A live attenuated VZV vaccine is used in 70-year-olds to prevent herpes zoster.

- The antiviral nucleoside analogue ACV can be used prophylactically and therapeutically for HHV-1, -2, and -3.

Further reading

Ansari, A., Li, S., Abzug, M.J., and Weinberg, A. (2004). Human herpesvirus 6 and 7 and central nervous system infection in children. *Emerging Infectious Disease* **10 (8)**, 1450–4.

Berges, B.K. and Tanner, A. (2014). Modelling of human herpesvirus infections in humanized mice. *Journal of General Virology* **95**, 2106–17.

Gershon, A.A. and Gershon, M.D. (2013). Pathogenesis and current approaches to control of varicella-zoster virus infections. *Clinical Microbiology Reviews* **26**, 728–43.

Kang, M-S. and Kieff, E, (2015). Epstein-Barr virus latent genes. *Experimental & Molecular Medicine* **47**, e131.

Kennedy, P.G.E., Rovnak, J., Badani, H., and Cohrs, R.J. (2015). A comparison of herpes simplex virus type 1 and varicella-zoster virus latency and reactivation. *Journal of General Virology* **96**, 1581–1602.

Yuen, K.S., Chan, C.P., Wong, N.H., Ho, C.H., Ho, T.H., Lei, T., *et al.* (2015). CRISPR/Cas9-mediated genome editing of Epstein–Barr virus in human cells. *Journal of General Virology* **96**, 626–36.

 Questions

1. Write notes about herpes latency reviewing the clinical and molecular aspects.

2. Describe HHV-3, varicella-zoster, and the biology of the infection.

3. Can herpes infections be prevented or cured?

Smallpox: human disease eradicated but zoonotic pox virus infections common

24.1 Introduction

The smallpox epic provides at least three 'firsts': the first vaccine, the first disease to be totally eradicated (Fig. 24.1), and the first virus infection against which chemotherapy was clinically effective.

Although smallpox itself is now extinct, other poxviruses infect humans; and as we shall see in Chapter 31, one of them may be used in a rather surprising way to immunize against quite different diseases. Because of the threat of bioterrorism or accidental release of previously unrecognized freezer stocks, some countries are stockpiling vaccinia virus vaccine against smallpox and safer vaccines are being developed to be held in a strategic reserve.

Leslie Collier, our senior author, made a significant contribution to the smallpox eradication campaign when as Director of the Lister Institute he developed a new stabilizer for the vaccine to be used in tropical and temperate countries of the world.

24.2 Properties of the viruses

24.2.1 Classification

Most of these viruses do not infect humans. Others are pathogenic both for animals and humans and two, smallpox and molluscum, only for humans (Table 24.1). The genera are distinguished on the basis of morphology, genome structure, growth characteristics, and serological reactions; there is close serological relationship between the viruses within each genus and good antigenic cross-protection between genera.

24.2.2 Morphology

These are the largest viruses of all and have 100 proteins in the virion; the orthopoxviruses are brick-shaped, whereas orf and molluscum tend to be ovoid (Fig. 24.2). The image reconstruction shows the virions to be barrel-shaped, 360×270–250 nm in size with a single lipid membrane bilayer and a complex internal structure of a dumbbell-shaped core and two accompanying protein lateral bodies, so named after their location in the virion. The core has a bilayer core wall and contains a large number of virus-encoded enzymes as well as the viral DNA.

24.2.3 Genome

The nucleic acid is a linear double-stranded DNA ranging in size from 134 kbp (parapoxviruses) to 300 kbp (avipoxviruses) (Fig. 24.3). The DNA has 10kb inverted repeats at each end of the genome and terminal loops that covalently close and connect the two DNA strands at the genome ends (Fig. 24.4). Several genes lie within these repeats. The hairpins and flanking sequences have important functions in DNA replication. This large genome encodes for 100–200 polypeptides. Centrally located genes for DNA replication, transcription, and morphogenesis are conserved and are essential for virus replication, whereas genes near the two termini affect host range and virulence.

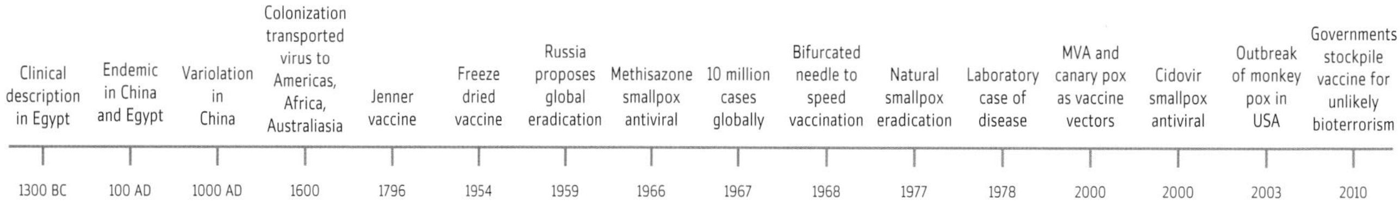

Clinical description in Egypt	Endemic in China and Egypt	Variolation in China	Colonization transported virus to Americas, Africa, Australiasia	Jenner vaccine	Freeze dried vaccine	Russia proposes global eradication	Methisazone smallpox antiviral	10 million cases globally	Bifurcated needle to speed vaccination	Natural smallpox eradication	Laboratory case of disease	MVA and canary pox as vaccine vectors	Cidovir smallpox antiviral	Outbreak of monkey pox in USA	Governments stockpile vaccine for unlikely bioterrorism
1300 BC	100 AD	1000 AD	1600	1796	1954	1959	1966	1967	1968	1977	1978	2000	2000	2003	2010

Fig. 24.1 Timeline for smallpox.

Table 24.1 Poxviruses that infect humans

Genus	Virus	Primary host(s)	Clinical features in humans
Orthopoxvirus	Variola	Man	Smallpox
	Vaccinia	Man	Vesicular vaccination lesion
	Cowpox	Cattle, cats, rodents	Lesions on hands
	Camel pox	Camels	As for cowpox
	Monkeypox	Squirrels, monkeys	Resembles smallpox clinically
Parapoxvirus	Pseudocowpox	Prairie dogs (USA), cattle	Localized nodular lesions ('milker's nodes')
	Orf	Sheep, goats	Localized vesiculogranulomatous lesions
Yatapoxvirus	Tanapox	Monkeys	Vesicular skin lesions and febrile illness
Molluscipoxvirus	Molluscum	Man	Multiple small skin nodules

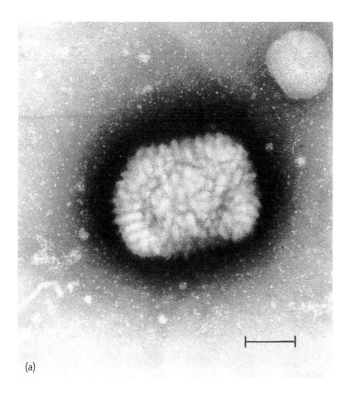

(a)

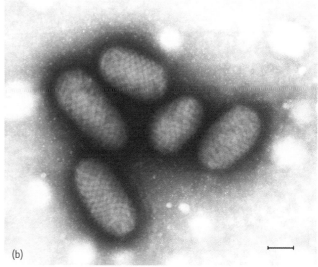

(b)

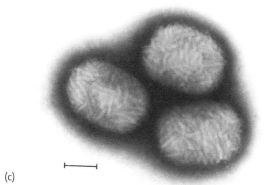

(c)

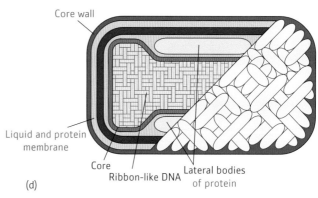

(d)

Fig. 24.2 Electron micrographs of poxvirus. (a) Vaccinia; (b) orf; (c) molluscum contagiosum; (d) model of poxvirus showing lateral bodies, ribbon-like DNA, and dumbbell core structure. Scale bar = 100 nm.

24.2.4 Replication

Unlike other DNA viruses, the poxviruses replicate only in the **cytoplasm**, and at least in the laboratory can replicate in cells without a nucleus. Since most of the cellular enzymes for DNA replication and RNA synthesis are located in the cell nucleus, poxviruses are obliged to provide their own! Virus entry into cells is a complex process and involves both low affinity interactions with general cell-surface molecules, such as glycosaminoglycans, and specific interactions between groups of virus and cellular proteins. Indeed, a putative entry complex of eight virus proteins has been identified in vaccinia virus. The viral core enters the cytoplasm of the cell and acts as scaffolding for subsequent replication events. The virus core transports the many virally encoded essential enzymes, such as viral transcriptase, transcription factors, capping and methylating enzymes, and a poly(A) polymerase, into the cell. Transcription of viral DNA is therefore initiated rapidly and 'early' viral genes

are activated, coding for enzymes involved in viral DNA replication. All poxvirus genes contain a single ORF and there is no evidence of virus mRNA splicing.

A 'rolling hairpin' model of DNA replication is generally accepted, similarly to parvovirus. 'Intermediate genes' have a peak transcription around 2.5 h and 'late genes' peaking at 4 h post infection. Intermediate genes encode for proteins for the regulated expression of late genes. 'Late genes' encode structural proteins and the factors required to start early gene expression that are packaged into the maturing virion.

Poxviruses interfere with cytokine and chemokine signalling and complement production, and confer resistance to IFN by interfering with IFN-induced antiviral proteins, cell signalling pathways, or apoptosis.

Virus assembly takes place in particular areas of the cytoplasm and immature viruses can be visualized quite easily as crescent-shaped membrane structures. The crescents develop

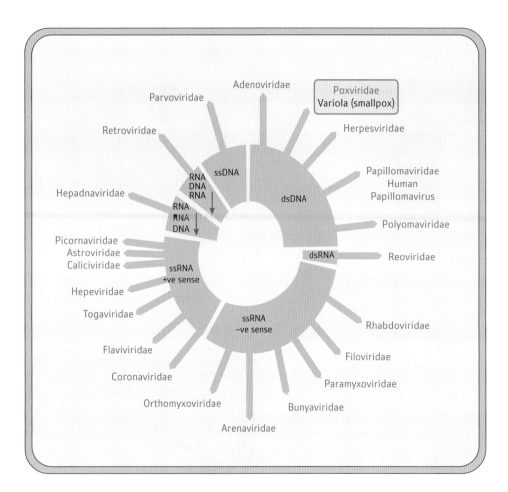

Fig. 24.3 Baltimore classification scheme and the poxviruses.

into spheres as the viruses begin to self-assemble. Lipids are delivered to the crescents from the ER as well as vesicles containing viral envelope proteins. During maturation the virions move to the Golgi complex, where they are enveloped before being released by budding or by cell disruption. Most of the virions remain inside the cell but others are wrapped in two extra membranes and transported to the plasma membrane. Adjoining cells can be infected via cell membrane projections, which contain mature virions on their tips.

24.3 Clinical and pathological aspects of smallpox

24.3.1 Historical note on the disease and vaccination

The Anglo-Saxon word 'pokkes' meant a pouch and refers to the characteristic vesicular lesions. The term 'small pox' was introduced during the sixteenth century to distinguish it from the 'great pox', or syphilis. The Latin term, *variola*, means a spot. It seems likely that smallpox has been with us for a very long time: the mummy of the Pharaoh Rameses V (1100 BC) bears lesions

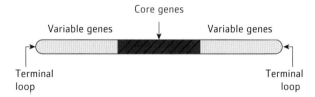

Fig. 24.4 Genome structure of poxvirus.

highly suggestive of this infection (Fig. 24.5) and there are many later accounts of its ravages. Indeed, smallpox was so widespread that it was often regarded as the norm, rather than the exception, and few people—if they survived—escaped its disfiguring scars. In India it was thought to be a divine visitation and even had its own goddess, Kakurani.

The story of Edward Jenner's discovery in 1796 that inoculation with cowpox would prevent smallpox is well known, but it is not so widely appreciated that others had made similar observations, notably Benjamin Jesty, a Dorset farmer, who inoculated his own family. To Jenner, however, goes the credit for showing that, following the inoculation of young James Phipps with cowpox, deliberate inoculation with smallpox material failed to induce the disease.

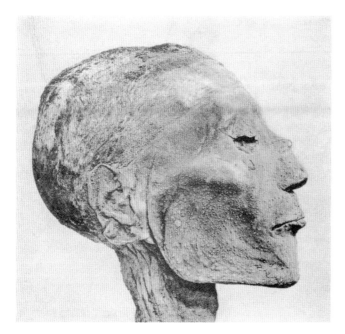

Fig. 24.5 Mummy of Rameses V. The lesions of the face are suggestive of smallpox. (Reproduced, with permission, from Dixon, C.W. (1962). *Smallpox*. J&A Churchill, London.)

The early vaccines were derived from cowpox (Latin *vacca* = cow) and propagated by arm-to-arm inoculation. During the twentieth century, other strains of poxvirus (vaccinia) that could be grown in quantity on the skins of calves and other animals were used. Vaccinia can now be cultivated in cell culture for vaccine production allowing a vaccine free of animal protein and bacteria.

The global eradication of smallpox

With improved methods for preparing vaccine in bulk, and for testing its safety and potency, vaccination was widely practised around the world; nevertheless, wide regional differences in uptake ensured that the disease continued to smoulder on, flaring up at intervals into epidemics. In 1966, faced with this situation, at a meeting the Alma Alta in the Soviet Union WHO voted US$2.5 million for an immunization campaign designed to eradicate smallpox completely within a decade (Fig. 24.6).

This vast enterprise is referred to again in Chapter 31, but needs a book to itself. Its success may be judged by the fact that the last case of naturally acquired infection was recorded—in Somalia—in October 1977, only months beyond the target date fixed ten years previously.

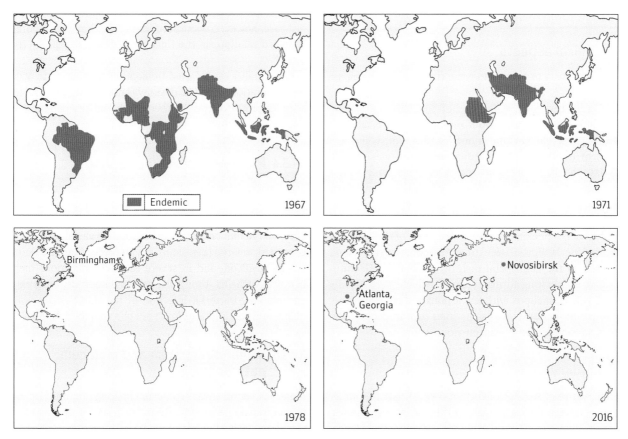

Fig. 24.6 The global eradication of smallpox by the World Health Organization. (Taken with permission from Fenner, F., Henderson, D.A., Arita, L., Jezek, Z., and Ladnyi, L.D. (1988). *Smallpox and its Eradication*, World Health Organization, Geneva.)

In *Health for all by the year 2000* Dr Mahler, Director General of WHO, said 'There is little joy in life nor any kind of justice for a child condemned to disease or early death because of the accident of birth in a developing country. Smallpox eradication is a sign, a token, of what can be achieved'.

In 1966, at the start of the campaign, 46 countries recorded 131 697 smallpox cases (perhaps 1% of the real number). This virus was endemic in Africa, Asia, Indonesia, and Brazil. By 1973 80% of the vaccine was produced in endemic countries. In 1968 the bifurcated needle was introduced and a diagnostic test. Alongside the technique to stabilize the vaccine virus developed by L. Collier, the crucial elements for success were now in place. Vaccination in each country was combined with surveillance and containment. Therefore valuable resources were concentrated in areas where there were cases. In 1984 for example there were 8403 outbreaks, 11 000 cases of smallpox. Flying squads removed cases to infectious disease hospitals and the neighbours of a case were vaccinated. This is now called 'ring vaccination'.

A major task was overcrowding, ignorance, and prejudice. For example the smallpox Goddess Shitala Mata was popular in India. In the final four years of the campaign in West Africa alone there were 23 changes of government in 18 countries, which seriously disrupted the campaign. A team of 800 000 national staff alongside 2000 million doses of vaccine and 700 international staff spelled the demise of the virus. The cost was US$300 million and one can compare the cost of placing a man on the moon at US$4000 million.

24.3.2 Clinical aspects

It is very gratifying to be able to write this section in the past tense!

There were two main categories of smallpox, caused by slightly different viruses: **variola major** had a mortality of about 30%, whereas **variola minor**, or **alastrim**, killed less than 1% of its victims. The incubation period was usually 10–12 days; a febrile illness of sudden onset lasting 3–4 days was followed by the appearance of a rash progressing from macules to papules, vesicles, and pustules (Fig. 24.7), which then formed crusts. Surviving patients were often left with unsightly scars or pockmarks. The distribution of the rash was **centrifugal**, i.e. it affected the extremities more than the trunk, as opposed to the centripetal rash of chickenpox. Haemorrhagic and fulminating forms occurred, which were rapidly lethal. Modified smallpox with few lesions and comparatively little constitutional upset was sometimes seen in people who had been vaccinated some years previously.

24.3.3 Epidemiology

The main source of smallpox infection was the patient's upper respiratory tract in the early stages of the disease, but fomites, such as bedding and clothing, were also of some importance. Spread from country to country was facilitated by increases in the volume and speed of international travel, and importations

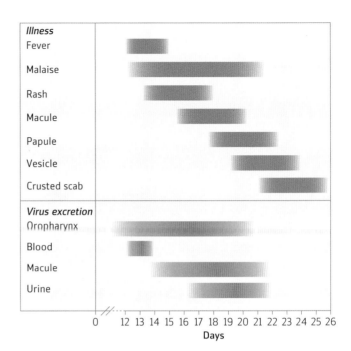

Fig. 24.7 Clinical and virological events during infection with smallpox.

into Europe and North America occurred with some frequency well into the twentieth century. The last natural community outbreak in the UK was 1962, due to imported smallpox; the last case, however, was a secondary infection from laboratory-acquired smallpox in the University of Birmingham in the UK in 1978.

Case fatality rates varied tremendously from <1 to 40% but the Generation Time (12 days) was not so variable, and at this time lesions developed in the mouth and the patient became infectious for others.

24.3.4 Pathogenesis and pathology

The pattern was that of an acute generalized infection (see Chapter 1, Section 1.1.1). The skin lesions were caused by direct viral invasion and local virus replication and the vesicles contained large numbers of virions. Although variola is primarily a dermotropic virus (Fig. 24.8), organs other than the skin were always involved, and in severe cases complications included keratitis, arthritis, bronchitis and pneumonitis, enteritis, and encephalitis. Bacterial infection of the skin vesicles was frequent, and was a serious matter in pre-antibiotic days.

24.3.5 Immune response and molecular pathology

An attack nearly always conferred lifelong immunity, mediated by neutralizing antibody. Recovery from infection was, however, largely effected by cell-mediated responses.

Poxviruses undermine host cell defences in many and varied ways. After infection the virus produces a protein, K3L, which

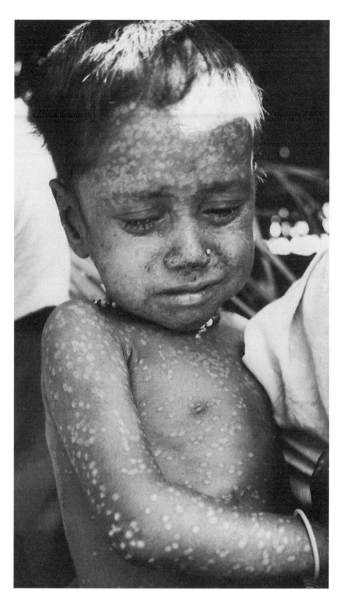

Fig. 24.8 Smallpox. This 2-year-old Bangladeshi girl was the last case of smallpox to be seen on the Asian subcontinent, and the last case in the world of variola major, the more severe form of the disease. (Photograph from World Health Organization.)

closely mimics the substrate of protein kinase R (PKR), an important component of the innate immune system. Viral infection blocks the interferon-induced inhibition of protein synthesis. Furthermore the ESL gene product binds to dsRNA and blocks both 2′-5′ oligoadenylate synthetase and PKR antiviral pathways. Similarly the function of cytokines such as TNF-α and interleukin-1 are blocked via a serine protease inhibitor and by producing decoys which bind to TNF-α and IL-1.

24.3.6 Laboratory diagnosis

The differential diagnosis most often needed was between smallpox and chickenpox, which could on occasion resemble each other clinically. **EM**, which readily and rapidly distinguished

between pox and herpes virions in vesicle fluid, proved most valuable in this respect and still remains an important method for diagnosing all poxvirus infections in specialized laboratories alongside the more generally used workhorse PCR.

24.3.7 Control

New anti-smallpox drugs

Coincident with the eradication of smallpox with the live attenuated vaccine, one of the first antiviral drugs was field tested and shown to have therapeutic and prophylactic activity—methisazone. Most recently, other inhibitors have been discovered and licensed including cidofir and a newer version called brincidofovir, which inhibits the viral DNA polymerase. Another drug, Tecovirimat (or Arestyvir), blocks exit of the virus from cells.

Vaccinia and smallpox vaccine

Two other examples of vaccine strains both derived from original vaccinia stocks by passage in cell culture are Imvamune and Lc16m8.

Variola as a biological weapon

Following tragic episodes of laboratory-acquired infection, it was internationally agreed that samples of the virus should be held only in two maximum security Category 4 laboratories, one in Atlanta, USA, and the other in the Novosibirsk, Russia; furthermore, large stocks of vaccine are held in store for use in the unlikely event of a reappearance of the disease. Even so, in view of the depredations of smallpox, the herculean efforts needed to eradicate it from the planet, and its possible use in biological warfare, the World Health Assembly recommended in 1996 that all remaining stocks of variola virus should be destroyed. At the time of writing (2015) this decision has still not been implemented. The complete destruction of this virus is controversial, as, however dangerous, every living species embodies a unique fund of biological information and the obliteration of any one of them represents an irreplaceable loss to future generations. In this case, however, the complete nucleotide sequences of several smallpox viruses are known, and the viral DNA can be cloned into bacterial plasmids in such a way that it cannot be expressed, but still embodies the genetic information as a permanent record of one of the great viral plagues of mankind.

But smallpox could be seen to be a poor choice for a bioweapon. The R_0 at 2 is low and the generation time of 12 days relatively high. The means, as with SARS CoV, MERS CoV, and Ebola, that suspected contacts of a case can be quarantined and even vaccinated during the incubation period. Furthermore, most persons over the age of 50 have already been vaccinated. Governments stockpile traditional vaccinia virus vaccine alongside the more weakened cell-culture versions such as MAVA (more attenuated vaccine Ankara).

In view of the ability of poxviruses to survive for long periods in the dried state, it has been suggested that infective smallpox virus might still exist in preserved cadavers; this seems unlikely, but it is interesting that in 1986 Italian workers claimed to have serological and electron microscopic evidence of the virus in a Neapolitan mummy dating from the sixteenth century AD. Exhumation of frozen graves in Siberia has also uncovered well-preserved bodies with smallpox lesions where virus has been visualized by EM. In the 1990s near our own hospital in London, a lead coffin-encased baby from the nineteenth century had smallpox lesions with EM visualized virus.

24.4 Other poxvirus infections

As we have seen, poxviruses are widespread in nature. Here, we shall briefly describe only those that infect humans. They are also the subject of intense research from non-virological biologists who wish to analyse the enzymes involved in mRNA metabolism.

24.4.1 Cowpox

Like most of the agents in this family, cowpox is, as its name implies, a zoonosis (Table 24.1); infection of humans is rare, but may be seen in rural practices. Although recent evidence points to cats and rodents rather than bovines as the main reservoir of infection, most cases in humans seem to be acquired from cows with sores on their teats. The lesions usually present on the hands or face and resemble severe vaccinia infections.

Treatment with anti-vaccinia immunoglobulin may be helpful if given early.

24.4.2 Monkeypox

This orthopoxvirus zoonosis, which despite its name is predominately an infection of rodents which can spread to monkeys, occurs in western and central Africa; it is of some concern because it causes an illness in humans very similar to smallpox, with a significant mortality rate. Importation into the USA of exotic pets from Africa, such as Gambian giant rats, is thought to be responsible for the subsequent transmission of virus to other USA pet animals, including prairie dogs. In a spectacular and embarrassing breakdown of public health surveillance in the USA, an outbreak of monkeypox from an imported pet from Africa caused typical pox lesions in children, which remained undetected in 2003 for many weeks. As of July that year, 71 cases were reported to CDC: a number were severe, but fortunately there were no deaths. The main clinical features were rash, fever, respiratory symptoms, and lymphadenopathy. Prairie dogs were an important source of infection, but elsewhere squirrels seem to be the main reservoir, infection occurring mainly in children, who can acquire it by playing with captive animals. Human-to-human transmission is more frequent than was at first thought, but control by vaccination is not difficult. Two clades of the virus are known, West African and Congo Basin monkeypox, and the latter is more virulent in primates.

Could monkeypox mutate to smallpox or occupy the same ecological niche? The question was raised as soon as infections in humans were recognized; but fortunately the poxviruses, like others with DNA genomes, are genetically very stable, so that this dire possibility can, one hopes, be ignored.

24.4.3 Parapoxviruses

Pseudocowpox infects various species of cattle, whereas **orf** is found in sheep and goats, in which it causes contagious pustular dermatitis; the viruses are very similar. In humans, lesions occur on the hands and face after contact with infected farm animals; the lesions of pseudocowpox are nodular ('milker's nodes'), whereas those of orf are granulomatous. Parapoxvirus lesions are characteristically painless and resolve over a period of weeks without specific treatment.

24.4.4 Tanapox

This virus takes its name from the Tana River in Kenya, where it was first diagnosed. It is prevalent in monkeys in Kenya and Zaïre, but unlike monkeypox appears to be spread by insect bites. There is usually only one vesicular lesion, but its appearance is preceded by fever and quite severe malaise. Recovery is uneventful.

24.4.5 Molluscum contagiosum

Molluscum is the sole member of the *Molluscipoxvirus* genus; its study is not helped by our inability to grow it in the laboratory. **Molluscum contagiosum** affects only humans; it is a comparatively common skin condition, characterized by **multiple small (1–10 mm) nodular lesions** mostly on the trunk (Fig. 24.9). They become umbilicated and contain caseous material in which 'molluscum bodies' can be readily demonstrated. These are quite large (30 μm long) ovoid structures containing many virions (Fig. 24.10). The diagnosis is usually obvious from the clinical appearance, but in case of doubt, a simple test is to identify the molluscum bodies in expressed material stained with Giemsa or Lugol's iodine. The virions themselves can readily be found by EM.

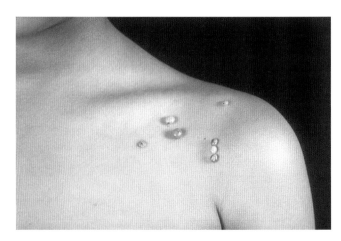

Fig. 24.9 Molluscum contagiosum. (Courtesy of the Department of Dermatology, The Royal London Hospital, London.)

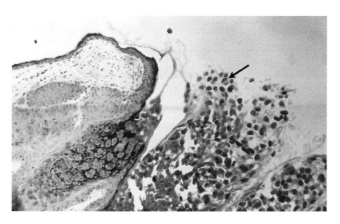

Fig. 24.10 Section of a molluscum contagiosum lesion of the skin. The small, ovoid molluscum bodies (arrowed) are packets of virus particles.

Transmission is by contamination of skin abrasions through contact; the infection can be sexually acquired. A molluscum lesion at the lid margin often results in severe **conjunctivitis** and **keratitis**, which resolve when the nodule is treated.

Disappearance of the lesions is often spontaneous, but can be helped along by cryotherapy, curettage, or treatment with caustic agents such as salicylic acid or phenol.

24.4.6 Ectromelia

Ectromelia, or mousepox, does not infect humans, but is mentioned here for two reasons. First, it provided the model for Dr Frank Fenner's research on the pathogenesis of acute viral infections in humans (Chapter 4); and secondly, it may become endemic—and very difficult to eradicate—in stocks of laboratory mice.

 ## Reminders

- The Poxviridae are comparatively large, brick-shaped or ovoid viruses with a **complex** structure containing many virus-encoded proteins. Their genome is **dsDNA**, ranging in size from 186 kbp (variola) to 230 kbp (cowpox).

- **They replicate only in the cytoplasm** and transport into the cell a multitude of enzymes for replication including transcriptases, capping and methylating enzymes, and poly(A) polymerase.

- Early viral genes code for enzymes involved in DNA replication. Intermediate and late genes are transcribed at the onset of DNA replication and code for structural proteins and proteins, many of which interfere with the host response to infection.

- Most poxviruses cause zoonoses, but some are pathogenic for humans. Of these, smallpox was the most important until eradicated in 1977 by the WHO vaccination campaign. It was an acute generalized infection: the mortality of variola major was about 30% and that of a milder form, variola minor or alastrim, less than 1%.

- Monkeypox is a similar infection and occurs in West Africa as a natural infection of rodents; it is occasionally transmitted to humans, but person-to-person spread is rare. There was a recent outbreak in the USA linked to pet shops that had imported mammals from Africa, and to prairie dogs.

- **Cowpox**, **pseudocowpox**, and **orf** affect farm animals and occasionally cause local lesions, which resolve spontaneously, on the fingers and faces of people coming in contact with them. Cowpox may be treated with anti-vaccinia serum.

- **Molluscum contagiosum** causes multiple small nodular lesions containing packets of virions known as molluscum bodies. A lesion at the lid margin may cause **conjunctivitis** and **keratitis**. The lesions may eventually disappear spontaneously, but can be treated by cryotherapy, curettage, or caustic chemicals.

- A combination of PCR and EM is used for diagnosing all poxvirus infections and units able to rapidly diagnose smallpox have been set up again to counter bioterrorism or natural emergence of monkeypox.

- A new generation of safer, more highly attenuated vaccines such as 'more attenuated vaccine Ankara' (MAVA) have been prepared in cell culture and are stockpiled.

- Smallpox would be a poor choice as a bioweapon, because of low R_0 and high generation time, but the storage of vaccine continues as a 'just in case scenario' and a sensible public health strategy.

- New anti-smallpox drugs are stockpiled and an example is the virus DNA polymerase inhibitor brincidofovir.

 ## Further reading

Fenner, F., Henderson, D.A., Arita, I., Jezek, Z., Ladnyi, I.D. (1988). *Smallpox and its Eradication. Geneva*: World Health Organization, Available at: http://www.whqlibdoc.who.int/smallpox/9241561106.pdf

Ferguson, N.M., Keeling, M.J., Edmunds, W.J., Gani, R., Grenfell, B.T., Anderson, R.M., *et al*. (2003). Planning for smallpox outbreaks. *Nature* **425**, 681–5.

Reardon, S. (2014). Smallpox watch—frozen mummies and envelopes of scabs could contain remnants of one of history's most prolific killers. *Nature* **509**, 22–33.

Weinstein, R.S. (2011). Should remaining stockpiles of smallpox virus (Variola) be destroyed? *Emerging Inf Dis* **17**, 681–682.

Questions

1. What are the features of the natural history of smallpox that would make the virus a poor choice as a bioweapon?

2. Describe the key steps of virus replication and pinpoint those stages where intervention by antivirals can happen.

3. Outline the WHO programme 'eradication of smallpox'.

Adenovirus: respiratory, eye, and gastroenteritis viruses

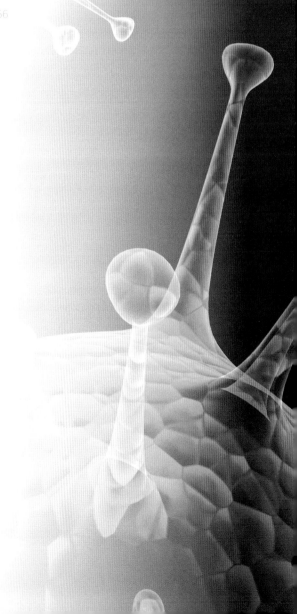

25.1 Introduction

These viruses were first isolated from human adenoids, hence their name: they cause widespread disease, ranging from colds to pneumonia and diarrhoea to bladder infections, and have an extraordinary range of pathology. Adenoviruses infect the eye, gut, and respiratory tree. Scientifically the study of adenovirus replication led to important discoveries of messenger RNA splicing.

25.2 Properties of the virus

25.2.1 Classification

Members of the family *Adenoviridae* are icosahedral, non-enveloped linear dsDNA viruses some 80 nm in diameter (Fig. 25.1). Thus far, adenoviruses have been isolated only from vertebrates. There are five recognized genera: *Mastadenovirus*, isolated from mammals, including humans; *Aviadenovirus*, isolated from birds; *Atadenovirus*, isolated from birds, mammals (including a marsupial), and reptiles; *Ichtadenovirus*, isolated from fish; and *Siadenovirus*, isolated from reptiles and birds. Fifty-seven serotypes of human adenoviruses are now known, distinguished originally on the basis of cross-neutralization tests, but this number will undoubtedly creep slowly upwards as time passes. The human adenoviruses fall into seven groups, A–G. In general, their host range is confined to one species.

25.2.2 Morphology

When the first EM pictures of adenoviruses were published by Horne and his team at the National Institute for Medical Research in London, they caused a minor sensation, revealing as they did an architecture of remarkable beauty and precision (Fig. 25.2). The capsid is formed from 252 capsomeres, which are arranged in an icosahedron with 20 sides and 12 vertices. A unique morphological feature is a slender fibre projecting from each of the 12 vertices of the icosahedron, giving the virus something of the appearance of an orbiting satellite (Fig. 25.3).

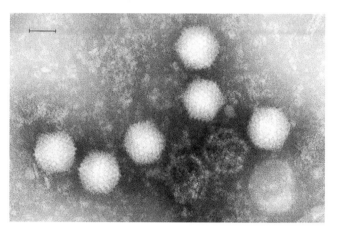

Fig. 25.2 Electron micrograph of adenovirus.

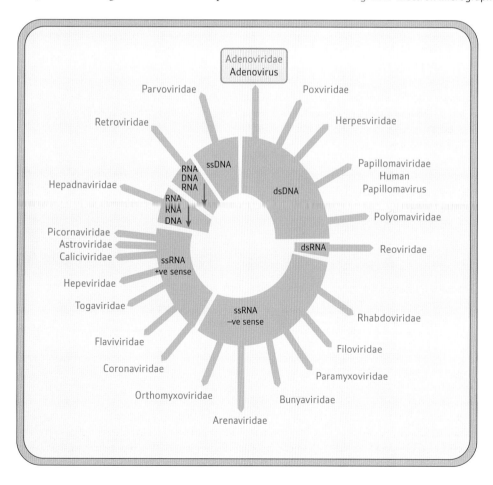

Fig. 25.1 Baltimore classification scheme and adenovirus.

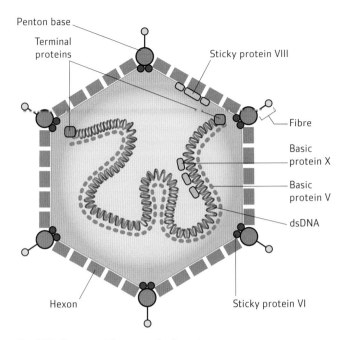

Fig. 25.3 Structural features of adenovirus.

At the fibre base is the penton protein, whilst the other structural proteins are called hexons. Most immune response during infection is directed towards these two structural proteins.

25.2.3 Genome

The genome is linear dsDNA, 36–38 kbp, thus Baltimore class I (Fig. 25.1), with inverted terminal repeats and a protein primer at each 5' terminus. The inverted terminal repeats differ in size, ranging from 20 to 200 base pairs, and are essential for the replication of the virus genome. The human adenovirus genome encodes for five transcription units expressed early (E) in infection of a cell (E1A, E1B, E2, E3, and E4), three delayed early transcription units (IX, IVa2, and E2 late) and one late (L) transcription unit that is processed into five mRNAs (L1–L5). Each transcribed gene gives rise to multiple forms through mRNA splicing, and, although the proteins translated from such splicing are different, many have related functions.

25.2.4 Replication

The replication cycle is divided into two phases separated by the onset of virus DNA replication. Although it is tempting to divide the replication cycle along the lines of early and late gene expression the functional distinction between such genes is often blurred, with early genes continuing expression and functioning at late times. In cell culture 5–6 h of infection is required before DNA replication is detected and the infectious life cycle is completed within 24–36 h. To initiate infection the virus binds to the host-cell receptor via the fibre (Fig. 25.4). The receptor used by most adenoviruses is the plasma membrane protein CAR, a component of epithelial tight junctions. Following CAR binding, the penton base of the adenovirus binds to integrins on

the cell surface, causing the detachment of the fibres and subsequent entry into the cell of fibreless virions by endocytosis. The endocytosed virions are released from the endosome and transported by the microtubule network of the cell to the nucleus. Virus uncoating occurs at the nuclear pore complex leading to the entry of the virus DNA associated with virus protein VII into the nucleus where replication takes place.

To successfully complete the virus replication cycle, adenoviruses alter the cellular environment to induce the cell to enter S phase, counteract antiviral defences, and synthesize new viral proteins needed for virus DNA replication. Transcription of both strands of the viral DNA by cellular RNA polymerase II leads to sequential formation of early and late mRNAs and proteins. RNA is transcribed in a precisely controlled order from nine promoters. E1A is the first transcription unit to be expressed and processed as two spliced mRNAs encoding for small and large E1A proteins. E1A proteins are strong, promiscuous transcriptional activators, inducing the expression of other adenovirus transcriptional units and cellular genes. E1A does this by binding to a variety of cellular transcription factors and regulatory proteins. In particular, both small and large E1A bind to the crucial cell cycle regulatory protein 'Retinoblastoma' (Rb) causing the displacement of the transcription factor E2F. The binding of Rb to E2F negatively regulates the entry of cells into S phase. When E2F is free from Rb it acts as a transcription factor inducing cellular genes for DNA replication. E2F also activates adenovirus promoters. Therefore, E1A binding to Rb and releasing E2F not only induces virus gene expression but also ensures virus DNA replication occurs in a cellular environment suitable for DNA synthesis, namely the S-phase.

In total about 12 adenovirus non-structural (NS) proteins are expressed before the genome is replicated. Two of these, E1B and E4orf6 are required to counteract the negative effects on the cell of E1A. Abnormal stimulation of the cell cycle leads to the activation of the cellular protein p53, and this protein functions to block inappropriate cell cycles and induces apoptosis. E1A expression therefore induces p53. E1B and E4orf6, along with various cellular proteins, target p53 for polyubiquitination. This addition of ubiquitin to p53 marks it for degradation by the cellular proteosome thereby removing it from the cell and preventing cell cycle arrest and apoptosis activity.

The cell environment is now ready for virus DNA replication and synthesis of viral structural proteins. Virus DNA replication takes place in two stages. Firstly, sequences in the double-stranded ends of the genome in the inverted terminal repeats are recognized by the virus replication proteins including the E2B DNA polymerase (AdPol) and new single-stranded DNA genome copies are made. Because of the inverted terminal repeats these single-stranded DNAs can base pair the ends of the genome together to form 'pan handle'-like structures. This structure allows the second stage of virus genome replication, where the double-stranded inverted repeat is recognized by the virus replication proteins as before, and the single-stranded genome region is copied to make full length double-stranded virus genome.

Expression of adenovirus late genes occurs at the onset of virus DNA replication. At the same time the virus proteins E1B-55K

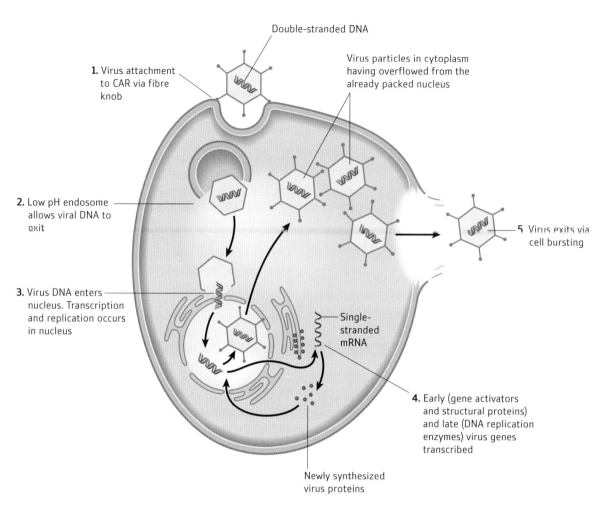

Fig. 25.4 Replication cycle of adenovirus.

and E4orf4 block the export of cellular mRNA from the nucleus, leading to the rapid accumulation of virus late RNAs in the cytoplasm and therefore the preferential translation of virus late mRNA into protein. New virus fibres and penton capsomers are assembled in the cytoplasm, and imported into the nucleus, where new virus particles are assembled, often in such vast numbers that they form crystalline aggregates. This process involves the cleavage of the virus proteins VI, VII, VIII, and terminal protein by the virus protease that is also part of the assembling new mature virion. The new virus particles are released from the cell aided by a number of virus proteins including the E3 11.6kDa 'adenovirus death protein'. Not unexpectedly, with this amount of viral synthesis and alteration of cell cycle control, the infected cell dies.

25.3 Clinical and pathological aspects

A very wide range of tissues can be infected by members of this family, including the eye, the lung, and the gastrointestinal system (Table 25.1). Some members cause cancer in laboratory animals but fortunately not in humans.

Table 25.1 Some clinically important human adenoviruses

Disease	Predominant serotypes
Infantile gastroenteritis	40, 41
Upper respiratory tract infections	1, 2, 3, 5, 7, 11
Lower respiratory tract infections	3, 4, 7, 21
Pharyngoconjunctival fever	3, 4, 7
Epidemic keratoconjunctivitis	8, 9, 37
Acute haemorrhagic conjunctivitis	11
Acute haemorrhagic cystitis	7, 11, 21, 35
Genital ulcers; urethritis	2, 19, 37
Gastroenteritis and pneumonia in immunocompromised patients	Many serotypes especially 40, 41

25.3.1 Clinical features

Endemic respiratory infections in children

Most children are infected with adenoviruses early in life, but probably fewer than half these infections result in disease, the frequency of symptomatic illness depending on the type of virus; adenovirus 2, for example, causes comparatively little illness. In young children, the symptoms include a stuffy nose and cough, whereas in older children pharyngitis is common. In some colder areas of the world, for example, China and Canada, adenoviruses type 3 and 7 can cause pneumonia in infants younger than 2 years (Fig. 25.5).

A good proportion of these infections result not from aerosols, but from faecal–oral spread on cups and utensils. This mode of spread is not uncommon with other respiratory viruses.

Adenoviruses types 4 and 7 are notorious for causing outbreaks of upper and lower respiratory tract infections in military recruits, probably because of mixing together of people from different geographical locations and crowding together as recruits. The illness usually lasts for about 10 days and pneumonitis is not uncommon.

Immunosuppressed patients

Severe infections, including pneumonias, encephalitis, and fulminant hepatitis, may occur in AIDS patients, and in immunosuppressed persons receiving organ transplants.

Pharyngoconjunctival fever

This syndrome is characterized by pharyngitis and conjunctivitis, mainly in children and young adults. Many outbreaks have been associated with swimming pools, and the possibility of an adenovirus infection should be considered in any patient with conjunctivitis developing about a week after using a communal pool.

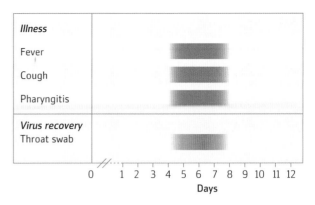

Fig. 25.5 Clinical and virological events during respiratory infection with adenovirus.

Epidemic keratoconjunctivitis ('shipyard eye')

Why 'shipyard eye'? This form of keratoconjunctivitis, which may leave permanent corneal scars and impaired sight, first occurred in epidemic form during the Second World War in American shipyards, which were, of course, very busy. Foreign bodies (e.g. flakes of rust) often had to be extracted from the workers' eyes and the infection was spread both within and without the clinics that they attended. In one shipyard, a particular employee acquired a high reputation for removing foreign bodies more skilfully than the local doctors; unfortunately, he never sterilized his home-made instruments and thus spread adenovirus type 8 to many hundreds of his workmates. Any criticism of this amateur operator should be tempered by the knowledge that, on a number of occasions, adenovirus infections of the eye have been transmitted by medical staff using infected solutions or inadequately sterilized instruments, notably tonometers.

Eye infections (conjunctivitis) in the community

Adenovirus induced conjunctivitis is certainly not restricted to shipyards as the following case study illustrates.

Medical case story: Conjunctivitis

Patients do not always fully engage with current best practice in health care. They may decline to attend for medication reviews, proper investigation, or even to listen to simple advice, much to the discomfort of their attending doctors.

You are just back from 2 weeks' leave and are not quite into the swing of things yet. Consequently, you are running 25 minutes late.

Mrs G flurries into your room carrying an 8-month-old baby. As she crosses the threshold you can see the baby's right eyelashes are covered with a yellow, crusted discharge.

Mrs G hurriedly explains that the baby's right eye became sticky yesterday. 'The nursery wants me to get Lily some eye drops before they will let her come back,' she says.

It just so happens that you have been doing some reading about conjunctivitis recently. You have been annoyed before about nursery insistence on antibiotics for conjunctivitis, and so you took the trouble to look up Public Health England (PHE) advice on the topic. The PHE advise that children do not have to be kept off nursery with conjunctivitis because it is not that infectious. A person with conjunctivitis is unlikely to infect another; R_0 is 0.25. R_0 is the basic reproduction number—that is, the average number of cases caused by one infected person. As a comparison R_0 is ~14 for measles and ~2–4 for influenza.

So here is an excellent opportunity to explain these facts and develop a new conjunctivitis 'script'. You know you are running late, but you are willing to spend a bit longer with this mum to explain fully.

You first examine the baby's eyes. The conjunctivas are both quite inflamed. There is discharge on the lashes more on the right than the left. You look for signs of periorbital cellulitis, but there is no redness around the eyes or on the eyelids. You check the baby is essentially well by asking about temperature and appetite. The mum agrees the baby is fine and in good spirits, eating and drinking normally.

The mother is twitching in her seat, obviously in a hurry. 'Do you know how to wash the eyes?' you ask.

'Yes, yes,' says the mum. 'I boil up water and pour it on cotton wool so it's all sterile, and then I take a piece of cotton wool and wipe it across her eyes once and discard it. I've got some with me in a sterile Tupperware container.'

'And wash your hands before and after', you add. She nods in agreement.

'It's not possible to tell simply by looking at the eye whether it is a viral or a bacterial infection', you begin ponderously. 'So it's always difficult to know if antibiotic drops will help or not'.

Here she interrupts. 'I'm awfully sorry; my parking meter is running out. I would really like the eye drops. I've found them very good: excellent, in fact. They always work quickly.'

And as you hand over the prescription, she puts the baby under her arm and says 'Thanks so much', makes for the door, and leaves.

You sit dumbly for a moment before you recover with a wry smile to laugh at the reversal of the usual pattern. Normally, it is you levering people out of the door. You then reflect on what else you wanted to say.

You wanted to convince her to have a trial period during which she simply cleaned the eye. You could have given her a 'delayed prescription'—that is, a prescription that she should not use unless the eye infection was not improving with cleaning after 3 or 4 days, or was getting worse. Research on delayed prescriptions has found they are an effective way of cutting down antibiotic use. Patients are usually very satisfied with this compromise and most manage to avoid using the prescription.

The dosage of chloramphenicol eye drops is four times a day. Fortunately, this is written on the script as you did not have time to tell the mother. If the mother embarks on treatment she should continue treating until the infection has been better for 48 h. Usually, it is needed for about a week in total.

In fact, chloramphenicol eye drops are available over the counter at pharmacies for children who are 12 months or older. It would have been helpful for the mother to know this for future occasions.

You have found that treatment failure is often an issue in conjunctivitis. Sometimes the infection is viral and so the antibiotics don't work. Sometimes the parent uses the treatment intermittently, or the eye drops don't even reach the eye (for example, if the child is not co-operative, which tends to be the rule). For this reason, two people are usually needed to administer eye drops in a child.

You wanted to advise her to continue taking the child to the nursery, and to get her to refer the nursery manager to the PHE website so that they might change their protocols for conjunctivitis, and stop pressuring for inappropriate antibiotics.

Most importantly, you failed to safety-net this child, by asking the mother to return if the child's eyes got worse or were not improving, or if the skin around the eyes became red. This is your responsibility; you feel your medical protection society would take a dim view of you missing it out.

You didn't get time to say any of these things to Lily's mother. Next time you will just have to resist the pressure to hurry, so that you can offer advice in a safe and timely manner.

★ **Learning Points**

- **Virological:** Conjunctivitis is a common minor presentation. If linked with an upper respiratory tract infection it makes it more likely to be virological. The commonest causes are adenovirus and *S. aureus*. Other possible causes are:

Viruses	Bacteria
Adenovirus	*Staphylococcus aureus*
Herpes simplex (causes keratitis)	*H. influenzae* (vaccinated against)
Varicella zoster (ophthalmic shingles)	*S. pneumoniae* (vaccinated against)
Enteroviruses (tend to be tropical)	*Chlamydia trachomatis* (newborn)

Viruses	Bacteria
Measles (as part of the illness. Vaccinated against.)	*N. gonorrhoeae* (newborn)
Other; acanthamoeba (tropical parasite)	*Ps. aeruginosa* (after trauma)

- **Clinical:** If the conjunctivitis has been present since birth then it is essential to take an eye swab to exclude Chlamydia.

- **Personal:** You will find that you gradually develop a 'script' or package of information for common complaints, which will help you inform patients consistently and efficiently.

Gastrointestinal infections

Symptomatic adenovirus infections of the gut occur mainly in infants and are caused by types 40 and 41.

Other syndromes

Adenoviruses have been implicated in acute intussusception in infants, necrotizing enterocolitis, haemorrhagic cystitis, and meningoencephalitis. They may also cause life-threatening pneumonia and other infections in immunocompromised patients, including those with AIDS.

25.3.2 Pathogenesis

Infections in humans are rarely lethal and the pathogenesis of adenovirus infections is difficult to study. However, infected cotton rats develop a similar pulmonary syndrome, characterized by replication in the bronchiolar epithelium and infiltration with lymphocytes. In this model, disease appears to be due more to the immune response than to direct effects on the pulmonary tissues; one should, however, be cautious in extrapolating these results to the disease in humans.

25.3.3 Epidemiology

All 57 serotypes are endemic in the community and some may cause explosive outbreaks of disease, usually respiratory, but also involving the eye. Another feature of the epidemiology of adenoviruses is the degree of seasonal variation. As an example, most outbreaks of pharyngoconjunctival fever in school-age children occur in the summer, perhaps because they use swimming pools more often. Eye infections may also be acquired following infection of the respiratory or alimentary tracts and, as we have seen, by inadequately sterilized instruments or other fomites. Epidemics of respiratory disease in military recruits occur almost exclusively in the winter. In general, of the 57 adenovirus serotypes, infections with types 2, 3, 5, and 7 are most common throughout the world. Type 1 and 2 infections occur in early childhood, whereas types 3 and 5 predominate late in life. The virus is known to infect persons by aerosol and by direct contact, but is probably also spread by the faecal–oral route, especially where hygiene is poor.

The vigorous humoral and cell-mediated **immune responses** account for the generally mild diseases caused by adenoviruses, except of course in immunocompromised patients, in whom they may be lethal.

25.3.4 Laboratory diagnosis

Rapid molecular testing for virus DNA is the method of choice. In contrast viral isolation from faeces, pharyngeal swabs, conjunctival swabs, or urine is slow, taking at least a week and perhaps as long as a month with some serotypes.

25.3.5 Prophylaxis

Because of the disruptive effects of large outbreaks of respiratory disease in army camps, the US military as long ago as the 1960s encouraged the development of a vaccine. Live preparations of adenovirus types 4 and 7 were enclosed in a gelatine capsule and swallowed, bypassing the stomach and being released in the intestine. Here, the virus replicates and induces immunity, but causes no overt disease. However, the problems for widespread use of such a vaccine in the community are formidable, not least of which being the variety of serotypes causing respiratory disease. Another worrying thought is that certain adenoviruses have oncogenic effects in animals, although there is no evidence of these in humans.

Molecular biologists have now taken a special interest in these adenovirus vaccine viruses as potential viral vectors and are beginning to use them to carry other genes into the body, thus inducing immunity not only to adenoviruses but also to influenza, HIV, and Ebola.

 Reminders

- Some of the viruses have an icosahedral structure embellished by a projecting fibre at each of the 12 vertices.
- Adenoviruses have a linear dsDNA genome some 36–38kbp in size.
- The genome has a short inverted terminal repeat sequence whilst the 5′ ends have a virus coded terminal protein which primes DNA synthesis.
- Genes are arranged into transcription units. There are six early transcription units, two intermediate units, and one late unit.

- The 57 human viruses are grouped as species A–G.
- Alternative RNA splicing results in 50 or so virus proteins.
- 'Shipyard eye' is an epidemic keratoconjunctivitis caused mainly by serotype 8; many of the remaining 51 serotypes cause respiratory symptoms.
- Some adenoviruses cause outbreaks or small epidemics of respiratory infections in the winter, and others, gastrointestinal illness.
- Adenoviruses are used as vectors for vaccines for carrying other virus proteins (e.g. Ebola, influenza).

Further reading

Branton, P. and Marcellus, R.C. (2011). Adenoviruses. In Acheson, N.,
(ed.), *Fundamentals of Molecular Virology*, p 274–84. John Wiley &
Sons, New York.

Questions

1. Briefly outline the clinical range of infection caused by
adenoviruses. Can members of this family be used for
recombinant vaccines in humans?

Group 5 Single-stranded DNA viruses

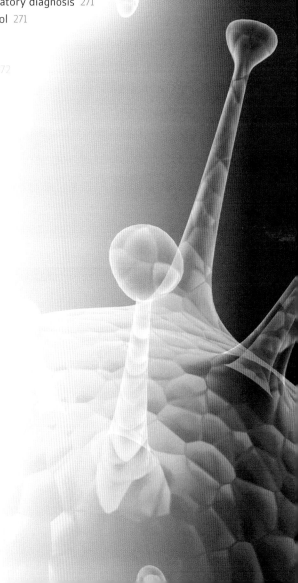

Parvoviruses

26

26.1 Introduction

It might be thought that viruses as a class represent the ultimate in parasitism, reliant as they are on their host cells to provide most of the machinery for replication. The parvoviruses (Latin *parvus*, meaning small), however, show a still further degree of dependence, as they can replicate only in the presence of active DNA synthesis in rapidly dividing host cells or under the influence of another virus infection, that is they depend on a helper virus.

Their genome of 5kb is small but larger than hepatitis B virus (3kb) and the same size as polyomaviruses. However, the DNA is single-stranded (Fig. 26.1), allowing compact packaging and hence the 'small' virion. The human viruses were not discovered until 1966 (Fig. 26.2).

These viruses have a complex relationship with the host whilst at the same time there is only restricted ability to cultivate them in the laboratory and there are no animal models. Their wide circulation and self-limiting clinical course have led to a diminished appreciation of their clinical importance.

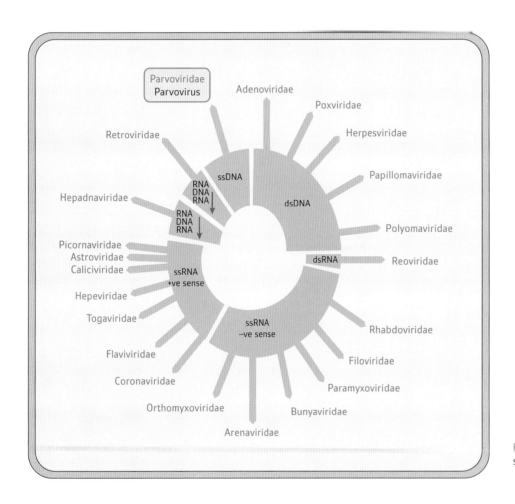

Fig. 26.1 Baltimore classification scheme and *Parvoviridae*.

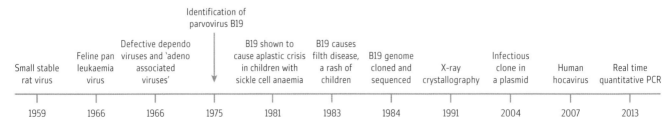

Fig. 26.2 Timeline for *Parvoviridae*.

26.2 Properties of the viruses

26.2.1 Classification

The taxonomic classification of the family *Parvoviridae* was updated in 2014. The family is divided into two subfamilies: *Parvovirinae*, which infect vertebrates, and *Densovirinae*, which infect invertebrates. The subfamily *Parvovirinae* has 8 genera whereas the subfamily *Densovirinae* has 5 genera. Very few members of the *Densovirinae* have been studied and sequenced but they infect mostly insects and shrimps, whereas the *Parvovirinae* genera infect humans, birds, and other animals. To date, viruses that infect humans belong to 5 *Parvovirinae* genera: *Bocaparvovirus* (human bocavirus 1–4), *Dependoparvovirus* (adeno-associated virus 1–5), *Erythroparvovirus* (parvovirus B19), *Protoparvovirus* (Bufavirus 1–2) and *Tetraparvovirus* (human parvovirus 4 G1–3) (see Table 26.1).

26.2.2 Morphology

Parvoviruses are small, non-enveloped viruses with icosahedral symmetry, whose simple structure has been unveiled by X-ray crystallography. They are 18–26 nm in diameter (Fig. 26.3). The capsid has 60 copies of the capsid proteins. Each subunit has the same β-barrel shape of other icosahedral viruses. The single-stranded DNA is tightly compacted within the capsid structure.

26.2.3 Genome

The *Parvoviridae* are the only viruses with linear single-stranded DNA genomes of approximately 5 kb in size. All parvovirus genomes have a similar structure where the ends of the genome have terminal repeats that allow the formation of base-paired hairpin structures at each genome end (Fig. 26.4). This length of the self-complementary sequence ranges from 115 to 300 nucleotides. The 3′ terminal hairpins act as

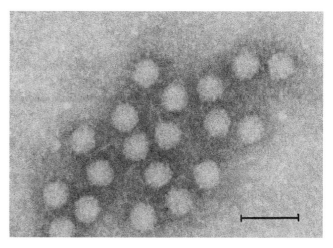

Fig. 26.3 Electron micrograph of parvovirus in serum. The virions are clumped together by antibody molecules. (Courtesy of Dr Ian Christie.) Scale bar = 50 nm.

primers to initiate replication of virus DNA. The virus non-structural proteins are located to the left half of the genome and the structural proteins to the right. However viruses of the family use very different strategies to control gene expression. Adeno-associated viruses use both P4 and P19 promoters within the left-hand reading frame to generate mRNAs. B19 uses a single promoter for transcription of the entire genome. Differential splicing, in all cases, produces virus-specific mRNAs. The parvoviruses were initially thought to be double-stranded until it was discovered that each virion contains either a positive- or a negative-sense genome strand, a unique situation.

26.2.4 Replication

The limited coding capacity of the virus genome ensures that these viruses are very dependent on host-cell functions. They can only productively infect cells that are going through their cell cycles, unlike polyomaviruses and papilloma viruses which are able to induce cells to enter DNA synthesis (S) phase.

A number of different cell-surface proteins can modulate parvovirus binding and entry. The cellular receptor for parvovirus B19 is globoside (erythrocyte P antigen receptor). Integrin αVB5 are co-receptors for infection with adenovirus associated virus type 2.

The virus is endocytosed and the capsids migrate to the vicinity of the nucleus where the capsid is believed to enter the nucleus through the nuclear pore complex. Replication of virus DNA and transcription of mRNA take place in the nucleus but only when the cell reaches S phase or is induced by a helper adenovirus. The virus genome is transcribed into a number of different alternatively spliced mRNAs depending on the type of parvovirus. For example, B19 produces nine different mRNAs (R1–R9) that encode for the proteins NS1 and NS2, the structural VP1 and VP2, and the additional proteins 7.5 kDa and 11 kDa. Parvoviruses do not encode for a DNA polymerase and,

Table 26.1 Parvoviruses infecting humans

Genus	Viruses	Diseases
Erythroparvovirus	B19	Erythema infectiosum (fifth disease). Foetal infections. Aplastic crisis
*Bocaparvovirus**		Recently discovered respiratory and enteric virus
Dependoparvovirus	Adeno-associated viruses (AAV) types 1–5	AAV-2 possibly implicated in foetal infections

*Classification in this genus is tentative.

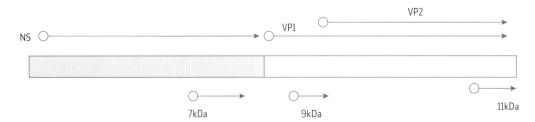

Fig. 26.4 Genome structure of B19 and functional mapping.

therefore, the ability to replicate the DNA genome is completely dependent on the host cell DNA polymerases. The parvoviruses only replicate in rapidly dividing host cells, in which S-phase functions provide the necessary help. The parvovirus DNA is not capable of independent DNA replication in the nucleus as the virus protein NS1 is required to 'nick' or cut part of the newly synthesized genome DNA to allow synthesis of a new genome strand to complete. The parvoviruses that can replicate without helper viruses are termed autonomous parvoviruses and, as noted above, are mainly animal viruses.

The dependoparvoviruses used to be called 'adeno-associated viruses' as they were seen by EM in association with enteric adenoviruses (Chapter 25). These viruses, although having a genome structure the same as autonomous parvovirus (terminal repeat hairpins), actually encode for two different proteins called Rep and Cap. The Rep protein has multiple functions, one of which is to nick virus-replicating genome in the same way as NS1. However, these viruses depend on either the induction of cellular DNA polymerases by a helper virus such as Adenovirus, or the delivery of another virus DNA polymerase as is the case with a Herpes simplex helper virus. The dependoparvovirus genome produces three transcripts that also undergo differential splicing, encoding for different forms of Rep (Rep78, 68, 52, and 40) and VP1, 2, and 3. Rep 78/68 are the main virus regulatory proteins and, for example, activate transcription, prevent cells from moving through the complete cell cycle, and have sequence-specific endonucleases to activate Adenovirus. Rep 52 is thought to be involved in virus encapsidation.

Dependoparvoviruses can establish latent infection of cells, a property normally associated with herpesviruses. In latency the virus genome is maintained in the cell without extensive virus gene expression. Adeno-associated virus (AAV) achieves this by integrating its virus genome into the host-cell DNA at a specific site on chromosome 19. The integrated genome becomes active only when the cell becomes super-infected with a helper adenovirus that can provide the enzymes essential for replication. AAVs, however, have not been shown to cause disease, although they are widespread.

All parvovirus capsid proteins are translated in the cytoplasm and then transported to the nucleus where new virus particles are assembled, and then either retained in the nucleus or transported out of the cell.

Perhaps not unexpectedly for a family of viruses with a predilection for fast-dividing cells, virologists have enquired whether cancer cells could be targeted with parvoviruses as a therapy. Certainly parvovirus infection has been shown to have an amelioratory effect on some animal cancers. Cancer cells by definition go through the S phase more often than 'normal' cells; moreover they are more sensitive to the toxic effects of NS1 and more NS1 is produced in cancer cells. This appears at first sight to be a 'perfect storm' situation. There is even another factor. The adenovirus-associated DNA appears to the infected cell as 'damaged' DNA and initiates a damage response involving P53, which normally arrests the cell cycle. But cancer cells do not possess P53 and so continue in their cell cycle leading to cell death in mitosis.

26.3 Clinical and pathological aspects

Members of the genus *Erythroparvovirus* such as B19 cause the most problems to us (Fig. 26.5). The virus was discovered by accident during screening of blood donations and can cause serious illness. The disease spectrum (foetal abnormalities and aplastic anaemia) results from the predisposition of the virus for infection and damage to dividing cells.

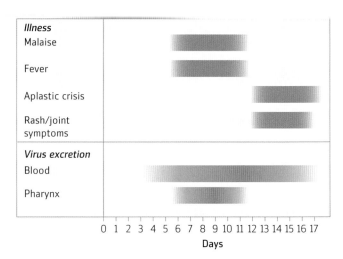

Fig. 26.5 Clinical and virological events during infection with parvovirus B19.

26.3.1 Clinical features of B19 infection

The parvoviruses such as B19 infecting humans cause three main syndromes.

Aplastic crisis in anaemia patients

The need of parvoviruses for actively dividing cells in which to replicate is illustrated by their lytic effect on red cell precursors: this is not clinically obvious in otherwise healthy people, but in those with a chronic haemolytic anaemia, e.g. sickle cell anaemia or thalassaemia, parvovirus infection may precipitate an **aplastic crisis** with very low haemoglobin and disappearance of circulating reticulocytes. This is because the virus replicates in and damages the rapidly dividing late erythroblasts which are erythroid progenitor cells in the bone marrow. Such events may be fatal if not treated rapidly by blood transfusion.

Erythema infectiosum

This 'infectious rash' is sometimes known as 'fifth disease', which is less of a mouthful. (The original classification of the exanthematous illnesses of childhood comprised six diseases: the other five are measles, rubella, scarlet fever, exanthem subitum, and Duke's disease, a rash of obscure aetiology.) A more picturesque name is 'slapped cheek syndrome', which helps you to remember that, especially in children, the presenting feature is often **erythema over the malar areas**, followed within the next 4 days by a maculopapular rash on the trunk and limbs, which may persist for 2 or 3 weeks. Volunteer and field studies show the incubation period for the rash to be 13–18 days. There may be some fever and malaise in the early stages, and mild febrile illness without rash is common. Virus is not detectable at this time and the current opinion is that of an immune-mediated pathology.

This is predominantly an infection of children, but adults may also acquire it; in them, and **particularly in women**, the **joints** are much more likely to be involved. Those of the hand and fingers are most often affected, but there may also be arthropathy of the arms, legs, and spine. Arthralgia may persist for a few weeks.

Differential diagnosis

Erythema infectiosum may be diagnosed by the characteristic rash, especially during an epidemic, but differential diagnosis from other illnesses with fever and rashes is not always easy. The syndrome of fever, rash, and arthropathy may be clinically indistinguishable from rubella, but it is essential to establish the correct diagnosis with absolute certainty in pregnancy; fortunately, laboratory tests can give the answer.

Meningitis and encephalitis following B19 infections have been described, but are very rare.

Foetal infection

B19 infection in pregnancy may result in foetal death or birth defects if the foetus is infected during pregnancy.

Persistent infection

Persistent infection with B19, associated with chronic anaemia, has been observed in a variety of conditions associated with immunodeficiency, including acute lymphocytic leukaemia, HIV infection, and therapeutic immunosuppression. Persistence may be aided by the continual appearance of new mutated strains.

26.3.2 Epidemiology

Infections with B19 occur worldwide, and tend to occur in cycles with peaks about every 5 years. There are three genotypes with distinct evolutionary lineages. Epidemics occur during late winter and early spring, and mainly affect young schoolchildren. Spread is by the **respiratory route**, although occasional transmissions by blood transfusion have been reported. These viruses can also be transmitted through the placenta. In schools attack rates can reach 60% and the virus spreads easily in families.

Infections can be acquired from blood where it can be present in quantities as high as $10^{10}/\mu l$ and blood products because both the virus is heat stable and its small size means it can pass filters used to purify these products.

26.3.3 Pathogenesis and pathology

After primary infection of the upper respiratory tract there is an intense viraemia lasting about a week. The dominant feature of infection is the high lytic activity of the virus in rapidly dividing cells, which accounts for its ravages both in foetal cells and in the adult haemopoietic system.

26.3.4 Immune response

Clearance of virus from the blood coincides with a sharp IgM antibody response, followed shortly by the appearance of IgG antibody. The rash and arthropathy in erythema infectiosum are probably mediated by immunological factors, including the deposition of immune complexes.

26.3.5 Laboratory diagnosis

These viruses cannot readily be grown in the laboratory, but can be found in the blood by **EM** during the viraemic stage. Virus DNA can also be identified by various tests, including the PCR and dot-blot hybridization. **Detection of IgM antibody by ELISA** indicates a current or recent infection, whereas the much more prolonged presence of IgG is a sign of past infection.

26.3.6 Control

There is at present no specific vaccine, although experimental vaccines in the form of VP2 and VP2 expressed as virus-like particles have been developed for parvovirus infections. Prophylaxis with antibody is directed mainly at prevention of spread to highly susceptible people, e.g. those with severe anaemias, and to laboratory staff who may be exposed to high concentrations of virus. Immunoglobulin can be used to block persistent infection with B19 virus.

Human bocavirus

Human bocavirus (genotype 1) has been found in children with upper respiratory tract illness all around the world and is an important pathogen. Bocavirus genotype 2–4 causes diarrhoea in children. The virus can be detected in samples of blood, faeces, and respiratory secretions, often in association with other respiratory viruses. Predominant symptoms are cough, fever, hypoxia, and wheezing.

Reminders

- The *Parvoviridae* are the smallest viruses and the only ones containing ssDNA. Virions contain either positive- or negative-sense strands. They need help from other viruses which can induce S phase or from independently rapidly dividing (S phase) host cells in order to replicate.

- The parvovirus causing the most significant disease in humans is B19, the sole member of the genus *Erythroparvovirus*. It causes:
  - **Erythema infectiosum**, a febrile illness with rash spread by the respiratory route, mainly in young children. Cure is spontaneous. In adults, especially women, there may be arthropathy;
  - **Foetal infections**;
  - **Aplastic crises** in people with pre-existing chronic haemolytic anaemia, e.g. thalassaemia or sickle cell disease, or who have immune deficiencies.

- Human bocavirus genotype 1 causes significant respiratory symptoms and genotypes 2–4 cause gastrointestinal disease in children worldwide.

- The differential diagnosis of erythema infectiosum may be difficult clinically, but can be made by appropriate laboratory tests. In pregnancy, it is particularly important to distinguish it from rubella which in the first trimester can result in nearly 100% foetal abnormalities.

Further reading

Gallinella, G. (2013). Parvovirus B19: achievements and challenges. *Virology*. http://dx.doi.org/10.5402/2013/898730.

Hao, R. *et al.* (2013). Correlation between nucleotide mutation and viral loads of human bocavirus 1 in hospitalized children with respiratory tract infection. *Journal of General Virology* **94**, 1079–85.

Manning, A., Willey, S.J., Bell, J.E., and Simmonds, P. (2007). Comparison of tissue distribution, persistence, and molecular epidemiology of parvovirus B19 and a novel human parvovirus bocavirus. *J Infect Dis* **195**, 1345–52.

Questions

1. Outline the pathology of viruses of this family and identify any correlation between virus characteristics and disease.

2. How important are bocaviruses?

Group 6 Single-stranded positive sense RNA with an RT

Retroviruses: HIV-1 and -2 and HTLV

27.1 Introduction

The first discovery of a retrovirus was made as long ago as 1910 by Peyton Rous (Fig. 27.1), working at the Rockefeller Institute for Medical Research in New York. This agent, avian sarcoma virus, induced tumours in muscle, bone, and other tissues of chickens. Rous received the Nobel Prize for this discovery, but it was not until the 1930s that other retroviruses, causing tumours in mice (mouse mammary tumour virus and murine leukaemia virus), were discovered. However, these viruses were regarded as laboratory curiosities until the description of feline leukaemia virus, a gammaretrovirus, first discovered in the 1960s, which appeared to spread naturally in household cats. Perhaps as a portent for the later discovery of HIV-1 in humans, Visna Virus of sheep, discovered in 1954, was shown to persistently infect sheep and could lead to ovine progressive pneumonia.

Together with feline immunodeficiency virus (FIV), discovered in 1986, and HIV-1, discovered in 1983, it became clear that a group of retroviruses called lentiviruses could infect animals and cause an immune deficiency in these infected animals.

Retroviruses possess a defining enzyme, reverse transcriptase (RT), which uses the viral RNA genome as a template for making a DNA genome copy, which then integrates into the chromosome of the host cell and there serves either as a basis for viral replication or as an oncogene. Howard Temin and David Baltimore both received Nobel Prizes for their spectacular discovery of this RT enzyme, which overturned a central dogma of molecular biology—that genetic information flows in one direction only, from DNA→RNA→protein. However, it should be noted that, subsequently, hepadnaviruses (Fig. 27.2) like hepatitis B virus were also shown to have a RT and are able to reverse transcribe their genome.

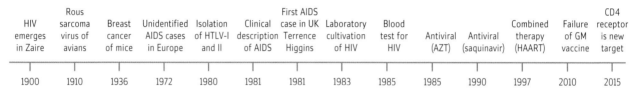

HIV emerges in Zaire	Rous sarcoma virus of avians	Breast cancer of mice	Unidentified AIDS cases in Europe	Isolation of HTLV-I and II	Clinical description of AIDS	First AIDS case in UK Terrence Higgins	Laboratory cultivation of HIV	Blood test for HIV	Antiviral (AZT)	Antiviral (saquinavir)	Combined therapy (HAART)	Failure of GM vaccine	CD4 receptor is new target
1900	1910	1936	1972	1980	1981	1981	1983	1985	1985	1990	1997	2010	2015

Fig. 27.1 Timeline for retroviruses.

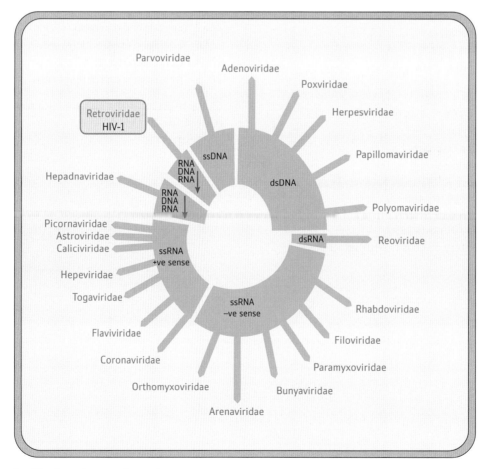

Fig. 27.2 Baltimore classification scheme and retroviruses.

The discovery of HIV-1, the causative agent of AIDS, only 3 years after the first human retrovirus (HTLV-1) from leukaemia patients resulted in a huge surge of scientific and medical interest in this previously rather obscure family of viruses.

27.1.1 The discovery of HIV-1 as the cause of AIDS

In 1981 a new clinical syndrome, characterized by profound immunodeficiency, was recorded in male homosexuals and was termed AIDS (Acquired Immune Deficiency Syndrome). An unusual prevalence of **Pneumocystis carinii pneumonia** was noted in a group of young, previously healthy, male homosexuals. Before then, this parasite pneumocystis had been associated with disease only in patients whose immune systems had been seriously impaired as a result of drug therapy or by congenital cellular immune deficiency. At the same time came reports of previously healthy young homosexuals in New York and San Francisco who had developed a rare cancer, **Kaposi's sarcoma**. The first isolation of a retrovirus (**HIV-1**) from an AIDS case was made by Luc Montagnier and Françoise Barré-Sinoussi at the Pasteur Institute in Paris early in 1983 and quickly confirmed by Robert Gallo in the USA. The first case of AIDS in the UK was diagnosed in late 1981 in a homosexual from Bournemouth, after he returned home from Miami. An AIDS research trust was established in his name, Terrence Higgins. Now 5000 new cases occur daily in the world and more than 35 million people have been infected.

27.2 Properties of HIV

27.2.1 Classification

The family *Retroviridae* is so named for its possession of a RT enzyme (Latin *retro* = backwards) (Table 27.1). Of the seven genera now recognized—alpharetroviruses, betaretroviruses, gammaretroviruses, deltaretroviruses, epsilonretroviruses,

lentiviruses, and spumaviruses—only two cause disease in humans:

- *Lentivirus*, containing HIV-1 and 2 (Latin *lentus* = slow). The lentiviruses are distinguished by the presence of a vase-or cone-shaped nucleocapsid, absence of direct oncogenicity, and the lengthy and insidious onset of clinical signs.
- *Deltaretrovirus*, which contain HTLV-I and -II. They are distinguished by their characteristic genomes and their ability to cause tumours, rather than immunosuppression.

The spumaviruses (Latin *spuma* = foam) cause a characteristic foamy appearance in infected primate cell cultures. As far as we know, they are not pathogenic.

27.2.2 Morphology

The typical HIV particle is 100–150 nm in diameter and is spherical (Fig. 27.3). The virus has an outer envelope of lipid penetrated by 72 glycoprotein spikes (Fig. 27.4), the envelope (env) protein. The env protein is a trimer and each monomer is composed of two subunits, the outer glycoprotein knob (gp120) and a transmembrane portion (gp41), which joins the knob to the virus lipid envelope. The receptor binding site for CD4 is present on gp120, as well as very important antigens such as the V3 loop.

There are abundant cellular proteins, notably MHC class I and class II molecules in the lipid envelope. The lipid envelope encloses an icosahedral shell of matrix protein (p17), within which is a vase- or cone-shaped protein core (p24) and virion protein R (VPR) and contains two molecules of single-stranded positive-sense RNA in the form of a ribonucleoprotein. Bound to the genome are several copies of the RT, integrase (IN), and protease (PR) enzymes. Retroviruses, through the packing of two genomes are the only known diploid viruses.

Table 27.1 Classification of primate retroviruses

Genus	Virus	Disease caused	Natural hosts
Deltaretrovirus	Human T-cell leukaemia virus (HTLV-1)	Adult T-cell leukaemia/lymphoma; tropical spastic paraparesis	Humans
	Human T-cell leukaemia virus (HTLV-2)	Hairy cell leukaemia (very rare)	Humans
Lentivirus	Human immunodeficiency virus (HIV-1)	Immune deficiency, encephalopathy. Virus can infect chimpanzees, but causes no clinical signs	Humans and chimpanzees
	Human immunodeficiency virus (HIV-2)	Immune deficiency. Less pathogenic than HIV-1	Humans and monkeys
	Simian immunodeficiency virus (SIV-1)	Immune deficiency. No disease in wild African green monkeys, but AIDS in rhesus monkeys	Monkeys
Spumavirus	Human spumavirus	Inapparent persistent infections	Primates and other animals

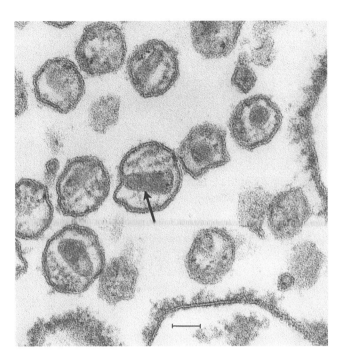

Fig. 27.3 Electron micrograph of HIV-1, vase-shaped core of the virion is arrowed. (Courtesy of Dr David Hockley.) Scale bar = 50 nm.

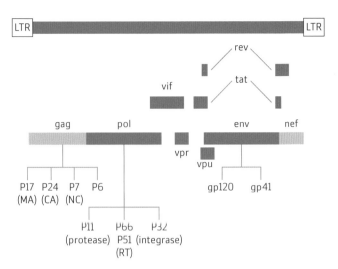

Fig. 27.5 Genome structure and encoded proteins of HIV-1.

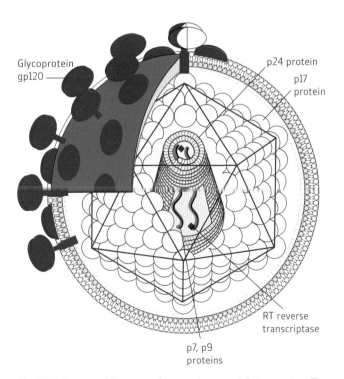

Fig. 27.4 Structural features of human immunodeficiency virus. The glycoprotein (gp120) spikes protrude though the lipid membrane. An icosahedral shell (p17) underlies the membrane and itself encloses a vase-shaped structure. The diploid RNA is enclosed in the 'vase'.

27.2.3 Genome

The organization of the positive-sense single-stranded RNA genome of HIV-1, approximately 9 kb in size, is complex (Fig. 27.5). It has no oncogene, but has some unique features, particularly the possession of control genes which can enhance viral replication, such as Rev (regulator of expression of virus proteins), Vpr (viral protein R), Tat (transactivator of transcription), Vpu (virus protein unique), and Vif (viral infectivity factor), and a repressor gene, Nef (negative effector). The genome therefore is more complex than other retroviruses. The genome is flanked at each end by Long Terminal Repeats (LTRs). The 3′ LTR has the polyadenylation signal while the 5′ LTR has the enhancer promoter sequences for viral transcription.

The HIV-1 primary transcript produces multiple mRNAs of three size classes by alternative splicing . The largest is unspliced 9 kb full length RNA (Gag and Gag-pol proteins). The polymerase protein (Pol) coded by this RNA has three constituents, the largest being RT; another is an integrase with the important function of integrating the HIV proviral genome into cellular DNA; and the third is a viral protease, which has an important cleaving function, after release of the virus from the cell. All these enzymes are the targets of novel antivirals.

The second mRNA class is the singly spliced 4 kb mRNAs which encode Vif, Vpr, Vpu, and Env whilst the smallest class 2kb mRNAs are doubly spliced and code for Tat, Rev, and Nef.

27.2.4 Replication

HIV-1 binds specifically to the CD4 receptor, which is expressed on the surface of certain T lymphocytes (the T-helper cells) and also macrophages. In contrast, in sexual transmission, moving across mucosa and aided by local abrasions and inflammation the virus interacts with dendritic cells where HIV-1 gp120 binds to a cell-surface protein DC-SIGN. Dendritic cells also move extensively and will present bound virus to CD4+ T cells.

In common with most viruses, the complete replication cycle takes only 24 h although the replication in macrophages is much slower. Virus entry is achieved by 30 minutes after its introduction, RT activity by five hours, and integration by 13 hours. The latter is an important 'extra' for the virus: the ability to integrate virus DNA into the chromosome of memory T cells and other cells. The intricacies of cell biology such as 'cell-to-cell' contact and cell movement are important factors. T cells, particularly, travel great distances daily, and dendritic cells, for example, contact thousands of T cells hourly.

As well as the CD4 receptor HIV also requires co-receptors to allow cell entry and these belong to the chemokine receptor family. T-cell tropic HIV-1 viruses (also called X4 viruses) use the chemokine receptor CXCR4 as a co-receptor, whereas macrophagetropic (M tropic or R5 viruses) primary isolates use CCR5. Individuals with defective CCR5 alleles exhibit some resistance to HIV-1 infection, which also suggests that CCR5 has an important role in HIV-1 replication. In addition both CXCR4 and CCR5 receptors help promote efficient infection of the CNS by HIV-1. Both forms of HIV may be present in the initial infecting droplet, although if the method of infection is sexual then the R5 strain is dominant. The virus genetic variation occurs on the SU protein in the variable domain 3.

After attachment, the virus penetrates the cell by 'fusion from without' (Chapter 3). Interaction with the CD4 receptor and gp120 causes conformation changes in the Env protein and exposure of the fusion peptides of gp41, and also allows binding to the secondary receptors. The nucleocapsids, now in the cell cytoplasm (Fig. 27.6), associate with cellular microfilaments as a scaffold. This is termed the 'pre-integration complex', which then moves into the nucleus via the nuclear pores. When T cells have been activated the virus matrix (P17) protein encodes a nuclear localization signal which interacts with a host cell family of protein importins. Large numbers of other host cell factors contribute to entry into the nucleus. The viral RT enzyme directs the synthesis of a cDNA strand (minus strand) using the host cell-derived transfer RNA (tRNA) as a primer and the viral RNA as template. Then viral RNAase enzymatically removes the original viral RNA, while the RT now synthesizes a second DNA strand (the plus strand). Note that these few sentences could be expanded into a chapter of its own on the complexity of this process! Integration of double-stranded viral DNA into the host chromosome then occurs to form a proviral DNA. The proviral DNA is integrated at random into the host cell genome using the virus integrase enzyme. In essence the enzyme binds to each end of the linear virus DNA and brings them together alongside the cellular DNA by a well-organized cleavage and ligation. Several host cell enzymes are involved. Once the integration is complete the proviral DNA behaves like a portion of the host

cell chromosome and is replicated via the normal mechanisms of host cell DNA replication as the cell divides. The proviral DNA may reside quietly in the chromosome for years or may be activated immediately. Such latently infected cells are sleeping 'fifth columnists' and are missed by the immune system. Moreover, memory CD4 T cells are long lived. Around one thousand million virions are made and destroyed daily in a patient and only a minute fraction is integrated. Still, this warns us that if a real cure for HIV is to be achieved then the quiescent proviral DNA has to be dealt with. No more than 1% of the CD4 T cells have the integrated provirus and approximately 1% of these cells replicate the virus.

Post-integration, both viral and cellular factors are needed to activate HIV transcription, with particularly important elements sitting in the HIV LTR. External stimuli, such as activation of T cells via their T-cell receptors or virus infection and the presence of cytokines, result in stimulation and transport of cellular transcription factors such as NF-kB, AP-1, Spl, and NFAT, which transactivate the virus 5′ LTR. The primary RNA transcript is then spliced to give different viral mRNAs.

Early viral mRNA transcripts encode Tat, Rev, and Nef. Tat can dramatically increase the quantity of viral RNA and is known to increase viral transcription by recruiting cellular factors and promoting efficient and genome-length transcription. Rev is important to aid export of viral RNAs to the cytoplasmic ribosomes, where the so-called accessory proteins Vif, Vpr, and Vpu are synthesized as enzymes. Without Rev the virus mRNAs remain in the nucleus. The Rev protein shuttles between the nucleus and the cytoplasm. Even when HIV mRNAs reach the cytoplasm Rev can still exert control by stimulating their translation. The Vif protein is located in the cytoplasm and because it physically interacts with virus genome RNA it is also present in the HIV virion. Deletion mutants of HIV without Vif show reduced infectivity in cell cultures. APOBEC3G is a cellular defence mechanism and Vif binds to this protein and causes its degradation, thus neutralizing its effect. The accessary 100 amino acid protein Vpr binds to gag and so is also found in virions as well as in the cell cytoplasm. It has an important role in the movement of the pre-integration complex into the nucleus in non-dividing cells and also stops infected cells at the G2 stage of their cycle, which may help virus replication. The Vpu 81 amino acid protein binds to cellular CD4 and triggers a degradation. CD4 would otherwise bind to virus gp160 and so impede its exit from the cell and incorporation into new virions. It also stimulates release of virions from the cell surface, much akin to the NA of influenza. Finally, Nef is a virulence factor working in a tripartite manner. Firstly, Nef decreases MHC-1 on the cell surface by blocking the normal movement of MHC-1 from the Golgi to the plasma membrane. The normal function of MHC-1 in presentation of virus peptides is thus reduced and the immune system becomes less aware of viral invasion. Secondly, the Nef mutants which have been studied are less able to infect cells. Finally, Nef can activate T cells but these cells have a reduced function.

Late gene products for HIV are Gag, Pol, and Env proteins. The different virus structural proteins begin to assemble at the plasma membrane as the polyproteins Gag and Gag-Pol. The Gag-Pol protein is produced by ribosomal frame shifting on the gag-pol RNA transcript. This is a 'stuttering' mechanism where the ribosome is retarded or alternatively jumps forward one nucleotide base to commence reading the triplet in a different frame. Hence two viral proteins can be translated from a single mRNA. This appears rather complex but does produce Gag-Pol in the correct ratio for virus assembly, especially for Gag which

is an important structural component. The viral genome and Gag-Pol associate with the Env proteins which have matured in the ER and Golgi, and also migrate to the plasma membrane. The virions bud from the plasma membrane as immature particles. At this stage, called post budding, the viral protease cleaves the Gag-Pol polyprotein into the individual virion proteins, namely MA, CA, and NC and the enzymes RT, IN, and PR. The proteins rearrange spontaneously to form the characteristic sarcophagus-shaped core and the virus becomes a fully fledged infectious HIV.

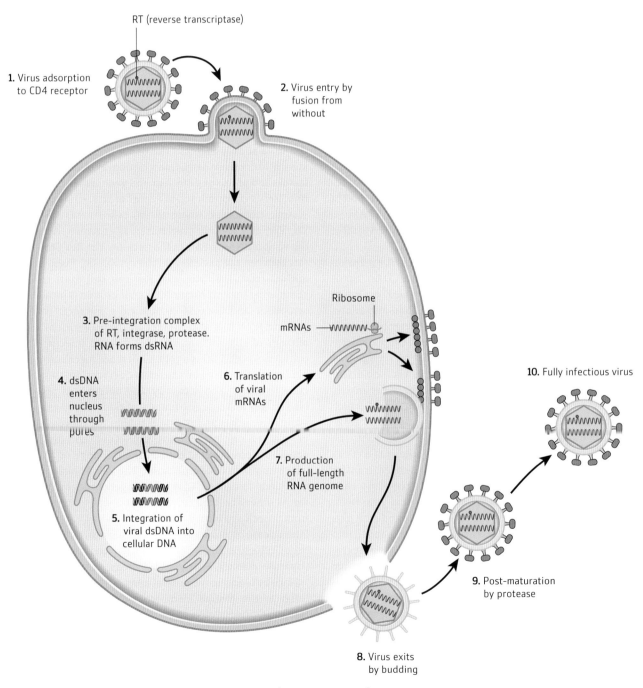

Fig. 27.6 Simplified version of replication cycle of retrovirus (for details see text).

27.3 Clinical and pathological aspects of HIV-1

Medical case story HIV

It is important to be on the lookout for HIV seroconversion illness and also to be vigilant about testing people with risk factors for HIV. The most important HIV issue in the UK and elsewhere is the number of undiagnosed cases. Firstly, because these people are at risk of HIV-related disease and death, and secondly, because they can carry high levels of virus in their blood, genital fluids, and semen that, combined with lack of knowledge of their infection, means they are much more likely to spread the virus to other people.

Peter Manly is a 42-year-old man. He is tall, thin, and has pale-looking skin—although this pallor is probably normal for him. He says he feels tired all the time, and this has been going on for 6 weeks. Initially, he had a sore throat, aches and pains, a temperature, and a headache. He also had a blotchy rash on his back. Since then he has just felt tired.

Because of the initial illness you are thinking of viral causes for his symptoms—for example, glandular fever or flu-like illness. You have been taught to link glandular fever and HIV in your mind, so now that you have thought, 'non-specific illness: glandular fever?', then you must also think, 'HIV seroconversion?'.[1] The blotchy rash on his back is a typical symptom of HIV seroconversion illness. Or it could be a minor condition irrelevant to his tiredness, such as pityriasis versicolor, a fungal infection.

You have thought of HIV, so you now have some awkward questions to ask. You wonder if he is a homosexual, which increases the likelihood of HIV. Asking questions about his sex life will come out of the blue for Mr Manly.

First you say, 'I don't know very much about you. Could I ask you a few background questions?'

He is very gentle and agrees, smiling briefly; 'Sure'.

'Do you have a partner at present?'

'Yes,' he says.

You say, 'Is your partner male or female?'

'Female, yes, we've been married for five years now.'

'And is your partner from this country?'

'Yes, we are both English,' he agrees.

'And before that partner, did you have any other partners?'

'No, no other partners,' he says. And you think, 'hmmmm,' because you have found patients do not always feel like sharing everything with their doctors.

'And have you had any foreign travel recently?' you press on.

'Yes, we were in Africa living in a township for 6 months. We got back 6 months ago.'

Actually, you were not expecting him to say this. You think 'Africa, lots of HIV,' and you say, 'Can I ask what work you were doing out in Africa?'

'I work as a pastor,' he answers.

Ah. Now that you have broached all these questions you feel exhausted and your suspicions have been neutralized. You lean back a little. 'So both you and your wife came together without having any previous partners?' you ask curiously. 'That's right,' he says.

You proceed to your final check, '... and have you ever had any tattoos, or piercings, or injected drugs?'

'No, no.'

'So can we conclude you are low risk for sexually spread viruses such as hepatitis B or HIV?'

'Yes,' he agrees.

Now you only have to deal with the 'tired all the time' side of things. Your heart does not sink when you see patients with this condition, because you have a strategy to deal with it.

First, you systematically run through first his daily work, sleep, and eating routine.

Next you run through his life, by which you mean occupation, partner, accommodation, and parents, to identify any stresses.

Then you check his mood: is he depressed or anxious, does he look forward to things, is he hopeful about the future?

Finally, you get out a blood form to exclude medical causes. You tick glucose (diabetes), full blood count (anaemia and immune status), iron (low iron stores and haemochromatosis), thyroid function tests (hypothyroidism), monospot (glandular fever), serology (HIV, hepatitis B, and C), liver function tests, and urea and electrolytes. In this instance, you also request a blood film for malaria.

Although he appears low risk for HIV, you still request this test, after asking his permission. You no longer have to do pre-test counselling, so you do not have to spend time on a long discussion about the implications. Apart from mentioning the 3-month window period, and advising him how he will get the results of the test and when, there is no need to go into more depth. The pre-test counselling was a barrier to HIV testing and so it has been largely abandoned.

Now that effective treatment is available with highly active antiretroviral therapy (HAART), life expectancy for people with HIV is greatly improved. Ticking the fasting glucose test on the form, or the hepatitis B box can also reveal life-changing diagnoses but don't involve special counselling, and HIV should be seen in a similar light.

The important thing is to be alert—for patients with a higher risk of HIV: for possible HIV seroconversion illnesses: or for symptoms or signs suggestive of HIV disease. An early diagnosis means early treatment, and this increases life expectancy by 10 or 20 years and reduces the likelihood of further spread.

Mr Manly is satisfied you have been thorough. You feel the most likely reason for his symptoms is convalescing from a flu-like illness. His tests come back normal and he doesn't present again.

Notes

[1] Madge, S. *et al. HIV in Primary Care.* Medical Foundation for AIDS and Sexual Health. Available at: www.medfash.org.uk. (accessed 29 July 2010).

★ **Learning Points**

• **Virological:** Human immunodeficiency virus (HIV) is a retrovirus. It is spread by sexual intercourse, IV drugs, transplacentally, and by breast feeding. Transmission is reduced by early diagnosis and treatment with highly active antiretroviral treatment (HAART), and prognosis with treatment is now up to 20 years and counting.

The seroconversion illness is a non-specific illness similar to glandular fever or flu-like illness, with aches and pains, sore throat, temperature, and sometimes a blotchy rash on the trunk. Without treatment, the infection is quiescent for 2–4 years, and then progresses to HIV disease including opportunistic infections, cancer, neurological manifestations, and death.

• **Clinical:** Have a low threshold for testing for HIV. If you are considering the diagnosis of glandular fever, then also consider HIV. Also consider testing in patients with risk factors for HIV or multiple/recurrent minor illnesses, such as oral thrush, molluscum contagiosum, shingles, herpes, seborrhoeic dermatitis, and genital warts.

• **Personal:** There is no doubt that it can be difficult and draining to broach these sensitive topics with a patient. Mr Manly took these enquiries very calmly, but the same routine will send other patients into a spin, making you question your decision to think of HIV in the first place

---

27.3.1 The clinical staging of the disease

Table 27.2 summarizes the generally accepted three-stage CDC scheme for the disease: primary acute infection (or asymptomatic infection), symptomatic conditions, and advanced disease (AIDS). Particular emphasis is placed on CD4 cell counts (of < 200 cells per µl) and HIV symptoms. We should note that there is a parallel classification from WHO which is broader and encompasses world experience; it is not dependent upon CD4 cell counts since the laboratory back up may be poor in many less wealthy countries. The WHO scheme has a fourfold staging classification.

After a mild acute influenza-like illness with fever, headache, malaise, weight loss, and pharyngitis with an incubation time of about 3 weeks, a viraemia spreads virus throughout the body. This acute primary stage A lasts 4 weeks. There can be a vigor-ous immune response resulting in dramatic fall of virus in the next weeks. A correlate of disease progression is the steady-state viral load (the important set point) at 6 months after this primary infection. Low virus genome counts of 12 000 or so correlate with slow clinical progress, with the disease becoming quiescent, and with CD4 cells slowly dropping by 30–60 cells per year. Many patients feel fatigue (stage A3).

Approximately 60% of asymptomatic cases from stage A move into stage B of the disease within the next 2–4 years. This is characterized by **fever, weight loss (<10% of body weight), persistent lymphadenopathy, night sweats**, and **diarrhoea** (Fig. 27.7). The most important factor here is accelerated loss of CD4 cells. Virus replicates continuously. There is tremendous variation in the length of this period between patients.

The clinical symptoms illustrated in the three blocks are termed A, B, and C, but in reality tend to merge together. The period from infection to onset of AIDS at stage C may extend to 8 years or more. Viraemia is most intense at the acute primary phase (A) and the virus RNA 'set point' or load at this time gives an indication of the problems ahead. Early intervention with antivirals will modify each stage of the disease and reduce virus infectivity. As well as blood, virus is also present in seminal fluid and vaginal fluid.

Untreated patients then proceed inexorably to clinical **AIDS (stage C)** commonly heralded by thrush, herpes zoster, and, more seriously, **Pneumocystis carinii pneumonia** and, in many parts of the world such as Africa and Asia, tuberculosis. In untreated patients the time from infection to death may be as long as 10 years and is inevitable in 70% of infected untreated persons. In contrast, patients well treated with antiviral drug combinations have a significantly improved life expectancy.

Table 27.2 Classification system for HIV infection

	Clinical categories		
CD4 cells/ mm³	A acute (primary) HIV	B Symptomatic	C AIDS
≥500	A1	B1	C1
200–499	A2	B2	C2
<200	A3	B3	C3

Data from Centers for Disease Control and Prevention (updated 2013)

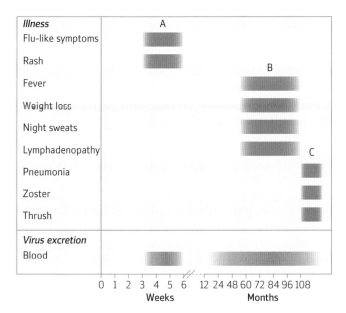

Fig. 27.7 Clinical and virological events during infection with HIV-1.

Remember that current estimates are that 1 in 10^6 CD4 cells harbours latent virus and that passage of 70 years would be needed to eradicate these by natural cell senescence.

The signs and symptoms of AIDS may vary somewhat in the different categories of persons infected with the virus. Kaposi's sarcoma is less common, for example, in haemophilia patients, who have been infected as a result of transfusion of virus-contaminated factor VIII preparations. In Africa, most patients die from tuberculosis, whereas superinfection with clinical pneumocystis is more important in the USA and Europe. The clinical manifestations of AIDS are protean, but can be summarized as malignant, infective, and neurological.

The WHO clinical staging of HIV/AIDS progresses from stages 1–4 with stage O being the primary infection, and stage 1 the asymptomatic state perhaps with PGL (persistent generalized lymphadenopathy). Stage 2 includes patients with unexpected weight loss, herpes zoster, and fungal nail infections. In clinical stage 3 patients have severe weight loss (10% of body weight), chronic diarrhoea, TB, and persistent fever. By stage 4 the predominating symptoms are wasting syndrome, Kaposi's sarcoma, toxoplasmosis, lymphomas, candida, CMV, and disseminated non-TB bacteria.

There are 'long term survivor' groups and, in contrast, so called 'fast progressors' and both are the subject of intense clinical and virological research.

Malignant disease

The most common of the malignant tumours is Kaposi's sarcoma, but aggressive B-cell lymphomas, non-Hodgkin's lymphoma, and genital cancers may also appear.

Infective manifestations

In the absence of HAART (highly active antiretroviral therapy) these are frequent, and include tinea, gingivitis, oral and oesophageal candida, chronic sinusitis, and other infections of the skin and mucous membranes. In countries where HAART is used these clinical conditions can be the first sign of non-compliance with drug therapy. Several latent viruses are activated, including herpes simplex and zoster, and papillomaviruses. Tuberculosis is a frequent complication, even in the USA. Common opportunistic infections are *Pneumocystis carinii*, EBV interstitial pneumonitis, cryptosporidia and microsporidia, CMV retinitis, enteritis and infection of the brain by *Toxiplasma gondii*, and *Cryptococcus neoformans*. These opportunistic diseases can be suppressed by using specific anti-HIV drugs to keep CD4 cells above 500 cells/mm³. In general, without treatment, the 6-month risk of AIDS for a young patient with CD4 count of 350 cells/mm³ ranges from 0.6% with a viral load of 3000 copies (µl) to 2.5% with a load of 300 000 per µl. Older patients develop disease faster. No differences have been noted between men and women.

Neurological sequelae

These were common before the advent of HAART, and included dementia, severe encephalopathy, myelopathy, and motor dysfunction. There may be diminished memory, tremors, and loss of balance as well as signs of peripheral neuropathy.

HIV in children

Without antiretroviral therapy 13–40% of babies born to HIV positive mothers acquire infection. There is clear evidence of transplacental HIV infection; indeed, this is now recognized as the second most common mode of transmission. Transmission can also take place during delivery or from breast milk. The course of disease in infants is accelerated compared to adults; 20% of infected infants develop AIDS in the first year of life and approximately one third die in the first 5 years. Use of HAART in the pregnant mother can dramatically reduce the chance of an infected baby.

A child is considered to have ARC if interstitial pneumonitis, persistent candidiasis, or parotid swelling is present for 2 months together with two or more of the following:

- persistent generalized lymphadenopathy;
- recurrent bacterial infection; e.g. *Strep. pneumoniae, H. influenzae, Salmonella*, and *P. jiroveci* pneumonia;
- hepatic or splenic enlargement;
- chronic diarrhoea;
- growth failure.

In the absence of HAART the mortality of children with HIV is high, but the outcome in those with less severe disease or symptomless infection is less certain. The risk of paediatric HIV is certainly higher in babies born to mothers who are symptomatic, rather than simply being seropositive.

27.3.2 Pathogenesis

HIV-1 enters the body either during **sexual intercourse, needle drug abuse, transfusion with contaminated blood products**, or **via the placenta** (maternal–foetal transmission). There have been a number of instances of patients infecting healthcare staff or vice versa, during surgical or dental procedures. Virus-infected lymphocytes are present in sperm and may infect via microscopic breakages in the endothelial lining of the vagina or rectum. It is assumed that, initially, CD4+ lymphocytes, macrophages, and dendritic cells are infected by the virus. However, as viral antigen has been found in only about 1/10 000 lymphocytes, the reason for their ultimate dysfunction and death is still a subject of some speculation. HIV-1 is strongly cytopathic in the laboratory and can cause cell-to-cell fusion resulting in the formation of giant syncytia and cell death. Viral glycoprotein present on the surface of an infected cell can interact with the CD4 receptors on many adjacent uninfected cells, thus multiplying the effect. However, the virus seems to be less aggressive to cells when actually in the body because there have been no reports of syncytia, for example, in AIDS patients. Moreover, many of the CD4+ lymphocytes that die have not been infected by HIV; they are innocent 'bystander' cells. Some immunologists have, therefore, proposed that there is an autoimmune component in the pathogenesis. Certainly host genetics are important, with HLA class I alleles B*27 and B*57 associated with low viral loads and long-term progression to disease.

Virus replication occurs mainly in peripheral lymphoid organs, spleen, lymph nodes, and gut-associated lymphoid tissue. High rates of viral replication occur early after infection and ensure future destruction of CD4+ lymphocytes. Massive lymphocyte destruction occurs daily, accompanied by equally massive cellular regeneration.

The replication of the HIV genome is enhanced in antigen-stimulated T cells, and it is assumed that persons with concomitant infections that stimulate T-cell replication have a greater chance of succumbing to AIDS. Macrophages are also infected with HIV and are themselves similarly stimulated by other antigens. They act as a reservoir of the virus as integrated proviral DNA, as do T lymphocytes and memory cells in the lymph nodes and microglia in the brain. As infection progresses, B lymphocyte functions are affected through their regulation by CD4+ (T_h) cells (Chapter 5). The destruction of the CD4+ helper cell subset is particularly damaging to the overall orchestrated immune response of the host. This malfunction of the immune response leads to the appearance of opportunistic organisms which are normally held in check by immune T cells. The size of the viral load at the acute phase (set point) can give an important indication of clinical progression or otherwise. Viral loads of up to 10 million viral RNA copies per ml can drop to 12 000 within the year reflecting the good news of slow progression, whilst viral RNA copy numbers of 60 000–100 000 at this time suggest rapid progression to AIDS.

27.3.3 Immune response

The dynamics of the antibody response are important for serological diagnostic purposes and most diagnostic kits take advantage of the high levels of anti-env (gp120) antibodies and their longevity. Antibodies to the Env protein develop slowly and remain at high levels throughout the infection. Antibodies to the internal p24 protein have a different temporal pattern and rise during the early stages of the infection, only to decrease in parallel with the onset of serious signs of the disease. CD4 cells dip during the primary infection, then slowly decline without treatment with antiviral drugs, and then rapidly decline as AIDS sets in.

27.3.4 Epidemiology

Origin of HIV-1 and genetic variation

HIV-1 is a zoonosis, infecting primates in the wild, often asymptomatically, and then opportunistically crossing the species barrier. The cut/slash method whereby chimpanzee hunters were bitten during capture of these chimps for food is generally accepted as the point of transfer. A retrospective genetic molecular clock analysis when matched to a timeline of historical events in Central Africa suggests that the virus first infected humans in South-east Cameroon in the late 1800s. Local circulation of virus was likely followed by spread to Kinshasa where by the 1920s 300 000 or so workers were travelling by rail to work in the mineral mines. Migrant workers are thought to have spread the virus to Angola and Zambia. By the 1960s the sex industry in Kinshasa had increased dramatically and included gay men from California and Europe, which gave an opportunity for the virus to spread internationally. There may have been eight independent transmissions to humans at varying times resulting in HIV-1 subtypes A–H.

The actual number of persons infected is a matter for conjecture but may reach 35 million worldwide. Of course, the numbers with active disease are much fewer, around 3 million.

In the absence of effective vaccines, infection can be prevented by individual action (Table 27.3) combined with screening of potentially contaminated blood and blood products to prevent iatrogenic infections. There is evidence of modified sexual behaviour in the USA and certain African countries, which leads to a reduced incidence of infection. Transmission most commonly occurs during sexual intercourse with exchange of virus-contaminated semen, genital secretions, or blood. Unprotected receptive anal intercourse is the most risky activity. The risk is probably 0.5%. The second most risky behaviour is via direct inoculation with blood by reusing contaminated needles. The risk here is 0.3%. The third primary mode of transmission is from an infected mother to her child where the risk is 10–40%. Treatment of the mother during pregnancy with antiviral drugs and the newborn at birth can significantly reduce this risk, which rises again if the mother breast-feeds.

There is considerable genetic variation between isolates of HIV-1 virus. In the case of the Env protein, the degree of variation is comparable with that between subtypes of influenza A virus, namely amino acid sequences differing by up to 25%. Three distinct genetic groups of virus are now recognized, namely the major (M) and the outlier (O), and N. Group M has more than 95% of all isolates. Group M is further divided into genetic subtypes or clades A–K and a further 40 circulating recombinants. These recombinants are described as patchwork mosaics with very diverse genetic segments from many clades. These clades have different geographical locations; for example,

Table 27.3 Individual actions for preventing HIV transmission

			Estimated risk of transmission
Needle-borne transmission	Needles or syringes must not be shared		1 in 300
Sexual transmission	A reduction in the number of sexual partners to decrease the chance of being exposed to an infectious person. The following guidelines can be used to decrease the risk of infection:		1 in 200
	Absolutely safe	Mutually monogamous relationship	
	Safe	Sexual intercourse using a condom containing spermicide; non-insertive sexual play	
	Risky	Oral, vaginal, or anal intercourse. Virus is present in vaginal secretion and seminal fluid, and in minute traces in saliva	Female to male 1 in 1000 Male to female 1 in 500 Male to male 1 in 200
Prenatal transmission	Those at possible risk of HIV infection should be tested for antibody. Antibody-positive women should not become pregnant		1 in 4
	Those who do should be treated with AZT to reduce the risk of transmission to the neonate		1 in 10
Blood transfusion	Avoid blood transfusion in high-risk countries with contaminated blood		9 in 10

clade B is spread by homosexual activity in Europe and the USA, whereas clades C and E have a predilection for heterosexual spread in Africa. By far the most infections in the world are now caused by group C.

The most explosive outbreaks of HIV-1 have occurred in the so-called Group III countries in Asia (including India and Thailand), Eastern Europe, the Middle East, and Africa, caused predominantly by clade A spreading heterosexually. The intermediate nations include Central and East Africa, where the virus has spread for the last 25 years, mainly by heterosexual intercourse and by infection of babies. Finally, in Western Europe, USA, and Australia, the virus has spread mainly by homosexual intercourse and the number of cases is expected to decline. Clade B viruses predominate in the UK.

27.3.5 Effective HIV vaccines have not been developed to date

As well as abstinence from sexual intercourse or the use of condoms, an effective vaccine would be very useful in preventing the further spread of HIV-1. However, complete protection is unlikely to be achieved against a genetically hypervariable virus infecting a mucosal surface, although it must be acknowledged that vaccines against a similarly diverse virus, namely influenza, have been used for over 50 years.

A chemically inactivated whole virus vaccine (a so-called therapeutic vaccine) has been tested in individuals already infected and induces a weak immune response. Some progress has been made with recombinant DNA techniques whereby the gene for HIV Env has been transferred to yeast cells or *E. coli*, and large quantities of pure viral spike protein (gp160 or gp120) have been produced by deep fermentation. However, the first

such vaccine tested on a high-risk group of male homosexuals failed to give satisfactory protection against the virus.

An experimental live attenuated HIV vaccine is under investigation, in which the viral *nef* gene has been mutated or altered. The vaccine gave protection in animal models but worryingly the virus was able to reconstitute a mutagenized Nef and became somewhat virulent once more. This gene normally enhances viral infectivity and, therefore, any interference with functioning would reduce viral replication and spread.

Probably most vaccine research is directed towards a prime boost approach in which the HIV *env* gene has been cloned into a harmless poxvirus, such as canary pox. Injection into the arm and subsequent replication of the poxvirus DNA containing the HIV *env* gene primes the immune system of the individual. A boost to the immune system then follows by later injection of HIV Env protein, itself produced by recombinant DNA technology. The two separate antigens stimulate T- and B-cell immune responses. Unexpectedly, a recent molecular vaccine where an HIV protein (gp120) was cloned into adenovirus has been shown to enhance HIV infection in some patient subgroups.

There is a big research focus on the clinical use of human monoclonal antibodies which cross-neutralize the clades of the virus. Also very recently a CD4 molecule has been fused with a mimetic of the amino terminus of the co-receptor CCR-5. This synthetic compound has broad neutralizing activity against most HIV isolates.

27.3.6 The success of highly active antiretroviral therapy (HAART)

No complete cure yet exists for this retroviral infection, and it is difficult to envisage the development of any such antiviral because, although very successful drugs have been developed which will repress viral replication, it is not yet possible to excise an integrated viral genome. We would note here that recent claims from Australia and elsewhere of a complete cure in a child given higher doses of HAART have not been substantiated. A number of HIV positive individuals are known who have lived into old age and not developed the symptoms of AIDS.

The mathematics of infection are daunting: in a patient there are 10^5 or more HIV viruses per ml of plasma with a turnover rate of 10^{10} per day! Also, the virus exists as a quasi-species, with every nucleotide of the HIV genome being able to be mutated each day. Thus drug resistance will always be an issue. On the positive side there are now a plethora of inhibitors that can block all stages of virus replication and more are being developed (Fig. 31.2). This is a fast-moving field of basic chemistry and clinical work.

Clinical experience with HIV over the last two decades, plus the long accumulated knowledge of the successful use of drug combinations versus TB, has pointed away from monotherapy. But a very practical world intervenes here—economics. In the USA, for example, HIV treatment of a single patient costs $20 000 per year, with a total care bill for HIV patients of $15 billion per year. In the UK the lifetime cost of treatment is £360 000 per person. However, this cost will reduce as a number of the drugs come off patent, so allowing cheaper 'generic versions' to be sold.

Such intensive regimens of several drugs are referred to as highly active antiretroviral therapy (HAART) and reduce the chances of drug-resistant virus emerging. In the developed world mortality has been reduced significantly since 2000 and this combined therapy can reduce virus load to 50 copies per ml or less in plasma.

There is broad consensus that antiretroviral therapy should be started in patients with a confirmed diagnosis with CD4 cell counts below 500 cells/mm^3 regardless of symptoms, and also in persons with AIDS and severe symptoms, in all pregnant women with HIV, and children younger than five years. A combination regimen for adults, including pregnant and breast-feeding women, includes a fixed dose of efavirenz plus TDV plus 3TC or FTC. A combination with a single tablet once per day (Atripla) has acceptable side-effects and virological and treatment response and has a combination of emtricitabine, efavirenz, and tenofovir. It must be appreciated that the antiretroviral drugs can sometimes interact unfavourably together and interact with other drugs as well. Recently a drug has been licensed (Truvada) as a prophylactic taken daily (also called pre-exposure prophylaxis) for persons who have unprotected sex with an HIV positive partner.

Since this is not a textbook of clinical medicine, and also since the possible drug combinations are almost endless, we would refer students to the WHO website for up to date information (http://www.who.int/hiv/pub/guidelines/en/).

Close laboratory monitoring of HIV gene copies commonly shows a marked (10-fold or more) reduction within 14 days of the start of treatment. Any failure to detect such reductions implies a lack of adherence by the patient to the drug regimen or, less likely with HAART, the emergence of a drug-resistant virus. Over the first year of HAART therapy the CD4 cell count should rise by 200 cells/mm^3 or so, whilst the patient should improve clinically with fewer superinfections. However, there can still be problems of drug toxicity, including peripheral neuropathology, lipo-atrophy, and pancreatitis and rashes.

Virus decontamination in the home and hospital

Fortunately, HIV is not a physically robust virus and can be easily destroyed on surfaces and equipment by 35% ethanol for 5 min, or hypochlorite (25 000 ppm), or by glutaraldehyde (0.5%).

27.3.7 Laboratory diagnosis of HIV infection

It cannot be overemphasized that laboratory investigation is needed for a definitive diagnosis of HIV infection. The diagnosis cannot be made on clinical grounds alone. Most laboratories screen blood for antibodies to viral proteins using a rapid **ELISA** method. In view of the great personal and social implications of a positive test for anti-HIV antibody, **it is important to carry out a confirmatory test** by a completely different technique. Modern third generation tests are 99–100% specific, but false negatives do occur. There may be a period of several months when the virus is present, but the infected person has not yet produced detectable antibodies. If, therefore, a person is a member of the high-risk groups, and is concerned about a recent sexual or other risk event, serological testing will have to be performed again after 6 or even 12 months before a negative result is acceptable. Dependent upon the level of laboratory experience RT PCR can be used to detect the virus directly. **Viral genome load assays** are now routinely performed in many countries, but certainly not all, to quantify the number of viral genome RNA copies in plasma by RT PCR. The results correlate well with clinical prognosis; for example, a high number of RNA copies in an early phase of the disease indicates a potentially rapid onset of AIDS.

Gloves and a disposable plastic apron or gown must be worn when taking blood or other specimens from ARC or HIV patients. Eye protection is recommended. Obviously, **safety procedures for taking, packing, and transporting clinical specimens must be strictly observed.**

It must be noted that in the USA home testing is used, whereby a skin prick is made and blood dried on to a paper filter, which is then mailed to a laboratory under code. **Personal accidents involving potentially HIV-infected material must be reported and skin pricks treated immediately** by encouraging bleeding, and washing with soap and water. Hospital pharmacies retain stocks of nucleoside and nucleotide RT inhibitors and protease inhibitors for immediate administration to staff who have been potentially infected via skin pricks. The chance of becoming infected after a 'needlestick' with infected blood is low at 0.3%.

27.4 Characteristics of the human lentivirus HIV-2

The second human lentivirus, **HIV-2**, was first isolated from immunosuppressed patients in West Africa and while transmitted in similar ways to HIV-1 is less pathogenic. Many fewer people succumb to HIV-2 than HIV-1 and moreover the incidence of HIV-2 is declining.

The virus morphology, replication strategy, and genome, as well as preventative measures, closely resemble the much more studied HIV-1.

The virus has its origins from a sooty mangabey monkey in West Africa, probably from a hunting accident as this species is consumed locally and often kept as pets. There are eight distinguishable groups of HIV-2 in humans and each originated independently from cross-species transfer. Countries with a past colonial or trading link with Portugal, including south-west India, Angola, Mozambique, and Brazil, all have significant numbers of infected people, presumably transferred by the activities of the early Portuguese explorers. Portugal still has the highest prevalence of HIV-2 in Europe, accounting there for 4.5% of HIV cases.

27.5 Characteristics of the human delta retrovirus HLTV-I

27.5.1 Introduction

In 1978, Robert Gallo, in the USA, isolated a retrovirus from the lymphocytes of an African American leukaemia patient that had been maintained in culture in his laboratory by a new technique involving stimulation of the cells with the cytokine IL-2.

Japanese workers had earlier noticed clustering of cases of adult T-cell leukaemia/lymphoma in the southern islands of Japan. To epidemiologists this clustering hinted at an infectious aetiology. There are now known to be four lineages (1–4).

27.5.2 Genome and replication

Similarly to HIV-1 this virus infects CD4 T cells and integrates into the cell chromosome. The genome of 9 kb has genes for retroviral proteins Gag, Pro, Pol, and Env, flanked by LTRs (Fig. 27.8).

The virus uses alternative splicing and internal initiation codons for regulatory and accessory proteins as well as structural proteins. An important gene product is Tax, which can immortalize T lymphocytes, suggesting that it is important for virus-induced leukaemia.

27.5.3 Origin and epidemiology

HTLV-I is presumed to have originated in Africa, where it infects Old World primates such as baboons and pygmy chimpanzees (bonobos). After zoonotic transmission to humans, it may have reached the Americas along with the slave trade, and then the southern islands of Japan with early Portuguese explorers who had previously been in Africa. The virus has proved to be endemic not only in southern Japan, where there are over 1 million carriers, but also in parts of Central and South America, the Caribbean, and Central Africa, and in Australian aboriginals.

Clusters in families show that vertical transmission or close contact is needed for spread. The important mode of transmission is sexual intercourse. Prevalence in blood donors may reach 5% in epidemic areas and 15–20 million people have been infected worldwide.

27.5.4 Clinical aspects

Most persons infected with HTLV-I remain asymptomatic for life, but in about 5% overt disease may appear after an incubation period of 10–40 years: there are two distinct clinical manifestations.

HTLV-I presents in middle-aged adults as acute **aggressive lymphoma** of the skin and most viscera, including particularly the liver, spleen, and lymph nodes. A further important diagnostic feature is hypercalcaemia, with or without bone lesions. The leukaemic phase of the disease is not always evident. Typical symptoms are malaise, fever, jaundice, weight loss, and the symptoms of opportunistic infections. Variants of what is usually

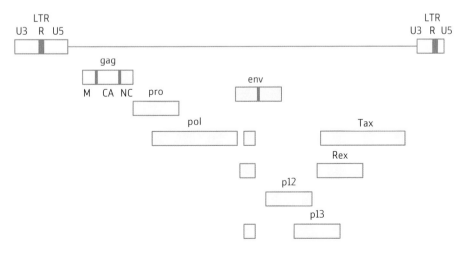

Tax interacts with cellular transcription factors. Rex is a post-transcription regulator increasing the expression of viral mRNAs whilst p12 is required for replication in new dividing cells. P13 induces apoptosis in infected cells.

Fig. 27.8 Genome structure of HTLV-I.

an acute and aggressive disease occur and include a chronic T-cell lymphocytosis. A form of 'smouldering' adult T-cell leukaemia/lymphoma is also seen, in which the patients present with persistent lymphocytosis, with or without lymphoma.

The second distinctive clinical form is where most patients have the chronic neurological disease, **tropical spastic paraparesis**, **TSP**, sometimes called **HAM** (HTLV-associated myelopathy). Progressive demyelination of the long motor neurone tracts in the spinal cord results in spastic paraparesis of both lower limbs. Women are affected more often than men. Magnetic resonance imaging reveals multiple white matter lesions in both the spinal cord and brain.

Analysis of CSF can show high levels of pro-inflammatory cytokines and also activated lymphocytes. Treatment with antiretroviral (HIV-1) drugs or corticosteroids is not wholly successful.

27.5.5 Transmission and pathogenesis

HTLV-I is transmitted by **intravenous drug abuse, sexual intercourse**, and **blood transfusion**, and enters the body inside infected CD4+ lymphocytes in semen or blood, as well as vertically from mother to infant via breast milk, and possibly via the placenta. Interventions in endemic areas to prevent transmission by breast feeding may be possible.

The prime suspect for initiating disease must be the HTLV-I-infected lymphocyte, transformed in adult T-cell leukaemia/lymphoma and presumably chronically activated in TSP. The product of the virus regulatory gene *tax* transactivates transcription from the viral LTR and also from cellular oncogenes. Neuronal damage and demyelination are probably consequences of the inflammation, so that HTLV-I may not be truly neurotropic. As only a few cases of both diseases in the same patient have been reported, despite the many hundreds of each in Japan and the Caribbean, infected lymphocytes seem to be committed to produce one disease or the other, or, much more often, neither.

27.6 Characteristics of the human delta retrovirus HTLV-II

A second virus, HTLV-II, was isolated in Seattle, USA, from the cells of a patient with a rare 'hairy cell' leukaemia, but little is known about it at present. It is prevalent in intravenous drug users in the USA, but is not known to cause disease in them. HTLV-II is known to cause a 'hairy T-cell' leukaemia, but very few cases have been described. Even less is known about HTLV-III and -IV.

 Reminders

- Retroviruses are enveloped spherical viruses and contain positive-strand **diploid RNA** of approximately 10 kb in size. There is an icosahedral matrix shell, which encloses a vase-shaped protein core containing RNA. They possess a reverse transcriptase (**RT**) able to catalyse transcription of viral RNA into DNA (proviral DNA) and also the enzymes integrase and a protease.

- The genome is flanked by LTRs having polyadenylation signals and promoter sequences for viral transcription.

- Viruses enter the cell by fusion from without: a so-called pre-integration complex (PIC) is transmitted through nuclear pores where genome RNAs are transcribed by the RT into a dsDNA copy which is randomly integrated into the host cell chromosome.

- A zoonotic origin in Africa is postulated, entering humans from chimpanzees around 1900.

- CD4 T cells have a receptor for the virus and are the most characterized reservoir of the virus, but monocytes, dendritic cells, astrocytes, and macrophages can also be privileged sites. Destruction of the CD4 T cells causes disorganization of the cellular immune system, thus making the patient vulnerable to otherwise mild and common infections.

- The three stages of HIV-1 disease in untreated patients are:

  - the asymptomatic phase;

  - ARC with persistent lymphadenopathy, night sweats, and diarrhoea;

  - full-blown AIDS with a plethora of opportunistic infections.

- Laboratory studies are required to confirm infection with HIV-1. **ELISA** detects specific antibodies whilst quantitation of **viral RNA** load, as measured by number of viral RNA copies by real time RT PCR, is an important prognostic test.

- The life of HIV patients is prolonged by treatment with **combinations of three of the antiretroviral drugs (HAART), such as non-nucleoside and nucleoside inhibitor** and a **protease and fusion inhibitor**, to reduce the emergence of drug-resistant mutants.

- Worldwide, 80% of HIV infections are transmitted by heterosexual intercourse. Perinatal and blood transfusions account, respectively, for 10 and 5% of infections.

- HIV-2 is the second human lentivirus with a zoonotic origin but much is less widespread than HIV-1 and less pathogenic.

- The delta human retrovirus HTLV-I causes cancer (**adult T-cell leukaemia lymphoma**) and also a chronic neurological disease, TSP, or HTLV-I associated myelopathy (HAM). There are 20 million infected persons, mostly asymptomatic. The related virus HTLV-II causes a rare hairy cell leukaemia. Still less is yet known about HTLV-III and -IV.

Further reading

Arts, E.J. and Hazuda, D.J. (2012). HIV-1 antiretroviral drug therapy. *Cold Spring Harb. Perspect Med* **2**, 1–23.

Coghlan, A. (2014). Hitching a ride on trains helped HIV spread. *New Scientist*, Oct 11.

Costiniuk, C.T. and Jenahian, M.A. (2014). Cell to cell transfer of HIV infection in implications for HIV viral persistence. *Journal of General Virology* **95**, 2346–55.

Dahiya, S. (2012). Deployment of the human immunodeficiency virus type I protein arsenal: combating the host to enhance virus transcription and providing targets for therapeutic development. *Journal of General Virology* **93**, 1151–72.

Deng, K. *et al*. (2015). Broad CTL response is required to clear latent HIV-1 due to dominance of escape mutations. *Nature* **517**, 381–5.

Faria, N.R., Hodges-Mameletzis, I., Silva, J.C., Rodés, B., Erasmus, S., Paolucci, S.,*et al*. (2012). Phylogeographical footprint of colonial history in the global dispersal of human immunodeficiency virus type 2 group A. *Journal of General Virology* **93**, 889–99.

Haigwood, N.L. (2015). Tied down by its own receptor. *Nature* **519**, 36–7.

Koff, W.C. (2010). Accelerating HIV vaccine development. *Nature* **466**, 161–2.

Maxmen, A. (2012). Generic HIV drugs will widen US treatment net. *Nature* **488**, 267.

WHO (2013). Consolidated guideline on the use of antiretroviral drugs in treating and preventing HIV infection. IBBN 97892415005727, pp 112–115.

Questions

1. Discuss the origin of HIV-1 and HIV-2 and how the viruses emerged and became global.

2. What are the problems developing a vaccine versus HIV-1?

3. Discuss points of attack in the HIV-1 life cycle of antiretroviral drugs.

28

Group 7 Circular double-stranded DNA viruses with an RT

Hepadnaviruses: hepatitis B and D

28.1 Introduction

Although the worldwide incidence of hepatitis A must run into millions, as a cause of serious disease and death this infection pales almost into insignificance beside hepatitis B. WHO estimates that 2 billion persons have been exposed and 350 million people worldwide are chronically infected. In some countries more than 10% of the population is afflicted. In surveys of the global burden of disease HBV ranks as the tenth leading cause of death and so is a major public health priority. Fortunately, there are excellent recombinant vaccines and over 180 countries of the world have started a Universal Vaccination Campaign for neonates to blunt the clinical effect of the virus. New Directly Acting Antivirals (DAAs) are used in combination therapy (Fig. 28.1), and the transmission routes of the virus—blood, sex, and birth—are extremely well defined.

From a scientific viewpoint the virus has the most compact and unique genome and shares with HIV the use of reverse transcriptase (RT) as a key step in the replication cycle, whilst a whole range of clinical effects have been documented and the virus is a major cause of liver cancer.

28.2 Properties of hepatitis B virus

28.2.1 Classification

The family name (Fig. 28.2) is *Hepadnaviridae* (HEPAtitis DNA viruses). There are two genera, *Orthohepadnavirus* and *Avihepadnavirus* that, respectively, affect certain mammals (woodchucks and squirrels) and birds (Peking ducks).

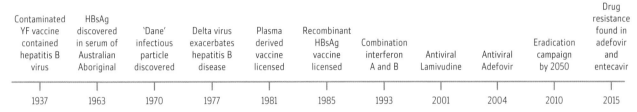

Fig. 28.1 Timeline for hepatitis B.

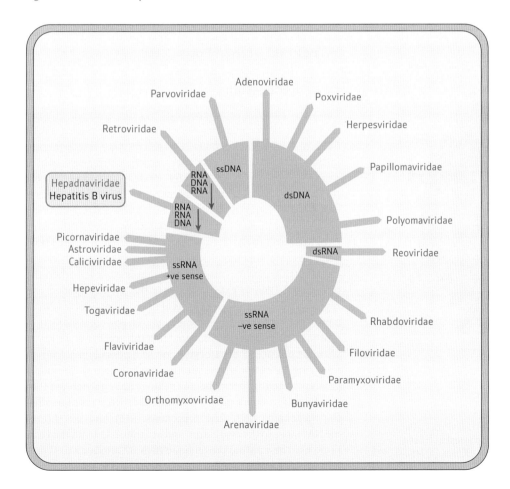

Fig. 28.2 Baltimore classification scheme and hepatitis B virus.

28.2.2 Morphology

Electron microscopy of the blood of acute and some chronic cases of hepatitis B reveals many particles differing in shape. Most of these particles are spheres and filaments and are 'empty' non-infectious particles; there may be as many as one million relative to a true infectious virion. The latter is 42 nm in diameter, and appears double-shelled by electron microscopy; this is the complete and infectious virion (Fig. 28.3b), sometimes called the Dane particle after its discoverer.

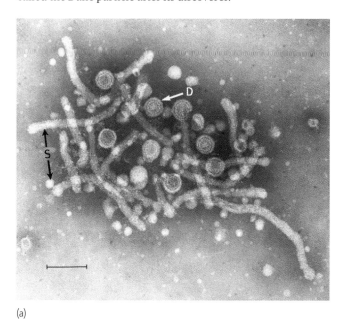

(a)

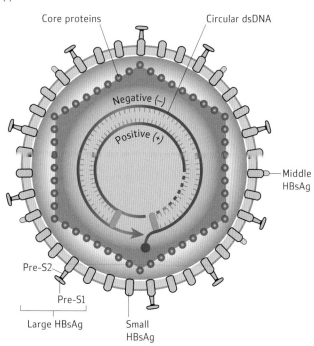

(b)

Fig. 28.3 (a) Electron micrograph of hepatitis B virus. D, Dane particles; S, large (filaments) middle and (spheres) of HBsAg. (Courtesy of Dr David Hockley.) Scale bar = 100 nm. (b) Model of the Dane particle of HBV.

The core of the infectious Dane particle is an icosahedral nucleocapsid, and contains the DNA genome, a DNA-dependent polymerase that can repair the gap in the virion DNA template and have reverse transcriptase activities involved in genome replication (P), the hepatitis B core antigen (HBcAg; the icosahedron itself), and the hepatitis B e antigen (HBeAg). The core is surrounded by glycoprotein spikes: large S, middle S, and small S. The large S protein has a pre-S1 and pre-S2 region, the middle S protein has pre-S2, and the small S protein has neither.

28.2.3 Genome

The virus has one of the smallest viral genomes of circular dsDNA some 3.2 kbp in size (Fig. 28.4). The organization of the genome is complex, with four overlapping open reading frames (ORFs). The four ORFs are called C (capsid protein), Pol (polymerase that also has RT activity), S (surface antigen), and X (regulatory protein). Because these gene regions overlap, the virus genome can code 50% more protein than would be expected. In fact translation yields seven different viral proteins. The genome also contains two enhancer regions for the regulation of virus gene transcription (Enh1 and 2), a polyadenylation signal, a packaging signal, and direct repeats (DR1 and DR2) which are involved in reverse transcription. The genome conformation is unique in that for most of its length the DNA is double-stranded but one strand ('short or positive strand') has a gap about 700 nucleotides in length. The 'long' negative strand has a nick near the 5′ end and is also linked to the viral DNA polymerase P (with RT activity) at the 5′ end.

28.2.4 Replication

The virus attaches to a hepatocyte via the liver cell sodium-dependent taurocholate co-transporting polypeptide (NTCP) using the virion pre-S protein. This cell receptor is only present in liver cells. The S protein then undergoes cleavage, which exposes a fusion motif. The latter catalyses fusion of the host plasma membrane and the virus envelope. The virus nucleocapsid is released into the cytoplasm and the interaction of virion core and the nuclear pore of the cell allow entry of the virus DNA to the cellular nucleus. Here the relaxed circular (RC) DNA genome containing the nick is converted into covalently closed circular (CCC) DNA. This conformation is achieved when the plus-strand DNA is extended across the single-stranded gap region using 5′ terminal remnants of the packed genome RNA as a primer. Essentially the genome is now an episomal mini-chromosome. It should be remembered that this cccDNA is the key to virus persistence in hepatocytes and also possible reactivation following, for example, cessation of antiviral therapy. The minus genome strand is transcribed by cellular RNA polymerase II to give four subgenomic mRNAs plus a 3.4 kb RNA transcript called the pregenome RNA. The pregenome RNA and the shorter mRNAs move to the cytoplasm, and are translated into the virus proteins. The unusual pregenomic RNA is translated to C protein, which itself assembles into particles, and to the virus DNA

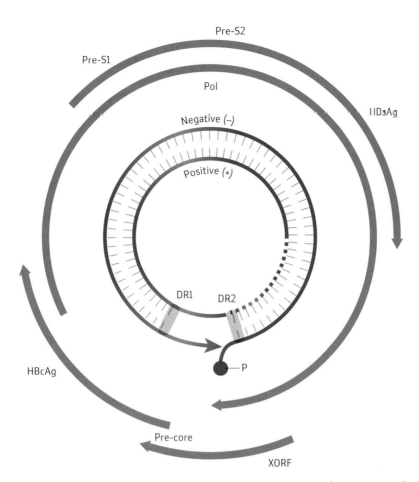

Fig. 28.4 Genome structutre and encoded proteins of hepatitis B. Labelled proteins are HBsAg (surface antigen), HBcAg (core antigen), P (polymerase), pre-S1 and pre-S2 core proteins, and the enigmatic X protein.

polymerase/reverse transcriptase, and is then itself incorporated as RNA into the new virus core structure.

The essential DNA polymerase/reverse transcriptase enzyme attached to the pregenomic RNA complex initiates the synthesis of the first DNA (minus) strand in the virus core. The original RNA strand is degraded as the enzyme moves along its length and the plus (incomplete) DNA strand is synthesized within the core structure. The virus core interacts with the pre-S domain of the S protein in the endoplasmic reticulum to form a Dane particle (Fig. 28.5). The new virions themselves are exported in multi-vesicular bodies and emerge by budding, without cell lysis, as with many other enveloped viruses. Excess pre-S domains are translocated and released from the cell and form the empty non-infectious spheres and filaments noted in Section 28.2.2. The hepatitis B core antigen (HBcAg) and hepatitis e(early) antigen (HBeAg) are alternative translation products of the core gene, with HBeAg translation requiring an upstream precore region ATG codon. The e antigen is so named because of its early appearance in serum during acute hepatitis. The HBeAg may also suppress the host immune system and so protects cells that contain the virus itself.

The replication strategy is unique for a DNA virus, namely using an RNA intermediate and a reverse transcription step. The use of RT with its lack of proofreading leads to relatively high mutation rates, but the extremely small genome size helps to prevent a large degree of genetic variability. Most variation occurs in the pre-S region, which is important for virus attachment, and entry. Vaccines contain the S protein, which induces a protective anti-S response. Therefore mutations in the S protein can abrogate vaccine protection. There are in fact three S proteins but all have a common C-terminal portion. The small S protein (Australian antigen) is most abundant. The middle-sized (MS) protein has an extra 55 terminal amino acids (encoded by the pre-S2 domain). The largest (LS) protein contains the pre-S2 and S domains plus 119 N-terminal amino acids from the pre-S1 region.

The X protein (HBx) is required for efficient infection and replication in vivo, but also interacts with many cellular proteins and weakly transactivates promoters of different cellular genes. Studies have suggested that HBx may play a role in HBV-induced liver cancer.

28.2.5 Antigens and antibodies

The virus core itself has its own antigenic specificity, referred to as HBcAg, and, as mentioned in Section 28.2.4 above, through use of an alternative AUG also encodes the HBeAg. The small

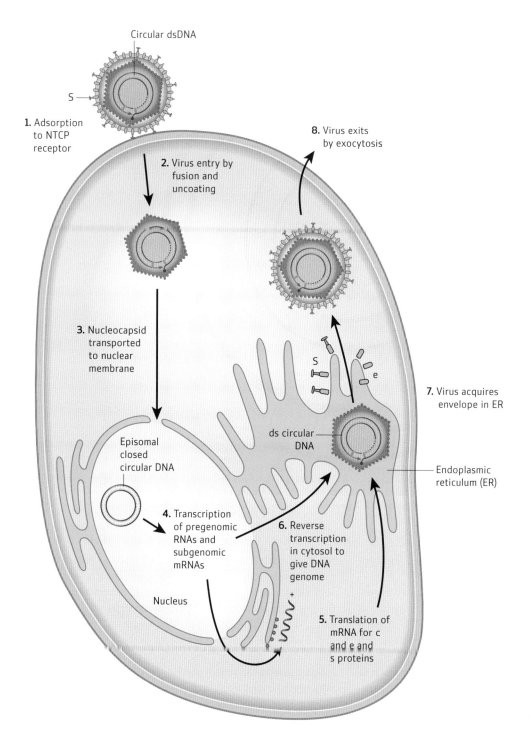

Fig. 28.5 Replication cycle of hepatitis B virus.

surface antigen (SS) is named HBsAg and was, in fact, the one originally found in the blood of the aborigine; for this reason it was at first referred to as 'Australia antigen', but this term is no longer used.

HBV markers

The main antigens **HBsAg**, **HBcAg**, and **HBeAg** each induce corresponding antibodies (Table 28.1). With the exception of HBcAg, all these antigens and antibodies, together with the viral DNA polymerase (encoded in the *pol* gene), can be detected in the blood at various times after infection and are referred to as 'markers', because their presence or absence in an individual patient mark the course of the disease, and also give a good idea of the degree of infectivity for others (Table 28.2). HBcAg is readily detectable only in the hepatocyte nuclei. More recently this antigen and its corresponding antibody have become more important to provide evidence of ongoing 'occult' HBV infection. In this subgroup of patients there is no detectable HBsAg and yet they are infected. Therefore in some countries blood donors are now tested for both HBsAg and HBcAg.

Table 28.1 Serological markers of hepatitis B infection

Marker	Remarks	Presence
HBsAg	Surface antigen, not infective. In three configurations: S, SM, and SL	Acute and chronic infections, including antigenaemia
HBeAg	Found in core of virion as the pre-core protein. Presence in blood indicates infectivity	Acute and chronic hepatitis
Viral DNA polymerase	As for HBeAg, above	As for HBeAg, above
Antibodies		
Anti-HBs	Indicates recovery; protects against re-infection	Convalescence
Anti-HBe	Presence indicates little or no infectivity	Convalescence
Anti-HBc	In IgM form, indicates recent infection	The first antibody to appear. Persists in IgG form for life

NB HBcAg, the core antigen, is not readily detectable in blood and is not used as a marker.

28.2.6 Hepatitis B virus subtypes and genotypes

As well as the main antigens already mentioned, HBcAg, HBeAg, and HBsAg, the surface antigens are endowed with serological specificities that enable us to define subtypes of the virus (Fig. 28.6). There is a group-specific antigenic determinant, *a*, associated with various combinations of subtype determinants *d*, *y*, *w*, and *r*. These combinations are themselves grouped into 10 genotypes labelled A to J, which are useful epidemiologically as their geographical distributions differ. Genotypes have a nucleotide sequence divergence of 7–15%. Genotype A is frequent in the USA, Northern Europe, and Africa. Genotypes B and C are prevalent in the Far East. Genotype D is detected mainly in Mediterranean countries whereas E occurs more often in sub-Saharan Africa. In combination with HBV DNA sequencing they are also helpful in deciding whether a particular carrier—e.g. a surgeon or dentist or a sexual partner—is the source of infection of another person. The finding of identical subtypes would not, of course, confirm the possibility, but differing subtypes would rule it out.

Hepatitis-like viruses existed in primates before the New World monkeys branched off 26 million years ago. However there is no indication of zoonosis with hepatitis B itself. Analysis of human population movements and evolutionary analysis of virus genome mutations indicate that genotype C, for example, existed uniquely in Australian Aborigines for 50 000 years. Some genotypes have spread more recently. Descendants of black slaves in America have genotype A similarly to European immigrants to the USA.

There is increasing evidence that the clinical picture, long-term progression, and even response to drug therapy may relate to the virus genotype. Thus genotypes C, D, and F are more pathogenic whilst genotypes A and B respond more to interferon therapy.

28.3 Clinical and pathological aspects of hepatitis B virus infections

28.3.1 Clinical features

Patterns of infection are variable and are influenced by age, sex, and the state of the immune system. We shall start with acute infections acquired in adulthood and then consider the very special and globally important problems posed by perinatal infection.

Table 28.2 Infectivity of HBV carriers

Markers					
	Antigens		**Antibodies**		
Degree of infectivity	HBsAg	HBeAg	Anti-HBe	Anti-Hbs	Anti-HBc
High	+	+	−	−	+
Intermediate	+	−	−	+	+
Low	+	−	+	+	+

Subtypes of HBsAg

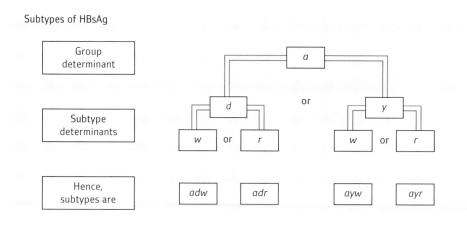

Fig. 28.6 The antigenic subtypes of HBsAg.

Adult infections

The incubation period is usually within the range 60–80 days (Fig. 28.7). Typically, there is a **prodromal phase** similar to that of hepatitis A, but sometimes marked by a transient rash and arthropathy, probably due to virus/antibody interaction. This is followed by overt **jaundice**, after which 90% of patients recover uneventfully within a month or so. In others, however, the outcome may be chronic infection or even rapid death.

Figure 28.8 shows the course of a typical acute infection in an immunocompetent adult. Even before the appearance of jaundice there is a rise in **serum transaminases** and **HBsAg** is detectable in the serum, followed soon afterwards by **HBeAg** and virus **DNA genomes** (not shown). **Anti-HBc** is the first antibody to appear. The next is **anti-HBe**, a good prognostic sign as its production heralds the disappearance of HBeAg and thus of infectivity. Although HBsAg is the first antigen to appear, **anti-HBs** is the last antibody to do so; its arrival indicates complete recovery and immunity to re-infection.

At the other end of the spectrum (Fig. 28.8) one patient in a thousand, usually a female, develops fulminant **hepatitis** and dies within 10 days in hepatic coma; this is the result of an abnormally active destruction of infected hepatocytes by cytotoxic T lymphocytes.

We are now left with the 10% or so of patients who do not come into either of these categories, but who become **chronic carriers**. This diagnosis is made when the serological profile has not reverted to the normal post-recovery pattern within 6 months of onset. There are three varieties.

In **chronic antigenaemia** the patient fails to form anti-HBs and the appearance of anti-HBe may be delayed. Although HBsAg persists in the blood for many years, liver function is normal, the patient is well, and is of little or no danger to others. This picture is often seen in those with impaired immunity. The serological pattern is similar in chronic persistent hepatitis, in which, however, there is a mild degree of liver damage.

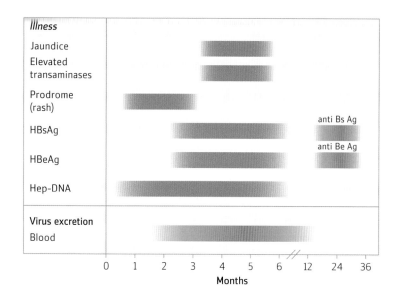

Fig. 28.7 Clinical and virological events during acute infection with hepatitis B virus.

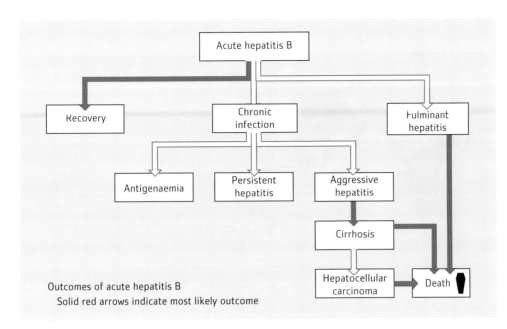

Outcomes of acute hepatitis B
Solid red arrows indicate most likely outcome

Fig. 28.8 Clinical outcomes of hepatitis B infection.

Chronic aggressive (or active) hepatitis is a different story, and these patients create the great bulk of the problems surrounding hepatitis B. They fail to produce either anti-HBs or anti-HBe. As a result, they continue to carry both HBsAg and infectious virions in their blood, and are thus **infectious for others**; they are sometimes referred to as 'super carriers'. There is significant damage to the liver parenchyma and raised transaminase levels indicate impaired function. These patients are liable to repeated clinical episodes of hepatitis and are at risk of developing **cirrhosis**; some may eventually succumb to malignant disease of the liver.

Hepatocellular carcinoma (HCC) may result from integration of viral genome into the DNA of the hepatocytes, but we acknowledge that the pathogenesis is not fully understood. It is clear, however, that HCC arises only after a chronic infection with continuing production of complete virions that has been in progress for at least 2 years. It is often, but not always, preceded by cirrhosis of the liver. Co-factors, such as aflatoxin contaminating groundnuts, may be contributory causes, but there is now no doubt that HBV itself is the primary carcinogen (Fig. 28.9).

Perinatal infection

This is the most important route of transmission of the virus in countries with high prevalence. Perinatal vaccination to break the infection chain has now started in over 180 countries and will be the key to the eventual global eradication of hepatitis B, which has been targeted for 2050.

Infants born to a mother with acute hepatitis B may themselves become acutely infected. By contrast, infants born to mothers who are HBeAg-positive carriers do not develop the acute disease, but 95% of them become infected. Of these, 10% are infected in the uterus of mothers acquiring infection during the first trimester of pregnancy. More often, however, it is transmitted either from **contact with blood and body fluids at birth** or within the next few months by close contact with the mother and siblings. The acquisition of maternal anti-HBc across the placenta seems to impair the normal immune response; unfortunately, nearly all such infants proceed themselves to become HBeAg-positive carriers, and many die later in life of cirrhosis or liver cancer. This is the main acquisition route of the virus in high endemicity countries. The outlook for boys is worse than for girls: 50% of males eventually die of these complications, compared with only 15% of females, who seem to mount a better immune response. The preponderance of surviving females favours perpetuation of the carrier state in populations with a high prevalence of the disease, as it is they who pass the infection to the next generation. As infection acquired early in life does not present as an acute infection with jaundice, the clinical pictures of hepatitis B in areas of high and low endemicity differ considerably.

28.3.2 Epidemiology and phylogeography

Mode of transmission

In modern societies the three ways of contracting the disease, namely neonatal infection, sexually, or via contaminant blood or needles, are closely linked with prevalence of the virus. In zones of high endemicity most infections result from perinatal spread from mothers to their babies. In low endemicity regions of the world sexual transmission is most important whilst transmission from faulty blood products and incompletely sterilized needles is mainly restricted to developing countries.

Concentrations of virus are high in blood and serum, moderate in semen, vaginal fluid, and saliva, and low or undetectable in urine, faeces, sweat, tears, and breast milk. Because the titres of virus are so high in some body fluids (between 1 and 100 million virions/ml), invisibly small quantities—0.00001 ml or even less—can transmit the infection. It is therefore easy to understand that minor abrasions or cuts can serve as portals of entry. HBV is readily spread by **sexual intercourse**, particularly among male homosexuals, and in **intravenous drug abusers** by the sharing of needles and syringes. It may even be transmitted by splashes of minute

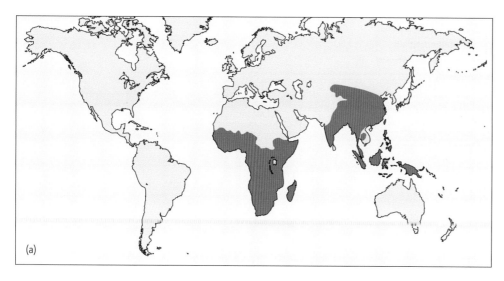

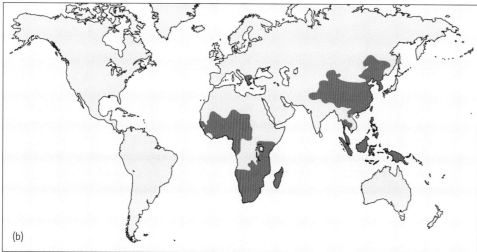

Fig. 28.9 Hepatitis B and liver cancer. Geographical distributions of: (a) HBsAg carriers (prevalence 5% or more), and (b) primary hepatocellular carcinoma (annual incidence 10 cases or more per 100 000 population).

droplets into the eye. There are now quite a number of reports of healthcare workers becoming infected from acute or chronic carriers of HBV, and of patients acquiring the virus from their surgeons or dentists or tattooists. Transmission is possible in homes in the absence of sexual activity but where hygiene is low and there is sharing of towels and toothbrushes or electric shavers.

The worldwide impact of HBV infections—some facts and figures

The prevalence of hepatitis B in a community can be estimated by the presence of anti-HBs or, more readily and informatively, by the proportion of the population who are HBsAg-positive carriers or carry virus DNA. There are 350 million carriers in the world.

- The high prevalence areas (10–20%) are East and South-east Asia, the Pacific Islands, and tropical Africa. However universal HBV vaccination of neonates is so effective that nations with high vaccination rates have already joined the medium prevalence group.

- Russia, the Indian subcontinent, parts of Africa, eastern and south-eastern Europe, and parts of Latin America are areas of medium prevalence (2–20%).

- The prevalence is low (<1%) in the rest of Europe, Australia and New Zealand, Canada, and the USA.

- In the UK, the HBsAg carrier rate in the general population is less than 1%. For this reason the UK has not introduced universal HBV vaccination of neonates. Higher rates in the various ethnic minorities reflect to some extent the prevalences in their countries of origin.

The following figures will give some idea of the devastating effect of HBV infections throughout the world:

- It is estimated that there are 350 million HBV carriers; of these, 75% were infected at birth.

- There are about 950 000 new cases every year in the WHO European region, of whom 90 000 become carriers; of these, 19 000 eventually die of cirrhosis and 5000 of liver cancer. In the USA, the annual incidence of new cases is

Medical case story Hepatitis B case study: view from the hospital ward

Your duty as a doctor is to first 'do no harm', but also to make sure you are untroubled by ill health so that you can do your job effectively. As part of an application to medical school you are asked about your vaccination status. Have you received vaccinations against previously well-known childhood diseases like measles, mumps, and rubella? Have you got a BCG scar? Have you been vaccinated against hepatitis B, and can you prove it?

Doctors sometimes perform procedures on patients that have the potential to spread blood-borne viruses such as hepatitis B and C. Imagine that you are a surgeon and you are doing an open cholecystectomy. You have made a large laparotomy incision and you have both hands inside your patient's abdominal cavity trying to remove the gall bladder. Although you are careful, for a split second you lose sight of your hands and before you know it you have accidentally nicked a hole in your glove and slightly pricked your finger with a scalpel.

What if your patient has hepatitis B? Are you now at risk? What if you have hepatitis B and you've just bled from your finger into the patient's abdominal cavity? Did you know that there could be as many as 100 000 live virus particles in a microdrop of blood?

This sounds like a dramatic and extremely unlikely scenario, but in fact, there have been cases of medical practitioners—doctors, surgeons, and dentists—who have transmitted blood-borne viruses to their patients, whether knowingly or not.

Such was the case of an orthopaedic surgeon who had emigrated from a country of high HBV endemicity to the USA. He had been working as a surgeon for 9 months before it was discovered during evaluation of a needlestick injury that he had chronic HBV infection with a high viral load. The surgeon had previously completed two hepatitis B vaccination series without achieving adequate protection, and thereafter had declined further vaccination.

A team was set up to trace all patients operated on by the surgeon and offer them hepatitis B testing. The surgeon and members of his operating team were interviewed separately to try and identify potential lapses in surgical technique or infection control practices. A total of 13 interviews were conducted, and 328 patients contacted.

Of the 232 patients who consented to be tested, two patients were found to have active hepatitis B infection, a strain that was genetically identical to that of the surgeon. Six patients with no further lifetime risk factors for acquiring hepatitis B were shown to have had the virus and cleared it. It is not known if this virus came from the surgeon. It is not clear from the interviews how the virus could have been transmitted; the surgeon had an excellent track record and always reported double gloving. It was hypothesized that microperforation or unknown laceration of the glove may have occurred.

This case study highlights the importance of adequate vaccination against hepatitis B, but also the adoption of safe practices to prevent transmission of blood-borne viruses, especially those we cannot vaccinate against like hepatitis C and HIV. In doing so, we protect patients against harm, but also the medical staff remain healthy.

Reference

Enfield, K.B., Sharapov, U., Hall, K.K., Leiner, J., Berg, C.L., Xia, G.L., *et al.* (2013). Transmission of hepatitis B virus from an orthopaedic surgeon with a high viral load. *Clinical Infectious Disease* **56** (2), 218–24, doi: 10.1093/cid/cis869.

---

200 000, with corresponding mortality rates for cirrhosis and HCC.

- The global death rate from HCC is estimated at 250 000 per annum.

High-risk groups

This term applies to people who by reason of their country of birth, way of life, or type of work are at higher than average risk of acquiring HBV infection or of passing it on.

We have already described one such group, babies born to carrier mothers in high endemicity areas. Table 28.3 gives examples of high-risk groups in areas of low endemicity; all, in one way or another, are liable to come into contact with infective blood or body fluids. At this point, we recall that the first clue to the nature of HBV was the reaction between surface antigen (HBsAg) in the serum of an Australian aborigine and what we now know to have been anti-HBs in the blood of a haemophiliac, both of whom would have been in high-risk groups.

28.3.3 Pathogenesis and pathology

HBV is even more difficult to propagate in the laboratory than HAV and many aspects of pathogenesis remain obscure, in particular, exactly what happens to the virus during the long incubation period. What is clear, however, is that very many complete (Dane) virions and even more HBsAg particles are liberated into the blood stream. As many as 10^9 virions are present in blood, an extraordinarily high infectious titre of virus.

A vigorous cellular immune response blocks virus replication and most but sometimes not all hepatocytes containing HBV are destroyed. Cytotoxic T lymphocytes (CTLs) act on infected cells only if the viral antigen is presented in association with MHC

Table 28.3 Groups at higher than average risk of HBV infection

Category	Risk factors or group
General community	Sexually promiscuous people
	Intravenous drug abusers
	Partners of HBeAg-positive carriers
	Infants of HBeAg-positive mothers
	Families adopting children from high-risk areas
	Travellers who reside in high-risk countries
Patients	Repeated blood transfusions
	Long-term treatment with blood products, e.g. haemophiliacs
	Chronic renal failure patients
Healthcare staff	Work in mental institutions
	Tours of duty in high endemicity areas
	Surgical and dental operations
	Some pathological laboratory work, including autopsies
	Work in STD clinics. Prisoners and staff in contact with them
	Medical and dental students, nurses, and doctors

class I. Hepatocytes are normally not well endowed with this antigen, but HBV infection stimulates production of IFN-α, which in turn increases the display of MHC class I on the liver cells and thus permits their lysis by the cytotoxic lymphocytes (Chapter 5). Combination of virus-induced lysis of hepatocytes along with destruction by immune cells is responsible for the pathological changes in the target organ, the liver, and the clinical picture of acute hepatitis. However, complete eradication of HBV is not usually achieved because some genomes remain as cccDNA in 'occult' form in liver cells. In such an occult infection with extremely low expression of virions there are no detectable signs of pathology.

In chronic aggressive hepatitis B (Section 28.2.1), there may be cirrhotic changes, which, in a proportion of those thus affected, are a precursor of primary liver cancer.

Since **immunopathological damage is a major component** of the body's response to HBV, it is no surprise that many of the clinical features are notably diminished in those with defective immunity, e.g. those with Down's syndrome or AIDS or in immunosuppressed recipients of human organs.

28.3.4 Oncogenicity

The mechanism of oncogenicity is not well understood. A theory is that hepatocarcinoma starts from clonal expansion of single cells where fragments of HBV DNA have become integrated in several chromosome sites. The protein X sequence may be a preferential site for integration into the human genome. Truncated pre-S and S proteins have oncogenic properties *in vitro* and, along with X protein, activate deregulated cell replication leading to tumours. Antiviral chemotherapy, by suppressing copies of HBV DNA, is expected to stop the progression to HCC.

28.3.5 Laboratory diagnosis

Specific tests

Most larger laboratories now test for HBV DNA as a measure of virus replication. The full marker profile is established by testing, by ELISA, for the three antibodies, **anti-HBs**, **anti-HBe**, and **anti-HBc**. It must be said that the sensitivity of these latter tests could be improved.

Once the diagnosis is confirmed quantitative data on the number of virus genome copies in the blood using real time PCR, for example, is essential to monitor progress of the patient, especially should there be intervention with antiviral drugs. The modern PCR kits allow detection close to the lower detection limit of one HBV DNA molecule per reaction mix along with a huge dynamic range up to 10^7 molecules or more.

Non-specific tests

These are useful for assessing liver function, but obviously do not distinguish between the various forms of hepatitis. In the acute phase or in exacerbations of chronic aggressive hepatitis **the levels of alanine aminotransferase and other liver enzymes are raised**, and prothrombin is depressed. Serum bilirubin is also increased.

28.3.6 Specific treatment with interferon and/or antivirals

Interferon therapy

The acute infection does not normally need treatment, but the threat of an HBeAg-positive carrier state demands action. Some at least of those who become carriers in adult life are naturally deficient in IFN, and some cures have been obtained with large doses of **IFN-α**. There are now established guidelines to select particular patients who would most benefit from drug therapy. For example the American Association for the Study of Liver Disease recommends treatment for those with HBV DNA copies of $>10^5$ copies per ml and ALT (alanine aminotransferase) more than twice the normal upper limit.

An advantage of IFN therapy is that the molecule itself enhances the body's own innate immune system and, for example, may enhance apoptosis of virus-infected cells. Patients with active inflammation, moderate viraemia $<10^8$/ml, and elevated transaminases are also suitable candidates for treatment. Treatment should be prolonged for at least 48 weeks and a decrease in virion DNA level in plasma predicts long-term success.

Some at least of those who become carriers in adult life are naturally deficient in IFN, and some cures have been obtained with large doses of IFN-α. Treatment should be prolonged for at least 6 months. As with Direct Acting Antivirals not all genotypes are equally susceptible and genotypes A and B are more responsive than C and D. The best candidates for IFN-α therapy are those with active inflammation (elevated transaminases) and moderate viraemia. However, side-effects can be quite severe and only a minority of patients show a sustained response.

Directly Acting Antivirals (DAA)

A new era opened when the nucleoside analogue lamivudine, a thio derivative of deoxycytosine, an HIV RT inhibitor, was shown to inhibit the RT of HBV and moreover decreased HBV DNA levels by 1000–10 000-fold. Histological improvement was noted in half the patients. However, after cessation of drug therapy a viral rebound was found and drug-resistant mutants were detected in 70% of treated patients after five years of therapy. Today lamivudine is not used as a first-line drug because of these issues of drug resistance.

Drug combinations of other nucleoside analogues are now being used including the acyclic nucleotide analogue adefovir and the guanosine analogue entecavir. The incidence of drug resistance is low in treatment of naïve patients but unfortunately with pre-existing lamivudine usage cross-resistance problems arise in more than half the patients.

A new era has opened with the discovery of more DAA drugs. Tenofovir is now the drug of choice for chronic hepatitis B, being well tolerated and, at least to date, resistant mutants have not been detected. These newer nucleoside analogues, but also including lamivudine, act by direct inhibition of RT enzyme activity through competitive binding with endogenous substrates and also through direct incorporation into the viral DNA to act as chain terminators.

For prevention of virus re-infection after liver transplantation a combination of RT inhibitors with HBV immune globulin can be used.

Clinical consideration is now given for antiviral therapy of mothers with high viraemia during the third trimester of pregnancy to reduce virus transmission to the newborn baby.

28.3.7 Vaccines and the Universal Vaccine Campaign

Vaccination is the most important way of reducing the worldwide burden of this disease. We are fortunate in having highly effective **vaccines** prepared from HBsAg, which are both safe and effective; and to a lesser extent human **immunoglobulin** with a high titre of anti-HBs (HBIG), which can be used in combination with vaccine to provide immediate passive protection.

Recombinant hepatitis B vaccine

The introduction of **genetically engineered vaccines** superseded the use of first generation vaccines prepared from the chemically treated plasma of infected individuals. The gene coding for HBsAg is cloned into yeast cells, which are cultured on a mass scale rather cheaply and produce large quantities of the antigen. The dosage of the vaccine employed in the UK, for example, is 20 μg of HBsAg given intramuscularly at 0, 1, and 6 months, with boosters at 5-year intervals for those at special risk. The third dose is essential to secure full protection.

Hepatitis B vaccines are highly effective with around 99% sero protection rates in adolescents and children. However there are a small minority of mutant virus strains against which the vaccines do not protect and also 'non responders' to vaccine. These are mainly adults and vaccinated but unprotected non responders (viz. do not have detectable anti-HBS antibody) can amount to 7% of vaccinees. Risk factors for non-response include old age, obesity, smoking, and impaired immune responses because of diabetes and haemodialysis. Some vaccines, unfortunately rather expensive, contain a pre-S region of the S protein and are especially reserved to immunize 'non responders'. This approach is sound because the pre-S is the most effective target of neutralizing antibodies and is highly conserved with fewer escape mutants.

Candidates for vaccination

In areas with high prevalence of HBV carriers, the first priority is immunization of newborn infants (called 'Universal HBV Vaccination'), with the ultimate aim of reducing the incidence of liver cancer. This initiative is particularly crucial because the risk of progression from acute to chronic infection is 90% in infants. Universal HBV vaccination was started in Taiwan as long ago as 1984 and has reduced the childhood carrier rate from 10% to 0.5%; the incidence of liver cancer has dropped by 70%. The Universal Vaccination Campaign has been so successful that some countries such as China will move from high HBV endemicity (10–20% prevalence) to intermediate prevalence (2–10% incidence). The vaccine is credited with protecting

80 million children from HBV. The vaccine is given in three doses and costs less than US$30 for the course.

In the developed countries, candidates for vaccination are as follows:

- The high-risk groups listed in Table 28.3.
- Neonates born to mothers who are carriers, or who had acute hepatitis B in pregnancy (see also 'Perinatal infections', in Section 28.3.1).
- People who have undergone a significant exposure to risk of infection. In this context, 'significant exposure' means: a penetrating injury with a potentially contaminated sharp object; mucocutaneous exposure to blood; or unprotected sexual exposure.

The schedules for post-exposure prophylaxis depend on various factors, including the HBV status of the source and of the person exposed, and the period elapsed since exposure. They are summarized in Table 28.4.

A booster dose of vaccine may be given at 12 months to those at continuing risk of exposure to HBV.

General measures to prevent infection

Horizontal spread of infection is blocked by preventing blood or body fluids from an infected person gaining access to the circulation of someone else. Degrees of infectivity can be assessed from the marker profile (Table 28.2).

Healthcare staff, as well as being vaccinated, must take the obvious personal precautions, such as keeping cuts and abrasions covered, and **wearing gloves** when injecting or operating

upon actual and potential high-risk patients. All hospitals should have detailed **codes of practice** for use in wards, theatres, clinics, and laboratories, and it goes without saying that these must be meticulously observed. They include instructions for the use, whenever possible, of disposable equipment, sterilization by heat, chemical disinfection with hypochlorites or glutaraldehyde, and for the action to be taken in the event of spillages or injury of staff.

Special care is taken to exclude HBV carriers from entering **renal dialysis units**, where, because of the depressed state of immunity of the patients, HBV infection readily becomes established with risks both to patients and staff. Stringent control measures have eliminated hepatitis B from dialysis units in the UK, but it is still prevalent in some other countries.

28.4 Properties of hepatitis D (delta virus)

28.4.1 Classification

This curious little agent was first detected in Italy in 1977, in people undergoing exacerbations of chronic HBV infections. It is known as hepatitis D virus or HDV, and has been provisionally recognized as a separate genus, *Deltavirus*. It is a unique virus. The virus is 40 nm in diameter and contains **ssRNA** only 1.7 kb in size. This is much too small to code for the usual virus functions and so its replication depends on that of HBV itself. Delta or D agent is thus an incomplete virus, reminiscent of the

Table 28.4 HBV prophylaxis for reported exposure incidents (reproduced with permission from *Immunization Against Infectious Disease*, p. 176. HMSO, London, 2006, updated 2010)

	Significant exposure			Non-significant exposure	
HBV status of person exposed	HBsAg positive source	Unknown source	HBsAg negative source	Continued risk	No further risk
<1 dose HB vaccine pre-exposure	Accelerated course of HB vaccine* HBIG×1	Accelerated course of HB vaccine*	Initiate course of HB vaccine	Initiate course of HB vaccine	No HBV prophylaxis Reassure
≥ 2 doses HB vaccine pre-exposure (anti-HBs not known)	One dose of HB vaccine followed by second dose 1 month later	One dose of HB vaccine	Finish course of HB vaccine	Finish course of HB vaccine	No HBV prophylaxis Reassure
Known responder to HB vaccine (anti-HBs >10 mIU/ml)	Consider booster dose of HB vaccine	Consider booster dose of HB vaccine	Consider booster dose of HB vaccine	Consider booster dose of HB vaccine	No HBV prophylaxis Reassure
Known non-responder to HB vaccine (anti-HBs < 10 mIU/ml 2–4 months post-immunization)	HBIG × 1 Consider booster dose of HB vaccine	HBIG × 1 Consider booster dose of HB vaccine	No HBIG Consider booster dose of HB vaccine	No HBIG Consider booster dose of HB vaccine	No prophylaxis Reassure
	A second dose of HBIG should be given at 1 month				

*An accelerated course of vaccine consists of doses spaced at 0, 1, and 2 months.

dependoparvoviruses described in Chapter 26. Its outer coat is, in fact, formed from HBsAg (S, pre-S1 and pre-S2), the specific delta antigen (HDAg small and HDAg large) itself being within the core.

28.4.2 Morphology

The small virions appear to be spherical without visible projecting spikes and consist of a single-stranded RNA genome surrounded by 70 copies of the delta virus-coded antigen (HDAg). There is an outer coat of HBsAg made up of the three hepatitis B proteins (HBS) (Fig. 28.10).

28.4.3 Genome

The negative-sense ssRNA genome is 1.7kb in size and is unique among animal viruses in being **circular**, in this respect resembling certain plant pathogens, the viroids. There appear to be a number of forms of the ssRNA: the genome, a complimentary copy or antigenome, and a linear RNA with a 5′ cap structure and 3′ poly (A) tail form that mediates translation of the HDV antigen. During replication the HDV genome becomes very abundant in cells, reaching levels up to 300 000 copies per cell.

28.4.4 Replication

The virus enters hepatocytes via interaction of the hepatitis B S antigen incorporated into its envelope with the cellular receptor NTCP. The rest of the virus replication is independent of hepatitis B. The virus is uncoated in the cytoplasm and the nucleocapsid transported to the nucleus. There is a dearth of knowledge about any requirement for endosomal compartments and

acidification. The genome is transcribed in the nucleus into a full length circular positive-sense RNA and a shorter linear transcript that acts as an mRNA for the delta antigen. The exact mechanism of virus genome transcription and replication are not understood, but host RNA polymerase I, II, and III are involved. From a molecular virology viewpoint an RNA directed RNA synthesis by a host RNA polymerase is totally unique and not without scientific controversy.

Packaging of the viral genome, 70 molecules of the delta antigen, and the three components of HBsAg (L, M, and S) takes place in the cytoplasm and the complete virion with the hepatitis B S protein is released from the cell. Since virus assembly requires the three envelope (S) proteins of the helper hepatitis B virus the assembly pathway of the two viruses must closely overlap.

28.5 Clinical and pathological aspects of hepatitis delta virus infections

28.5.1 Clinical features

This agent is **transmitted along with HBV**, either at the time of first infection with the latter (co-infection) or during a subsequent exposure (superinfection). It may have little or no effect on the associated hepatitis B infection, but can cause **fulminant hepatitis** and death. Mortality rates of co-infected individuals are 10-fold higher than for hepatitis B alone. Development of cirrhosis is three times higher than for hepatitis B alone and 80% of patients develop this in 5–10 years.

The incubation period is 3–7 weeks, after which there is a prodromal phase with malaise, nausea, and loss of appetite. Jaundice is accompanied by biochemical evidence of liver damage and fulminant hepatitis may develop at this point. HDV infections superimposed on chronic hepatitis B are also liable to become chronic: titres of anti-IgM and IgG remain high indefinitely. Most cases are clinically important.

28.5.2 Epidemiology

Phylogenetic analysis points towards an African origin of the virus with a common ancestor around 1930. The co-infection is still prevalent in Italy where it was first discovered, and also in other parts of the Mediterranean and in the Middle East, Asia, Russia, and Latin America. However, a decline in prevalence is occurring because of universal vaccination of children against hepatitis B. Those mainly at risk are **intravenous drug abusers**, and epidemics among them have been documented in Scandinavia and elsewhere. It is estimated that 15–20 million people worldwide are infected with hepatitis D alongside their hepatitis B infection.

On the basis of sequence differences, HDV isolates have been divided into eight genotypes, which vary in their geographical distribution. Thus type 1 is widely distributed in Europe, the USA, North Africa, and the Middle East, whereas types 2 and 3

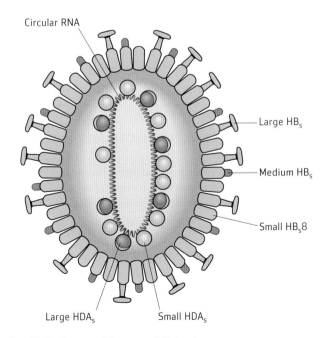

Circular RNA

Large HB$_s$

Medium HB$_s$

Small HB$_s$8

Large HDA$_s$ Small HDA$_s$

Fig. 28.10 Structural features of *Deltavirus*.

are prevalent in the Far East and Latin America, respectively. Recent nucleotide sequence analysis is detecting more virus clades, especially in Africa.

28.5.3 Pathogenesis and pathology

As far as we know, HDV, along with HBV, replicates *in vivo* only within hepatocytes. It is difficult to assign HDV a specific role in pathogenesis when its replication is so closely intertwined with that of the accompanying HBV. There is nothing special about the pathological changes in the liver, which are those to be expected of hepatitis B. Concurrent or superinfections with HDV tend to be more severe than those caused by HBV alone leading to increased chance of severe hepatitis, fulminant hepatic failure, chronic liver disease, cirrhosis, and hepatocellular carcinoma.

28.5.4 Immune response

Both IgM and IgG antibodies are formed. Little is known of any cell-mediated response specific to HDV.

28.5.5 Laboratory diagnosis

Diagnosis depends mainly on **tests for IgG and IgM antibodies**. Both the delta antigen and genomic RNA can also be detected in the blood by RT PCR.

28.5.6 Control

Immunization

Immunization against hepatitis B also protects against co-infection with HDV. Second cases of hepatitis D infection are not known.

General measures

These are similar to those for hepatitis B, the most important being the elimination of syringe and needle sharing by drug addicts and the use of gloves along with high levels of hygiene. Current policies of excluding hepatitis B infected blood donors successfully reduce chances of transmission during transfusion. Like hepatitis B, the virus can be present in blood in very high numbers. The only therapy at present is with alpha interferon.

 # Reminders

HBV belongs to the family *Hepadnaviridae*, genus *Orthohepadnavirus*. It is a spherical enveloped virus with a circular dsDNA genome. The infectious virion is 42 nm in diameter (Dane particle). Uniquely the reverse transcriptase enzyme is used to replicate progeny DNA from an RNA intermediate.

- Seven proteins are coded by four overlapping reading frames and are LS, MS, and SS (the surface proteins), C and E core proteins, the regulatory X protein, and the DNA-dependent polymerase including reverse transcriptase activity.

- Transmission is by **blood** or **body fluids**, very small amounts of which may be infective for others. Neonatal infection from the mother is the predominant method of infection globally in high incidence regions (10–20% prevalence) such as South-east Asia and tropical Africa.

- Sexual transmission and sharing needles of drug addicts are important methods of spread in medium and low incidence countries and regions (<1% low incidence in USA, Western Europe, UK, Australia, New Zealand; 2–20% prevalence in India, S. America, and Eastern Europe).

- During active infection, following a long incubation period of 30–180 days, the blood contains numerous infective virions (Dane particles), and as many as a million times excess non-infectious smaller spheres and tubules of surface antigen (**HBsAg**).

- The antibodies **anti-HBc**, **anti-HBe**, and **anti-HBs** appear in that order and indicate recovery. People who fail to form anti-HBe or anti-HBs become chronic carriers.

- HBe-positive carriers are at risk of **cirrhosis of the liver** and **HCC**. Babies acquiring the infection at birth nearly all come into this category. The high carrier rate in some countries, e.g. in Africa, is correlated with a high incidence of liver cancer.

- **Treatment** of chronic hepatitis acquired in adulthood with IFN-α and lamivudine is successful in 50% of patients.

- A new generation of Directly Acting Antiviral (DAA) nucleoside analogues such as adefovir and entecavir, and also tenofovir, are used in combination to ameliorate clinical symptoms and prevent or at least reduce the emergence of drug-resistant mutants.

- **Prevention** of hepatitis B infection depends on blocking person-to-person transmission by avoiding needle sharing from addicts and, importantly, sterilized needles in hospitals. Blood supplies are screened to remove hep B in many but not all countries.

- The vaccine is highly effective at prevention of infection. All healthcare workers are obliged to be vaccinated and there are extensive Universal Vaccine Programmes to immunize new-borns in countries with high endemicity. In countries with lower endemicity only so called high-risk persons and healthcare workers are vaccinated.

- **Delta agent (hepatitis D)** is an incomplete RNA virus dependent on HBV for replication.

- The genome of delta is a unique circular ssRNA 1.7kb in size and virus replication is restricted to hepatocytes.

- It may be transmitted along with hepatitis B, or may come as a superinfection. It exacerbates the normal clinical course of

HBV. The virus is prevalent in **intravenous drug abusers**. Diagnosis depends mainly on tests for IgM and IgG antibodies. Hepatitis B vaccine also protects against HDV, whilst increased screening of blood banks is also reducing its prevalence.

Further reading

Fung, J. *et al.* (2011). Nucleoside/nucleotide analogues in the treatment of chronic hepatitis B. *J.A.C* **66**, 2715–25.

Gerlich, W.H. (2013). Medical virology of hepatitis B: how it began and where we are now. *Virology Journal* **10**, 239.

Hughes, S.A. *et al.* (2011). Hepatitis delta virus. *Lancet* **378**, 73–85.

Lai, M.M. (2005). RNA replication without RNA dependent RNA polymerase: surprises from hepatitis delta virus. *Journal of Virology* **79**, 7951–8.

Taylor, J.M. (2006). Hepatitis delta virus. *Virology* **344**, 71–6.

Trepo, C. *et al.* (2014). Hepatitis B virus infection. *Lancet* **384**, 2053–63.

Questions

1. Outline the replication strategy of hepatitis B virus and identify stages amenable to intervention with new drugs.

2. How successful is the hepatitis B vaccine? Suggest improvements dependent upon knowledge of virus structure.

3. Discuss pathogenicity of hepatitis B and D viruses.

Practical aspects

The clinical virology laboratory

29

29.1 Introduction

As detailed knowledge of the genomes of viruses has increased, and with the discovery of nucleic acid amplification methods such as the polymerase chain reaction (PCR), the use of virus genome detection for clinical diagnosis has increased dramatically. In general, virus genome detection is more rapid and sensitive than cell culture and can be used before a detectable antibody response has developed. Further, with the discovery of potent antivirals for hepatitis C, hepatitis B, HIV, and influenza there is now huge demand for the diagnostic laboratory to monitor levels of virus in the appropriate clinical sample of the patient using rapid molecular tests. These tests can also give the first indication of drug-resistant virus mutants, which could compromise antiviral therapy. This change has happened during the lifetime of this textbook. At the time of the first edition of *Human Virology*, in 1993, the clinical diagnostic laboratory mainly isolated viruses using classical cell culture techniques. Viruses were grown in the laboratory over a period of 4–15 days and identified by cytopathic changes in the cells. Retrospective serological diagnosis was also commonly used and in this manner a huge database was established and clinical syndromes were linked to viruses. But this was of little use to the ill patient.

Molecular techniques for virus diagnosis have now taken over the laboratory almost universally and are highly commercialized in the form of kits, giving sensitive and reproducible results. Whole virus genomes can be sequenced in days and at large scale using new 'Next Generation Sequencing' (NGS). Importantly the particular genotype of the virus may be identified within the broader family and the potential virulence and antiviral susceptibility can be predicted from the virus genotypes. There are also some bedside tests for diagnosing influenza or RSV and norovirus, which take 15 min and resemble the use of a dipstick for diabetes. Serological diagnosis of HIV in a patient can be done at home in 2–3 min. In a first for virus genetics, hepatitis C was identified only by molecular detection of its genome without any ability to culture the virus. The power of RT PCR and virus genome nucleotide sequencing was demonstrated in a very practical manner in China in 2009 in the face of a swine influenza A (H1N1) outbreak. In early 2009, when a new influenza-like illness not related to known human influenza viruses was reported, the rapid virus genome sequencing of the new influenza virus showed it was derived from several viruses circulating in pigs, with the initial transmission to humans having occurred several months before recognition of the outbreak. Twenty million travellers at airports were screened by thermal imaging and for clinical symptoms. Throat swabs from 20 000 persons with a high temperature were sent to 63 regional laboratories who returned data within 3 h. Seven hundred and fifty persons were found to be positive for swine flu and quarantined for 7 days in an attempt to stop the spread of the virus in China.

More recently, in 2014, samples taken during the Ebola outbreak in West Africa were rapidly virus genome sequenced, showing they were all genetically similar and allowing the conclusion that West Africa was dealing with a single source outbreak rather than multiple infectious sources. This was reassuring and easier to control than a multi-source outbreak. However, continued genome sequencing of Ebola shows that multiple lineages of the virus arose in humans and some of those were still circulating in 2015. In contrast, for MERS CoV in Saudi Arabia, the same type of virus genome analysis showed that the virus causing human infections was genetically more diverse than predicted and, given a lack of large-scale human infection, suggested that different virus lineages were the result of different infections from an unknown source, a source later identified as camels.

Most diagnostic virology laboratories operate to level 2 of biological safety and can handle typical epidemic community viruses such as influenza, herpes, HIV, and hepatitis A, B, and C. In contrast, emerging viruses such as pandemic influenza, SARS Cov, MERS CoV, and Ebola require the use of level 3 and level 4 (the highest category) laboratories and highly trained staff. However, once a clinical sample has been placed in disruption medium for molecular amplification, by definition the virus has been destroyed and is safe to use in a category 2 laboratory, although care must still be taken, as some viruses have infectious genomes if the virus nucleic acid remains intact and can be introduced into cells. These laboratories, therefore, are carefully monitored by authorities for quality control and usually operate to an internationally agreed system called Good Laboratory Practice (GLP) or Good Clinical Practice (GCP).

The smaller virology laboratories have become totally dependent on molecular diagnostic kits and real time PCR, whereas the larger establishments retain a variety of methods at their disposal, including cell culture and electron microscopy.

29.2 Collection and transport of clinical specimens to the laboratory: vital steps

From the point of view of the new clinical virologist and also the scientist performing the test and perhaps undertaking research into viral pathology this is probably one of the most important sections in the textbook, and careful thought must be given—based, one hopes, on a knowledge of viral pathology—as to which clinical samples should be collected. Whichever techniques are used in the clinical or even research laboratory, no results will be forthcoming without good quality samples from the patient.

To perform its task properly the laboratory must be provided with:

- the right specimens;
- taken at the right time;
- stored and transported in the right way (for example, some viruses can be completely destroyed by freezing).

Table 29.1 summarizes the types of specimen to be collected when viral infections of various body systems are suspected.

Table 29.1 Specimens required for isolation of virus or detection of antigen or viral genome

Disease	Specimen
Respiratory infection	Nasal or throat swabs; post-nasal washing
Gastrointestinal infection	Faeces and diarrhoea (rectal swab not so satisfactory)
Vesicular rash	Vesicle fluid, throat swab, faeces
Hepatitis	Serum, faeces
Central nervous system	Cerebrospinal fluid, throat swab, faeces
AIDS	Unclotted blood and/or plasma for genome quantification

NB. In addition to the above, 5–10 ml of clotted blood for serological tests is always required for determination of 'resting' levels of antibody. A sample of blood is needed at the acute phase of the illness and around two weeks later when antiviral antibodies will have developed. A comparison of the level of antibodies in the two samples may often provide a definite diagnosis of infection with a particular virus.

29.2.1 Clinical sampling

Swabs

The amount of material collected must be adequate. In particular, throat or skin swabs must be taken fairly vigorously. Swabs are broken off into a vial of transport medium, usually tissue culture medium of amino acids and vitamins containing antibacterial and antifungal antibiotics to inhibit contaminants, a protein stabilizer (such as bovine serum albumin) to protect sensitive viruses, and a buffering solution at pH 7.0. For molecular diagnosis, the clinical material can be placed in a special transport medium, which denatures virus protein, inactivates the virus, and releases viral RNA or DNA. This material, if processed correctly, is safe to handle afterwards, even for viruses such as Ebola needing level 3 security for virus cultivation.

Nasopharyngeal aspirates

These are very useful in the diagnosis of upper respiratory tract infections of young children, e.g. RSV, but obtaining a satisfactory specimen with minimum distress to the child requires skill and practice. Bedside molecular diagnostic kits are now available for influenza and RSV.

Vesicle fluid

Vesicle fluid for electron microscopy (EM), for example for a poxvirus or herpes, is collected on the tip of a scalpel blade, spread over an area about 3–4 mm in diameter on an ordinary microscope slide, and allowed to dry. A swab can be used for molecular diagnosis.

Faeces, to identify enteroviruses or rotaviruses

These should be placed in a dry sterile container; they are preferable to rectal swabs for virus isolation. A bedside diagnostic kit for rotavirus gives rapid answers.

Clotted blood

As a general rule 5–10 ml blood is taken. A syringe, rather than a vacuum tube, is used to take blood; the needle should be removed before expelling the blood to avoid haemolysis.

Unclotted blood

EDTA-treated blood to stop clotting is used for detecting HIV, hepatitis B, and hepatitis C genomes by RT PCR and sequencing.

29.2.2 Storage and transport

Specimens should be placed in secure plastic bags and labelled in accordance with local practice. They should go to the laboratory as soon as possible after collection; if kept overnight, they should be held at 4 °C, rather than being frozen, which tends to destroy viruses with a lipid envelope.

Much time will be saved if request forms are properly completed. Brief indications of the **date of onset** of illness, **clinical signs**, and **suspected diagnosis** are much more important than a specification of the tests required.

29.3 Rapid diagnostic methods

Many clinical laboratories can give diagnostic and also quantitative results within a working day, particularly for quantitation of genome copies of HIV and hepatitis C and B. Antibody techniques have a longer (2 day) turnaround time.

29.3.1 Detection of viral genomes by nucleic acid amplification methods

These molecular methods now dominate viral diagnosis and are used in most clinical laboratories to detect and quantify the DNA genomes of CMV, hepatitis B, and HPV in clinical samples.

RNA viruses such as HIV, hepatitis C, and influenza can be detected, and also quantified by incorporating an initial step of reverse transcriptase (RT) to transcribe the viral RNA to a piece of DNA (RT PCR).

Viral genome quantification in patients' samples ('viral load') using real time quantitative PCR is now a major activity in clinical laboratories, particularly for patients treated with the newer antiviral drugs, to measure the reduction in viral load following chemotherapy of HIV, and hepatitis B and C viruses.

To avoid cross contamination, which is possible with a few molecules of DNA or RNA, laboratory staff need to be trained with detailed standard operating procedures (SOPs). For laboratories which still prepare mixtures of reagents there must be physical separation of reagents and equipment. Four completely separate laboratories are needed and each laboratory will have its own gowns and equipment. These laboratories are a reagent preparation laboratory for the PCR master mix, a clinical sample processing laboratory where nucleic acid is extracted, a target loading laboratory where the clinical sample is pipetted into the PCR master mix, and finally a thermocycling laboratory. The samples should flow in the direction of laboratory 2 to 4. Staff must change gowns and thoroughly wash and disinfect hands before moving laboratories.

Polymerase chain reaction (PCR)

Two distinct oligonucleotide primer sequences carefully chosen to match a common nucleotide sequence of the virus under investigation, one on each strand of the target viral DNA are added to a clinical sample that has been treated with heat (94 °C) and detergent to denature and, therefore, physically separate the strands of viral DNA. The primers specifically hybridize with the homologous nucleotide stretches on the viral DNA genome. A DNA polymerase termed Taq polymerase (from *Thermophilus aquaticus*), which acts at high temperature, is added. After 1 min the temperature is reduced to 52 °C for 20 s to allow annealing of primers and the temperature is then raised to 72 °C for 5 min to allow DNA polymerization to occur. Under these conditions, and only if the oligonucleotide primers have hybridized, the Taq enzyme generates multiple copies of the nucleotide stretch between the two primers. Multiple cycles of DNA denaturation, annealing of primers, and polymerization can be programmed in the microprocessor-driven heating block. In this manner, a portion of a single molecule of viral DNA can be amplified a millionfold in a few hours to give a quantity of DNA that can be separated in a polyacrylamide gel, and then even visualized by addition to the gel of ethidium bromide and exposure to UV light. 'Nested PCR' is even more sensitive. Following initial amplification of a unique stretch of viral DNA, a further set of 'internal' primers is added that anneal to DNA within the original fragment, allowing a smaller stretch to be amplified.

Multiplex PCR

A technique whereby several primer sets are incorporated into the reaction mixture, so allowing the detection of genes from multiple viruses which can cause the same clinical symptoms. An excellent example is a respiratory syndrome whereby primers can be designed for influenza, RSV, adenovirus, and common cold, and so allow precise virus causation to be established. As new antivirals for respiratory viruses are developed, multiplex PCR will become even more important. As a word of warning these multiplex techniques sound straightforward but in practice there are often problems of the primer sets giving spurious interactions. Overall this method is less sensitive than the use of individual primers for a particular virus.

Branched chain techniques

The method uses highly sensitive branched DNA (bDNA) probes to detect and quantify viral RNA sequences. The sensitivity derives from signal amplification, rather than the target amplification that provides the basis for PCR.

For example, to measure HIV RNA copy number in plasma, the clinical sample is first centrifuged to pellet the virus particles. The virus is lysed and the viral RNA mixture is added to microtitre plate wells coated with oligonucleotide probes, which match conserved sequences in the HIV genome. The viral RNA forms double-stranded duplexes with the probe sequences and is thereby captured. After the well is washed, bDNA amplifier molecules are hybridized to the bound HIV RNA, and then alkaline phosphatase probes that bind to the bDNA amplifier molecules are added. The HIV-specific bDNA probes are thus linked to an enzyme that catalyses the release of a chemiluminescent molecule from its substrate, and the amount of light produced is proportional to the quantity of viral genome. The method is faster, less laborious and expensive, and requires less technical ability than PCR.

The NASBA amplification technique

This method targets RNA viruses or mRNA transcripts of DNA viruses and uses three enzyme systems at the same time to amplify a particular viral genome sequence. It can be quantitative. The three enzymes are RT, phage T7 DNA-dependent RNA polymerase, and RNase H.

A viral genome specific primer is added, which also incorporates the T7 promoter and hybridizes to the viral genome. This is extended by the RT enzyme. The RNase degrades the RNA strand and the RT, utilizing a second primer, produces template DNA. Multiple copies of RNA are produced from this DNA template by the T7 DNA-dependent RNA polymerase.

Quantification of viral genomes by real time polymerase chain reaction

Undoubtedly, this technique has become a laboratory workhorse. The best known system is TaqMan. Essentially, this PCR method does not wait for an end quantitation, but detects and follows genome amplification as it goes along. A specific probe binds to the viral amplicon under investigation, and is hydrolysed to produce fluorescent molecules, which are immediately detected and quantified. Alternatively a dye is encouraged to intercalate into the dsDNA being produced in the first reaction, and as more dye is trapped the fluorescence increases. Often real time PCR reactions finish within minutes, giving an instant clinical diagnosis.

Isothermal methodology

This method does not require precision thermal equipment but rather a simple water bath, and the two favourite techniques in use are loop mediated isothermal amplification (LAMP) and helicase dependent amplification (HDA).

Next Generation Sequencing (NGS)

There is an annual growth of viral genome sequences deposited in GenBank of around 20%, with around 1 billion viral sequences now on deposit. Most of these sequences are derived from the human viruses discussed in this textbook. There are over 400 000 sequences of HIV, 100 000 of hepatitis C, and 30 000 of influenza A virus. With such enormous data resources virus genome sequencing has been used to support diagnostics, and with the advent of high-throughput next generation sequencing (NGS) it is predicted that such methods will become essential for many areas in clinical virology, including virus discovery and metagenomics (viromes), molecular epidemiology, and pathogenesis, as well as supporting decisions on vaccine design and antiviral drug resistance. Up until recently, many viral genes on deposit were sequenced by first generation methods involving DNA amplification by PCR and DNA sequencing using chain terminating inhibitors (Sanger sequencing). This method allowed up to 1000 bases per day to be generated per sequencing run. The new massively parallel and high-throughput methods of NGS enable many millions of bases of sequence to be studied in the same sequencing run. Virus genomes can be produced by NGS from the same clinical specimens as used for other molecular diagnostics. DNA sequencing technologies have changed dramatically since the mid-2000s; however, technology has reached what is likely to be a temporary plateau, with three so-called NGS sequencing methods platforms—Illumina, Ion Torrent, and PacBio—representing most of the currently purchasable machinery. Essentially, all platforms require a DNA sample to be prepared as an NGS library for sequencing, or for RNA viruses for cDNA to be made first. For non-cultured clinical samples, however, the majority of the nucleic acid in a sample is derived from the host genome (or transcriptome), meaning that if sequenced in its totality only a small percentage (often less than 1%) of the total sequence produced from the sample is pathogen. The solution for large-scale sequencing of virus genomes direct from clinical samples is to enrich the virus genome, either by whole genome PCR amplification or via genome capture and enrichment by hybridization. Through

this, the number of samples that can be sequenced per run of the sequencing machine is increased, as NGS sample libraries can be barcoded and pooled, allowing 10s to 100s of samples to be sequenced at the same time. However, what is still lacking is a good set of computer programs for the non-expert assembly of virus genomes and the use of the considerable information that can be gained from them. This is improving, but at the time of writing there is no commercial solution for virus NGS in clinical virology labs. We predict this will be solved by the next edition of this book. Nevertheless the case study presented here shows the power of NGS especially in outbreaks of new virus infections.

Monitoring the clinical effects of antivirals by quantitation of viral load

Laboratory help is essential for clinical care of AIDS patients being treated with drug combinations, HAART (Chapter 31). Patients need to be monitored every 2 months and expectations are to detect fewer than 50 HIV genome copies per ml of plasma after antiviral drug treatment, compared with a typical figure of 10 000 RNA genome copies at the start of antiviral therapy, say 3–4 months previously. The test also needs to quantify virus from HIV virus groups M, A–G, and N and O with equal efficiency.

The laboratory is also essential for the continuing care of patients chronically infected with hepatitis B and treated for example, with a combination of lamivudine, famciclovir, and adefovir. One 100- to 1000-fold reduction of viral DNA load would be typical in the months following antiviral therapy.

Similarly, a rapid and sustained reduction in RNA genome copies of HCV following therapy with IFN-α and ribavirin or the

Case study Use of Next Generation Sequencing in Middle East Respiratory Syndrome Coronavirus (MERS CoV)

When a patient developed severe respiratory illness in Saudi Arabia in 2012, all known tests for human respiratory viruses were negative. Following tissue culture of a sample from the patient a portion of the culture media was sequenced by NGS and a new coronavirus sequence was discovered, closely related but nevertheless genetically distinct from SARS CoV that infected 8273 people with 775 deaths in 2002–3. The virus genome led to the rapid development of PCR diagnostic tests and the identification of more people infected with the virus in the Arabian peninsula. Mathematical modelling of the new virus epidemic suggested the virus was possibly just able to sustain human-to-human transmission but that most infections were only recognized in hospital settings. In the first use of NGS in a virus outbreak, virus genome sequencing and phylogenetic analysis showed that the hospital transmissions were short chain trans-

missions and the virus genetic diversity was consistent with multiple new infectious introductions into humans from an unknown source, followed by short transmission chains in humans. This was reassuring, as, although the potential for sustained human transmission exists, the virus is so far not well adapted for human spread. Serology of animal samples suggested the camel was the source of MERS CoV, and again sequencing the virus genome in camels and humans with respiratory disease and close contact with camels showed the virus was the same, confirming camels as a source of infection. Recently, from May to July 2015 MERS CoV caused multiple infections in South Korea. Rapid virus genome sequencing showed the MERS CoV in South Korea was genetically related to the virus circulating in 2015 in the Arabian peninsula, and as such reassured that infection control procedures could still contain the virus successfully.

new directly acting antivirals such as telaprevir and boceprevir (protease inhibitors) and also daclatasvir (polymerase inhibitor) would predict successful treatment. With hepatitis C the laboratory can also help by identifying which of the five types has infected the patient, because these respond differently to antiviral therapy.

The quantitative PCR methods described earlier in this section give the first warning of a resurgence of viraemia in a patient following long-term therapy. The first possibility to be investigated is point mutations in the RT or protease genes of HIV, or in the polymerase or protease of hepatitis C and the neuraminidase of influenza, indicating emergence of a drug-resistant virus. Obviously, HIV, hepatitis C, and hepatitis B patients are treated with drug combinations to minimize this possibility. However, rapid clinical action may have to be taken to replace one of the drugs in the combination.

Usually viruses avoid the action of antiviral drugs by a point mutation in the target gene and these mutations are well known. Therefore, a number of molecular tests can be used to search for these particular mutations. In particular, a so-called point mutation assay utilizes PCR primers synthesized so as to hybridize to the drug-sensitive or drug-resistant virus only. However, direct nucleotide sequencing is now so automated that this is the method of choice, sometimes used in conjunction with so-called chip technology, whereby literally thousands of pre-synthesized oligonucleotides arranged on a microchip are allowed to interact with a PCR-amplified DNA from the virus in question. Computer analysis of these interactions can pinpoint a dominant drug-resistant mutant in a viral population.

29.3.2 Enzyme-linked immunosorbent assay (ELISA) and radioimmunoassay (RIA) to detect viral antigens

ELISA has become, alongside PCR, the workhorse of most diagnostic virology laboratories as it is automated, kits are commercially available, and it can be adapted to the identification of many viral antigens, e.g. the p24 antigen of HIV-1, hepatitis B, rotavirus, etc., and the corresponding antibodies. For detection of a virus the clinical sample is added to a plastic plate to which is bound antivirus antibody (Fig. 29.1). After washing, an enzyme-labelled specific antibody is added and if the virus antigen has bound then fluorescence is detected even after washing because of tight antibody binding.

29.3.3 Latex agglutination tests

Latex particles are coated with viral antigen and agglutinate when mixed on a slide with specific antiserum. The test is rapid, easy to read, and does not require complicated equipment. It is, however, liable to prozone effects, giving false negative results at low dilutions of serum. It is useful for the diagnosis of rotavirus and norovirus in samples of diarrhoea.

29.3.4 Electron microscopy and immunoelectron microscopy

Samples are negatively 'stained' with phosphotungstic acid, so that the virions, which are not penetrated by the stain, stand out as white particles on a dark background. However, virus particles must be present at a concentration of at least 10^6/ml to stand a chance of being identified; it is therefore sometimes nec-

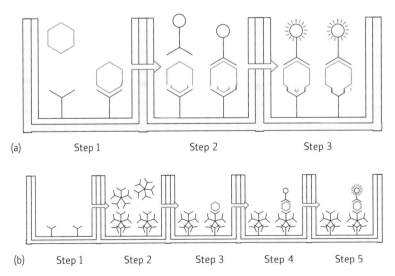

Fig. 29.1 (a) Direct identification of antigen by capture and ELISA. Step 1: Addition of specimen containing antigen that combines with the specific 'capture' antibody on a plastic surface. Step 2: Addition of enzyme-labelled specific antibody. Step 3: Substrate is added, reacts with bound enzyme, and undergoes colour change. (b) Identification of specific IgM antibody by capture and ELISA. Step 1: Plastic surface coated with antibody to IgM. Step 2: Patient's serum added; IgM molecules are captured by the anti-IgM. Step 3: After washing to remove unattached IgM, test antigen is added and combines with any captured IgM of the same specificity. Steps 4 and 5: as steps 2 and 3 in (a). Note that the captured IgM molecule on the left, having no specificity for the test antigen, does not react.

essary to use simple virus concentration methods such as a high-speed microfuge (18 000 rpm). Needless to say an experienced microscopist is needed if the results are to be reliable. The method is used less and less nowadays, but was still used to identify the SARS coronavirus. The 'small round' viruses causing diarrhoea are often present in diarrhoea at levels of 10^9 per ml and are easily spotted by this method.

The value of the test may on occasion be increased by using immune electron microscopy (**IEM**), which is the addition to the specimen of specific immune serum that agglutinates a particular virus, thus making the virions easier to find and adding serological specificity to their identification.

29.4 Virus isolation in cell cultures

This original technology of the virology laboratory is being used less and less, but still represents a 'gold standard' for molecular tests. It was used with great success, in conjunction with electron microscopy, to identify the novel coronavirus causing SARS in 2003. Cell culture methodology is used nowadays in pharmaceutical and research laboratories for the production of virus vaccines. Molecular cloning methods and the use of plant cells and yeast cells will surely overtake mammalian cell culture for vaccine production in the next decade.

Some large diagnostic laboratories will use three cell lines such as Vero (monkey kidney), MDCK (dog kidney), and MRC-5 (human diploid) for screening and cultivating for viruses mainly from the respiratory tract, CSF, or faeces. These mammalian cells are grown in a single layer (**monolayer**) on the surface of a plastic container and inoculated with the test specimen. They are then observed daily by light microscopy for virus-induced cytopathic changes (**CPEs**), which are sometimes sufficiently characteristic to be reported without further investigation.

29.4.1 Propagation of cell substrates

The **growth medium** used to cultivate cells contains a solution of salts at physiological concentrations, glucose, amino acids, essential vitamins, and antibiotics to inhibit bacterial and fungal contaminants; it is buffered at pH 7.2–7.4. Foetal calf serum is added to a concentration of 10–20% to provide supplements essential for cell growth. When the cells have formed a confluent monolayer the growth medium is replaced by a **maintenance medium** containing only 2–5% serum, which permits little or no further cell multiplication.

When test cells are needed, a stock monolayer is treated with trypsin or versene to disperse it into a suspension of individual cells, which are then diluted in growth medium to a concentration of one million per ml and distributed into other vessels for use. Multiwell plastic dishes are now widely used.

29.4.2 Types of cell culture

Semi-continuous cell lines

Such cells are derived from human or animal foetal tissue. They have the normal diploid karyotype and, hence, can be used for

vaccine production. Some lines are also used for diagnostic work. Classic cells are called WI-38 (from the Wistar Institute in Philadelphia) and MRC-9 (from the Medical Research Council laboratories).

These cells are termed 'semi-continuous' because they have a limited lifespan and can be subcultured through only 50 or so generations. Even so, millions of cultures can be obtained from a single foetal organ. In practice, several thousand vials of cells suspended in dimethyl sulphoxide are frozen in liquid nitrogen after only a few passages in the laboratory. Each of these vials can be used to generate further cultures (the 'seed lot' system). For measles, mumps, or rubella vaccine production, only cells at a given low passage level, e.g. the fourth, are used and each batch of cells is carefully tested for karyotype and other properties to exclude the possibility of malignant change. Such strict criteria are not necessary in the diagnostic laboratory, where cells from several successive subcultures may be used for virus isolation.

Continuous cell lines

These are the most widely used for diagnostic and research work. They are derived either from a tumour or from normal cells which after repeated culture have become transformed so that they behave like tumour-derived cells, i.e. they have an abnormal number of chromosomes. They can be propagated indefinitely. Some cell lines such as Vero (monkey kidney) or MDCK (dog kidney) are now licensed to cultivate viruses for vaccines such as polio and influenza respectively.

Lymphocyte cultures

B lymphocytes will divide and continue to do so indefinitely if infected with the herpes virus Epstein–Barr (EBV). This characteristic, known as '**immortalization**', is an example of cell transformation by a virus.

Importantly, T lymphocytes will grow in the presence of a lymphokine, IL-2, formerly known as T-cell growth factor. This finding proved essential to the study of the human retroviruses (HIV and HTLV, Chapter 27), which can be propagated in such cultures with the formation of syncytial giant cells.

29.5 Detection of antiviral antibodies

Investigation of post-infection antibodies in a patient can be a very important test. ELISA methods are easily automated and have replaced the previous rather specialized techniques of virus neutralization and HI tests (for influenza) except in specialized research laboratories. It must be emphasized again that isolation of a particular virus, although suggestive of a diagnosis, does not always prove a casual association between it and the patient's illness. This is especially the case with the molecular tests detecting virus genomes which can be too sensitive and may not indicate an infectious virion. As we saw in Chapter 4, some viruses are shed from clinically normal people. This uncertainty factor can be greatly diminished by the use of serological tests, in which (1) a **rising titre of antibody** to a particular virus is sought, or (2) serum is tested for the presence of **specific IgM antibody**. Method (1) depends on testing paired samples of serum, the first being taken as soon as possible after onset and the second 10–14 days later. **A fourfold or greater rise in titre** of the relevant antibody is considered significant. Method (2), more widely used, has the advantage of rapidity in that specific IgM antibody is detectable a few days after the onset of illness. Some techniques detect newly secreted antibodies from B cells. The so-called ELISPOT technique and ELISA methodology is used to pinpoint and quantify individual antibody-producing B

cells which have been prior diluted to approximately 100 cells per plastic cell.

29.5.1 Class-specific (IgM) antibody tests

We have already described the rapid ELISA method used for detecting viral antigens or antibodies. The ELISA-type 'capture' method (Fig. 29.2) in particular is adaptable to the detection of

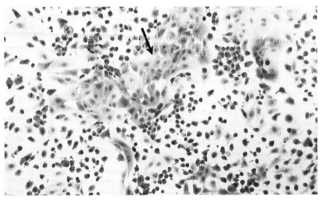
(a)

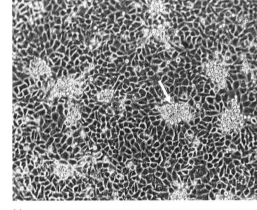
(b)

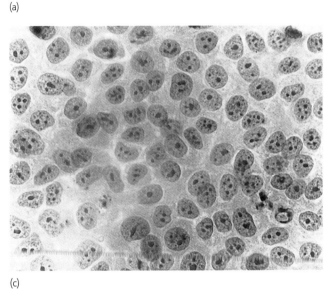
(c)

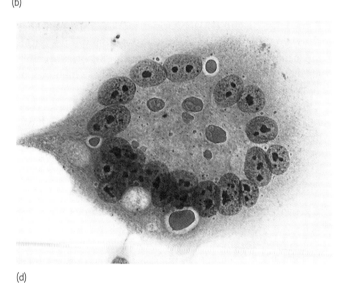
(d)

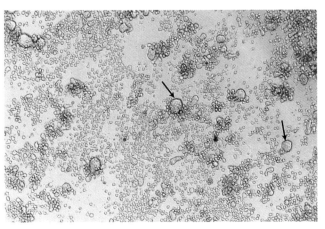
(e)

Fig. 29.2 CPEs of viral infection. (a) Enterovirus infection of a continuous line of human embryo lung cells (HEL). Areas of rounded, dead, or dying cells lie between islands of normal cells (arrowed). (b) HSV type 2 infection of baboon kidney cells (phase contrast). The affected cells (arrowed) are swollen and refractile. (c–d) High-power view of a continuous line of human epithelial cells (c) and a multinucleate giant cell (d) resulting from infection with respiratory syncytial virus. (e) Infection of a line of human lymphocytes with human immunodeficiency virus (HIV-1). Multinucleate giant cells are arrowed. (Parts (a) and (c) are reproduced with permission from Robinson, T.W.E. and Heath, R.B. (1983). *Virus Diseases and the Skin*. Churchill Livingstone, London.)

specific antibodies, of which IgM is the most useful diagnostically. It is detectable within days of infection and remains so for 3–9 months, so that its finding is good evidence of a current or recent infection.

29.5.2 Immunoblotting methods

The Southern blot method has no geographical connotation, but was named after the worker who invented a widely used method for DNA hybridization. A similar method used for **RNA** hybridization inevitably became known as **Northern blotting**, and **Western blotting** refers to its application for identifying **proteins**. It is still used in some laboratories for diagnosis of HIV infection.

As the correct diagnosis of HIV-1 infection has such important personal and social implications, it is necessary to confirm a positive result from ELISA (see Chapter 27) by a different technique, usually Western blotting (or immunoblotting). Virus proteins are separated as bands according to their molecular weights by electrophoresis through a polyacrylamide gel. The bands are eluted ('blotted') on to chemically treated paper, to which they bind tightly. The test human serum from the patient is added to the paper strip and any specific antibody attaches to the viral proteins. As in some other sandwich-type tests, an antihuman antibody labelled with an enzyme is added, followed by the enzyme substrate; the paper is then inspected for the presence of stained bands, which indicate the presence of complexes of specific antibody with antigen (Fig. 29.3).

We can conclude that the virology diagnostic laboratory, like its counterpart in bacteriology, holds a key spot in the mind of general practitioners and hospital doctors alike. Not to be forgotten is the patient, who often understands as well as anyone what difference a tenfold reduction in viral genome can mean whilst undergoing antiviral therapy.

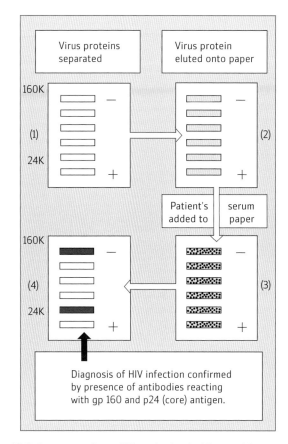
Fig. 29.3 Detection of anti-HIV antibodies by Western blot.

As a final footnote, the diagnostic laboratory has a niche role in forensic phylogenetics to sequence samples from different individuals, perhaps infected by mistake by a doctor with hepatitis C or even deliberately by a partner with HIV. By constructing a family tree it can be seen whether two infections are likely to be interrelated if a likelihood ratio is calculated.

Reminders

- Provision of **appropriate patient specimens** and clinical information is essential.
- Laboratories use molecular techniques (PCR, branched chain, NASBA) for quantifying viral genomes in patients treated with antivirals and analysing viral genomes for drug-resistant mutations.
- TaqMan (Roche) is a typical real time PCR technique able to give instant quantitative data on viral genome levels.
- Next Generation Sequencing (NGS) technologies are used for understanding virus outbreaks, transmission chains, and searching for 'new' viruses but only from already existing virus families.

- Electron microscopy has had a resurgence of interest for identification of pox viruses and was also important for diagnosing the first SARS infections. The specificity of electron microscopy is enhanced by the use of monoclonal antibodies.
- **Virus isolation in cell cultures** has been superseded by the more rapid molecular methods in many laboratories. It is still useful for the identification of new viruses. Viruses cause particular cytopathic changes in cells, which may give a good indication of the family of a new virus concerned.
- ELISA techniques can be automated and are used to detect virus-reacting antibodies for diagnosis and to check antibody levels in pregnant patients who have been vaccinated with rubella.

Further reading

Bhattacharya, S. (2014). Infectious forensics. *Nature* **506**, 424–6.

Cabo, F. (2012). Application of molecular diagnostic techniques for viral testing. *The Open Virology Journal* **6**, 104–14.

Peretti, A. *et al*. (2015). Hamburger polyomaviruses. *Journal of General Virology* **96**, 833–9.

Radford, A.D. *et al*. (2012). Application of next generation sequencing technologies (NGS) in virology. *Journal of General Virology* **93**, 1853–68.

Tang, Y.W. and Ou, C.Y. (2012). Past, present and future molecular diagnosis and characterisation of human immunodeficiency virus infections. *Emerging Microbes and Infection* **1**, e19.

 Questions

1. Explain the scientific basis of:

    a. RT PCR.

    b. ELISA methods.

    c. Next Generation Sequencing (NGS).

    d. cell cultures.

2. What are the main uses of the modern virology laboratory?

Control of viral diseases by immunization

30.1 Introduction

Not until the germ theory of disease was accepted in the late nineteenth century and concepts of immunity were developed was it possible to develop the scientific framework for the practice of immunization against microbial diseases. We now know that vaccines contain what in modern terminology are virus proteins with antigenic areas called **epitopes**. Viral epitopes induce both B-cell and T-cell immunity; they may be contained in a continuous length of protein, or may have a discontinuous structure whereby two separate pieces fortuitously come together because of the tertiary folding of the polypeptide or protein. Vaccine-induced immunity may last several years or a lifetime, the ability to make these long-term responses being mediated by 'memory cells' (Chapter 5, Section 5.4.2).

The first vaccines to be prepared in the laboratory were those against smallpox and rabies. Other milestones of virus vaccine development include: attenuated yellow fever (YF) vaccine, manufactured from infected chick embryos; poliomyelitis and other viruses propagated in cell cultures; and, most recently, experimental vaccines made from viral proteins and even DNA.

In this chapter, we aim only to cover the general principles of immunization. More detailed information on individual vaccines will be found in the relevant chapters in Part 2. Figure 30.1 plots vaccine developments over five generations.

30.2 The technology and practicalities of virus vaccine production and development

30.2.1 Virus vaccine production and standardization

Table 30.1 lists virus vaccines currently licensed around the world, most of which are prepared from live virus strains which have had their virulence weakened or attenuated in the laboratory by various methods. These vaccine viruses are selected on the basis of their good **immunogenicity** and **lack of pathogenicity**. They must also be **highly stable genetically**,

i.e. with a minimal chance of reversion to the pathogenicity of the original 'wild-type' strain.

The preparation of all microbial vaccines demands the highest standards of laboratory practice, and vigorous efforts are made by international organizations such as WHO to ensure that virus vaccine seed material and production facilities in different countries conform to given standards. Thus, vaccine manufacturers use well-characterized stock vaccine viruses and cell cultures for preparing vaccines. A seed virus technique is used whereby a large batch of vaccine virus is frozen at –70 °C. From this, the actual vaccine is produced by infecting cells with virus that is only two passages removed from the seed. This procedure reduces the opportunity for unwanted mutations to occur in the vaccine virus, which might alter its virulence or antigenic characteristics.

The amount of infective virus in a live vaccine is quantified in cell cultures to ensure an adequate infectivity titre. For inactivated or subunit vaccines, such as influenza or hepatitis B, the amount of viral antigen is adjusted to about 10 µg protein per dose. Needless to say, rigorous quality assurance checks, including of course tests for microbial contamination, are carried out from start to finish of the production process.

30.2.2 Choice of cell substrates

Although we live in an age of biotechnology almost all the current virus vaccines, with the notable exception of hepatitis B and human papilloma virus (HPV), are produced using the traditional technology of cell culture developed half a century ago. However, cloning techniques and expression of viral proteins such as Hep B SAg in yeast or HPV as virus-like particles in mammalian cells are already key techniques.

John Enders and his colleagues achieved a major scientific breakthrough in the 1940s when they reported that a strain of polio virus could be cultivated in monkey kidney cells, which, unlike neural tissue, can readily be grown in the laboratory. They were awarded the Nobel Prize for their work. The result of this observation (together with the discovery of antibiotics, which prevented bacterial contamination of cell cultures) was the development of effective vaccines, first against polio and then other viral infections.

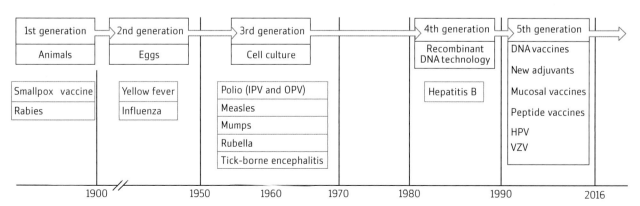

Fig. 30.1 Five generations of virus vaccines.

Table 30.1 Some virus vaccines currently licensed

Vaccine	Source of virus	Live vaccine?	Route of administration	Comments
Vaccinia	Vero cells	Yes	IM	Stocked for emergency use only; US and Russian armies immunized
Yellow fever	Eggs	Yes	IM	Long 30-year immunity induced. 500 million doses per year
Influenza	Eggs or Vero or MDCK cells	Yes	IM (killed) or intranasal (live)	Antigenic variation of virus outdates vaccine yearly; 70% effective. 5 million doses per year in total
Polio	Monkey kidney or human diploid cells	Yes	IM (killed) or oral (live)	Both vaccines are highly efficacious and will enable polio to be eradicated
Measles	Chick embryo cells	Yes	IM	Component of the successful triple vaccine (MMR)
Rubella	Human diploid cells	Yes	IM	With at least 90% efficacy
Mumps	Chick embryo cells	Yes	IM	Long-lasting immunity
Rabies	Human diploid cells	No	IM	Pre- and post-exposure prophylaxis
Hepatitis A	Human diploid cells	No	IM	
Hepatitis B	Yeast (recombinant)	No	IM	Recombinant DNA vaccine
Human papilloma virus	Pure virus-like protein (VLP)	No	IM	First vaccine against a viral-induced cancer. New vaccines have 9 HSV types

IM, intramuscular.

Hot topic Delivery of polio vaccine in a patch of bioneedles

There are innumerable migrating immune cells in the epidermis and these are missed when a vaccine is given intramuscularly. New techniques include bioneedles which are biodegradable alongside highly heat stabilized vaccine. In a recent study (Kraan, H. et al. Vaccine **33**, 2030–37, 2015) the scientists formulated polio serotypes 1, 2, and 3 into bioneedles which were lyophylized. The virus antigenicity was retained for one day even at 45 °C and imaging showed that the inactivated poliovirus vaccine (IPV) was retained at the inoculation site for 3 days. Moreover the antibody response was similar to intramuscular injected vaccine. These hollow mini-implants made of extruded starch degrade and release antigen in the epidermis. Advantages for field use are the avoidance of needles and heat stability of the vaccine, so obviating the need for a cold chain. Indeed as part of the experiment IPV bioneedles were taken on a three-week trip in the Middle East with temperatures averaging 26 °C and reaching a high at 46.5 °C. Recoveries of the poliovirus D antigen were in excess of 70%.

In general, the criteria for selection of cells for vaccines are ready availability, the lack of potential oncogenicity, genetic stability, and freedom from demonstrable contamination with extraneous viruses. Most vaccine viruses are now cultivated in human cells or a stable cell line of monkey kidney called Vero. These human cells are diploid, with a limited lifespan in culture, as in nature. Commonly used cells, such as WI-38 or MRC-5, were derived from aborted foetuses and this may raise ethical problems for some people. The cells can be passaged only 40–50 times in vitro before dying out, so , in practice, large batches of cells are frozen in liquid nitrogen at an early passage level and are made available for manufacturers to produce virus vaccines.

A more recent technique, still experimental, involves the use of plant cells, whereby a gene coding for a virus protein is cloned into the plant cell using a plant bacterium. The plant cells, such as tobacco cells, produce the new antigen.

Even with a high degree of attention to safety other viruses can occasionally creep into the vaccine production. The rotavirus vaccine cultivated in Vero cells was shown by deep sequencing to contain live porcine circo virus type 2. This had entered the vaccine production in contaminated porcine trypsin which is an important enzyme used to separate cells in culture. The enzyme was accessed from pig intestines. Fortunately the porcine virus is not pathogenic for humans.

30.2.3 The virological and genetic basis of killed or attenuated virus vaccines

Chemically killed vaccines

Chemical agents such as formalin or β-**propiolactone** inactivate viruses grown in massive quantities in eggs or cells by chemically cross-linking base pairs in virion RNA or DNA. Classic killed vaccines are rabies, influenza, polio, and hepatitis A. This inactivation is not always straightforward and clumps of viruses need to be avoided.

Live attenuated vaccines

Most attenuated or 'weakened' virus vaccines are made from RNA viruses, such as polio, measles, mumps, and rubella. This is probably no coincidence, because the virus RNA replicase has a low fidelity of transcription (Chapter 3). This means that in every cycle of infection, lasting say 8 h, an RNA virus might throw off 500 different mutants compared with the 10 000 or so viruses per cell identical to the parent strain. However, advantage can be taken of the observations, dating back even to Pasteur, that some of these mutants may be less virulent for humans than the parent virus. They may be selected or enriched in the laboratory by rapid passage; namely, transfer of virus from cell culture to cell culture; this is the so-called 'attenuation by passage'. Another method, used alone or in combination, is to perform the passages at a low temperature (28–30 °C). This allows the emergence of temperature-sensitive (ts) or cold-adapted mutants that replicate preferentially at these lower temperatures, at the same time losing their pathogenicity. Such strains, with many mutations in the genome, are termed attenuated; Dr A. Sabin used such techniques to produce the attenuated live polio viruses used in the Oral Polio Virus vaccine (OPV).

Great advantage was taken of virus passage at low temperature to produce the two master strains of influenza A almost 50 years ago in St Petersburg and in Ann Arbor. Another candidate master strain is A/PR/8/34 (H1N1), which is attenuated for humans by extensive animal passage. All the master strains have mutations in all 8 genes and the master strain provides 6 such genes to construct a suitable vaccine virus whilst the wild-type virulent virus provides only the current HA and NA. Each year authorized WHO laboratories use genetic reassortment to produce these 'starter' viruses for vaccine production units around the world.

The genetic basis of attenuation

When such attenuated viruses are analysed, a surprisingly limited number of specific mutations are noted in key genes. Comparison with the genome of wild-type virulent polio type I genome shows that the attenuated strain has undergone 55 substitutions out of 7441 bases; 21 of these substitutions result in actual amino acid changes. With the important type 3 polio vaccine strain, which is known to revert to virulence more easily than type 1, only 10 nucleotides are mutated, of which only three result in amino acid changes in the structural protein. New live attenuated influenza vaccines have about 10 mutations, with at least one in each of the eight genes.

For the first time a new technology of reverse genetics is being applied to negative-strand RNA viruses. A negative-stranded genome such as influenza can be excised, transcribed to DNA, mutated regions introduced, and reinserted into a cell with a reconstituted RNA transcriptase that allows infection of the cell and incorporation of the genetically modified gene. We can anticipate a new generation of GM vaccines and an excellent example is the pandemic influenza A vaccine for avian influenza (H5N1). The HA and NA genes have been modified by reverse genetics and then the six other genes have been added.

30.2.4 Problems with viral vaccines

Although the development of effective virus vaccines is one of the major successes of biomedical research, some serious—but fortunately rare—problems of safety have come to light.

As an example of a totally unexpected problem, an inactivated vaccine against RSV (see Chapter 17) resulted in some immunized children developing a more serious infection than their non-immunized classmates, when they were in contact with virulent virus. It has been reasoned that chemical inactivation of the virus vaccine distorted the immune response allowing excessive production of IgE against one of the viral spike proteins, but this was 50 years ago. A modern AIDS vaccine incorporating an adenovirus as a vector has recently shown a hint of a similar fault.

Another problem encountered with some attenuated vaccines, such as those for polio and influenza, is reversion to parental-type virulence. The incidence of post-vaccination polio is about 1 in 300 000 doses of attenuated vaccine. Such events provide a warning that we still do not fully understand the underlying genetic mechanisms determining virulence or attenuation of most viruses and the degree to which any live vaccine virus can recombine its genome in nature and revert back to virulence.

Finally the transfer of the cow prion (BSE) to humans has raised serious questions about the wisdom of using bovine trypsin enzyme for cell culture or bovine albumin for stabilizing virus vaccines for fear of contamination with the heat-stable prion. Synthetic alternatives are now used and even pig trypsin, but as we have noted above with the rotavirus vaccine, this can itself be contaminated with pig viruses.

All at once or one at a time? The measles, mumps, and rubella (MMR) vaccine controversy

In the 1980s a combined MMR vaccine prepared from attenuated strains of these three viruses was introduced for large-scale community vaccination. Although even a single dose conferred immunity in about 90% of those receiving it, it was recommended that the vaccine should be normally be given in two doses at 12–15 months of age, but could be given at any age, say 13–14 years, thereafter as a booster.

The widespread use of the MMR vaccine use was followed by dramatic reductions in the incidence of all three infections. However, it was contended by some that the vaccine was responsible for various complications, ranging from febrile reactions soon after its administration to more serious illnesses, in particular autism, and Crohn's disease of the lower bowel.

The heated arguments surrounding the validity of these allegations have generated an enormous literature, reminiscent of the earlier debates about the complications of pertussis vaccine, and indeed about the introduction of vaccines per se in the late nineteenth century when there was a religious view that 'God's will was being overruled by science'. There is certainly not enough space or the desire here to discuss the issues in depth. The main contention of the anti-MMR lobby is that administration of triple vaccine, added to the others recommended in early childhood, is in danger of 'overloading the immune system'. Their assertion is that the vaccine should therefore be given as three spaced single vaccines. This extended schedule is not, however, approved officially because it leaves vulnerable children unprotected for too long against one or other of the three infections; and, indeed, there is evidence that use of the single vaccines has been followed by increases in their particular disease incidences.

It has to be said that there is much misleading information about this topic on the internet, which is often the first port of call for parents who are sometimes mistrustful of advice from official sources. This is a pity, because the strong conclusion from many well-conducted observations in the field is that there is no evidence for an association between triple vaccine, on the one hand, and autism and bowel disease, on the other. Furthermore, it is now widely accepted that the early researches on which this conclusion was based were flawed.

Further internet debates revolve around the HPV vaccine, which, given the evidence of effectiveness (90%) at preventing cervical cancer, might be expected to be free of controversy. But here small groups in the population object on the basis that the vaccine could lead to promiscuity in the young girls receiving the vaccine.

Some contraindications and possible side-effects of viral vaccines are listed in Table 30.2.

30.3 Virus vaccines and public health

As a result of mass immunization campaigns during the last two decades, childhood infections such as polio, measles, mumps, and rubella are well controlled in many of the more wealthy countries, so well, in fact, that some parents now withdraw their children from vaccination schemes, not wishing to accept a very small risk of side-effects when they perceive that the incidences of the illnesses in question are now extremely low. However, it should also be remembered that in a global context infectious diseases still take a heavy toll both in mortality and general suffering. Paralytic polio is still present in Asia and Africa (although the WHO plan will see the virus eradicated by 2020), and measles still causes very serious problems in children in these countries. These viruses can be, and are, imported by holiday makers into the USA and Europe. In 2014, 644 persons were diagnosed with measles in the USA, nearly all imported. This is an indication of low vaccination in some regions in the USA, such as Orange County, CA. As a comparison measles was virtually eliminated by 2000.

Table 30.2 Main side-effects and contraindications for licensed viral vaccines

Vaccine	Potential side-effects	Main contraindications for vaccination
Inactivated vaccines		
Influenza	Local reactions, including redness at inoculation site. Guillain–Barré syndrome is exceedingly rare as is narcolepsy associated with the adjuvant used in one of the pandemic 2009 vaccines	Egg allergy
Rabies	Mild local reactions with modern human diploid cell vaccine	None
Polio	None	None
Hepatitis A and B	None	None
HPV	Local reactions	None
Live attenuated vaccines		
Measles	Mild. Malaise, rash, fever, headache in a low proportion of vaccinees	Pregnancy; the immunocompromised; serious egg allergy (except rubella and polio)
Rubella	Mild lymphadenopathy and joint pain in a low proportion	
Mumps	Mild. Fever and parotitis. Rare post-vaccination meningitis with Urabe strain which is now no longer used	
Yellow fever	Mild. Malaise, headache in a low proportion of vaccinees	
Polio	Vaccine-associated paralysis; exceedingly rare. Production of live vaccine will now cease for routine use and killed vaccine will be used worldwide for the final global eradication	

Hot topic Should childhood vaccination be compulsory?

Even given the fact that, after clean water, vaccination is the most effective public health intervention, averting 2 million deaths worldwide, some parents decline some or even all of the ten childhood vaccines. Since immunization is so beneficial to society and vaccines are effective and safe, and community protection comes from 90% vaccination figures, should vaccination be compulsory?

However satisfactory this sounds, compulsory vaccination threatens public trust. Obviously there can be conflicts of rights and ethics, in this case the right of a child to protection from ill health versus the rights of a parent to make choices for their child and the rights of a community or society to be protected from ill health.

The UK has a long history of opposition, albeit in a minority of parents, to vaccination, dating from the time over 200 years ago that smallpox vaccination was compulsory. In modern times parents resist vaccination because of worries about mercury in the vaccine as a preservative, or as a cause of autism, or alternatively they have religious attitudes. In the USA vaccination rates vary significantly from state to state. There were 9000 cases of pertussis in 2010 in California, the largest outbreak since 1947. Although in theory a child has to be vaccinated against measles before school entry, there has been lax enforcement in certain states in the USA.

In Australia incentives have been offered to parents to encourage higher vaccination rates, including tax rebates, reduced insurance rates, or direct payments. It is also clear that many of the pro vaccine messages created by public health authorities, especially in the USA, fail to increase vaccination rates in children and may even increase concerns about communicable diseases. For example although the CDC website successfully corrected misconceptions about MMR causing autism, the respondents recalled other concerns about vaccine to defend their anti-vaccine attitudes. Most analyses of the economic impact of vaccination focus on rubella and measles which show billions of dollars of net benefit. In countries such as South Africa, Gambia, and Côte d'Ivoire national measles and/or rubella immunization is highly cost-effective and confers a net benefit.

The scientific problems faced in the development and production of safe and efficacious vaccines for some diseases have been overcome. Overall, only the political will and economic resources to use them in developing countries are now needed. As long as reservoirs of any of the viruses remain, importations into apparently 'virus-free' countries will be a constant problem. Influenza vaccines are well established to negate early death, stroke, and pneumonia in the over 65s, pregnant women, and diabetics, but vaccination rates even within the EU vary from 187 per 1000 in the Netherlands and the UK to 19 per 1000 in Latvia.

Smallpox eradication is the best example of a vigorous international approach to the elimination of an important infectious disease. It was initiated by the USSR (Russia) at a WHO meeting in Alma Ata in the late 1960s and supported fully by the USA;

Hot topic Could a universal influenza vaccine be developed to give protection against all subtypes?

Influenza vaccine is reformulated by manufacturers yearly according to the scientific advice from WHO. Viruses in the vaccine are required, quite naturally, to reflect dominant variants of influenza viruses in the community. However, the vaccine production time span extends to 6 months or so with WHO deciding which virus strains to use in February, and vaccine production starting in May and continuing throughout the summer for the autumn vaccination campaign. To give an example of the size and complexity, 15 million embryonated hen's eggs are needed to produce the 10 million or so doses of inactivated vaccine used in England alone.

But are there epitopes in the virus external HA and NA and the internal NP and M which are shared by all influenza A virus or even between influenza A and B viruses? Several such 'universal' peptides have been described in the internal M2e protein, the internal NP, and also the stem region of the HA, which could form the basis of a 'Universal Influenza Vaccine'. There are several experimental vaccines under scrutiny at present. Essentially a useful vaccine would extend for 24 months or more. The vaccine would need to give >20% reduction in pneumonia and a reduced hospitalization rate. Ideally a shelf life of 5 years would be required and the administration route would be IM or intranasal (IN).

Two such vaccines have been tested in the HVivo quarantine unit in London. In one vaccine the NP was cloned into vaccinia virus Ankara. There was a reduction in clinical illness upon deliberate infection with influenza a few weeks later and there was also a reduction in virus titre in the nasal wash. A second vaccine containing peptides to PB1, NP, and NA induced primarily a T-cell response and clinical illness after challenge was inversely correlated to the number of peptide-reactive CD8 T cells. But such vaccines now need to be tested more widely in field clinical trials in the EU.

with the necessary funds and scientific expertise the world vaccination campaign was pushed forward by WHO in some of the poorest nations on Earth with dramatic success. Similar international co-operation could now result in the vanquishing of polio and even hepatitis B (by 2040). Global production of 500 million doses of influenza A/Swine (H1N1) vaccine (in 2009–2010) has helped to blunt the medical and social problems experienced in previous pandemics but, as already noted, the health departments of many countries do not treat the disease seriously enough.

30.3.1 Cost-effectiveness of vaccines

The increasing availability of effective vaccines inevitably imposes costs on the budgets of national and international health authorities, but there are few more splendid examples of cost-effectiveness than immunization, which produces massive savings on hospitalization and medical treatment. This is particularly the case with polio, measles, and rubella, which sometimes give rise to serious long-term sequelae.

Apart from the vital global strategy of the international WHO programme, most national health authorities organize their own childhood immunization schedules. The details vary from country to country; the UK schedule is summarized in Table 30.3.

It is most important that these national campaigns are conducted in a vigorous manner to achieve immunization rates greater than 90% in childhood. Sometimes it is impossible to predict at the outset whether a vaccination campaign will produce enough herd immunity to prevent spread of the natural virus. Figure 11.11 in Chapter 11 shows the results of two successive strategies to prevent rubella epidemics, by immunizing either adolescent girls only or babies of both sexes. The latter approach proved by far the more successful.

In the UK and Europe, influenza vaccine is administered to individuals at special risk, e.g. the **elderly** and those with **cardiac or respiratory disease**, and, when an epidemic is threatened, to certain groups such as **healthcare staff** and people in public services. In the face of the A/Swine H1N1 pandemic, priority groups for vaccination were the under-fives, younger persons with medical problems, pregnant women, obese persons, and front-line NHS staff. Elderly groups in this pandemic were not at risk because they had immunity to this pandemic strain, which had spread in the community between 1918 and 1956.

Immunization of all medical, paramedical, and dental staff against hepatitis B is now obligatory. In some European countries universal immunization of babies with hepatitis B vaccine is the rule. It is quite likely that the UK and USA will also adopt this approach.

Hot topic A vaccine against Ebola

For most of 2014 an Ebola outbreak preoccupied the health departments in Sierra Leone, Liberia, and Guinea Bissau, and over 10 000 persons died. During this time high levels of hygiene were difficult to achieve because of poverty and lack of safe water. The virus spread in hospitals and amongst families who had buried their own relatives and had touched the body before burial. But several groups in the EU, USA, and Canada started work on 'embryonic' Ebola vaccine projects. This illustrates the power of molecular biology. High-security laboratories to handle virus are only a necessity at the start because thereafter most of the work is DNA cloning and nucleotide sequencing.

At unprecedented speed in the field of vaccinology two experimental vaccines were produced and are being investigated in large phase II clinical trials within 12 months of the Ebola outbreak starting in West Africa. Both approaches depend upon cloning a glycoprotein gene from Ebola into existing live 'carrier' viruses, namely a chimpanzee adenovirus and vesicular stomatis virus (VSV). There is no natural immunity to the latter two carrier viruses to interfere with vaccine 'take'.

To progress to a clinical trial of efficacy in a community, even with a life-threatening virus like Ebola, a 'candidate' vaccine has to pass vigorous safety tests in animal models to demonstrate freedom from toxic molecules and induction of a measurable immune response. In the case of the Ebola vaccines an animal model has already been established in monkeys. These two candidate vaccines protected monkeys from an otherwise lethal injection of Ebola.

The next stage in vaccine development is a 'first in man' experiment usually involving 10 or so volunteers in a special phase I clinical trial unit where informed consent to the injection is given and the experimental protocol carefully explained. These volunteers are clinically examined twice daily and carefully monitored for any untoward physiological or haematological changes from the new vaccine. Very importantly, the researchers study immune response in the succeeding 2–3 weeks to the injected vaccine including antibody and immune T cells to the viral epitopes. At this stage it may be obvious that the volunteer should be given a second booster dose of vaccine, perhaps 10 days later. After careful analysis of the clinical trial and scientific data a decision is taken whether or not to design a phase III trial whereby volunteers, perhaps 10 000, are vaccinated and will be followed in the community in West Africa, alongside an unvaccinated control group, for signs of Ebola. Even at this stage it cannot be assumed that such a vaccine is protective. Vaccination could even induce an aberrant immune response making the vaccinee more susceptible to the disease, as is the case in nature with another haemorrhagic virus, dengue virus.

Optimistically, if the vaccine gives a degree of protection against a previously defined disease endpoint, such as reduction in clinical disease, symptoms, duration of symptoms, or reduced height of the infectious titre of the virus at the acute phase, then a committee of experts, not associated with the vaccine development group, could recommend a larger clinical trial or suggest submission of the complete data dossier to NIH in the USA and EMA in Europe for a license to utilize the vaccination in the community. Recently a novel clinical trial design has been used to show vaccine efficacy, called a 'ring trial', a novel cluster randomized trial. When an Ebola case is identified epidemiology is used to identify the contacts of that case, a so-called ring of possible exposed people around the index case. A number of

such cases and rings are found, and split into two groups of clusters. In half the groups, all the contacts are given a vaccine, in the other half of the groups the vaccine is delayed for a number of weeks. When the data was analysed for this type of trial, where 90 clusters totalling 7651 people were involved, it showed that in the 48 clusters (4123 people) immediately vaccinated, no new cases of Ebola arose, whereas in the 42 cluster (3528 people) in the delayed vaccine group there were 16 new Ebola cases in 7 of the clusters. This showed the vaccine, used in this way, was 100% effective in preventing new cases of Ebola. In the future of Ebola the most likely recipients of a vaccine will be healthcare nurses and doctors in hospitals in West Africa, as well as cluster vaccination, like the clinical trial, in the community.

Table 30.3 Schedule of routine immunizations in the UK

Vaccine	Age
DTP and Hib, Polio, PCV, rotavirus, Men B	2 months
Men C, rotavirus, DTP, Hib, Polio, rotavirus, Men B	3 months
DTP, Hib, Polio, Men C, PCV, Men B	4 months
Hib, Men C, Men B	12 months
Measles/mumps/rubella (MMR), PCV	13 months
DTP, Polio, MMR, IPV, Influenza	2–5 years
HPV	12 years (girls)
DTP, IPV, MMR, Men C, and Polio	14 years
Men ACWY	19–25 years

Children should therefore have received the following vaccines:	
By 4 months:	3 doses of DTP, IPV, Hib
	2 doses of PCV, rotavirus, Men B and 1 dose Men C
By 14 months:	First dose of measles/mumps/rubella
	Booster dose of Hib/Men C, PCV, and DTP and Men B
By school entry:	4th DTP, IPV; second dose of MMR
Before leaving school:	5th dose of DTP, IPV
	Human papilloma virus (girls), 2 doses Men C, Men ACWY

Adults should receive the following vaccines:	
Women seronegative for rubella:	Rubella
Previously unimmunized individuals:	Polio, tetanus, diphtheria
Individuals in high-risk groups:	Hepatitis B, hepatitis A, influenza, pneumococcal vaccine
Elderly over 65:	Influenza and once off PVC
Elderly over 70:	VZV

Key; D, diphtheria; tetanus; P, pertussis (whooping cough); IPV, intramuscular polio; Hib, *Haemophilus influenzae* Type B; PCV, pneumococcal conjugate vaccine; Men C, meningitis C; MMR, measles, mumps, rubella; HPV, human papilloma virus.

D, diphtheria; T, tetanus; P, pertussis; Hib, *Haemophilus influenzae* Type B.

PCV, pneumococcus; IPV, inactivated polio; Men C, meningitis type C; Men B, meningitis type B.

Reproduced with permission from Department of Health (1996, updated 2010). *Immunization Against Infectious Disease*. HMSO, London, pp. 46–7. Updated on www.nhs.uk

Table 30.4 Examples of human immunoglobulins used for passive prophylaxis

Preparation	Comments
Normal human immunoglobulin (HNIG)	For prevention or modification of measles in persons at special risk after contact with an infection, e.g. immunocompromised children or adults For prevention or modification of hepatitis A in travellers to endemic countries excluding Europe, USA, and Australasia
Hepatitis B immunoglobulin (HBIG)	Can be co-administered with vaccine to provide rapid protection. Administered to persons with needlestick injuries. Not available for travellers
Human rabies immunoglobulin (HRIG)	To provide rapid protection after exposure to virus until vaccine immunity develops
Zoster immunoglobulin (ZIG)	For immunosuppressed or leukaemia patients; neonates or pregnant contacts of cases
RSV human monoclonal antibodies	For children with life-threatening pneumonia
Lassa convalescent plasma	Used therapeutically in a few patients
SARS and MERS convalescent plasma	Used therapeutically in a few patients
Ebola convalescent plasma	Used therapeutically in a few patients
Human monoclonal antibodies to Ebola	Used therapeutically in a few patients as a cocktail of 3 monoclones produced in tobacco plants

30.3.2 Overseas travel

Before going abroad, especially to the less developed countries, travellers should seek advice on immunization, which may change from time to time according to local circumstances. Visitors from more developed countries often forget that they may be at risk of influenza in the winter period. Decisions as to which inoculations should be given may depend on the type of travel and accommodation: the requirements of backpackers in the interior may differ from those staying in luxury hotels on the coast.

YF may be contracted during visits to endemic areas, and immunization is obligatory for entry to these regions and subsequent travel to uninfected countries. Prophylactic **rabies vaccine** is not recommended unless the traveller is visiting an endemic area for an extended period or is travelling and working in remote country regions. Travellers to developing countries may opt for hepatitis A vaccine or the combined **hepatitis A and B** vaccine.

30.3.3 Vaccine storage and usage

All vaccines must be kept at about 4°C and not allowed to freeze. Live vaccines are particularly susceptible to heat inactivation, and some are provided as freeze-dried powder to be reconstituted with sterile water at the time of use. Vaccines that cannot be freeze-dried, e.g. oral polio, must be transported in special cold boxes containing sensors that give a warning if the internal temperature rises beyond the safety limit. This is known as 'the cold chain'.

Live polio vaccines are administered as oral drops, usually on a sugar lump; all the other current viral vaccines, whether live or inactivated, are administered by intramuscular injection.

30.4 Passive immunization

Injection of **human immunoglobulin** preparations containing appropriate antibodies gives immediate partial or complete protection against infection by certain viruses. Table 30.4 summarizes the preparations currently available. Such protection is immediate upon injection of the antibodies, but is not sustained beyond about 4 months because of decay of antibodies and the production in the recipient of antibodies against the preparation. This is by no means a universal protective method because administration of antiviral antibodies to persons already infected with certain viruses could actually make the infection worse; with dengue, for example, antibodies may form complexes and provoke an untoward reaction (Chapter 12), but for certain infections it may be a useful adjunct to active immunization.

It has found a niche recently for treatment of emerging infections caused by SARS CoV, MERS CoV, and Ebola where no other therapies were available. Experimental human monoclonal antibodies are under clinical investigation for RSV and influenza. Indeed, this idea of finding highly active human antibodies to a virus and either using them singly or in a cocktail as a form of immunotherapy like antiviral treatment has offered considerable promise recently as a new way of treating diverse virus diseases.

30.4.1 The practicalities of immunizing children and adults

Two essentials are the training of nurses and doctors in both the clinical aspects of injection and safe use of needles, and also the cold chain to store and transport vaccines. The latter is vital and domestic refrigerators should not be used, refrigerators should be regularly de-iced, and vaccines kept in their original packets so retaining information of batch numbers and expiry dates.

Medical case story MMR

Your next patient enters. She has her 13-month-old daughter with her, who has a minor ailment. As she gathers up her rice cakes and raisins, which the child has scattered on the floor (despite your assurances that clinic rooms are not good places for children to eat), she mentions her concern about the MMR vaccine and the supposed link with autism.

You are amazed that people still bring up this topic. After all, it is well over 15 years old and has long since been disproved. You order your thoughts to confront what you consider to be her potentially selfish intentions.

People who decline the MMR vaccine are relying on everyone else to continue having it. As long as high levels of children have been vaccinated, these illnesses cannot circulate. This is known as herd immunity. To keep the herd immunity intact, over 95% of the population has to have been vaccinated. However, the uptake in your area is so low (64%) that there is a real chance of an outbreak. In fact, children registered at your practice are advised to have their first MMR at 13 months and their second 3 months later, rather than waiting for the preschool booster as is customary.

First you ask her what she knows. This is always a good starter when there is a difficult conversation ahead. You also ask her if she has any specific worries about the MMR vaccine and bear in mind how many conflicting reports she will have been bombarded with both from the media and from friends and family.

Presenting your most important points in a friendly and clear way, you outline the following.

* First, the MMR is important because rates of vaccination are so low in your area that her child will be put at personal risk if she does not have it.

* Secondly, some people are not able to have the vaccine, or have not yet had it. This includes children with immunological problems such as leukaemia, or who are on transplant rejection medicines, and also all babies under 12 months in this country. These children depend on the herd immunity.

* Thirdly, her own unborn grandchildren will be at risk. If her daughter were to catch rubella whilst pregnant the baby would be likely to suffer congenital rubella (and could be born blind, deaf, or be severely mentally disabled).

* Fourthly, the original paper published by Wakefield in The Lancet in 1998[1] was, in fact, a case report on 12 patients—the sort of 'evidence' that is regarded as anecdotal. Research done since, including retrospective cohort studies of over 500 000 children[2] and large prospective studies[3]

conclusively proves that the rate of autism in those who have had the MMR vaccine is the same as those who have not.

* Fifthly, Wakefield has been struck off the GMC register, and The Lancet has (at last) retracted his original paper.[4]

* Finally, if she was considering single vaccines she needs to know that the idea (of administering single vaccines as opposed to combined) was made up one day on the spur of the moment by a clinician at a press conference. It is not evidence based, and not offered by the NHS.

The single vaccines are administered over several months, rather than all at once, and this leaves the child exposed for longer. In many cases, the course is never completed, which also leaves the child at risk. The vaccines are expensive,[5] and the course involves six separate injections. Furthermore, there is no benefit to spreading out the vaccines on the grounds that children can only cope with one antigen at a time (as is sometimes argued). This premise is false: children come across hundreds of antigens a day: it is a normal part of everyday life.

You hand her a leaflet on MMR,[6] and direct her to a good website.[7] Good information sources can help people make an informed decision and increase the rate of uptake of the MMR vaccine.

You have done your best, and if this parent is still unconvinced, then it is best not to prolong the discussion.

Notes

[1] Wakefield, A.J., Murch, S.H., Anthony, A., Linnell, J., Casson, D.M., Malik, M., Berelowitz, M., et al. (1998). Ileal-lymphoid-nodular hyperplasia, non-specific colitis, and pervasive developmental disorder in children. Lancet 351, 637–41.

[2] Baird, G., Pickles, A., Simonoff, E., Charman, P., Sullivan, T., Chandler, S., et al. (2008). Arch Dis Child 93, 832–7, 1079. Doi: 10.1136/adc2007.122937.

[3] Marsden, K.M., et al. (2002). N Engl J Med 347, 1477–82.

[4] (2010) Retraction—Ileal-lymphoid-nodular hyperplasia, non-specific colitis and pervasive developmental disorder in children. Lancet 375, 445.

[5] https://www.gov.uk/government/publications/mmr-vaccine-dispelling-myths/measles-mumps-rubella-mmr-maintaining-uptake-of-vaccine (accessed 20 Feb. 2016).

[6] www.patient.co.uk

[7] Immunisation.nhs.uk/vaccines/MMR. Currently www.nhs.uk/conditions/mmr/Pages/Introduction.aspx

- **Virological:** MMR stands for measles, mumps, and rubella, and is a vaccine given at 13 months and, again, as a preschool booster from 3 years 4 months. Six to ten days after having the MMR children sometimes get a temperature and a mild measles-like rash.

- These vaccines have been given as a combined vaccine since 1988, when the number of measles notifications in the UK dropped from 82 000 and 16 deaths, to less than 10 000 with one death. The current incidence runs at around 2000 per year and usually no deaths.

- **Clinical discussions** with patients on immunizations crop up quite regularly. In the case of the MMR debate it is worth *reading* the Wakefield paper to have an informed opinion on the issue. There are always ongoing worries about vaccines, such as those about the adjuvants and preservatives used (i.e. mercury and thiomersyl), and there will surely be new vaccines (new variant influenza, human papilloma virus) causing further controversy in the future.

- **Personal:** Strongly held views about vaccines can be very aggravating, and the first thing to do is to note your own response to the situation, allowing you to handle a discussion with more detachment.

Boxes should be placed with sufficient space for cold air circulation in the main part and not the door of the refrigerator. Validated cool boxes need to be used if vaccines are to be moved to outlying clinics.

Nurses and doctors must have received specific training, including recognition and treatment of anaphylaxis. For this purpose, adrenaline must be immediately available. Vaccinators will make sure that there are no contra-indications and that the vaccinee or carer is fully informed and consents. The patient can give oral consent, but there must be a clear transfer of information to the patient including the nature of the vaccine itself, the risk of the disease in the unvaccinated, and the potential side-effects. Most viral vaccines are given intramuscularly, where they are least likely to cause local reactions. A suitable site is the deltoid area of the upper arm and antero-lateral aspect of the thigh (especially for infants under 1 year). The buttock is not a preferred site because of fatty tissue especially in nations (such as the USA) with the current epidemic of obesity. Visibly dirty skin should be cleaned, but most practitioners do not use isopropyl alcohol these days. For example, for intramuscular injection, a 25-mm needle is commonly used at a 90° angle to the skin, which is stretched and not bunched. Accurate records need to be kept of vaccine name, batch and expiry date, dose, site of inoculation, and signature of the vaccinator.

Common adverse reactions can be pain at the vaccine site (usually immediately) or, within a few hours, local side reactions such as swelling or reddening near the site. More systemic reactions, such as fever, myalgia, and headache can be recorded. Rare adverse reactions are anaphylaxis, at a rate of one per million vaccine doses.

In the UK, for example, there is further monitoring using the Yellow Card System. This voluntary scheme encourages GPs, pharmacists, nurses, and dentists to report adverse reactions to the government monitoring group at the MHRA (www.mhra.gov.uk). The MHRA works in conjunction with the Joint Committee on Vaccination and Immunization to assess vaccination strategies and success.

30.5 New approaches to vaccine development

30.5.1 Genetically engineered virus vaccines

Strong new ideas and the techniques of molecular biology have been introduced in the last two decades into vaccine technology. Thus, genes or portions thereof can now be **cloned in plasmid vectors** and transferred to yeast or mammalian cells and plant cells, as is already done in the manufacture of hepatitis B and HPV vaccines. In the example of cloning illustrated in Fig. 30.2, a viral gene coding for an immunogenic protein is excised from the virus and cloned into a plasmid vector, which is then inserted into a bacterial cell. The plasmid replicates alongside the bacterium and provides the necessary genetic information for synthesis of the viral protein. This protein may be produced in large quantities in a bacterial fermentation apparatus and purified from the bacterial culture. Of course, the vector may be cloned into eukaryotic cells, insect cells or larvae, or even into plants and plant viruses for efficient expression of the viral protein. A further example of the application of the technique is the virus-like particle human papilloma virus vaccine.

Perhaps the most exciting molecular technique is the cloning of viral genetic information into the genome of other large DNA viruses, such as vaccinia and the related canarypox, or adenovirus or the RNA animal virus vesicular stomatitis virus (VSV), which are used as 'Trojan horses' to carry the genes of the new vaccine virus into the recipient. Thus, immunogenic proteins of rabies, influenza (HA and NA), HIV (gp120), and hepatitis (HBSAg) have been cloned into the TK locus of the genome of vaccinia poxvirus or an even more innocuous pox virus called 'more attenuated virus Ankara' (MAVA). Moreover, insertion of new genetic information at this point in the vaccinia genome is expected to attenuate this virus further. The vaccinia virus genome is so large (150 kbp) that viral functions are not compromised by excision of a portion and reintroduction of a new

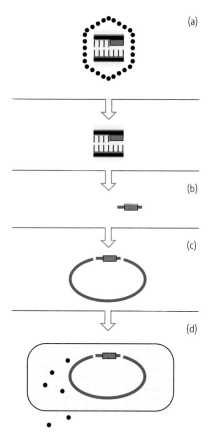

(a)

(b)

(c)

(d)

Fig. 30.2 The cloning and expression of a viral gene in a bacterium. (a) A viral gene (•), coding for an externally situated viral protein (•), is excised (b) and inserted into a plasmid vector (c). The vector is used to transfect a bacterium (d), in which the gene is translated to produce large amounts of viral protein. Such 'genetically engineered' proteins may be used for vaccine production.

gene. After inoculation into the skin, vaccinia virus replicates and, at the same time as producing its own proteins, codes for the new antigens specified by the cloned viral gene. Other viruses such as human adenovirus type 4, or a chimpanzee adenovirus to which there is no pre-existing immunity to complicate

vaccine 'take', can be used as vectors and can be given orally. Some RNA viruses, such as Venezuelan equine encephalitis (VEE), are used where a replicon system is designed whereby a vaccine virus only replicates when the additional RNA replicase is co-administered.

30.5.2 Short peptides as experimental vaccines

Our understanding of the nature of the B- and T-cell epitopes (antigenic determinants) on virus proteins is progressing rapidly. Some viral epitopes consist of only eight amino acids. Cattle can be protected against foot-and-mouth disease virus by immunization with short synthetic peptides of the correct amino acid sequence. In theory, many such epitopes for different viruses could be synthesized and linked together as a single immunogenic protein, able to induce B- or T-cell immunity to a wide range of infective agents. Vaccines to stimulate CD4 and CD8 T cells and directed towards conserved epitopes of virus internal structural proteins are being researched at present, especially for HIV and influenza.

30.5.3 Experimental DNA vaccines

Unexpectedly, researchers discovered that injected **viral DNA** would penetrate, albeit at low efficiency, mammalian epithelial cells, and be transcribed and copied into viral proteins by the cellular protein translating system. More importantly, these newly synthesized viral proteins would be processed into viral peptides and presented on MHC class I molecules exactly as if the cell had been infected with the virus itself. Viral DNA could thus act like an attenuated virus vaccine. Even the RNA genome of viruses, such as influenza or HIV, may be transcribed into DNA by reverse transcriptase enzyme and the DNA mutated by reverse genetics and subsequently utilized to infect cells. DNA vaccines would seem to have an important future and experimentation is particularly critical with HIV and also influenza.

Hot topic Broadly reacting neutralizing antibodies to dengue viruses

One of the first conventional vaccines against dengue has recently failed to induce immunity to all 4 serotypes of the virus. This partial failure has very practical consequences because in nature infection with one of the dengue serotypes can lead to enhanced disease after superinfection with one of the other 3 serotypes. Around 90 dimers of the virus glycoprotein E cover the virion surface and it is the major, if not the sole, target for neutralizing antibodies. A research group (Rouvinski, A., *et al.* (2015), *Nature* 520, 109–13) have now identified human antibodies that neutralize all four dengue serotypes and studied, using X-ray analysis, their interactions with a crystal structure of the E

protein. In this manner they identified a common antigenic site at the E dimer interface, incorporating what is called 'the E fusion loop'. But why should such a piece of E protein be shared by all four serotypes? The answer is that during virus replication in the cell a second glycoprotein (precursor M) binds to E at this site. Here it protects E from being prematurely changed by the low pH in the cell. Clearly the E protein has a vital role to play during infection and this explains the selection pressure to keep it unchanged among the four serotypes. It may yet be the Achilles heel of dengue. The practical outcome of this research is to formulate a vaccine composed of a pre-fusion E dimer.

30.5.4 Adjuvants

Adjuvants of various compositions can prolong and enhance the immune response to inactivated or subunit vaccines. Aluminium hydroxide has been widely licensed for use in humans. More recently, proprietary mixtures of oil and detergent have been used on a large scale with pandemic influenza A/Swine (H1N1) vaccine.

It must be acknowledged that the scientific basis of these molecules is still obscure. New research is focused on stimulating the innate immune system via Toll-like receptors, which are present on the outer membrane of macrophages and dendritic cells and can sense viral RNA and some viruses themselves.

Methods of presentation of viral antigens to the immune cells are important and immunogenicity can be enhanced if viral proteins are aggregated in novel ways with or without adjuvants. For example, aggregation of viral proteins by saponin molecules results in formation of **immune-stimulating complexes**; and viral proteins have been incorporated into lipid spheres (**liposomes**) containing muramyl dipeptides, which increase the immunological response and memory of the host.

But any new adjuvant must be tested very carefully because, as we have noted earlier (in Table 30.2), one of the adjuvanted pandemic influenza A/Swine vaccines produced narcolepsy in a minority of young vaccinees in Northern Europe and the UK in 2010.

 ## Reminders

- Smallpox has been completely eradicated and some other viral infections, such as polio, have been virtually eliminated. Humanitarian considerations apart, **immunization is one of the most cost-effective public health measures available**.

- Viral vaccines prevent infection by antigenic stimulation of the host and induction of memory T cells, resulting in the generation of neutralizing antibody and cytotoxic T cells. Immunity starts to develop some days after vaccination and memory cells may persist for decades.

- **Inactivated** (killed) vaccines consist of chemically inactivated whole virions (e.g. rabies, formalin-killed polio vaccine). Non-living vaccines are also prepared from fractionated virus containing immunogenic proteins, (e.g. 'split' influenza vaccine), or by recombinant DNA techniques (e.g. HBsAg vaccine).

- **Live vaccines** are prepared from viruses, which, by manipulation in the laboratory, are no longer pathogenic but retain their immunogenicity, e.g. yellow fever, oral polio, mumps, measles, rubella, and, most recently, for pandemic and epidemic influenza.

- Vaccines are not without serious side-effects, but these are extremely rare. Pregnancy and immunocompromised states are contraindications to live attenuated vaccines in case the virus crosses the placenta.

- **Passive prophylaxis with human immunoglobulin preparations** gives a measure of immediate protection against infection of measles, hepatitis A and B, rabies, VZV, Lassa, RSV, and Ebola. Immunoglobulin may be co-administered with inactivated vaccines, such as rabies and hepatitis B. Most recently human monoclonal antibodies to influenza, RSV, and Ebola have become available to treat seriously ill patients.

- Research is continuing into recombinant DNA techniques, the use of DNA itself as the immunogen, and the improvement of adjuvants and new adjuvants targeting the innate immune system. Peptide vaccines designed to stimulate CD4 and CD8 cells will undoubtedly contribute to the vaccine list in the near future. Replication-deficient VEE and adenoviruses are forming the basis of a new class of genetically modified (GM) vaccines.

 ## Further reading

Diekema, D.S. (2015). Improving childhood vaccination rates. *New Eng J Med* **366**, 391–3.

Nyhan, B., Reifler, J., Richey, S., and Freed, G.L. (2014). Effective messages in vaccine promotion: a randomized trial. *Paediatrics* **133**, 1–8.

Rouvinski, A., Guardado-Calvo, P., Barba-Spaeth, G., Duquerroy, S., Vaney, M.C., Kikuti, C.M., *et al.* (2015). Recognition determinants of broadly neutralizing human antibodies against dengue viruses. *Nature* **520**, 109–12.

Victoria, J.G., Wang, C., Jones, M.S., Jaing, C., McLoughlin, K., Gardner, S., and Delwart, E.L. (2010). Viral nucleic acids in live attenuated vaccines: detection of minority variants and an adventitious virus. *Journal of Virology* **84** (12), 6033–40.

Weiner, D. *et al.* (2010). Introduction to DNA vaccines. *Vaccine* **28**, 1893–6.

WHO (2013). Yellow Fever Vaccine Information. World Health Organization, Geneva, Switzerland. http://www.who.int/immunization/diseases/yellow_fever/en/ (accessed 1 Mar 2016).

? Questions

1. Discuss the relative merits of live versus dead virus vaccines, with examples.

2. What particular contribution to vaccinology have molecular methods made?

3. Write short notes on:

   a. Cost effectiveness of vaccines.

   b. Cell substrates.

   c. Adjuvants.

   d. Overseas travel vaccines.

   e. Universal flu vaccine.

Antiviral chemotherapy

31

31.1 Introduction

Certain important virus diseases, such as measles, poliomyelitis, yellow fever, and rubella, can be kept under very good control with live virus vaccines, and some viruses such as smallpox and, soon, polio even eradicated. However, as we saw in Chapter 30, it is difficult to imagine the development of successful vaccines for many other viruses, because of a multiplicity of serotypes, or variability or complexity of their antigenic structure. Other viruses such as hepatitis B and C have chronically infected millions and vaccines cannot break these established infections, only being able to prevent new infections in the case of hepatitis B. Finally some viruses such as SARS, MERS, Ebola, and Zika emerge and resurge very quickly, posing serious difficulties in timely vaccine formulation. Nevertheless, experimental vaccines for both MERS CoV and Ebola have recently been reported.

Instead of or complementary to vaccines, antiviral drugs have been successfully developed for a number of virus infections. There are currently 40 or so antiviral compounds licensed for use against herpesviruses, hepatitis B, hepatitis C, HIV-1, RSV, and influenza A virus (Fig. 31.1 and Table 31.1). However the majority of these molecules are anti-HIV drugs. If one observes the number of classes of these drugs for HIV and their success it is difficult not to conclude that this is the most successful drug development pipeline against any disease to date. Lessons learnt from this virus are being applied to hepatitis B and C, both of which have elements of chronicity and so need long term use of antivirals. A negative feature of all the known antivirals, with the exception of ribavirin, is their very narrow spectra of antiviral activity; thus, anti-herpes compounds have no effect against influenza and vice versa. However, some virus targets, such as the enzyme reverse transcriptase (RT), are common to such diverse families as retroviruses and hepadnaviruses, and a specific antiviral, such as lamivudine, can have an inhibitory effect against members of both families. Unexpectedly, a newly developed inhibitor of the RNA polymerase of influenza also inhibits the similar enzyme of the RNA filovirus Ebola.

All these drugs have been discovered by random biological screening in the laboratory, but chemists are now using three-dimensional structures of virus proteins, accurately determined by X-ray crystallography, to refine and even design inhibitory molecules that fit into essential sites on viral proteins. Examples are the four influenza NA inhibitors that have been redesigned using these techniques to bind more tightly to the active site of the viral enzyme than did the original molecule discovered many years earlier.

However, would it not be possible to devise a universal antiviral against all viruses, or does such a compound exist already as IFN? The answer is unfortunately, no. The great hopes raised over 50 years ago when IFN was discovered by Isaacs and Lindenmann in London have not been realized, and its applications in viral chemotherapy are still limited to certain hepatitis B and C infections, albeit with important clinical effects. However, the work of these pioneers has given rise to the discovery of a wide range of molecules—the cytokines—that have important effects in the innate arm of the immune system. Application of the new science of proteomics and searching databases of cellular proteins which aid virus replication may give us a new generation of virus blockers. The emergence of drug-resistant mutants is carefully monitored for the RNA viruses HIV, hepatitis C, and influenza, and often resistance can be attributed to a handful of mutations at or near the enzyme active site of the drug.

31.2 Points of action of antivirals in the virus life cycle

Scientists have searched for the last 58 years for molecules that would inhibit virus-directed events without interfering with normal cellular activities. The potential points of inhibitory action of antiviral drugs include:

- Binding to the free virus particle. Drugs may stabilize the free virus by cross-linking its structural proteins so that release of nucleic acid from its interior is impeded. An example is the class of drugs which bind to the capsid protein of rhinovirus.

- Virucidal compounds, which could destroy enveloped viruses, such as HIV or influenza A (H5N1) and norovirus, on contact on surfaces and on hands. The components of these 'disinfectants' are mild alcohols, mild acids, detergents, and phenolics. These 'disinfectants' are even being explored very carefully in the case of HIV as a pessary. However toxicity for cells lining the vagina could actually enhance the infection.

- Interference with virus adsorption or attachment to the receptor binding site on the cell. Synthetic cell receptors could act as 'decoys' and, hence, prevent infection of cells. An example is the attempted use of soluble CD4 and other molecules to prevent HIV infection. Most successful has been the use of drugs to block fusion of HIV with the cell membrane and thereby to stop virus entry. Another new

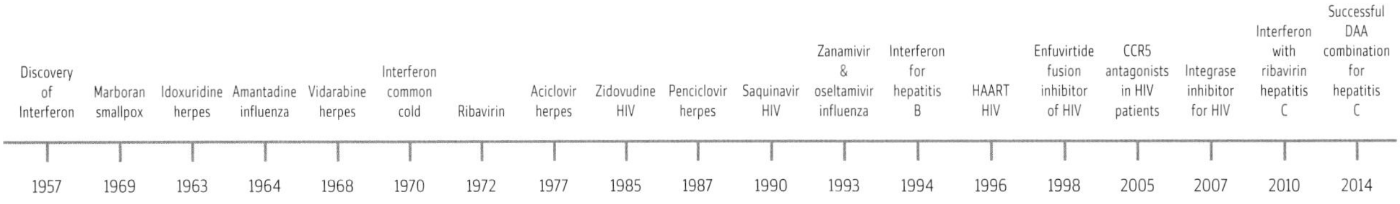

Discovery of Interferon	Marboran smallpox	Idoxuridine herpes	Amantadine influenza	Vidarabine herpes	Interferon common cold	Ribavirin	Aciclovir herpes	Zidovudine HIV	Penciclovir herpes	Saquinavir HIV	Zanamivir & oseltamivir influenza	Interferon for hepatitis B	HAART HIV	Enfuvirtide fusion inhibitor of HIV	CCR5 antagonists in HIV patients	Integrase inhibitor for HIV	Interferon with ribavirin hepatitis C	Successful DAA combination for hepatitis C
1957	1969	1963	1964	1968	1970	1972	1977	1985	1987	1990	1993	1994	1996	1998	2005	2007	2010	2014

Fig. 31.1 Timeline for the discovery of antivirals.

category of HIV drugs can cross-link the cell receptor with the HIV GP120 envelope protein.

- Inhibition of virus uncoating or release of viral nucleic acid from the virus in cytoplasmic membranes and vacuoles. The most investigated inhibitor in this group is amantadine, an antiviral against influenza A targeting the viral M2 ion channel.

- Inhibition of viral nucleic acid transcription and genome replication. This has been the favourite and most successful point of action of antivirals. Certain viruses code for specific enzymes of their own, such as hepatitis C and influenza RNA transcriptase, herpesvirus TK and DNA polymerase, and HIV and hepatitis B reverse transcriptase, integrase, and protease. These viral enzymes form the most important targets for inhibition.

- Interference with cellular processing of viral polypeptides, by preventing addition of sugar or acyl groups.

- Transcription of host-cell gene products which are essential components in the virus life cycle could be the targets for a new generation of virus blockers.

- Prevention of virus budding or interference with virus maturation. One of the most used antivirals, the inhibitor Tamiflu inhibits influenza virus NA, and normally acts at the virus cellular release stage. Some anti-HIV proteases block internal events catalysed by viral proteases immediately after release of virus and during its early extracellular maturation.

Table 31.1 gives examples of antiviral drugs affecting different steps in virus multiplication.

The current licensed antivirals are very effective inhibitors and, although few in number compared with antibacterials, are now widely used in hospital and in general practice. Typical chemical structures are illustrated in Fig. 31.2.

Table 31.1 Examples of antiviral drugs affecting different steps in virus multiplication (the list is not complete)

Target	Drug	Virus inhibited
Viral fusion	Enfuvirtide	HIV-1
Blocking CCR-5	Maraviroc	HIV-1
Penetration and uncoating	Amantadine	Influenza A
	Rimantadine	
Viral nucleic acid synthesis or Integration	Aciclovir (ACV) Penciclovir Cidofovir	Herpes simplex and varicella-zoster Small pox
	Ganciclovir	CMV
	Daclatasvir Sofosbuvir Ipasvir	Hepatitis C Hepatitis C Hepatitis C
	Lamivudine (3TC) NRTI	HIV-1 and HBV

Target	Drug	Virus inhibited
	Zidovudine (AZT) NRTI	HIV-1
	Didanosine (ddi) NRTI	HIV-1
	Stavudine (d4T) NRTI	HIV-1
	Emtricitabine NRTI	HIV-1
	Zalcitabine NRTI	HIV-1
	Abacavir NRTI	HIV-1
	Nevirapine NNRTI	HIV-1
	Efavirenz NNRTI	HIV-1
	Delavirdine NNRTI Nevirapine NNRTI	HIV-1 HIV-1
	Tenofovir (nucleotide analogue)	HIV-1 and hepatitis B
	Favipiravir (T705)	Influenza
	Interferon-α	Hepatitis B and C
	Foscarnet	HIV-1 and herpes simplex
Integrase inhibitor	Raltegravir and Elvitegravir	HIV-1 and HIV-2
Binding to intact virus particle	Disoxaril	Rhinoviruses
Virus release	Zanamivir	Influenza A and B NA
	Oseltamivir	Influenza A and B NA
	Peramivir	Influenza A and B NA
	Saquinavir	HIV protease
	Indinavir	HIV protease
	Ritonavir	HIV protease
	Lopinavir	HIV protease
	Nelfinavir	HIV protease
	Amprenavir	HIV protease
	Tipranavir	HIV protease
	Atazanavir	HIV protease
	Darunavir	HIV protease
	Telaprevir Boceprevir Asunaprevir Simeprevir	Hep C protease Hep C protease Hep C protease Hep C protease
Interference with cellular enzymes	Ribavirin	Hep B, Hep C, RSV

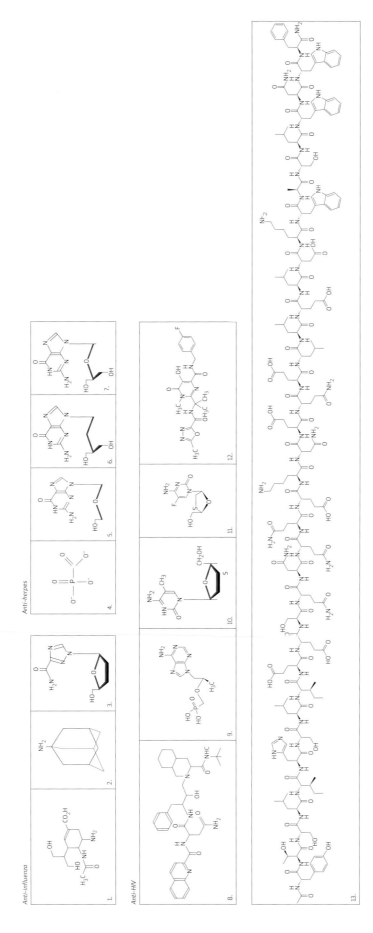

Fig. 31.2 The chemical structures of some antiviral compounds. (1) oseltamivir; (2) amantadine; (3) ribavirin; (4) foscarnet; (5) ACV; (6) penciclovir; (7) ganciclovir; (8) saquinavir; (9)Tenofovir; (10) lamivudine; (11) emtricitabine; (12) raltegravir; (13) enfuvirtide.

31.3 The use of antivirals: general considerations

A list of antivirals according to their clinical usefulness and effectiveness and indeed sales would be headed by ACV, the l-valyl ester prodrug valaciclovir, and the chemically related penciclovir against alphaherpesviruses. ACV cream is now licensed as an 'over-the-counter' drug for use against herpes cold sores; as the development of antivirals expands, this could be a portent of things to come, especially for mild respiratory viruses. Nucleoside analogues such as ddI, ddC, and 3TC, non-nucleoside analogue inhibitors, and protease inhibitors are widely used for treating HIV-1 infections especially combined as highly active antiretroviral treatment (HAART). Some combinations are now formulated into a single tablet. Treatment reduces virus load in the blood, stops decline of CD4 cells, and halts clinical problems. The NA inhibitors such as oseltamivir (Tamiflu) are being increasingly used to treat influenza in the elderly and special risk groups; they had an important role in the last swine influenza A (H1N1) pandemic of 2009 and are stockpiled for future pandemics. Treatment reduces complications such as pneumonia and quantities of virus in the upper airways. Less used are ribavirin against RSV infection in children and amantadine against influenza A. Pegylated IFNs are used to treat SARS patients and also hepatitis B and C patients, often alongside directly acting antivirals. A number of directly acting antivirals (DAA) have now been tested in the clinic against hepatitis C virus, successfully targeting virus replicase and protease enzymes and used without interferon.

31.3.1 Therapy versus prophylaxis

Antivirals are mostly used therapeutically and are administered as soon as possible after the first signs of infection. For example, ACV should be taken as soon as possible after the onset of herpetic lesions (i.e. at the early itching stage) or early in the onset of clinical signs of encephalitis, and amantadine or Tamiflu when the patient first begins to feel ill with influenza. Speed of use is essential, as if there is a delay even of 24 h the drugs are much less effective. Therapeutic use is also common with HIV, hepatitis B, and hepatitis C infection.

To a lesser extent antivirals are used to actually prevent a virus infection, e.g. amantadine or oseltamivir in the prophylaxis of influenza A in the home and hospital wards; see Sections 31.5.1 and 31.5.2. More recently, drug combinations have been used to prevent HIV infection developing in accidentally infected healthcare workers and also in partners of HIV positive persons (discordant couples).

Compared with vaccines, prophylaxis with a chemical antiviral has the advantage of speed of action, as some antiviral protection would be anticipated within an hour or so of drug administration. The converse is also true: when prophylaxis is discontinued the patient again becomes susceptible to infection.

A good example of antiviral prophylaxis is the long-term administration of ACV by mouth to prevent recurrent attacks of genital herpes, a condition that causes much discomfort and sexual disability. The NIs or M2 blockers are used to prevent infection with influenza in the family or workplace. The NI Tamiflu was used extensively to reduce virus transmission and prevent deeper infection in the respiratory tree during the swine influenza pandemic of 2009/10.

31.3.2 Pharmacology and side-effects

Following oral or intravenous administration, effective tissue concentrations of the antiviral drug may be achieved in minutes. The half-life of an antiviral drug is often only a few hours and, therefore, frequent dosing is required to maintain optimal levels in plasma and tissues. Penciclovir has the same antiviral spectrum as ACV against herpes simplex type I, but patients need only take three tablets a day, rather than five, because the former drug is more tightly bound inside the cell and its half-life is prolonged.

All drugs have side-effects and antiviral compounds are no exception. Even the very safe anti-herpes drug **ACV**, which has been used for more than three decades in millions of patients, can exceptionally cause gastrointestinal symptoms, such as nausea and vomiting. Particular care must be exercised in patients with renal failure, as even with normal kidney function about 70% of the drug is excreted unchanged in the urine. Severe renal malfunction can result in undesirably high concentrations of an antiviral drug in the blood. The structurally related anti-CMV drug, **ganciclovir**, induces more serious side-effects than ACV: for example, neutropenia occurs in one third of patients and thrombocytopenia, rash, and nausea may also occur.

Zidovudine (AZT) and nucleoside analogues: at the original dosage prescribed for AIDS patients AZT induced anaemia, neutropenia, and leucopenia in approximately one third of patients. These side-effects are now avoided by using lower doses of drug, which are, however, still effective against the virus, especially when used in a drug combination where synergy is common. HIV drug combinations can bring problems of unwanted side-effects and drug interactions as well as benefits of synergy.

Amantadine causes the slight neurological effect of 'jitteriness' in a few patients, which ceases immediately the drug is discontinued. Again, with careful adjustment and lowering of dosage, particularly for the elderly or for patients with renal problems, these side-effects can be avoided.

IFNs are certainly not free of side-effects, such as depression and other psychiatric changes, fatigue, influenza-like symptoms, and even severe somnolence.

Pregnancy is usually an important contraindication for the use of all these drugs because of potential damage to the foetus. Unexpectedly, however, nucleoside analogues administered to an HIV positive mother during pregnancy can have a beneficial effect by very significantly reducing the chance of spread of the virus from mother to foetus. Also the pandemic influenza A/swine virus (H1N1) caused most serious clinical signs in the late stage slightly immunosuppressed pregnant woman. The NI Tamiflu was used successfully during pregnancy to protect mothers and their foetuses from influenza.

31.3.3 The use of prodrugs

Some of the most practical pharmaceutical innovations are antiviral prodrugs with chemical side groups attached to the active antiviral molecule that enhance adsorption and tissue penetration. Host enzymes then cleave off the side chain from the prodrug to release concentrations of the active drug that are often higher than could otherwise be achieved. For example valaciclovir, the prodrug of ACV, is absorbed from the gastrointestinal tract and converted to ACV in the intestine and liver. Approximately 60% of the prodrug given by mouth is absorbed, compared with 20% absorption of ACV. Similarly, famciclovir is a diacetyl prodrug of penciclovir. Hydrolysis in the intestinal wall and metabolism in the liver remove both acetyl groups, and oxidation of this deacylated form converts famciclovir to penciclovir.

The widely used anti-influenza drug Tamiflu is a prodrug and the active antiviral molecule is produced by enzyme cleavage of the prodrug in the liver.

31.4 Herpes infections and antivirals

Members of this large family of DNA viruses cause a very wide range of diseases and moreover can take on a form of latency whereby recurrent infections can occur throughout life. There are no vaccines and so antivirals have a special role.

31.4.1 Aciclovir and related compounds

Mode of action

ACV and **penciclovir** possess an excellent combination of biochemical and pharmaceutical properties, which explains their unique anti-herpesvirus specificity. Indeed, they approach the ideal specification for an antiviral drug. First, the compounds are phosphorylated to the monophosphate only in herpes-infected cells, as this biochemical step requires a **herpesvirus TK** and cannot be achieved by normal cellular TK. The viral TK is less 'precise' than the corresponding cellular TK and so, unlike the latter, will accept 'fraudulent' substrates, such as ACV. Once ACV is phosphorylated in the cell, it cannot emerge because of the charged phosphate group that has been added by the TK enzyme. Moreover, as the pool of 'normal' unphosphorylated ACV in the cell is depleted, more ACV molecules move across the plasma membrane and these are, in turn, phosphorylated. In this manner, the drug accumulates only in virus-infected cells. Phosphorylation to the di- and tri-phosphate is then achieved by cellular enzymes.

Thus, the drug is likely to be of low toxicity as its tissue distribution is limited to virus-infected cells. The second specific feature of the drug is that the triphosphate, the active moiety, binds to and specifically **inhibits the herpesvirus DNA polymerase**. It has little effect on cellular DNA polymerase and, hence, is not toxic. These drugs can terminate DNA chain formation and so, at least in theory, might inhibit DNA replication in uninfected cells. However, if this happens at all, the effect must be very slight and the compounds are considered to be safe in clinical use: for example, patients with recurrent herpes simplex have been effectively treated with daily doses for many years without side-effects. However, a latently infected cell cannot be cleared of herpesvirus by ACV, which is thus unable to eradicate the infection.

The related antiviral, penciclovir, has a similar mode of action to that of ACV, but is retained even more tightly within the cell as the mono-, di-, or triphosphate, and therefore needs to be given less often. Penciclovir has approximately one-hundredth the potency of ACV in inhibiting herpesvirus DNA polymerase, but as it accumulates in very high concentrations and has an extended half-life, its clinical effect is similar.

Clinical application of ACV

ACV or penciclovir are used in the prophylaxis of herpes simplex and herpes zoster infections in bone marrow and heart transplant patients, and to prevent the spread of virus in those already infected. The drugs are also very effective in treating herpes simplex encephalitis, if administered early. Continuous treatment prevents recurrent HSV-1 and -2 infections, particularly those of the genital tract, and some patients have taken ACV orally for many years. Treatment of severe varicella-zoster (VZV) infections in the elderly and in immunocompromised patients requires higher dosage than herpes simplex infections. Fortunately, although drug resistance can occur, especially in immune compromised patients, this does not appear to cause a significant clinical effect.

A derivative, **ganciclovir**, has mild antiviral activity against **CMV** and is used to treat life-threatening CMV pneumonia after BMTs and eye infections in AIDS patients with serious CMV infection.

Nucleotide analogues such as **cidofovir** have recently proved effective in CMV eye infections.

31.5 Influenza and antivirals

This globally important virus has two modes of behaviour, epidemic and pandemic, and is, moreover, genetically unstable. Vaccines are used in some countries but are certainly not completely effective and have to be updated each year. Therefore antivirals play an important role in reducing hospitalization and serious complications especially in the elderly and other at-risk groups.

31.5.1 Inhibitors of influenza neuraminidase (NIs)

These drugs cannot be synthesized for large numbers of doses rapidly and, hence, have been stockpiled in tens of millions of doses by governments. For example the UK holds 30 million courses of Tamiflu and around 3 million courses of Relenza for use in a pandemic. A third and more recently developed NI is peramivir. Controversy has been kindled over possible side-effects versus efficacy of Tamiflu, but the most recent analysis showed large clinical benefits reducing hospitalization and death versus a very low incidence of side-effects such as nausea when the tablets are taken without food.

NA normally acts at the stage of virus release from the cell and inhibition of the NA enzyme by the drugs causes virus particles to aggregate at the cell surface, rather than releasing themselves to infect adjacent cells. The NIs have overtaken the M2 blockers as the drugs of choice against influenza, both because this drug class inhibits influenza A and B viruses and also because drug resistance is not an overwhelming issue. The recent pandemic of influenza A/swine H1N1 virus has witnessed the use of hundreds of millions of prescriptions for these two antivirals worldwide. It would be fair to say that the full clinical impact of this class of drugs has yet to be uncovered. Mutations in the NA around the enzyme active site result in drug resistance but the mutants have reduced ability to replicate and are often replaced by drug-sensitive fully replicative viruses.

31.5.2 Amantadine and rimantadine (virus M2 blockers)

These two drugs were amongst the first antivirals discovered 45 years ago. They are held in stockpiles around the world to counteract emerging pandemic influenza A viruses. For day-to-day use in epidemic influenza they have been overtaken by the neuraminidase inhibitors where there are fewer problems of drug resistance and which can target influenza B as well as influenza A.

These drugs act on the M2 ion channel across the viral membrane, which normally allows passage of hydrogen ions to the interior of the virus, particularly when it is within cellular lysosomes during early infection of the cell. Under these low pH conditions in the endosome two virus core proteins, NP and M, which bind closely with the viral RNA, dissociate and thus allow the viral RNA to leave the virion itself and to enter the cell nucleus to replicate. Blockage of the M2 ion channel by amantadine stops the induction of this low pH and therefore inhibits all subsequent events that would normally lead to virus replication.

Prophylaxis

Prophylactic administration of anti-flu drugs—the neuraminidase inhibitors (NIs) or **amantadine**—will prevent influenza A in 80% of individuals. The anti-flu drugs may be used prophylactically when the presence of influenza A virus in the community or a nursing home or indeed a household is confirmed. Prophylactic use can continue for five weeks or until the end of the epidemic is in sight. As with influenza vaccines, chemoprophylaxis is recommended only for the 'special risk groups', such as the over-65s, diabetics, and persons with chronic heart or chest diseases who have not been immunized or who wish to receive extra protection. Therapeutic use within 2 days of the first symptoms, particularly with the NA inhibitors, can reduce serious complications, hospitalization, and death particularly with the pre-pandemic influenza A virus of fowl such as influenza A (H5N1).

Favipiravir

This compound has been licensed in Japan and inhibits the RNA-dependent RNA polymerase of influenza. Surprisingly it has some antiviral effects against Ebola, at least when tested in the laboratory.

31.6 HIV infections and antivirals

More than 20 drugs are used clinically in this important and global infection and more are in development. There is no vaccine and hence antivirals occupy an important role. These drugs have transformed treatment and management of the disease away from specialist clinics to doctors in general practice. The disease is now viewed more akin to diabetes, where, similarly, adherence to therapy is of vital importance. It must be acknowledged that with HIV the contradictions of high science versus human behaviour and economics come strongly into play. Many infected persons live in developing countries in Africa and Asia where funding for health is low. In developed countries like the US and the EU adherence to daily prescribed drugs—even life-saving antivirals—can be very variable.

Medical case story AIDS-related meningitis

You are a medical student from the UK volunteering in a large hospital in Malawi, Central Africa. You are working alongside two Malawian junior doctors on the admissions unit for the General Medical Department, where it is your job to assess incoming patients and admit them onto a ward if necessary. Since you've only been here for a month, you cannot speak the local language very well and you require the help of a student nurse to translate.

You peer at the list of over 20 unseen patients, and quickly scan the documented blood pressure (BP) readings and temperatures to look for the sickest patient. As you look down the list a female patient called Ellena catches your eye. She has a low BP of 80/40, an elevated temperature of 39 °C, and an elevated heart rate of 130; she is evidently in septic shock.

With the help of the student nurse, you find this patient on a mattress on the floor. Blood cultures have already been taken by the specialist nurse following local protocol, and she is fixed to a drip hooked onto the neighbour's bed. Ellena is lying on one side and staring vacantly into space. As you look at her frail, emaciated body curled up under some sarong material you think for a moment that she is 12 years old, and not a 35-year-old mother as her notes suggest.

She is here with her mother, who will be required to look after her and nurse her during her time in hospital. As Ellena is too weak and breathless to do more than to confirm her name, you turn to the mother to ask details of her history through the translator. You are acutely aware that she needs oxygen, but the single oxygen cylinder on this ward is currently in use.

Ellena is from a small town a 2-hour bus ride from the hospital. For the past 10 days she has felt feverish and has been short of breath, coughing up white sputum, but no blood. Two days ago she developed a headache. When questioned directly, she admits she has a stiff neck, but no pain when looking into the light and no skin rashes. She has been losing weight for the last 3 months.

'Has Ellena ever been tested for HIV?'

'Yes,' said her mother. 'She has been taking antiretroviral therapy (ART) for a year now,' and she shows you some little bags containing stavudine, lamivudine, and nevirapine, the standard first-line ART in Malawi.

You think back to your student rotation on an Infectious Disease ward in the UK, where there are far more drugs available, as well as facilities to test for HIV strains and for pre-existing drug resistance. The British HIV Association (BHIVA) recommend a first-line treatment using efavirenz, and either tenofovir and emitricitabine (truvada), or abacavir and lamivudine (kivexa).[1] There is enough drug choice to tailor the regimen to the individual needs of the patient. This is not the case in Malawi.

You are not surprised that Ellena is HIV positive. About 80% of patients admitted onto the wards at this hospital have HIV and so many of the illnesses you see are complications of a severely suppressed immune system.

On examination of Ellena you note the following significant observations:

Generally

- Malnourished and weak
- Drowsy, but oriented

Airway

- Clear

Breathing

- Tachypnoeic with bilateral air entry and coarse crackles in both lung bases

Circulation

- Heart rate tachycardic, but regular, pulses weak

Development

- Glasgow Coma Score 14/15 (eye opening in response to speech)
- Marked neck stiffness, positive Kernig's sign
- Pupils equal and reacting to light
- Cranial nerves 2–12 intact, no papilloedema
- No peripheral neurological findings

Everything else

- No petechial rash noted
- Pale conjunctiva, but no jaundiced sclera
- Oral thrush, but no oral Kaposi's sarcoma

In summary, Ellena is in septic shock with neurological signs suggestive of meningitis. In addition, she has bibasal crackles suggestive of pneumonia, is malnourished, has oral candidiasis, and is probably anaemic. The findings are consistent with CDC stage 3 (see Chapter 27, section 27.3.1).

Having noted the absence of papilloedema, which would indicate raised intracranial pressure (the CT scanner is broken), you do a lumbar puncture whilst the patient lies on the mattress, and send off a sample of spinal fluid for analysis. Then you ask the Malawian doctors to review the patient and prescribe an immediate dose of Ceftriaxone 2g IV, the standard treatment in Malawi to cover suspected meningitis. You also exclude malaria by sending off a blood smear.

The Malawian doctor also prescribes additional antibiotics to cover for pneumonia. They ask you to sort out a chest X-ray to check for TB and PCP (*Pneumocystis jiroveci* pneumonia), and to have her sputum tested for AFB (acid-fast bacilli). Ellena is unlikely to receive oxygen due to limited resources.

You document all your findings, including your discussion with a named doctor, carefully in the notes, signing your name and student status clearly at the end. Ellena is then admitted onto the female ward.

As you move onto the next patient you reflect that your assessment of Ellena was far from complete, but as thorough as your limited resources would allow. You realize that you have just skimmed the surface of her problems, but you have 19 comparably ill patients waiting to be seen and you have to be realistic about what assistance you can offer given your limited knowledge and experience. You have set the ball rolling though, and you hope that your notes will be read by the experienced doctors on the ward and she will be given the help she needs.

Notes

[1] World Health Organization (2013). Consolidated guidelines on the use of antiretroviral drugs for treating and preventing HIV infection. Available at: http://www.who.int/hiv/pub/guidelines/arv2013/download/en/

★ Learning Points

- **Virological:** Ideally the patient's HIV strain and sensitivity should be assessed before selecting the most appropriate combination of ART.
- **Clinical:** The clinical presentation of meningitis can be atypical in patients with HIV, with subtle chronic symptoms of headache and neck stiffness.
- **Personal:** This patient's assessment was not complete. However, this is acceptable as the patient should be reassessed regularly anyway to check the effects of the therapy, and then gradually more of her problems can be addressed.

31.6.1 Nucleoside RT inhibitors (NRTIs)

Mode of action

The dideoxynucleoside analogues (Table 31.1) which have now supplanted AZT in clinical practice in many countries have a similar mode of action to that of AZT. The nucleoside analogue must be phosphorylated intracellularly to produce the active antiviral drug, which is the triphosphate. The latter is a very potent inhibitor of viral RT and prevents nucleotide chain elongation much in the same manner as ACV.

The 3′ positioning of the azido group of AZT, for example, blocks the essential phosphodiester linkage which would normally enable the next nucleotide to be added to the growing DNA chain. Furthermore, AZT triphosphate binds to the viral RT, rather than to the cellular DNA polymerase, giving some specificity of action. By contrast with ACV, cell enzymes rather than viral enzymes phosphorylate the AZT and this class of molecule and, hence, intracellular concentrations of the drugs increase in normal cells, as well as in virus-infected cells; this partly explains the toxic effects.

HIV-1 strains can quite quickly mutate in the *RT* gene, the point of action of AZT, to become resistant. The patterns of resistance are quite complex. Five mutations in the RT were dominant in AZT-resistant viruses. However, it was rather surprising to find that resistance to AZT could be reversed by other mutations induced in the RT by, for example, the dideoxynucleoside analogue ddI. Entirely new drugs can act synergistically to avoid drug resistance problems and can be used in lower dosages to avoid side-effects. Molecules such as ddI, ddC, and 3TC are used in **combination chemotherapy**, often with the addition of a viral protease inhibitor. This use of a combination of drugs is called HAART (highly active antiretroviral therapy). The NRTIs form the backbone of therapy for HIV in combination with non-nucleoside RT inhibitor (NNRTIs) and protease inhibitors.

31.6.2 Non-nucleoside RT inhibitors (NNRTIs)

The non-nucleoside RT inhibitors comprise several hundred antiviral compounds of very varied chemical structure, but only a handful have been tested in the clinic. They are powerful inhibitors of HIV in the laboratory, but drug-resistant HIV mutants appear almost immediately a patient is treated. There are complex patterns of cross-resistance between these drugs. Many of these compounds are highly selective for HIV-1 and bind tightly to the viral RT close to, but not overlapping, the polymerase active site, where they distort this region and reduce conformational changes and hence activity of the viral enzyme. They are used in combination with nucleoside and protease inhibitors.

31.6.3 Protease inhibitors

The virus-coded protease has the important function of cleaving certain HIV structural proteins at the post-release **maturation** stage of viral replication (see Chapter 3). Without such cleavages the newly released virus does not mature and is not infective. Thus the gag and gag-poly protein complexes cannot be processed without proteolytic cleavage into their mature form, namely structural proteins of the virus and the virus-coded enzymes RT, integrase, and protease.

At least nine new drugs have been found that inhibit HIV protease, but not mammalian cell proteases. These are powerful anti-HIV-1 drugs in the laboratory and the clinical data suggest that they are effective in AIDS patients. They are quite well tolerated, but are difficult to manufacture and patients require large doses, up to 2 or 3 g/day. Drug-resistant mutants are easily selected and the cross-resistance patterns are complex.

31.6.4 Nucleotide inhibitors

This class of drug, represented by tenofovir, also targets the viral reverse transcriptase enzyme and can be used in place of a NRTI drug in the HAART scheme.

31.6.5 Fusion inhibitors

These drugs stop the virus entering the cell and the most well tested molecule at present is enfuvirtide (Fuzeon), a synthetic 36-amino acid oligopeptide. Drug-resistant mutants arise quite quickly after monotherapy. Some strains of HIV even mutate to drug dependence whereby they actually require the drug for infection. Such a feature was observed with one of the first antivirals to be discovered, marboran, against smallpox half a century ago!

31.6.6 Antagonists to the virus co-receptor CCR5

Small molecules have been discovered which bind to the hydrophobic pockets within the transmembrane helices of the HIV co-receptor CCR5 and stabilize the receptor and antagonize biological activity. This new class of drug thus reduces the interaction between CCR5 and the V3 stem loop of the virion gp120 envelope protein. Because this drug binds to a cell protein drug resistance could only occur by altering virus co-receptor tropism such as switching to the CXCR4 co-receptor.

31.6.7 Integrase strand transfer inhibitors (InSTI)

The HIV integrase enzyme catalyses 3′ end processing of virus DNA and strand transfer when the provirus DNA integrates with the host cell DNA, and is the target of four new inhibitors with the lead taken by elvitegravir and raltegravir. These drugs have metal binding characteristics and react with magnesium ion co-factors on the integrase; they also have a hydrophobic group which interacts with the HIV DNA. This class of drugs also blocks the integrase of HIV-2. Also uniquely, virus drug-resistant mutants have a reduced replicative ability and so can be outgrown by the drug-sensitive wild-type virus in much the same manner as drug-resistant mutants to influenza NA. Cross-resistance patterns are common in this group of inhibitors and are complex.

31.6.8 Combination chemotherapy (HAART)

The most successful clinical strategy is to use a **combination** or 'cocktail' of drugs, e.g. efavirenz and tenofovir + emtricitabine or abacavir + lamivudine, (3TC). The precise combination used is somewhat dependent upon cost but also whether a patient has already been treated with one or more inhibitors previously and, where they may already exist, drug-resistant mutants, or alternatively is taking anti-HIV drugs for the first time as an adult, adolescent, or child.

Drug combinations can reduce the viral genome load to undetectable levels (<50 copies/µl) and, therefore, there is clinical evidence that HIV, like certain leukaemias, for example, may be held in check by such complex drug treatments. Patients are monitored carefully for viral genome load and should resistant mutants emerge to a particular drug in the combination this could be substituted. Cocktail antiviral chemotherapy in a single tablet (e.g. Atripla®) is easy to handle although expensive. Default of compliance of the patient can allow a very rapid rebound of drug-resistant virus, which is thus not totally eradicated and sequestered in so-called 'sanctuaries'. An example of such a virus reservoir is the brain where penetration of antivirals is likely to be poor.

Cost is not a small factor in antiviral chemotherapy of HIV, with cost estimates of US$700 billion globally over the next two decades. Nevertheless, many HIV patients have benefited from the effects of these new combinations of drugs and even return to work and lead a normal life.

In a wider sense there is a fortunate group of HIV infected persons called 'elite controllers' who have fewer than 50 copies HIV gene/µl without drug treatment. The so called 'German case' may be an example of eradication of the virus. This patient was being transplanted with stem cells from a donor who carried the 32 CCR5 mutation. HIV attaches to cells using both CD4 and CCR5 and CXCR4 receptors and persons homozygotic for a 32bp deletion in the gene coding for CCR5 are resistant to infection. In this patient the HAART treatment was stopped and the HIV RNA level remained below 1 copy/µl for four years. However it is also possible that the radiotherapy and chemotherapy for the transplant itself eradicated long lived virus reservoirs in certain subsets of cells. Unfortunately, HAART alone would not be able to remove cells infected latently with HIV. Resting CD4 cells as well as monocytes, dendritic cells, and macrophages and cells of the CNS as noted above have integrated HIV provirus which could form a huge virus reservoir.

Hepatitis C virus

This important and globally widespread virus causes both acute and chronic infection of the liver and no vaccine is available. Persistent infection can lead to chronic liver disease and hepatocarcinoma and therefore the virus is the leading cause for liver transplantation in developed nations. Antivirals are important interventions.

Until recently, approved treatments for patients were restricted to IFN-α (either in the native or pegylated forms) alone or in combination with the nucleoside analogue ribavirin. However, treatment failure is relatively high at about 60%. The three recombinant IFNs used are IFN-α2a, IFN-α2b (which is similar to 2a with a substitution of a single amino acid), and alfacon-1, a bioengineered consensus IFN-α, that has activity similar to natural IFN.

Directly acting antivirals (DAA) are now licensed for clinical use and especially useful are drug combinations such as telaprevir and boceprevir (both protease inhibitors) and alunaprevir and daclatoasvir. Also available is ombitasvir and sofosbuvir with or without ribavirin for less than 12 weeks of therapy.

As with HAART and HIV, these drugs in combination have different viral targets such as the hepatitis C virus RNA replicase and protease and significantly reduce virus load, which is carefully monitored. But similarly to HIV there can be problems of patient compliance and also expense. The most encouraging clinical data now indicates that these antiviral drugs can 'cure' an individual's HCV infection and that some combinations of DAA may be pan HCV genotype, namely blocking all strains of hepatitis C virus.

Hepatitis B virus

Despite the availability of vaccine, hepatitis B remains a major public health problem because of already existing chronic infection in up to 400 million carriers globally. Antiviral chemotherapy is the only option at present to control chronic HBV infection. Most attempts to eradicate or reduce body yields of HBV in a patient using chemotherapy have only had modest success to date. However, on a positive note, over the last decade a subgroup of patients infected with certain of the genotypes of the virus and with active liver disease and low-level viraemia have been found to respond to treatment with IFN-α given subcutaneously. Nevertheless this treatment gives rise to side-effects of the IFN.

A marked step forward from initial success with some patients given IFN-α subcutaneously was the discovery that nucleoside analogues originally developed to treat HIV or HSV, such as lamivudine (3TC), can be potent and well tolerated inhibitors of HBV. The drug acts by inhibiting the reverse transcription step during the replicative cycle of hepatitis B. It is effective in patients in combination with alpha interferon or in those who fail to respond to IFN-α and moreover can be given orally. Relapses can occur, however, in most patients when they discontinue therapy. It is also used as prophylactic treatment in liver transplants. Nevertheless, drug-resistant mutants emerge using monotherapy. A second nucleoside analogue, adefovir, an acyclic analogue of deoxyadenosine monophosphate has also been licensed for chronic hepatitis B patients.

More recently, clinical trials are in progress with famciclovir (dGTP competitor), adefovir (dATP), and entecavir (dGTP). More drugs are being synthesized and undoubtedly drug combinations will be favoured to lower genome copies and hence reduce the incidence of pathology of the liver and hepatocarcinoma, and to reduce the emergence of drug-resistant viruses.

Cellular proteins as targets for drugs to block virus replication

The existing antivirals described above have been found or even designed to target viral proteins, and thereby ensure specificity and safety. However, the biggest problem, as we have noted, for the RNA genome hepatitis C, influenza, and HIV is that these viruses have a 'pliable' genome and exist as a quasi-species with countless mutants, some of which are naturally resistant to the drug. A second problem is that the drugs that target viral proteins such as RT, RNA polymerase, and proteases have a narrow antiviral spectrum. Many cellular proteins are now being identified as essential for viral replication. Most of the drugs used in the wide pharmacopoeia target cellular proteins and so could these be a target for a new family of virus blockers avoiding the problem of virus drug resistance? Cellular topoisomerases, nuclear transport proteins, and protein kinases are all required over short periods of time for acute infections, and experimental inhibition of these proteins has already been shown to repress viral replication. It could be added that, although discovered many years ago, ribavirin may be a member of this class of drugs. It targets the cell inosine-5'-monophosphate dehydrogenase rather than a virus enzyme. Also, as expected for this class of drugs, no virus has been discovered which is resistant to ribavirin. At the moment ribavirin is used as an aerosol to treat children seriously ill with RSV. More recently ribavirin has been used in Ebola patients and in hepatitis C patients in combination with interferon.

31.7 Clinical use of interferons

Mammalian cells have evolved many systems to detect and then combat viral infection. For example, dsRNAs, a sign of viral infection in a cell, can activate cellular protein kinase-mediated responses that mediate the production of interferon as well as inhibit translation and even induce cell death. As a counterbalance, viruses may express their own dsRNA binding proteins that prevent interferon production by sequestering cellular dsRNA binding proteins. Also, short interfering RNAs or micro-RNAs down-regulate cellular gene expression by binding to complementary mRNAs arresting mRNA translation into protein. There are now the first indications that interference RNAs can be synthesized specifically to a viral mRNA sequence to block viral mRNA but not host mRNA. The mode of action of interferons and their properties are described in Chapter 5 and here we will only discuss their clinical use. In summary, though, interferon itself binds to the IFN receptor on cells, activates the JAK-STAT pathway, and up-regulates several IFN stimulated genes.

The possibility of using IFN in the clinic was limited for a long time by the small quantities available. With the development of cDNA cloning technology the situation changed and genetically engineered IFNs are now used quite extensively in clinical practice especially against hepatitis C and B infections. Five IFN preparations are licensed: α2a (Roferon A); α2b (Intron A®); αN3 (Alferon N®); β1B (Betaseron®); and γ1B (Actimmune®). Pegylated IFNs have a polyethylene glycol molecule attached to slow absorption and extend serum half-life. Thus, the standard dose of 3 MIU of IFN three times weekly can be reduced to one weekly for hepatitis B and C injection.

Clinical and laboratory responses to IFN are not uniform and can depend on the genotype of the virus. Thus approximately one third of patients with chronic HBV respond to treatment with 5 million units daily of IFN and remissions may be sustained. The most common side-effects are those of fever, chills, headache, and myalgia, which has led to the supposition that in many acute viral infections these symptoms may in fact be caused by IFNs stimulated by the virus itself.

31.8 The future: a combination of proteomics and genetics to identify host proteins as new targets

Although basic virological screening of extensive compound libraries of perhaps 400 000 molecules in pharmaceutical companies will continue to play a part in the discovery of new antivirals, knowledge of viral enzyme structure and functioning by a combination of genetics and X-ray crystallography is leading to a new generation of drugs. The interaction between an antiviral compound (disoxaril) and a rhinovirus protein was visualized at the atomic level by X-ray crystallography quite a few years ago and is a 'gold standard' study. This low molecular weight inhibitor of the common cold virus has been identified as binding in a 'cave' at the bottom of the receptor-binding pocket of the virus. The drug–virus interaction blocks pores in the virion through which ions would normally pass to its interior to aid uncoating. The virus is stabilized and uncoating blocked, and infection of cells is aborted. Such computer-aided studies at the atomic level can assist the design of brand new inhibitors or lead to addition of crucial side chains to existing drugs to improve target binding. The most exciting example of the use of atomic modelling is modification of a drug first discovered 40 years ago and known to block the NA function of influenza. The original drug was not active in animals, but computer-guided modification led to synthesis of a related molecule, which can prevent influenza infection in animals and in humans (Relenza and subsequently Tamiflu).

A vital enzyme for influenza is the polymerase which both transcribes the RNA genome into mRNAs and also replicates full length copies for the new genome. The atomic structure of the interaction of virus proteins PA, PB1, and PB2 of the polymerase enzyme and the viral RNA promoter has now been reported. The 5' extremity of the RNA promoter folds into a compact hook in a pocket of the virus proteins PB1 and PA. This complex is near the polymerase active site and appears integral to the structure, and definitely required to enhance and activate polymerase function. The crystal structure is high resolution and can now be used for structure-based design of inhibitors which target the active site of PB1 and the vRNA binding site.

New viruses, such as Ebola, SARS, MERS, Nipah, West Nile, and Hendra, have emerged and resurged recently and have become a focus for drug discovery. The worry about bioterrorism has renewed work on drugs against smallpox and polio.

There is a strong expectation that a combination of proteomics and molecular genetics will unravel the interaction of viruses with cellular proteins and, from the thousands involved in virus replication, the functions of some could be blocked, at least for a short time, and so abort infection. The replication of viruses such as influenza with small genomes (12kbp) is heavily dependent upon the cell enzymes, scaffolding, and structures. A recent clinical trial where 18 volunteers were infected in the London quarantine unit of hVIVO discovered strong transcriptional regulation of genes involved in inflammosome activation and genes encoding virus-interacting proteins, as well as cell-mediated antioxidant and innate immune responses. Other studies with influenza have identified cellular proteases, nuclear and general trafficking proteins, microtubule proteins, proteins involved in the recycling pathways of endosomes, and, perhaps most fruitful of all for new drugs, signalling pathways involved in virus replication and export of vRNAs from the nucleus.

RemInders

- Vulnerable points for attack with antivirals in the virus life cycle are adsorption to cells, virus **penetration, replication of viral genome,** and viral **budding and maturation.** Viral enzymes such as RNA-dependent RNA polymerases, DNA polymerase, protease, integrase, and RT are excellent targets for antiviral drugs.

- Antiviral compounds are usually used therapeutically to treat infections, but prophylactic use to actually prevent infection is possible with HIV and influenza.

- **Prodrugs,** namely famciclovir, the diacetyl version of penciclovir, and valaciclovir, the l-valyl ester of ACV, have better bioavailability when given by mouth and are used clinically. The NI prodrug Tamiflu similarly has excellent pharmacological characteristics. The dideoxynucleoside analogue inhibitors of HIV, such as AZT, are prodrugs, and are phosphorylated in both normal and virus-infected cells.

- Most antiviral drugs have a very narrow spectrum of activity. However, drugs selected for hepatitis C may also have effects against dengue, both being caliciviruses. Also some inhibitors of the RT of HIV, e.g. lamivudine, also block a similar RT step in the replication of hepatitis B virus.

- A range of cloned IFNs has unexpectedly opened up a new field of cytokine research, and is also used in the clinic against hepatitis B and C infections.

- A new generation of directly acting antivirals (DAA) has now replaced monotherapy of hepatitis C with IFN-α.

- **Combination chemotherapy** employing two or three antiviral drugs, each with a different molecular target, (HAART) is used to treat HIV patients and reduces the emergence of drug-resistant viruses. Similar strategies are emerging for the treatment of hepatitis C and hepatitis B.

- Viral nucleotide sequence data and X-ray crystallography of virus proteins and the science of proteomics may lead to a generation of 'designer' drugs.

- Emerging viruses, such as **SARS, MERS, West Nile,** Ebola, and Zika, or bio-threat viruses, such as smallpox, are new target viruses for chemotherapists.

Further reading

Bhattacharya, D. and Thio, C.L. (2010). Review of hepatitis B therapeutics. *Clin Inf Dis* **51**, 1201–08.

Huang, Y. *et al.* (2011). Temporal dynamics of host molecular responses differentiate symptomatic and asymptomatic influenza A infection. *PLOS Genetics* **7**, 1–8.

LeDouce, V. *et al.* (2012). Achieving a cure for HIV infection: do we have reasons to be optimistic? *Journal of Antimicrobial Chemotherapy* **67**, 1063–74.

Pawlotsky, J.M. (2011). Treatment failure and resistance with directly acting antiviral drugs against hepatitis C virus. *Hepatology* **59**, 1742–51.

Rehman, S. *et al.* (2011). Antiviral drugs against hepatitis C. *Genetic Vaccines and Therapy* **9**, 11–20.

Whitley, R. *et al.* (2013). Global assessment of resistance to neuraminidase inhibitors. *Clin Inf Dis* **56**, 1197–1205.

Questions

1. Describe the importance of the search for drug resistance mutations against antivirals for Hep B, HIV, and influenza.

2. What is HAART and could this strategy eradicate HIV from a patient?

3. How are new antivirals discovered?

4. Describe the mode of action of the anti-herpes nucleoside analogue drug aciclovir.

Index